Research Methods in Biomechanics

SECOND EDITION

D. GORDON E. ROBERTSON

University of Ottawa

GRAHAM E. CALDWELL

University of Massachusetts, Amherst

JOSEPH HAMILL

University of Massachusetts, Amherst

GARY KAMEN

University of Massachusetts, Amherst

SAUNDERS N. WHITTLESEY

HUMAN KINETICS

Contents

PART II KINETICS 61

PART III MUSCLES, MODELS, AND MOVEMENT   177

Preface

Biomechanics is a highly technical field, and its research methods change as rapidly as does technology. Research techniques are regularly replaced with new techniques because of the availability of faster and more sophisticated software and hardware. For example, 25 years ago many researchers used cinematography to record human motion; 10 years later cinematography was almost obsolete, having been replaced by VHS videography; and now, digital and infrared videography have become the preferred motion-capture technologies. Faster computers with essentially unlimited memory can process data using more complex analyses and more sophisticated statistical methods. Given these realities, this edition has added up-to-date research methods to existing chapters and includes several new chapters that outline advanced analytical tools for investigating human movement.

This text is organized into four parts. Parts I and II retain their structure from the first edition, with part I exploring planar and three-dimensional kinematics in research and part II examining issues of body segment parameters, forces, and energy, work, and power as they relate to two- and three-dimensional inverse dynamics analysis. Within the first two parts, chapters 2 and 7 have been extensively revised to reflect current research practices in biomechanics. Chapter 7 now reflects the role of software such as Visual3D in carrying out inverse dynamics analyses. A Visual3D Educational Edition is provided with this book purchase, so that the reader can experience the process of kinematic and kinetic analysis of human motion using Visual3D.

Part III of the text deals with the study of muscle activity and the mathematical modeling of human movement. Chapter 9, Muscle Modeling, has been updated and is bolstered by the addition of chapter 11, Musculoskeletal Modeling. Chapter 9 retains its emphasis on the Hill model and now includes more information on how to obtain parameters to allow the Hill model to represent individual muscles in a subject-specific manner. We have removed some of the musculoskeletal model material from the original chapter and include it in the new chapter 11, which is coauthored by Brian R. Umberger and Graham E. Caldwell. Chapter 11 explores the use of musculoskeletal models in analyzing human movement, an area of growing interest that per-mits the study of muscle forces beyond that allowed by inverse dynamics. Other chapters in part III address electromyographic (EMG) kinesiology and computer simulation of movement. EMG permits the monitoring and analysis of the active contractile characteristics of muscles, whereas computer simulations permit the study of motions without requiring that a subject perform the motion, which allows researchers, physicians, therapists, or coaches to test novel motions without placing people at risk of injury.

Part IV explores further analytical procedures that can be applied to biomechanical data, beginning with signal processing techniques and then moving on to two chapters new to the second edition. Chapter 13, Dynamical Systems Analysis of Coordination, coauthored by Richard E.A. van Emmerik, Ross H. Miller, and Joseph Hamill, outlines the theories and analytic methods used to investigate movement in complex systems with many degrees of freedom. This chapter focuses on how we assess and measure coordination and stability in changing movement patterns,

eBook available at HumanKinetics.com

VISUAL3D EDUCATIONAL EDITION

New to this edition is the access to the Visual3D Educational Edition software, created by C-Motion. The Visual3D Educational Edition can be used to display C3D and CMO data sets, but also provides the ability to manipulate sample data sets to help readers understand kinetic and kinematic calculations and to provide experience with professional biomechanical research software. To download the Visual3D Educational Edition and special sample data sets selected by this book's authors, visit **http://textbooks.c-motion.com/ResearchMethodsIn Biomechanics2E.php**. The sample data sets are downloadable from this page. Follow the Download Software link to download Visual3D Educational Edition. When opened using Visual 3D Educational Edition, these sample data sets will unlock additional functionality, allowing you to explore the full range of modeling and analysis capabilities of the professional version of Visual 3D.

For support, visit www.c-motion.com.

and it examines the role of movement variability in health and disease. Chapter 14, Analysis of Biomechanical Waveform Data, coauthored by Kevin J. Deluzio, Andrew J. Harrison, Norma Coffey, and Graham E. Caldwell, outlines statistical tools to identify the essential characteristics of any human movement. Biomechanists are faced with the sometimes daunting task of determining which variable or variables from thousands of possibilities (linear and angular kinematics, linear and angular kinetics) best characterize a particular motion. Techniques in this chapter can be used to select the best combination of these factors.

Human gait is used as one example of motion, but the techniques can be applied to any motion.

Each chapter includes an overview, a summary, and a list of suggested readings for those interested in learning more. In select chapters, sample problems are provided to serve as learning aids, and answers are provided in the back of the text. Sections titled From the Scientific Literature highlight the ways in which biomechanical research techniques have been used in both classic and cutting-edge studies in the field. The appendixes provide helpful mathematical and technical references, and a glossary provides a reference for terminology.

Biomechanics Analysis Techniques: A Primer

Gary Kamen

Every scientific discipline uses a unique set of tools that new scientists must master before they can contribute knowledge to that discipline. A molecular biologist could not begin to conduct research without knowing how to use and interpret information from a spectrophotometer or a gas chromatograph. A geologist intent on studying a volcano must select appropriate oscilloscopes, amplifiers, and seismographs; store the signals that these instruments record; and later analyze those signals using signal-processing techniques.

Just as knowledge of basic Newtonian physics and mastery of instrumentation and analytical techniques are prerequisites for conducting research in molecular biology and geology, so too are they required in the study of biomechanics. This book is a comprehensive resource on the tools needed to conduct research in biomechanics.

WHAT TOOLS ARE NEEDED IN BIOMECHANICS?

Some prerequisite knowledge is necessary to begin applying the principles of biomechanics in research. This text assumes that readers have a basic understanding of geometry, trigonometry, and algebra, including elementary vector algebra. Knowledge of basic **mechanics** according to Newton's laws is also necessary, and this text provides many examples of their application in the field of biomechanics. A good working knowledge of human anatomy is important: This text includes examples that apply to the human musculoskeletal and neuromuscular systems, and knowledge of the constraints imposed by the anatomical system is essential to acquiring a comprehensive understanding of the mechanics involved. Readers seeking additional information on these topics should consult the many textbooks that provide good reviews. A list of suitable readings and texts is included at the end of each chapter.

APPLICATIONS OF THE PRINCIPLES OF BIOMECHANICS: AN EXAMPLE

Of course, thoroughly understanding two- and three-dimensional kinematics and kinetics, anthropometrics, muscle modeling, and electromyography is useless without good research ideas or biomechanical problems to be solved. Consequently, before we begin detailing applied biomechanical principles, let us consider an example that illustrates how we can apply the knowledge to be gained from this text.

Locomotion is the hallmark that distinguishes organisms in the animal kingdom from plants, and animals have devised myriad methods to enable movement. Problems related to locomotion constitute a major area of focus for many biomechanists. Ants, despite their small body size, ambulate quickly by moving each leg at the correct velocity. Fish propel themselves efficiently even though they are subject to the large drag force of water. The anatomy of horses constrains their movement, yet they manage to select the right gait for the environmental conditions and the velocity of locomotion. Because human gait and postural control is a primary focus of this text, let us consider a locomotion and postural control problem that requires biomechanical tools to solve.

On returning home from a difficult day at the office, our subject must ascend the front steps to her home, open the door, and walk inside before setting her briefcase on a table. After climbing the stairs, she reaches for the doorknob, turns it, and finds it locked. She must retrieve her keys from her pocket while juggling her briefcase and keeping her balance. After entering the house, she places the briefcase on a table.

These seemingly simple tasks require the successful interplay of a complex system of musculoskeletal design and neuromuscular control. Consider some of the many subproblems to be solved:

▶ How does our subject climb the stairs? How can we describe the characteristics of the movement at each joint?

▶ How much muscular force is needed during the transition forward to the next step?

▶ Where does she decide to stop and reach for the doorknob?

▶ How does she maintain her balance while reaching for the door?

▶ How does she first turn the doorknob and then retrieve her keys and unlock the door?

▶ How does our subject walk into the house with her briefcase? At what point, for example, does the leg planted on the floor begin to move forward again?

▶ How is the briefcase placed on the table? What prevents it from being placed either too far forward or too far back?

These problems all require solutions gleaned using biomechanical tools. For example, we need to know about the displacements and forces produced at various joints. How do we refer to units of displacement or force so that other scientists can understand us? In the scientific literature, the preferred units of measure are part of an international system—the metric system—whose use has been agreed upon by all scientists. Virtually all biomechanical conferences and journals require the use of this system, so this text uses it exclusively. The system consists of seven fundamental quantities from which all other measurements are derived; for biomechanists, the most important ones are the kilogram, meter, second, and ampere. The basics of this system are outlined in appendix A.

Kinematic analysis describes the motions we see. As a person approaches stairs, we observe repetitive flexion and extension movements at the hip, knee, and ankle joints. These angular displacements change as the person begins to climb the stairs. Patients with a lower-limb injury may use a different pattern of angular joint displacement to perform these activities. We use displacement information to compute velocities and accelerations during performance of these tasks. The motions we observe may include linear as well as angular motion. Monitoring a point on the person's

trunk allows us to describe the instant-to-instant linear displacement and the resulting velocity of walking. We measure kinematic variables using an imaging system such as a film or video camera or instruments attached to the joint to measure displacement. Some motions may be too complex to describe using simple planar (two-dimensional; 2-D) coordinates. Overuse knee injuries are sometimes ascribed to inappropriate motions of the knee joint in the frontal or sagittal plane. Thus, more complex, three-dimensional (3-D) analysis may be required to adequately describe these motions. Methods for acquiring kinematic data and computing planar kinematics are covered in chapter 1, whereas chapter 2 considers 3-D kinematics.

Special computers can capture images from video data and compute the trajectories of reflective markers placed over a subject's joint centers and then analyze the motion patterns. These data are then processed to derive various kinematic measures, such as the range of motion of each joint, the velocity and acceleration of each segment, and the path of the center of gravity. Chapter 3 details methods for determining the body's center of gravity. These data can then be synchronized with the ground reaction forces to enable inverse dynamics analysis.

Kinematics describes the movements we see, but to understand why the motions occur as they do we must examine the kinetics, or underlying linear forces and rotational torques, that dictate the kinematic motion. External forces are those caused by interaction of a person with the environment. As our subject walks toward the stairs with her briefcase, she subconsciously forms a plan for how to transition from level surface walking to stair climbing. Each footfall generates a ground reaction force (GRF) that can be measured with appropriately placed instruments, such as force platforms. Force platforms are embedded wherever the researcher wants to measure the mechanical cost of performing the movement task. For example, a force platform embedded in a step will show GRFs that are slightly higher than those generated during walking. The peak vertical forces will be slightly lower than two times the subject's body weight while she is ascending the stairs but only 25% to 50% higher than her body weight when walking. As our subject opens the door, a force platform will show that her center of gravity has shifted to prevent imbalance. Chapter 4 describes methods of recording and analyzing forces, including how to use force platforms to measure ground reaction forces.

The internal forces and torques generated at each joint in the body can also be approximated by using GRFs. GRF patterns change as people ascend stairs because they must overcome the force of gravity (attraction by the earth) to raise their center of mass with each step. Measuring GRF patterns and segmental kinematics, in conjunction with inverse dynamics, can elucidate the strategies people use to maintain balance and keep internal forces within acceptable levels during movement.

To aid in performing the task of climbing and to help maintain balance, our subject might partially support herself with the handrail. Force transducers mounted between the handrail and its attachment to the wall or ground can quantify the amount of that support. Interaction with the door also requires force, this time in the form of rotational torque applied to the doorknob. Torque is measured by instrumenting the doorknob assembly with strain gauges. Our subject must also apply torque to the unlocked door to make it swing open on its hinges. Finally, placing the briefcase on the table requires of our subject both balance and gradual changes in her application of force to put it in the desired location. Appendix C sets out the basics of electricity and electronic instrumentation with strain gauges and other devices.

After measuring GRFs and computing 2- or 3-D motion patterns, we use inverse dynamics analysis to calculate for each joint the smallest possible force that is necessary to complete each action. Such analysis uses Newton's second and third laws to determine what forces and moments of forces exist at each joint. Chapter 5 introduces the theory and methods for performing inverse dynamics analyses for planar motions, whereas chapter 7 develops the techniques used for spatial (3-D) analyses.

The biomechanist can also compute the mechanical cost of the work done and the mechanical power required at each joint. As the velocity of movement increases, greater mechanical power is required. An elderly person may carry out all the necessary tasks for daily life at a much slower pace than a younger person, yet the energy cost to both will be similar and the mechanical power required at each joint will be maintained within the capacity of each joint. Chapter 6 explores how mechanical work, energy, and power are derived from kinematic and kinetic measurements.

The precise forces produced by the muscles and transmitted through the tendons, ligaments, and bones can be directly measured only with indwelling force sensors or estimated by modeling the musculoskeletal system. The actual forces may be larger if the person uses an inefficient technique to move the joint or because internal stabilizing forces from ligaments are used to prevent collapse in the face of unexpected perturbations, but at least the minimal required forces are determined. Alternatively, the activation patterns of the muscles can be measured to better quantify the role of the muscles during performance of the tasks.

Muscle activation can be studied using electromyography (EMG). Typically, sensors are attached to the skin to record the muscles' electrical activity. The input from these sensors is amplified by other instruments, and the output signal is digitally stored or viewed on an oscilloscope or other output device. EMG is frequently used in ergonomic evaluation to determine which muscles are under stress and at risk for injury and during sport performance to determine the phase of a motion during which a muscle group is the most active. Chapter 8 covers techniques for recording and interpreting EMG signals.

Why not rely on the information provided by the kinematic or kinetic analyses to predict which muscles are active? The techniques used to describe movement characteristics are not perfect predictors of the underlying activities in muscles. Part of the task of placing a briefcase on a table could be accomplished by allowing gravity to lower the briefcase rather than by activating specific muscles. Moreover, some muscles that are not directly involved in the task may be activated anyway. Walking up the stairs, particularly when carrying something, requires postural stabilization to prevent a fall. Activating other muscle groups stabilizes posture.

Other analytical tools allow scientists to answer questions not directly amenable to measurement techniques. Although kinematic, kinetic, and EMG analyses are indispensable for studying actual movements, the question that remains is whether stairs can be climbed in a more efficient or effective manner. After all, if the goal is simply to accomplish the motor task, it could be done in several different ways, including hopping on one or both legs from stair to stair or crawling up on hands and knees. But what is the optimal way for a person to ascend stairs? Forward dynamics models address such questions by simulating a movement given a set of internally applied forces and torques. Once the model is customized for a specific individual, optimal control techniques are used to find the best set of forces and torques needed to accomplish the task. If *best* is defined as the movement pattern that minimizes overall muscular effort, the optimization model finds the patterns of kinetics and kinematics necessary for the subject to climb the stairs with the least muscular effort. A different pattern is found if *best* is defined as the movement pattern that presents the subject with the least likelihood of falling while on the stairs. When results from optimization models are compared with the actual movement produced by our human stair climber, ways in which she can improve her performance may come to light. Muscle models that mimic the force-generating capabilities of actual muscles can be used to provide values for the internal forces in these forward dynamics models. Use of these models is essential to biomechanics research because the technology for measuring individual muscle forces is highly invasive and unsuitable for use in most research situations. Chapter 9 reports on the modeling of muscles to better understand their function, and chapter 10 discusses the topics of computer simulation and forward dynamics.

Many different types of data are required to perform these analyses. For example, to quantify the motions of reflective markers placed over joint centers, the researcher must digitize the video data using high-speed computers to obtain the positions of the body segments during the activities under study. To determine velocities and accelerations, researchers compute mathematical time derivatives using algorithms that require special smoothing techniques. The researcher must know which technique is appropriate for use and then evaluate whether the technique was successful. These topics are discussed in detail throughout the text. Chapter 12, in particular, describes various data-smoothing and data-processing techniques that result in reliable, noise-free data. Chapter 11 permits an even deeper analysis of walking by modeling the actual muscles and anatomical elements of the human instead of simplifying the kinetics as a net force and moment of force at each joint as was done in chapters 5 and 7.

Chapters 13 and 14 offer powerful tools to examine the considerable amount of data that modern **motion-capture** and **motion-analysis** software provide. Chapter 13 outlines the principles of *dynamical systems analysis*, which is used to study the underlying systems that coordinate complex motions such walking and running. Chapter 14 further introduces *principle component analysis* and *functional data analysis*, which can be used to extract the most important variables from the plethora of redundant and extraneous variables that typical motion-analysis software produces. For example, the analysis of walking can result in more than 1000 time-series when one includes the 3-D trajectories of each marker; the 3-D linear and angular kinematics of each segment and joint; and all the kinetic measures such as joint forces and moments of force, their powers, the 3-D angular and linear momenta of each segment, and the changes in segmental energy and the work done on each segment.

NUMERICAL ACCURACY AND SIGNIFICANT DIGITS

The following chapters have many sample problems with numerical solutions, but with the use of modern calculators, the number of significant digits in a numerical answer needs addressing. After performing computations on a calculator, you often have more digits in your display than you care to report. A rule needs to be applied to keep answers reasonable. One convention used in engineering (Beer et al. 2010) holds that to conserve an accuracy of 0.2% (historically based on the accuracy achieved by 10-inch slide rules) one needs three significant digits, unless the first significant digit is a one, in which case four significant digits are reported.

For example, the accuracy of the number 456 is 456 ± 0.5, or in percentages

$$\frac{\pm 0.5}{456} \times 100\% = 0.1096\%$$

In contrast, the number 105, which represents the numbers between 104.5 and 105.5, has a percentage accuracy of

$$\frac{\pm 0.5}{105} \times 100\% = 0.476\%$$

If we add a fourth decimal digit, however, the accuracy increases to ±0.0476%, which is well under the 0.2% accuracy threshold. Here are some examples of reporting numbers in this way:

1/6 = 0.1667

4/5 = 0.200

1/8 = 0.1250

56 300

145.5

237

945

1.000

5580

Notice that with large or small decimal numbers, spaces are added every third position from the decimal. This is the accepted SI format; however, it is permissible to leave out the space for numbers in the thousands, as shown in the third example. The following are incorrect forms:

76, 0.56, 6751, 25.05, 10.064, 932.0, and 22 456.56.

Correctly written these numbers are

76.0, 0.560, 6750, 25.1, 10.06, 932, and 22 500.

It is good practice to include a zero before the decimal in fractional numbers as illustrated by the second number in the examples above. In complex problems it is often necessary to break up a problem into several steps. Intermediate results must therefore be recorded before reaching the final answer. In such a case, to maintain 0.2% accuracy in the final answer, retain an additional significant figure in all intermediate results and then report the final result rounded to the required number of significant digits (i.e., 3 or 4).

Of course the physical (versus numerical) accuracy of an answer to any problem is based on the accuracy of the data used in the problem. If a measurement is only accurate to two significant digits, then any quantity derived from this measurement is also only accurate to two significant digits. In this text, however, we assume that all measurements reported in the problems are accurate to three or four digits and therefore all answers must follow the rule outlined above.

SUMMARY

Biomechanical analysis techniques allow us to solve many problems involving the interaction of humans and other animals with the physical environment. Studying the coordinated actions of limbed animals assists engineers in developing assistive devices, robots, and vehicles, such as those used to explore earth's moon. Understanding the types of angular motions that might put a joint at risk for injury allows researchers to develop knee braces that limit potentially hazardous joint motions while minimally restricting movement. The analysis of muscle activity during functional movements has contributed to the design of artificial limbs. Musculoskeletal models and motion simulation permit the testing of novel activities to determine their feasibility, physiological requirements, and safety. In the chapters that follow, readers will gain sufficient familiarity with these tools and techniques to begin applying them to real-world problems. This text uses a number of abbreviations for biomechanical terms which will be introduced as they are first used. For easy reference, here is a list of some of the frequently-used abbreviations.

LIST OF ABBREVIATIONS

2-D	two-dimensional
3-D	three-dimensional
A/D	analog-to-digital (converter)
CC	contractile component
COP	center of pressure
CNS	central nervous system
CRP	continuous relative phase
CSA	cross-sectional area
DOF	degrees of freedom
DRP	discrete relative phase
EMG	electromyogram, electromyography
FBD	free-body diagram
FDA	functional data analysis
FV	force-velocity (relationship)
FM	free moment
FPC	functional principal component
FL	force-length (relationship)
FΔV	force-extension (relationship)
GCS	global coordinate system
GRF	ground reaction force
IH	index of harmonicity
IRED	infrared emitting diode
LCS	local coordinate system
LED	light emitting diode
MUAP	motor unit action potential
PCA	principal component analysis
PCSA	physiological cross-sectional area
PEC	parallel elastic component
PRS	(force) platform or plate reference system
RMS	root mean square
SEC	series elastic component

PART I

KINEMATICS

Kinematics is the study of motion without regard to causes. Studying human motion in the past was a time-consuming, laborious, and expensive task because cinematography was employed and manual methods were needed to extract the trajectories of body parts from the film. Advances in technology have automated much of the processes of capturing motion data electronically and then extracting the two- or three-dimensional trajectories. Such technology is now commonplace in the motion picture industry, but biomechanists use additional software to obtain time derivatives of the various trajectories or combine the trajectories to reconstruct the motions of body segments and joints so that differences in motion patterns can be readily identified. Kinematics is also the first step to analyses by inverse dynamics (covered in part II) that estimate the causes of the motion. In this part, chapter 1 outlines how to record two-dimensional kinematics electronically, photographically, and videographically and how to extract digital data from the recordings. Chapter 2 outlines the additional mathematics and processing needed for three-dimensional kinematics. Note that chapter 12 outlines data smoothing techniques that are also important to the valid processing of kinematics, particularly accelerations. Several of the appendixes also have information concerning electronics (appendix C) and mathematical principles (appendixes D and E) that are required for data collection and analysis in kinematics. Note that text in boldface is a concept described in the glossary at the end of the book.

Chapter 1

Planar Kinematics

D. Gordon E. Robertson and Graham E. Caldwell

Kinematics is the study of bodies in motion without regard to the causes of the motion. It is concerned with describing and quantifying both the linear and angular positions of bodies and their time derivatives. In this chapter and the next, we

- ▶ examine how to describe a body's position;
- ▶ describe how to determine the number of independent quantities (called *degrees of freedom*) necessary to describe a point or a body in space;
- ▶ explain how to measure and calculate changes in linear position *(displacement)* and the time derivatives *velocity* and *acceleration*;
- ▶ define how to measure and calculate changes in angular position *(angular displacement)* and the time derivatives *angular velocity* and *angular acceleration*;
- ▶ describe how to present the results of a kinematic analysis; and
- ▶ explain how to directly measure position, velocity, and acceleration by using motion capture systems or transducers.

Examples showing how kinematic measurements are used in biomechanics research and, in particular, methods for processing kinematic variables for planar (two-dimensional; 2-D) analyses are presented in this chapter. In chapter 2, additional concepts for collecting and analyzing spatial (three-dimensional; 3-D) kinematics are introduced.

Kinematics is the preferred analytical tool for researchers interested in questions such as these: Who is faster? What is the range of motion of a joint? How do two motion patterns differ? An important application of kinematic data is their use as input values for inverse dynamics analyses performed to estimate the forces and moments acting across the joints of a linked system of rigid bodies (see chapters 5, 6, and 7). Thus, kinematic analysis may be an end in itself or an intermediate step that enables subsequent kinetic analysis. Whether kinematic variables are the primary goal of a research project

or merely the first step in a series of analyses, they need to be quantified accurately.

DESCRIPTION OF POSITION

To quantitatively describe the position of a point or body, we must first identify the tools we will use. Our main tool is the Cartesian coordinate system, within which we establish one or more frames of reference. One that is desirable but not always necessary to define is an *inertial* or *Newtonian frame of reference*, which is also called an *absolute reference system*, a *global reference system*, or a *global coordinate system* (GCS). This type of reference system is constructed from stationary axes that are fixed in their orientation so that the *X*-axis is parallel to the floor. The coordinate system is defined by

- ▶ an origin defined by the 2-D coordinates (0, 0) or the 3-D location (0, 0, 0) and
- ▶ two or three mutually orthogonal (at right angles to each other) axes, each passing through the origin.

In this chapter, we adhere to the GCS-axis convention adopted by the International Society of Biomechanics (ISB), shown in figure 1.1. In this definition, the *X*-axis direction corresponds to the principal horizontal direction of motion. The *Y*-axis is orthogonal to the *X*-axis, pointing upward vertically, whereas the *Z*-axis is a right-perpendicular to the *X-Y* plane (approximately medial-lateral to the subject). Note that an axis system can be right-handed or left-handed. The GCS described here is a right-handed system, with the third axis *(Z)* pointing to the right with respect to the plane formed by the *X*- and *Y*-axes (see figure 1.1). Right-handed axes are by convention the most common system.

In other texts, readers will encounter coordinate systems that do not follow the ISB convention. In principle, axis designations are arbitrary and easily understood as long as the researcher defines which one is being used.

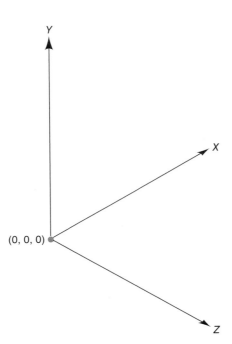

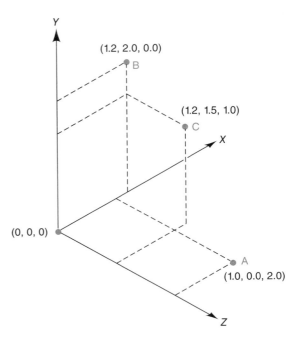

▲ **Figure 1.1** A right-handed GCS using the convention adopted by the ISB. The subject's movement progresses along the *X*-axis.

▲ **Figure 1.2** Example of point locations and Cartesian coordinates in the GCS depicted in figure 1.1. Points are located using their *(X, Y, Z)* coordinate triplets. Point A is on the *X-Z* "floor" of the GCS, 1 unit along the *X*-axis and 2 units along the *Z*-axis. Point B is on the *X-Y* "wall" of the GCS, 1.2 units along the *X*-axis and 2 units along the *Y*-axis. Point C is located 1.2 units along the *X*-axis, 1.5 units along the *Y*-axis, and 1 unit along the *Z*-axis.

Some mathematics and engineering textbooks, most force platform manufacturers, and many 3-D biomechanics applications (such as those in chapters 2 and 7) often use GCSs that differ from the ISB specification. In the field of 3-D biomechanics, for example, often the *Y*-axis corresponds with the principal horizontal direction of movement, the *X*-axis is orthogonal to the *Y*-axis in the horizontal plane (approximately medial-lateral to the subject), and the *Z*-axis is a right-perpendicular to the horizontal *X-Y* plane, pointing upward vertically. To familiarize readers with some of these differences, we chose to use various conventions in this text. In each chapter, the convention adopted is clearly stated. As stated earlier, the system in this chapter adheres to the ISB convention, and we use the *X-Y* plane to discuss sagittal plane movement. This convention is the most commonly used 2-D system in the biomechanics literature.

The designation of the origin is the cornerstone for our ability to quantify position within the GCS. Any point in the GCS can be described by its position in relation to the origin, given by its coordinates in 2-D *(X, Y)* or 3-D *(X, Y, Z)*, as shown in figure 1.2. Although the exact location of the origin is arbitrary, in biomechanics it is usually placed at ground level in a convenient location with respect to the motion studied. For example, when a force platform is used, the center or a corner of the force platform is a suitable location for the origin.

Now that we have established our reference system, we can use it to describe the location of any point of interest, such as the positions of markers affixed to the subject at palpable bony landmarks on or near joint centers. However, to describe the position of an object, or **rigid body**, rather than a specific point, we need to specify additional information. First, we must describe the location of a specific point on or within the object, such as the coordinates of its center of mass (see chapter 3) or its proximal and distal ends. In addition, because the object has a finite volume and shape, we must describe its orientation with respect to our established reference axes. To do this, we establish a second reference frame that has its origin and axes attached to, and therefore is able to move with, the body (figure 1.3). In general, this is known as a *relative* or *local coordinate system* (LCS); when applied to a human body segment, this system may also be called an *anatomical, cardinal,* or *segmental coordinate system*. Often, the LCS origin is placed at the segmental center of mass or at the proximal joint center, and the axes are aligned to roughly coincide with the GCS when the subject is in the anatomical position.

The relative positioning of the LCS axes with respect to the GCS defines the orientation of a rigid body or segment. At least three angles of rotation are needed to describe the LCS orientation in 3-D. Several different sets of angles can be used to specify the LCS orientation, but

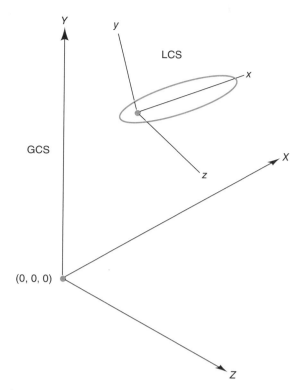

▲ Figure 1.3 LCS attached to an object located within the GCS.

each of these sets contains three independent angles. For example, the system used by the aircraft industry uses the terms *yaw, pitch,* and *roll.* Yaw is a left-right rotation of a plane in flight, pitch is a nose up–nose down motion, and roll is rotation about the plane's long axis. These rotations correspond to rotations about the *Y*- (vertical), *Z*- (lateral), and *X*- (anteroposterior) axes defined in the ISB's GCS convention. More complete descriptions of the LCS and 3-D angles of rotation are given in chapter 2.

DEGREES OF FREEDOM

From the preceding discussion, we see that the location of a given point in space can be described by the three pieces of information contained in its coordinate location *(X, Y, Z)*. The complete description of a rigid body, however, requires six pieces of information: the *(X, Y, Z)* location of its center of mass and the three angles that describe its orientation. The number of independent parameters (pieces of information) that uniquely define the location of a point or body is known as the object's degrees of freedom (DOF). Thus, a point has three DOF, whereas a rigid body has six DOF (figure 1.4*a*).

Although the complete description of motion involves spatial (3-D) movement, in many cases human motion can be described primarily in one specific plane. For example, walking and running involve relatively large excursions of segments within the sagittal plane defined by the *X*- and

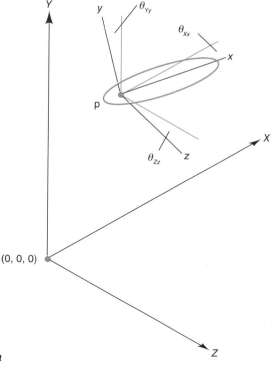

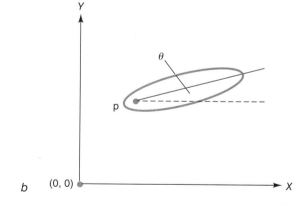

▲ Figure 1.4 *(a)* DOF in a 3-D coordinate system. The segmental endpoint p can be described by its *(X, Y, Z)* coordinates and thus has three DOF. To describe the position of the segment itself, three angles *(θ_{Xx}, θ_{Yy}, and θ_{Zz})* that describe the orientation of the LCS axes *x, y,* and *z* with respect to the GCS axes *X, Y,* and *Z* must be specified. Thus, in 3-D, a rigid body has six DOF *(X, Y, Z, θ_{Xx}, θ_{Yy}, and θ_{Zz}). (b)* DOF in a 2-D coordinate system. The segmental endpoint p can be described by its *(X, Y)* coordinates and thus has two DOF. To describe the position of the segment itself, an angle θ describing the orientation of the segment with respect to the GCS axes *X* and *Y* must be specified. Thus, in 2-D, a rigid body has three DOF *(X, Y,* and θ).

Y-axes of the GCS. Movements in the frontal and transverse planes exhibit less range of motion during walking or running. Therefore, many of the essential details of motion for walking and running can be determined from a sagittal plane analysis. This simplifies measurement, analysis, and interpretation for describing a movement and serves as an excellent starting place for understanding movements that are mainly planar in nature. The immediate advantage is a reduction in DOF from three *(X, Y, Z)* to two *(X, Y)* to describe the position of a point. For a rigid body in two dimensions, the DOF are reduced from six to three, with only two coordinates *(X, Y)* and one angle *(θ)* serving to locate the object (figure 1.4*b*) in the plane. The remainder of this chapter focuses on planar (2-D) kinematics. However, many of the concepts raised within the context of planar kinematics also apply to spatial (3-D) kinematics, as described in chapter 2.

KINEMATIC DATA COLLECTION

The most common method for collecting kinematic data uses an imaging or **motion-capture** system to record the motion of markers affixed to a moving subject, followed by manual or automatic digitizing to obtain the coordinates of the markers. These coordinates are then processed to obtain the kinematic variables that describe segmental or joint movements. The most common imaging systems use video, digital video, or charge-coupled device (CCD) cameras (e.g., APAS, Elite, Vicon Motus, Qualisys, and SIMI). They record motion using ambient light or light reflected by markers affixed to the body (figure 1.5). In laboratory settings, the cameras have their own lights and the markers have reflective tape that amplifies the marker's brightness compared with the skin, clothing, and background. Other video systems use infrared light or infrared cameras to identify marker locations. Some systems use reflective infrared light (as do Vicon Nexus and Motion Analysis Cortex), whereas others (e.g., NDI's Optotrak) use active infrared light-emitting diodes (IREDs). Active marker systems require a control unit that pulses the individual IREDs in sequence for correct marker identification.

To study planar motion, a single camera placed with its optical axis perpendicular to the plane of motion is sufficient. However, many laboratories use multiple cameras to record 3-D coordinates from both sides of

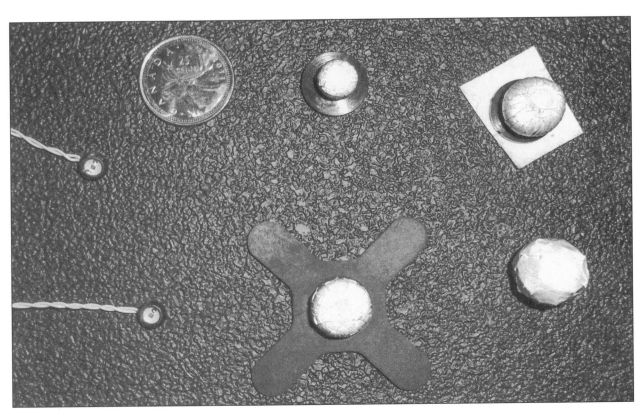

▲ **Figure 1.5** Markers used in the collection of kinematic data with imaging systems. Shown are typical passive reflective sphere markers used with video systems. On the left are active IRED markers that pulse infrared light to signify their locations to a base sensing receiver. A Canadian 25-cent coin (a quarter) is shown for size comparison.

the body (figure 1.6). Locating 3-D coordinates requires only two cameras. However, because markers may be blocked by a body part or may rotate out of the line of sight of either camera, multicamera systems grant a view of each marker by at least two cameras throughout the movement. Thus, a multicamera system is advantageous even for studying planar movements. In addition, with multicamera systems, the exact placement and orientation of each camera are not critical, as readers will see in the following discussion of calibration.

One of the advantages of modern imaging systems is that most of them have automated digitizing that quickly calculates and displays the coordinate position data from multiple markers throughout an entire movement sequence. Before the advent of such systems, 16 mm cinefilm was used to record human motion. Cinefilm has a number of advantages over video, including finer resolution and a wide range of shutter and camera speeds. Cinefilm analysis, however, requires time-consuming manual digitizing of coordinates, and a few seconds of film take hours to digitize. When this limitation is coupled with the delays associated with film processing, long turnaround times are the norm. Furthermore, because there is no way to view the film during or immediately after filming, errors in photogrammetry (focus, lighting, shutter speed) or camera alignment are discovered long after the subject has left the laboratory. Video systems permit real-time viewing of subjects and immediate replay so the user can check the accuracy of the recorded images.

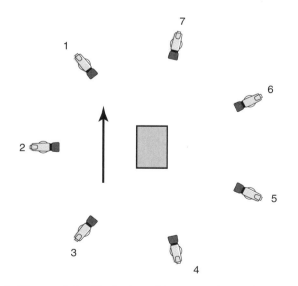

▲ **Figure 1.6** Typical multicamera (seven) setup found in a laboratory for studying human motion. This view is from directly overhead, looking down at the laboratory floor with its imbedded force plate (gray rectangle). The subject movement direction is indicated by the arrow.

Principles of Photogrammetry

According to the American Society for Photogrammetry and Remote Sensing (1980), *photogrammetry* is the "art, science, and technology of obtaining reliable information about physical objects and the environment through processes of recording, measuring, and interpreting images." Biomechanists mostly concern themselves with the factors essential for obtaining clear images on photographic or videographic media. Few biomechanists today use photography or **cinematography** to collect data because of the high cost and the long processing and manual digitizing times that these methods require. Videography is a much more common method, but with all of these mediums, many of the same principles are encountered in the quest to create usable photographic images. In this section, we discuss the factors most important to the biomechanics researcher. One issue, perspective error, is discussed in a later section, Calibration of Imaging Systems. More detailed information about photography, cinematography, and videography, such as the Additive Photographic Exposure System (APEX), must be obtained from specialized sources.

Field of view is defined as the rectangular area seen by the recording medium (film, video) through the camera's optics (see figure 1.7). Care must be taken to ensure that the movement falls within the camera's field of view. Motion before and after the period of interest also must be recorded to prevent inaccuracy near the ends of the coordinate data record that might result from the smoothing process (see chapter 12, as well as Signal, Noise, and Data Smoothing later in this chapter). Unfortunately, this may reduce the size of the subject's image, which should be as large as possible to improve the signal-to-noise ratio. A less obvious problem is that the digitizing system used to quantify the motion may reduce the recorded field of view through a slight magnification that effectively hides markers as they approach the edges of the field of view. Furthermore, motion at the edges of the image may be distorted as a result of poor optics or the use of a wide-angle lens. It is generally a good idea to ensure that the trajectories of markers do not pass near the edges of the field of view of each camera in the system.

One of the most important concerns when one is recording the motion of markers is the exposure time, which in a photographic system is related to the camera speed and shutter speed. *Exposure time* is the duration of time that the recording medium is exposed to light passing through the camera's lens. *Camera speed*, also called the *frame rate*, is how fast the camera records images on the medium. Typical video cameras record at 30 frames per second (fps) in NTSC format (North America) and 25 fps in SECAM or PAL format (Europe), depending on the line frequency of the electrical system (60 Hz in North America, 50 Hz in Europe). Standard

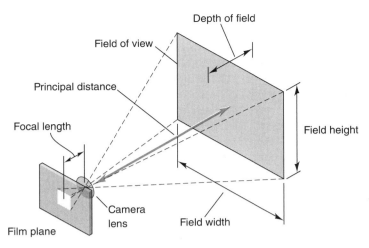

▲ **Figure 1.7** Photographic dimensions.

cine cameras used for filming movies record at 24 fps, but cameras typically used by biomechanists record from 100 to 500 fps. Specially designed video cameras record at rates such as 60, 120, and 240 fps, and expensive digital systems reach 2000 fps. Doubling the line frequency or recording two or more images per frame achieves these rates.

A technique used by some video digitizing systems is to separate interlaced video images. Analog or television video records images as two frames with only half of the lines on a video screen refreshed each 1/60th (North America) or 1/50th (Europe) of a second. Separating the odd and even screen lines into two fields reduces the image quality but doubles the frame rate, producing artificial frame rates of 60 or 50 fps. This also means that markers appear to shift up and down at the rate of 30 or 25 Hz. Low-pass filtering the data or digitizing a stationary marker eliminates this problem. This doubling technique is recommended for recording fast motions, but when slow motions are to be analyzed, use the full video frame for better picture quality.

Clearly, the faster a camera records images, the shorter the exposure time is. If the camera is running at 30 fps, the exposure time must be less than 1/30 s. Most cameras use a shutter that can change the exposure time to as little as 1/2000 s. When markers are poorly lit, these brief exposure times may reduce visibility and prevent automatic **digitizers** from locating the marker's coordinates. In contrast, a long shutter speed (e.g., 1/60 s) makes a fast-moving marker look like a streak across the screen instead of a single spot. In general, shutter speeds of 1/500 or 1/1000 s prevent streaking and make digitizing the coordinates of the marker more reliable. Sufficient light must be available to use these exposure times. Using a camera-mounted light and reflective markers will help to ensure visibility, as will using larger markers and brighter lights and moving the camera or, preferably, the lights closer to the markers. When recording new types of motion, run a pilot test using different shutter speeds, marker sizes, and lighting arrangements to determine the optimal setup.

The final major photographic consideration concerns *focus* and *depth of field*, which refer to the distance in front of and behind the subject that is considered to be "in focus." The camera's distance setting should be set equal to the distance between the lens and the object being filmed. This distance is called the **principal distance**. This number is marked on the camera lens and may be set manually or determined by autofocusing. Autofocusing generally is not useful in biomechanics because the camera may focus on the background or some other object if the subject is not in the field of view when recording starts. Then, as the subject enters the field of view, the optics refocus and temporarily blur the image until the camera can focus on the moving subject. Thus, when you are using autofocus, it is best to autofocus with the subject standing still in the middle of the field of view and then turn off the autofocusing system. Alternatively, measure the distance from the camera to the plane of motion and then set this distance manually on the lens.

Although generally not a factor with video cameras, a camera's aperture setting has several important influences on picture quality and depth of field. Modern video cameras have automatic irises that dilate and constrict like biological eyes, permitting more or less light through the lens to produce a correctly exposed image. If too much light enters the lens during a long exposure time, the image will be overexposed, making digitizing difficult or impossible. For example, if two markers are close together in an overexposed image, they may appear to be one marker. In contrast, if too little light is available because of poor lighting or brief exposure times, the markers will be underexposed and may be indistinguishable by the digitizing system. By changing the aperture of the lens, the user permits more or less light to reach the film. The *aperture* of a lens is the size of the opening of the lens's iris. The *iris* typically is a set of "leaves" that open and close to change the amount of light that passes through the lens. The *f-number* or *f-stop* of a lens is the ratio of the aperture of the lens divided by the *focal length*, which is the apparent distance between the front of the lens and the recording medium. As the iris closes, less light passes through the lens. Even with a completely open iris, however, some light is lost between the front of the lens and the back because of imperfections in the optical glass, filters, and lens coatings and the refraction and reflection of light within the lens barrel.

Each lens is marked with its maximum possible aperture. For example, a standard lens may be rated with an f-stop of 2, which means it will reduce the light by one-quarter with its iris fully open. Standard f-stops are based on powers of the root of 2 and rounded off to simplify their application. Table 1.1 shows standard f-stops and the amount of light each one restricts through the lens. At each higher f-stop, the amount of light permitted to reach the film is halved. Thus, increasing the f-stop from f/2 to f/2.8 (1 f-stop) reduces the light by 1/2. Conversely, each decrease in f-stop, such as from f/11 to f/8, doubles the amount of light that reaches the film. Zoom lenses, as a result of their complex structure, reduce the amount of light that passes through them by between 1/4 (2 f-stops) and 1/32 (5 f-stops).

Changing the aperture also affects the depth of field. As the iris closes down (is changed to a higher f-stop), there is less curvature in the part of the lens that focuses the image. This keeps the image in focus better and widens the depth of field. Some lenses show the relationship between the depth of field and aperture, making it possible to determine what region on either side of the focal distance will be in focus in the image. This usually is not possible on zoom lenses, however, because the focal length of the lens also affects the depth of field. For example, lenses with long focal lengths—telephoto lenses—magnify the image and reduce the depth of field. Conversely, wide-angle lenses reduce the image size, allowing larger fields of view, but increase the depth of field. However, wide-angle lenses should be avoided for biomechanics research because they distort the image and create artificial motions that must be corrected with complex transformations.

Calibration of Imaging Systems

For any system of collecting kinematic data, a suitable means of calibration must be used to ensure that the image coordinates are correctly scaled to size. For imaging systems, there are two basic methods. For 2-D systems with one camera, the simplest method is to use a calibrated ruler or surveyor's rod placed in the subject's plane of motion. The camera must be perpendicular to the plane of motion; otherwise, there will be linear distortions of the types shown in figure 1.8b. By digitizing the length of the ruler, a scaling factor (s) is determined for scaling in both directions (X, Y), that is,

$$s = \frac{\text{Actual length (meters)}}{\text{Digitized length (meters)}} \qquad (1.1)$$

$$x = su \qquad (1.2)$$

$$y = sv \qquad (1.3)$$

where u and v are the digitized coordinates of a marker and x and y are the scaled coordinates (see figure 1.8a).

The preferred method of calibration—essential for multicamera systems—is to establish a series of control points. Control points are markers, attached to a structure or affixed to the film site or laboratory, whose exact coordinates are known. For example, figure 1.9a shows a grid of 15 points on a board used to calibrate planar motion across a laboratory walkway. For 2-D analyses, at least four noncollinear points are required. At least six noncoplanar locations on a 3-D structure are needed for 3-D analyses. Figure 1.9, b and c, shows several 3-D structures with control points for calibrating in three dimensions. Other types of control-point structures are also used, such as Woltring's (1980) method, which uses several views of a plane of markers. Some commercial systems film a calibrated wand (figure 1.9d) that is wafted around the volume of the movement space (Dapena et al. 1982).

After the control points are filmed, equations are computed to scale the digitized coordinates into real metric units. In single-camera 2-D setups, image distortion resulting from misalignment of the camera's optical axis perpendicular to the plane of motion can be corrected. The common method of enabling the transformation of the data from digitized coordinates to real metric units is called *fractional linear transformation* (FLT) when applied to two dimensions or *direct linear transformation* (DLT) when applied to three dimensions

Table 1.1 Standard Photographic Apertures, Exposure Times, and Film Speeds

APEX value[a]	0	1	2	3	4	5	6	7	8	9	10
Aperture (f-stop)	1	1.4	2	2.8	4	5.6	8	11	16	22	32
Exposure time (s)	1	1/2	1/4	1/8	1/15	1/30	1/60	1/125	1/250	1/500	1/1000
Film speed (ISO or ASA)[b]	3	6	12	25	50	100	200	400	800	1600	3200

[a]APEX stands for the Additive Photographic Exposure System. Each increase in APEX value indicates a decrease in the light level by one-half.

[b]Each faster film speed requires half the amount of light that the previous speed needed for proper exposure.

(u, v) digitized coordinates   **(x, y) scaled coordinates**

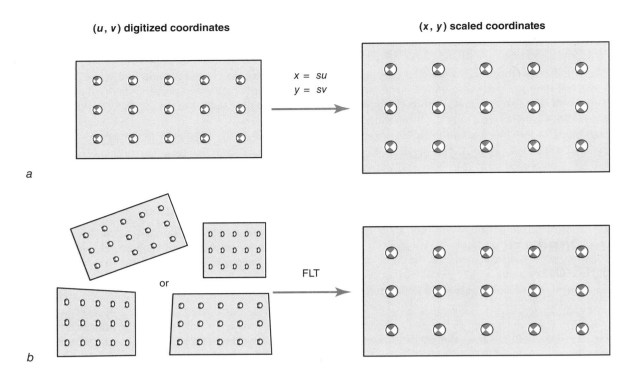

$$x = su$$
$$y = sv$$

FLT

a

b

▲ **Figure 1.8** Comparison of scale factor method *(a)* and fractional linear transform method *(b)* for transforming digitized images to real-life measurements.

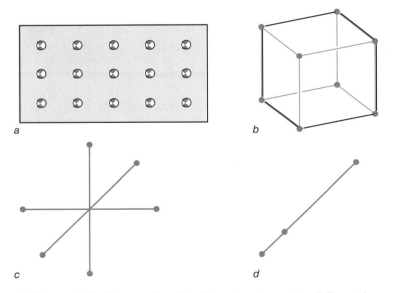

a   *b*

c   *d*

▲ **Figure 1.9** Types of calibration structures for 3-D motion calibration.

(Abdel-Aziz and Karara 1971; Walton 1981; Woltring 1980). The digitized coordinates that result are said to be **refined** rather than **scaled** because the data are altered in a more complex way.

When a camera is tilted with respect to the plane of motion, distances are distorted as illustrated in figure

1.8*b*. FLT and DLT correct these types of errors. In 2-D analyses, however, perspective errors may occur in which objects appear to shorten as they move farther from the camera. One way to minimize this effect is to use a telephoto lens and to zoom in on the subject to make the motion fill as much as possible of the camera's field of vision. This technique flattens the subject, reducing perspective errors. Of course, one needs a large enough room and clear lines of view to use this method. The FLT equation takes the form

$$x = \frac{c_1 u + c_2 v + c_3}{1 + c_7 u + c_8 v} \tag{1.4}$$

$$y = \frac{c_4 u + c_5 v + c_6}{1 + c_7 u + c_8 v} \tag{1.5}$$

where c_1 to c_8 are the coefficients of the FLT, u and v are the digitized marker coordinates, and x and y are the marker's refined coordinates in metric units. Note that when the camera's optical axis is very close to perpendicular to the calibration plane, the coefficients c_2, c_4, c_7, and c_8 are nearly zero. The coefficients c_1 and c_5 become scaling factors, similar to those that would have been digitized if a ruler was used. The reason these numbers are not equal is because most digitizers have different scales for horizontal and vertical. The coefficients c_3 and c_6 correspond to the differences between the origin of the calibration system and the origin of the digitization system.

To check the accuracy of these systems, compare the actual coordinate locations of the control points with their refined (i.e., transformed) coordinates. The standard error or root mean square (RMS) of the differences between the true coordinate locations and their digitized and refined locations reflects the accuracy of the system. Better still, film, digitize, and refine a second set of known coordinates to compute the system's accuracy.

Two-Dimensional Marker Selection

We have described how a Cartesian coordinate system locates a point in space and how imaging systems record the location of reflective markers. Logical questions for a biomechanist interested in studying human movement are as follows:

- ▶ Where do I place these reflective markers on my subject?
- ▶ How many markers should I use?
- ▶ Do I need other markers (not on the subject) in the camera's field of view?

The answers to these questions vary depending on the nature of the movement under study and the exact research questions being asked. An excellent starting point is to construct a model of the important anatomical parts involved in the motion. For planar motion, this model can be represented by a stick figure, as shown in

figure 1.10. For example, the model for a person running might include body segments representing the foot, leg, and thigh of each lower extremity; the upper and lower portions of each upper extremity; and a trunk segment. For a seated cyclist, a segment representing the bicycle crank arm might be needed along with body segments representing the foot, lower leg, and thigh of only one lower extremity, plus the trunk. The simplified model for the cyclist is possible because the bicycle crank imposes symmetrical motion of the legs. Also, if the cyclist is instructed to keep his hands stationary on the handlebars, upper extremity motion is minimal and can be excluded from the model.

Once a suitable model has been constructed, it can guide the researcher in marker selection and placement. At least two points must be quantified for each segment in which planar motion is to be modeled. Often, markers are placed at the estimated centers of rotation at each end of the segment or over proximal and distal anatomical landmarks (defined in chapter 3). Two points are needed to define the planar angular orientation of the segment (see Angular Kinematics later in this chapter). Therefore, for our cycling example, roughly 10 markers would suffice. For running, however, we would need more markers to properly represent the body (figure 1.11). Additionally, if we were recording the runner's GRFs, we would need another marker affixed to the force plate to correctly locate the position of the force vector on the runner's foot (see chapters 4 and 5).

The exact number of markers needed also depends on the nature of the 2-D motion at the joints between adjacent segments. A simple pin joint or hinge joint (e.g., phalangeal joints of fingers and toes) has only one degree of freedom—the freedom to rotate about the pin or hinge. Note that here, *degrees of freedom* has a slightly

▲ **Figure 1.10** *(a)* A simplified sketch (left) and stick-figure representation of a runner. *(b)* A simplified sketch (left) and stick-figure representation of a cyclist.

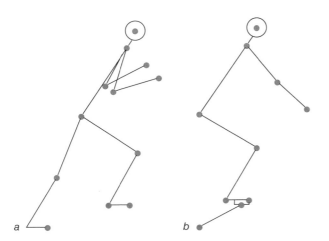

▲ **Figure 1.11** Markers needed for representing a runner (left) and a cyclist.

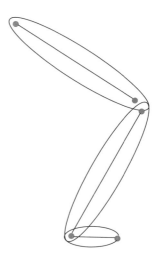

▲ **Figure 1.12** A joint's possible motions influence the selection of marker locations. The ankle and hip joints can be modeled as hinges and designated with one marker. Two markers can be used to designate the distal end of the thigh and proximal end of the leg, because at the knee both rotation and translation are possible.

different meaning than in our description of quantifying the position of a point or rigid body. The hip is also modeled as a hinge joint for planar analysis, despite its ball-and-socket construction that allows three different spatial rotations. For hinge joints, the endpoint marker for adjacent segments can be placed directly over the point representing the hinge (figure 1.12). Other joints are more complex; the knee joint, for example, permits flexion-extension and some translation across the tibial plateau (called *shear*). This means that the knee has two degrees of freedom (one for rotation and one for translation). In such situations, it is impossible to place a marker that will always represent the position of the moving hinge, and separate endpoint markers must be used for the adjacent segments (figure 1.12). In some cases, researchers choose to ignore small translational motions and simply model the joint as a hinge. They must select an appropriate level of detail for the specific research question being studied. See chapter 3, tables 3.1 and 3.3, for common marker locations used in 2-D studies and for measurements for computing the locations of "virtual" points without actual markers (e.g., segment centers of rotation and centers of mass).

Marker-Free Kinematics

In many situations, it is impractical, impossible, or undesirable to place markers on the subject performing the motion, such as athletes in competitions and patients in clinical situations who are unable to endure an extended experimental preparation period. In these situations, the researcher must digitize the movement record manually, using identifiable anatomical landmarks to locate the points necessary for the movement model. Efforts have been made, however, to develop automatic markerless **motion-analysis systems** (Corazza et al. 2007;

D'Apuzzo 2001; Trewartha et al. 2001). Such systems are not yet in widespread use but would be major breakthroughs in the analysis of human motion because they permit shorter data acquisition sessions in clinical and research laboratories and extend data collection beyond the laboratory to more ecological environments.

Marker-free systems use computer graphics techniques to match the shape of a body segment with a predetermined shape (Corazza et al. 2007; Trewartha et al. 2001). Initially, a model of the body that matches the general size and morphological characteristics of the subject is constructed. A video of the moving subject is recorded and converted to digital form. Computer software then attempts to align the computer model with the image of the actual performer in each frame of video data. If the computer can find an acceptable position for the model in every frame, that effectively replicates the motion of the subject. The software extracts the positions of individual points and segments from the model, effectively digitizing the locations of desired points. Corazza and Andriacchi (2009) used their system to track the total body center of mass for posturographic analysis, but such systems have not become standard methods for 3-D scientific kinematic or kinetic analyses of human motion.

LINEAR KINEMATICS

The Cartesian coordinate system permits quantification of the position of a point or rigid body either in

3-D space or on a 2-D plane. It is also the starting place for a complete kinematic description of any human motion. In this section, we introduce the kinematic variables of displacement, velocity, and acceleration, all of which describe the manner in which the position of a point changes over a period of time. The mathematical processes of differentiation and integration—the main concepts of calculus—relate these kinematic variables. **Displacement** is defined as the change in position. **Velocity** is the time derivative of displacement, defined as the rate of change of displacement with respect to time. **Acceleration** is the time derivative of velocity, defined as the rate of change of velocity with respect to time. Acceleration is therefore the second derivative of displacement with respect to time. These three kinematic variables can be used to understand the motion characteristics of a movement, to compare the motion of two different individuals, or to show how motion has been affected by some intervention. Sometimes, determining the time derivative of acceleration, called *jerk*, is desirable to assess the severity of head impacts during car crashes or collisions with surfaces or projectiles. Table 1.2 lists the commonly used kinematic measures and their associated symbols and units in the International System of Units (SI), including angular kinematic variables, which are discussed later.

In some situations, acceleration can be measured directly with a device aptly called an **accelerometer**. Integral calculus is then used to find the velocity and displacement data. Velocity is the first integral of acceleration with respect to time, whereas displacement is the integral of velocity with respect to time. Information on accelerometers is presented later in this chapter, and methods of integration are detailed in the section on the impulse-momentum relationship in chapter 4. The remainder of the present discussion concentrates on the process of differentiation.

Computing Time Derivatives (Differentiation)

There are several ways to compute the time derivatives of displacement once the position of a point has been established as a function of time for a particular movement. The starting point for kinematic analysis is the coordinate data that were digitized at equal time increments throughout the movement sequence. The exact number of data points and the time increment depend on the duration of the movement and the sampling rate of the motion capture system. For example, a 2 s movement captured at 200 fps yields an input data stream of 400 numbers, with each point separated by 5 ms (0.005 s). For the differentiation process, the X and Y coordinates are treated separately, meaning that these 2 s of motion produce 400 X positions and 400 Y positions for each marker digitized in the motion sequence; the earlier example of a runner with 20 markers would therefore generate 40 separate data streams, each one 400 numbers in length. These data are often listed in row-by-column format, with the columns containing the X and Y coordinates of different markers and the rows showing the incremental steps in time throughout the movement. Table 1.3 is a portion of such a data table.

There are three categories of methods for computing derivatives. Analytical methods involve the differentiation of mathematical functions and are taught in high school and college differential calculus courses. Graphical methods use the concept of the instantaneous slope (rise over run) of a graphed function. Finally, numerical methods apply relatively simple computing formulas to a set of data points that represent any varying function. All three categories are used in biomechanics, because each has strengths and weaknesses. To use analytical techniques, the positional data collected at equal time

Table 1.2 Kinematic Variables and Their SI Units and Symbols

Measure	Definition	Units	Symbols
Linear position, path length, or linear displacement		**m**, cm, km	x, y, z, s (arc), d (moment), r (radius)
Linear velocity	ds/dt	**m/s**, km/h	v
Linear acceleration	dv/dt, d^2s/dt^2	**m/s²**, g (=9.81 m/s²)	a
Linear jerk (jolt)	da/dt	**m/s³**, g/s	j
Angular position, plane (2-D) angle, or angular displacement		**rad**, °, r (r = revolution)	ω, β, γ, ϕ
Angular velocity	$d\theta/dt$	**rad/s**, °/s, r/s	ω
Angular acceleration	$d\omega/dt$, $d^2\theta/dt^2$	**rad/s²**	α
Solid (3-D) angle		**sr** (steradian)	Ω

Common unit is in **bold**.

Table 1.3 Sample Marker Data From a Walking Trial

Frame no.	Time	RIGHT SHOULDER X (cm)	Y (cm)	RIGHT HIP X (cm)	Y (cm)	RIGHT KNEE X (cm)	Y (cm)	RIGHT ANKLE X (cm)	Y (cm)	BALL OF RIGHT FOOT X (cm)	Y (cm)
1	0.000	−7.3	154.3	−6.1	92.4	−3.8	50.2	−8.6	13.7	1.3	1.1
2	0.020	−7.2	154.3	−6.3	92.2	−3.0	50.1	−8.6	13.8	1.3	1.1
3	0.040	−7.0	154.2	−6.7	91.8	−2.0	49.9	−8.5	13.8	1.3	1.1
4	0.060	−6.7	153.9	−7.2	91.3	−0.8	49.7	−8.4	13.9	1.3	1.1
5	0.080	−6.4	153.2	−7.8	90.6	0.6	49.4	−8.2	14.0	1.3	1.1
6	0.100	−6.0	152.0	−8.5	89.6	2.1	48.9	−8.1	14.0	1.3	1.1
7	0.120	−5.6	150.3	−9.4	88.4	3.5	48.5	−7.9	14.1	1.3	1.1
8	0.140	−5.2	148.1	−10.3	86.9	4.9	47.9	−7.7	14.1	1.2	1.1
9	0.160	−4.8	145.4	−11.3	85.2	6.2	47.4	−7.5	14.2	1.2	1.1
10	0.180	−4.3	142.4	−12.3	83.3	7.3	46.8	−7.3	14.2	1.2	1.1
11	0.200	−3.9	139.0	−13.3	81.3	8.3	46.3	−7.2	14.2	1.2	1.1
12	0.220	−3.4	135.5	−14.3	79.2	9.1	45.8	−7.1	14.2	1.2	1.1
13	0.240	−2.9	131.8	−15.2	77.1	9.8	45.3	−7.0	14.2	1.2	1.1
14	0.260	−2.5	128.2	−16.0	75.0	10.4	44.8	−6.9	14.2	1.2	1.1
15	0.280	−2.1	124.6	−16.7	72.9	10.9	44.4	−6.8	14.2	1.2	1.1
16	0.300	−1.7	121.1	−17.4	71.0	11.4	44.1	−6.7	14.3	1.2	1.1
17	0.320	−1.3	117.7	−17.9	69.1	11.7	43.8	−6.7	14.3	1.2	1.1
18	0.340	−1.0	114.4	−18.4	67.3	12.0	43.5	−6.6	14.3	1.2	1.1
19	0.360	−0.6	111.3	−18.8	65.6	12.2	43.3	−6.6	14.4	1.2	1.1

increments must first be fitted to an appropriate mathematical function. Once the position data are in equation form, analytical techniques produce equations that represent the corresponding velocity and acceleration patterns. A strength of this procedure is that it generates velocity and acceleration data that are free of numerical errors. Graphical methods are slow and do not lend themselves to either numerical or analytical data forms, but the ability to graphically differentiate is of great value in checking the validity of results from the other two computation methods. The most common approach is numerical differentiation, primarily because of the manner in which experimental data are collected. The columns of position coordinates equally spaced in time are in the precise format needed for applying numerical techniques. However, if not used carefully, numerical differentiation methods can result in computational errors. The following sections describe the steps necessary to minimize these computational errors.

There are a variety of different numerical derivative formulas. The equations presented here to obtain velocity and acceleration values are examples of *finite*

difference calculus. The method used is the *central difference* method.

$$v_i = \frac{s_{i+1} - s_{i-1}}{2(\Delta t)} \tag{1.6}$$

$$a_i = \frac{v_{i+1} - v_{i-1}}{2(\Delta t)} = \frac{s_{i+2} - 2s_i + s_{i-2}}{4(\Delta t)^2} \tag{1.7}$$

$$\text{or} \quad a_i = \frac{s_{i+1} - 2s_i + s_{i-1}}{(\Delta t)^2} \tag{1.8}$$

where v_i and a_i are the velocity and acceleration of the marker at time i, Δt is the sampling interval of the data (in seconds), and s is the linear position (x or y in meters). Notice that to obtain velocity and acceleration at a particular instant *(i)*, data from the time intervals before $(i - 1, i - 2)$ and after $(i + 1, i + 2)$ that instant are required. This is problematic at the start $(i = 1)$ and end $(i = n)$ of a data stream, because not all of these points are available. One way to obtain a velocity at time 1 or n is to use the following *forward* and *backward* difference equations (Miller and Nelson 1973):

$$v_1 = \frac{s_2 - s_1}{\Delta t} \tag{1.9}$$

$$v_i = \frac{s_{i+1} - s_{i-1}}{2(\Delta t)} \tag{1.10}$$

$$a_1 = \frac{s_1 - 2s_2 + s_3}{(\Delta t)^2} \tag{1.11}$$

$$a_n = \frac{s_n - 2s_{n-1} + s_{n-2}}{(\Delta t)^2} \tag{1.12}$$

A better way to obtain these first and last derivatives is to collect extra data, both before and after the motion of interest. If this cannot be done, *pad* the ends of the data with extra data, take the derivatives, and then discard the padding points. Padding points can be derived by mirroring the data at each end (Smith 1989) or linearly or nonlinearly extrapolating from the endpoints.

Signal, Noise, and Data Smoothing

Unfortunately, as a result of errors introduced by the digitization process, numerical calculations of velocity and, in particular, acceleration yield results contaminated with high-frequency noise. Figure 1.13 shows the acceleration pattern of a toe marker that was computed from position data digitized from video. Notice that an irregular pattern occurs even during the period when the marker is supposed to be motionless on the ground

(from 0.03 to 0.07 s). No amount of careful digitizing can eliminate this problem.

These noise spikes occur because small errors in the digitizing process represent large accelerations. Finding the second derivative of a pure sine wave analytically illustrates this phenomenon. Mathematically, this produces another sine wave that is phase shifted by 180°. However, if noise is present in the original sine wave, the second derivative is very different. Figure 1.14 compares the second derivatives of a pure sine wave and one to which random or white noise of amplitude ±0.01% (in other words, a signal-to-noise ratio of 10,000:1) was added before double differentiation. Clearly, the noise dominates the second derivative signal. For kinematic time derivatives, noise in the position data must be removed to obtain a valid acceleration pattern. There are a number of acceptable *smoothing* methods for removing the high-frequency noise induced by the digitizing process, including low-pass digital filtering, piecewise quintic splines, and Fourier series reconstruction (see chapter 12).

Accelerometers

One commonly used technology for directly measuring the kinematic variable of acceleration is the accelerometer. There are three types of accelerometers: strain gauge, piezoresistive, and piezoelectric. Piezoelectric accelerometers use the piezoelectric effect to measure acceleration. The **piezoelectric effect** occurs when certain crystals, such as quartz, are mechanically **stressed** causing a voltage. Piezoelectric accelerometers typically

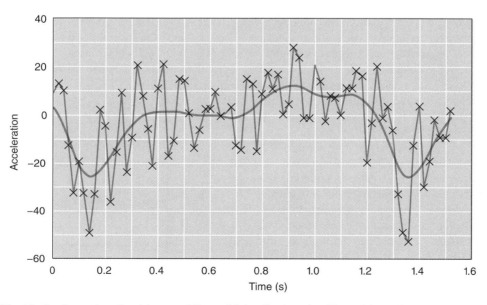

▲ **Figure 1.13** Vertical acceleration history of filtered (blue line) and unfiltered (gray line marked with × symbols) toe marker during walking. Notice that the unfiltered data vary irregularly around the filtered data.

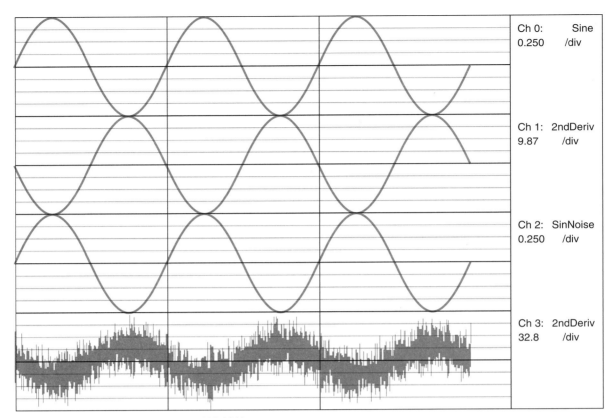

Sampling time: 3.00 s, sampling rate: 1000.00 Hz

▲ **Figure 1.14**   Histories of sine wave and its second derivative followed by the same sine wave with white noise added and its second derivative. The top waveform is the original sine wave sampled at 1000 Hz over 3 s. The second waveform is the second derivative of the sine wave. The third waveform is the sine wave with noise added (random numbers of range ±0.0001). The fourth waveform is the second derivative of the sine wave with the added noise.

have higher frequency responses than strain gauge accelerometers, but they do not have a true static response, so they should not be used for recording slow motions or periods of inactivity. Figure 1.15 illustrates the inner workings of a strain gauge accelerometer. Tiny strain gauges are attached to a cantilevered beam to measure the bending of the beam. If the beam is subjected to acceleration, the inertial mass at its free end causes the beam to bend in proportion to the imposed acceleration. Clearly, a sharp blow can easily damage such a fine element. Accelerometers are available as uniaxial units for measuring acceleration in one direction or in triaxial packages for measuring acceleration in three orthogonal directions. The calibration of an accelerometer depends on the type of sensing element it has. Strain gauge and piezoresistive units exhibit a static (DC) response and are calibrated by aligning their sensitive axis within the gravitational field. Because piezoelectric accelerometers lack this DC response, they must be calibrated at the factory using dynamic techniques. Padgaonkar and colleagues (1975) outlined an accelerometric system that can quantify the triaxial linear and angular accelerations (six DOF) of a single segment but requires nine uniaxial accelerometers to achieve stable results.

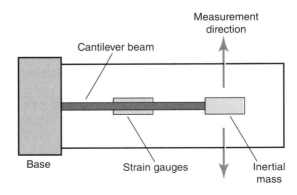

▲ **Figure 1.15**   Schematic diagram of a strain-gauge accelerometer.

FROM THE SCIENTIFIC LITERATURE

Mayagoitia, R.E., A.V. Nene, and P.H. Veltink. 2002. Accelerometer and rate gyroscope measurement of kinematics: An inexpensive alternative to optical motion analysis systems. *Journal of Biomechanics* 35:537-42.

These authors used four uniaxial seismic accelerometers and two rate gyroscopes to measure the kinematics of two lower-extremity segments during walking on a treadmill. They simultaneously quantified the motion with a well-calibrated optical system (Vicon). Although **accelerometry** usually is a real-time system, the complex mathematics required to derive the 2-D kinematics of the segments for comparison with the optical were done offline. The system performed favorably compared with a more expensive optical imaging system. The researchers noted that although their system was tested in a laboratory setting, it is capable, with an appropriate data logger, of collecting motion data in "almost any environment." Of course, it has the usual limitations of electronic systems, such as being unsuitable for measuring more than a few segments at a time rather than simultaneously measuring many segments during whole-body motions.

Graphical Presentation of Linear Kinematics

There are several methods of graphically presenting kinematic measures. Plotting formats for linear kinematic variables include **trajectories** (X-Y or X-Y-Z plots), in which the path of the object is displayed, and time series or *histories*, in which variables are graphed as a function of time. Figure 1.16 shows a trajectory and a time series for two segmental markers. Time series representation is the most prevalent plot used in biomechanical studies because it permits comparison of different kinematic features that occur simultaneously. It is also useful for comparison with kinetic variables such as force and biophysical signals such as muscle activity (recorded with electromyography [EMG]). In this way, the motion can be linked with the underlying forces or physiological events. Trajectory plots are useful when studying motion that occurs in both directions within the plane of motion (e.g., the motion of the knee joint during running, which has significant vertical and horizontal components). These plots also illustrate how a movement in a secondary direction accomplishes a prescribed movement in another direction. For example, Bobbert and van Zandwijk (1999) demonstrated how the center of mass moves in the anterior-posterior direction during a vertical jump by plotting the X-Y trajectories of the jumper's center of mass.

Ensemble Averages

In biomechanics research articles, authors often want to present data that are pooled across several trials of a particular movement condition. This involves combining the trials for an individual subject or combining the data of several subjects to get a sense of the kinematic patterns produced by the group. In both cases, the patterns produced by the pooling of data are known as *ensemble averages*. The first step in creating these average curves is to standardize the movement time for all the trials, because each one has a different duration. For example, a subject might complete the first trial in 0.8 s and the second in 0.9 s. This time normalization is accomplished by identifying a given portion of the motion (performed in all trials) as 100% and rescaling the sampling rate as a percentage of the duration. For cyclic motions such as gait, one complete cycle (right-foot toe-off to right-foot toe-off, for example) is normally the complete scaling period. Alternatively, the stance phase time alone could be used.

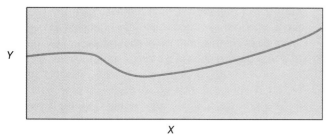

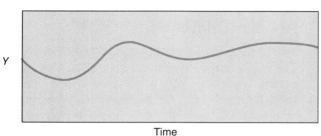

▲ **Figure 1.16** Comparison of *(a)* path trajectory graph and *(b)* a position history.

To accomplish time normalization, the original data for a given kinematic variable must be interpolated to find the values of that variable at specific times. This produces data equally spaced throughout the rescaled movement time (often each 1% or 2% of the motion). For example, if data were collected at a sampling rate of 200 Hz during a movement that lasted 0.9 s (900 ms), there would be 180 data points separated by intervals of 5 ms. If the 900 ms were rescaled to 100%, to identify data values at 1% intervals we would need to find 101 (0%-100%) data points separated by intervals of 9 ms. To do this, we interpolate between data that are 5 ms apart to estimate the values for data 9 ms apart. This can be done in several ways, with the two most common being linear interpolation and splining (with cubic or quintic splines).

For linear interpolation, it is assumed that the collected data were separated by such a small time increment that motion between any two consecutive times was linear in nature. For example, if a marker moved in a given direction from position 1.185 m at time 0.310 s to position 1.190 m at time 0.315 s, linear interpolation computes the following position-time data:

0.310 s	1.185 m
0.311 s	*1.186* m
0.312 s	*1.187* m
0.313 s	*1.188* m
0.314 s	*1.189* m
0.315 s	1.190 m

For data collected at sufficiently high sampling rates, this method works well. In cases in which movements occur rapidly and the sampling rate is not excessively high, a splining method can be used. The concept behind splines is the same as that with linear interpolation, although there is no assumption that the data within sampling intervals are linear. Cubic splining fits the entire data set (100% in rescaled time) to a series of cubic (third-order) polynomial equations, allowing the user to evaluate the equations at any chosen time, not just the time of the original data. For this reason, these equations are called *interpolating splines*. The quintic spline is similar to the cubic, except that fifth-order polynomials are used (Wood 1982). The advantage of quintic splines is that their second derivatives are cubic splines, whereas the second derivatives of cubic splines are lines. Thus, using quintic splines is preferred if the data are to be double differentiated. The resulting "acceleration" curves will therefore be continuous curves rather than a series of connected line segments.

After each trial's data are normalized, an ensemble average is created. From every trial to be included in the ensemble, the data representing the first time point (e.g., 0%) are summed and the mean and standard deviation (SD) are computed. This process is repeated for each time point of the movement, yielding a 101-point series for both the ensemble average and the SD about that average. The ensemble average data for different movement conditions within a study may be graphed and compared to illustrate variations in the motion. Alternatively, the ensemble average could be plotted along with ±1 SD bounds or the 95% confidence interval (±1.96 ± SD) to illustrate the degree of variation within the data (figure 1.17). This process is not unique to linear kinematics and can be applied to any type of signal, including angular kinematics, kinetics, and muscle activity variables (e.g., angular velocity, forces, moments of force, and EMGs).

ANGULAR KINEMATICS

Angular position measures can be divided into two classes. The first class concerns the angular position or orientation of single bodies. These are called *segment* or *absolute angles*, because they are usually referenced to an absolute or Newtonian frame of reference. The second class concerns the angle between two, usually adjacent, segments of the body. These are called *relative, joint*, or *cardinal angles*, because they measure the angular position of one segment relative to another.

Segment Angles

As stated earlier in the section on markers, at least two points must be quantified to describe the angular position of a human body segment in a 2-D plane. These absolute angles follow a consistent rule called the *right-hand rule* (figure 1.18), which specifies that positive rotations are counterclockwise and negative rotations are clockwise. Curving the fingers of the right hand in the direction of the angle or rotation and then comparing the direction of the thumb to the reference axes will indicate the sign of an angle or rotation about a particular axis. If the thumb points in the direction of a positive axis, then the angle or rotation is positive. For planar analyses, segment angles are often defined as the angle of the segment with respect to a right-horizontal line originating from the proximal end of the segment. Other conventions are possible, and researchers must identify the one they use.

Angular Conventions

Two conventions are used to quantify segment angles. The first measures angles like a compass, with angles ranging from 0° to 360°, whereas the second allows a range from +180° to −180° (figure 1.19). These conventions yield the same values for angles between 0 and +180° but different ones for angles between 180° and 360°. With the second convention (see figure 1.20), these angles range from 0° to −180°, making them easier to visualize.

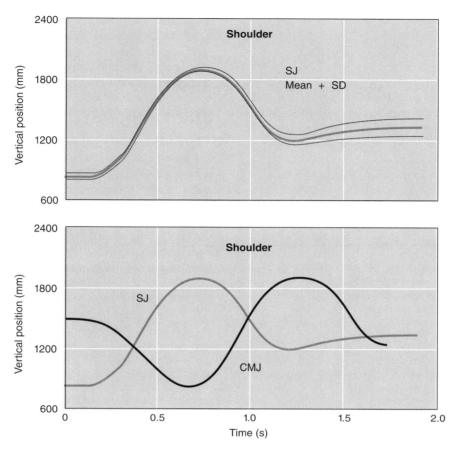

▲ Figure 1.17 Ensemble average plots of the vertical position of a marker attached to the shoulder of a subject while performing vertical jumps. The top panel shows the mean ±1 standard deviation of four squat jumping (SJ) trials. The bottom panel compares the ensemble from the squat jumps with the ensemble from 3 countermovement jumps (CMJ).

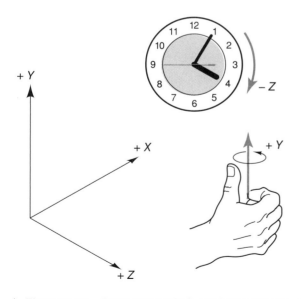

▲ Figure 1.18 Right-hand rule for defining directions of rotations.

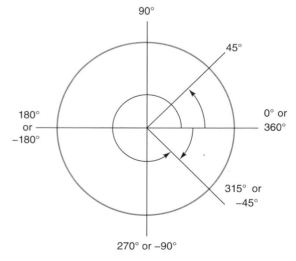

▲ Figure 1.19 Two conventions for defining the absolute angles of segments. One convention always measures angles in the range 0° to 360° whereas the other uses a range from 180° to −180°.

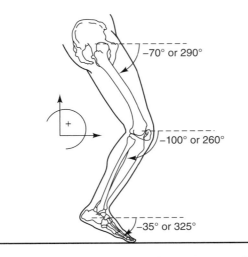

▲**Figure 1.20** Examples of absolute angles of the lower extremity. All angles are taken from a right horizontal from segment's proximal end.

Discontinuity Problems

With both conventions, a problem arises when a segment crosses the 0°/360° line or the ±180° line. For example, when a segment moves clockwise from an angle of 10° to 350°, the change is recorded as 350° − 10° = +340° instead of the correct value of −20°. Similarly, for the ±180° convention, whenever a segment moves counterclockwise from +170° to −160°, the difference is said to be −160° − 170° = −330° instead of the correct change of +30°. In both cases, it is possible that the segments rotated in the opposite directions through the larger angle changes, although this is unlikely if the data sampling rate was correct. To solve this dilemma, it is assumed that no angle changes by greater than 180° from one frame to the next. When one is graphing angular histories,

discontinuities also arise when a segment crosses 0°/360° or ±180°. This can be resolved by adding or subtracting 180° whenever an absolute angular change of more than 180° occurs. Of course, this correction can create angles that exceed ±180°.

Joint Angles

The human body is a series of segments linked by joints, so measurement and description of relative or joint angles are often useful. Quantification of a joint angle requires a minimum of three coordinates or two absolute angles, as shown in figure 1.21. When defining joint motions, we must remember that adjacent joints may have different directions for the same type of motion. For example, if knee flexion is a positive rotation according to the GCS, then flexion of the hip is a negative rotation (see figure 1.22). Notice that a biomechanical system that respects the right-hand rule is presented. Also shown is a medical system that is used by physiotherapists, anatomists, and the medical community to define the anatomical positions of joints. With the latter system, negative angles are avoided by specifying the type of joint motion (such as flexion, extension, and hyperextension).

Polar Coordinates

Angular motion of a rigid body also produces linear motion of individual points (such as markers) attached to the body. Furthermore, the amount of linear displacement of a point depends on its location with respect to the axis about which the body rotates. Consider the motion of the minute hand of a clock (figure 1.23). A marker placed at the center of the clock will undergo no linear displacement as the minutes tick by, but a marker at the end of the hand will sweep out a circular path.

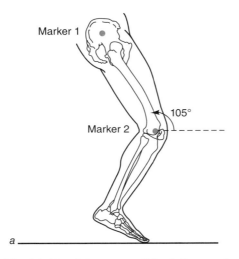

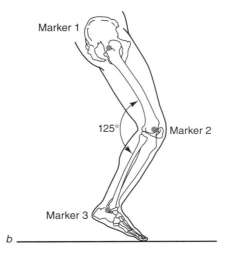

▲**Figure 1.21** *(a)* Absolute versus *(b)* relative angles.

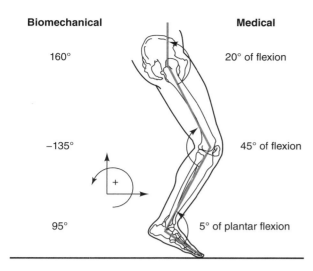

Figure 1.22 Examples of relative angles of the lower extremity.

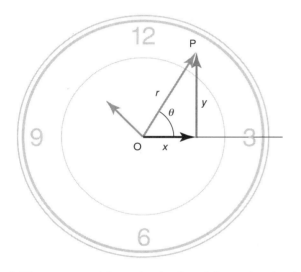

▲ **Figure 1.23** The polar *(r, θ)* and Cartesian *(x, y)* coordinates for the minute hand of a clock.

The mathematical description of such angular-to-linear motion is reflected in the use of *polar coordinates*, an alternative to Cartesian coordinates. In the polar system, as with the Cartesian system, two degrees of freedom describe the planar position of a point on a plane. A line is constructed from the origin of the axis system to the point (figure 1.24). The length *(r)* of the line represents one DOF, whereas the angle *(θ)* between the line and one of the reference axes (usually the axis right-horizontal from the clock center) describes the second DOF. These are called the polar coordinates of the point and are written *(r, θ)*. Note in figure 1.24 that drawing a line from the point to the reference axis forms a right

triangle. Using simple trigonometry, we can convert from polar to Cartesian coordinates as follows:

$$x = r \cos \theta \tag{1.13}$$

$$y = r \sin \theta \tag{1.14}$$

If we consider the center of the clock to be the origin, a marker farther from the center will have a greater length, *r*. Thus, for one complete rotation, a point at the end of the minute hand sweeps out a longer path than a point at the end of the shorter hour hand. Although Cartesian coordinates are used most often in biomechanics research, for some applications it is more convenient to use polar coordinates. It is always possible, given the Cartesian coordinates *(x, y)* of a point, to compute its polar coordinates *(r, θ)* using the following trigonometric relationships:

$$r = \sqrt{x^2 + y^2} \tag{1.15}$$

$$\theta = \tan^{-1}(y/x) \tag{1.16}$$

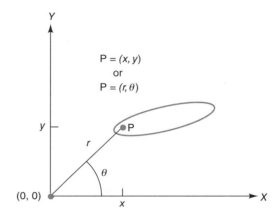

▲ **Figure 1.24** Polar coordinates can represent the position of a point as readily as Cartesian coordinates. Polar coordinates use the length of a line segment joining the point and the origin *(r)* as one coordinate and the angle between the line segment and a fixed axis *(θ)* as the second coordinate. The location of point *P* can be expressed either as *(x, y)* in the Cartesian system or as *(r, θ)* in polar coordinates. Note that in either system the point has two DOF.

Angular Time Derivatives

Just as with linear kinematic variables, differential calculus and integral calculus are also used to determine these angular variables. As stated earlier in this chapter, *angular displacement* is defined as the change in angular position. *Angular velocity* is defined as the rate of change of angular displacement with respect to time, and *angular acceleration* is the rate of change of angular velocity with respect to time. Conversely, angular velocity is the integral of angular acceleration with respect to time and angular displacement is the time integral of angular velocity. These three kinematic variables are used to describe the angular motion of rigid bodies during a motion sequence. The symbols representing units of measure for these angular variables were presented in table 1.2.

Once the continuity of the angular histories has been ensured, the angular velocities and accelerations can be computed using finite difference equations that are similar in form to those used for linear kinematics. The central finite difference equations for computing angular velocity *(ω)* and acceleration *(α)* are

$$\omega_i = \frac{\theta_{i+1} - \theta_{i-1}}{2(\Delta t)} \tag{1.17}$$

$$\alpha_i = \frac{\omega_{i+1} - \omega_{i-1}}{2(\Delta t)} \tag{1.18}$$

$$\text{or} \quad \alpha_i = \frac{\theta_{i+1} - 2\theta_i + \theta_{i-1}}{(\Delta t)^2} \tag{1.19}$$

where θ represents the angular position and Δt represents the time duration between adjacent samples (see Miller and Nelson 1973). The character *i* represents the particular instant in time that is being analyzed. As is the case with linear data, before these equations are applied, the raw angular positions must be smoothed to remove high-frequency noise (Pezzack et al. 1977). If the angular positions were derived from marker coordinates that were already filtered, no further smoothing is necessary.

Angular to Linear Conversion

In the earlier section on polar coordinates, we showed how linear and angular motion are related. When a rigid body undergoes angular rotation, it is possible to calculate linear velocity and acceleration from the known angular velocity and acceleration. Consider the example of a body rotating about a fixed axis of rotation at point Q in figure 1.26. Designate a marker attached at the far end of the body as point P so that points Q and P are separated by length *r*. As the body rotates, point P describes an arc (or circle) on the underlying surface. Now, attach to the body a 2-D reference frame with its origin at point P. One axis, called the **normal** axis, is at right angles to the curvature of the path, whereas the **tangential** axis is at a **tangent** to the path. Affixed to the body, this reference frame will rotate with the body so that the normal and tangential axes change their orientations within the GCS. The angular motion of the body is described by its angular velocity *(ω)* and angular acceleration *(α)* within the GCS. Note that such a system could represent a human body segment such as the thigh, where Q represents the stationary hip joint, P the moving knee joint, and *r* the length of the thigh.

The linear velocity of point P within the rotating reference system can be calculated from the equation $v_t = r\omega$, where v_t is the tangential velocity (i.e., in the direction of the tangential axis). Note that the normal velocity (v_n) is zero, because the length *r* is constant (because points Q and P are fixed relative to each other). Point P may

FROM THE SCIENTIFIC LITERATURE

Pezzack, J.C., R.W. Norman, and D.A. Winter. 1977. An assessment of derivative determining techniques used for motion analysis. *Journal of Biomechanics* 10:377-82.

This paper uses both direct kinematic measures of angle and angular acceleration and indirect kinematics of markers to assess two methods of smoothing digitized marker kinematics, particularly angular acceleration. First, data from an aluminum arm that could be rotated only in the horizontal plane (one DOF) were collected by filming the arm's motion from above while simultaneously measuring its angular position with a potentiometer (see appendix C) and its linear acceleration with a uniaxial accelerometer. The aluminum arm was manually moved in several different ways. The accelerometer was mounted so that it measured transverse acceleration (a_t), which was then converted to angular acceleration by dividing by the distance from the arm's center of rotation to the accelerometer $(\alpha = a_t/r)$. Next, the

film data were digitized and the angular positions of the arm were computed. Then, the angular acceleration of the arm was computed in three different ways—without data smoothing, after smoothing with Chebyshev least-squares polynomials of orders 10 to 20, and after digital filtering with a fourth-order, zero-lag Butterworth filter.

The results (figure 1.25) showed very good agreement between the two angular displacement measures (film vs. potentiometer), but significant differences existed among the three methods of computing angular acceleration. First, the unsmoothed accelerations were very noisy, although the waveform, on average, did follow the signal derived from the direct measure of acceleration. Second, the least-squares polynomial that accurately fit the dis-

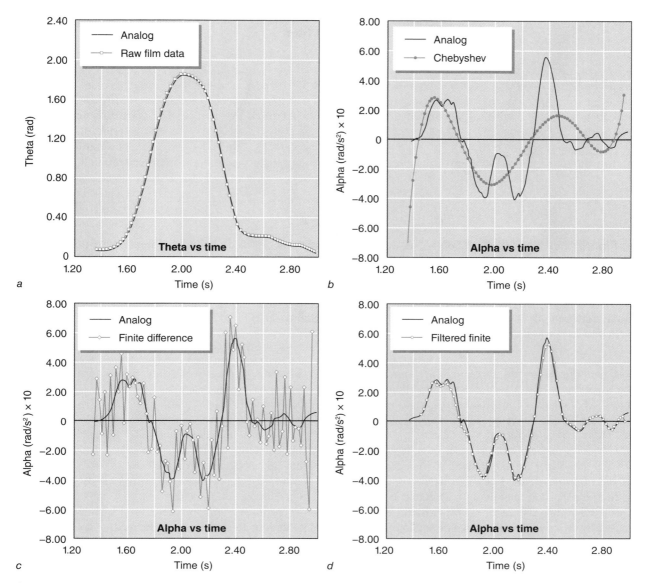

▲ **Figure 1.25** *(a)* Angular displacement of a humanly moved aluminum arm. *(b)* Angular acceleration of the arm after differentiation without data smoothing. *(c)* After least-squares curve fitting and differentiation. *(d)* After digital filtering and differentiation.

Reprinted from *Journal of Biomechanics*, Vol. 10, J.C. Pezzack, R.W. Norman, and D.A. Winter, "An assessment of derivative determining techniques used for motion analysis," 377-382, copyright © 1977, with permission from Elsevier.

placement signal did not follow the accelerometer signal after two derivatives were taken. Clearly, fitting a single polynomial to even a relatively simple human motion was not suitable. Third, after two time derivatives, the digitally filtered signal did closely match the true acceleration as measured by the accelerometer. Some attenuation of the signal occurred at the peaks, but these could have been corrected by increasing the cutoff frequency. See the following section and chapter 12 for more details on the use and implementation of digital filtering.

The importance of this paper cannot be overestimated. In addition to successfully reducing the effects of high-frequency noise in digitized film, the authors provided the data from one of their trials so that other researchers could evaluate their data-smoothing techniques (Lanshammer 1982a,b). Subsequently, two other techniques were also shown to have acceptable smoothing capabilities. Wood (1982) used piecewise quintic splines and Hatze (1981) used optimally regularized Fourier series to appropriately smooth human motion data.

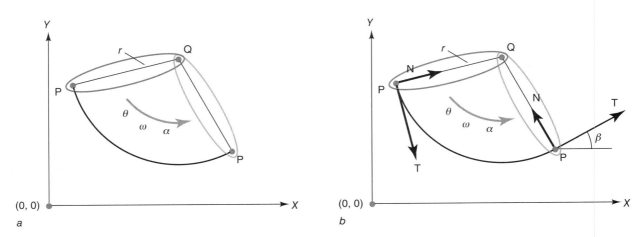

▲ **Figure 1.26** *(a)* Angular motion *(θ, ω, and α)* of a rigid body produces linear motion of points attached to it. Point P is located at a distance *r* from fixed point Q. As the body rotates through the angle *θ*, P scribes an arc such that its *(x, y)* coordinates are constantly changing. *(b)* LCS attached at P allows conversion from angular motion of the body to linear motion of the point P. The *tangential axis (T)* of the LCS describes a tangent to the arc scribed by P, whereas the orthogonal axis *N* is directed toward the axis of rotation point Q. Angle *θ* is the angle between the tangential axis and the right-horizontal and can be used to convert the tangential velocity *(v_t = rθ)* into its *x* and *y* components.

also undergo tangential acceleration, computed from the equation $a_t = r\alpha$. This acceleration will be nonzero if the body increases or decreases its angular velocity. If the body rotates at a constant angular speed, the angular and tangential accelerations will be zero.

Curiously, although the normal velocity is zero for circular motions, there must always be a nonzero *normal acceleration (a_n)* if the body is traveling in a circular path. The normal acceleration is calculated as $a_n = r\omega^2$. This acceleration, also called *centripetal acceleration*, is caused by the continual change of direction of the point P within the GCS. Readers should consult an engineering mechanics text such as Beer, Johnston, et al. (2010) for more detailed consideration of these accelerations.

The normal and tangential velocities and accelerations are defined within the reference system attached to the rotating body but can be converted to the GCS using knowledge of the body's angular orientation and simple trigonometric identities. In figure 1.26, the angle *β* is the angle formed by the tangential velocity and the right-horizontal, which is parallel to the *X*-axis in the GCS. The vector can be resolved into the components v_x and v_y, corresponding to the principal directions of the GCS, using the equations $v_x = v_t \cos \beta$ and $v_y = v_t \sin \beta$. Similar transformations can be applied to the accelerations a_n and a_t to express them in the GCS, as well.

Electrogoniometers

A **goniometer** is a manual device for measuring joint angles (figure 1.27*a*). It is essentially a protractor with two arms—one affixed to the protractor and one that rotates to measure angles. To measure joint angles electronically during motion, **electrogoniometers** are used. Typically, they are much less expensive than imaging systems and allow data to be collected and viewed immediately. Unfortunately, these devices do encumber movement because various electronics must be attached to the subject and most systems require cables to be run to a data collection system.

The most common type of electrogoniometer uses a potentiometer as the sensing element. A **potentiometer** is essentially a variable resistor (see appendix C). A constant voltage is applied across its terminals, and a wiper that turns with the potentiometer taps off voltage in an amount proportional to the amount of the turn (see figure 1.27*c*). One part of the potentiometer is attached to one segment of the joint and the other to the adjacent segment. Any angular motion of the joint causes the potentiometer to rotate and therefore change its output voltage. Other types of sensors include digital encoders, polarized-light photography, strain gauges, and fiberoptic cables.

One problem with goniometers is that not all joints act as pure hinges; any translational motion of the joint creates an erroneous angular rotation of the electrogoniometer. In principle, designing self-aligning mechanisms such as the four-bar linkage illustrated in figure 1.27*b* solves this problem. A similar self-aligning element was an integral component of the CARS-UBC (or MERU) electrogoniometer system developed by Hannah and colleagues (1978). This device simultaneously measured

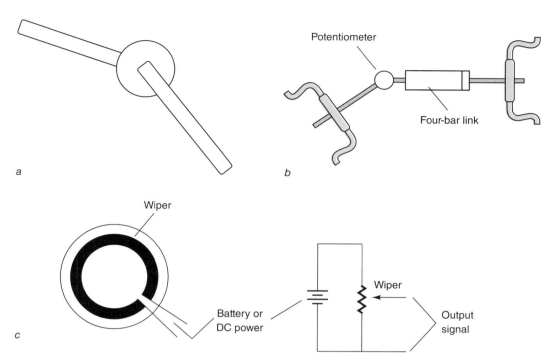

▲ **Figure 1.27** *(a)* Manual goniometer, *(b)* electrogoniometer, and *(c)* schematic of a potentiometer and electrogoniometer circuit.

triaxial angular motion of the ankle, knee, and hip of both legs in real time—a total of 18 signals. Although this system was clinically valuable for some patients, it is not as useful for examining people with severe disabilities because of the device's encumbrance. Another significant problem with all electrogoniometers is that they only measure joint angles. This prevents them from recording absolute motion of the segments with respect to a Newtonian frame of reference (in other words, a GCS), which is necessary for performing inverse dynamics analyses (see chapter 5). However, this type of system is useful in clinical environments where immediate joint kinematic information is required.

When selecting or building an electrogoniometer, use a sensing element that is continually differentiable. For example, as the wiper of most inexpensive potentiometers moves, it jumps from one loop of a coil of wire to another to vary its resistance. This contaminates the output position data with *discontinuity spikes* that disrupt the calculation of derivatives for obtaining angular velocity or acceleration. To eliminate these discontinuous steps, more expensive potentiometers have a continuous strip of conductive material. There is usually a break at one spot along the strip, and this portion must be placed outside the joint's range of motion (see figure 1.27c).

Another solution is to select a different type of sensor. One company, Measurand, devised a device, called

ShapeSensor, which uses a fine fiberoptic cable that transmits less light as the bend in the cable increases, allowing continuous measurement of the amount of joint rotation. Other systems (such as one made by Biometrics Ltd., formerly Penny and Giles) use strain gauges to quantify the degree of bending in a steel wire—and therefore angular position—about two axes (two DOF). For example, a transducer that crosses the elbow can simultaneously measure elbow flexion and forearm supination. Yet another solution is to use a polarized-light goniometer, which uses two sensors that are sensitive to polarized light (Chapman et al. 1985; Mitchelson 1975). One sensor is placed on the rotating segment, and the other is affixed to a stationary object. The relative positions of the sensors are determined by the orientation of each sensor within the polarized light plane.

Calibration of electrogoniometers is relatively straightforward. If the electrogoniometer design allows, it can be attached directly to a manual goniometer. Moving the manual goniometer from one known position to another while recording data from the electrogoniometer will provide a voltage measure equivalent to an actual angular displacement. From these measurements, a calibration coefficient is computed. If the electrogoniometer cannot be directly attached to a manual goniometer, a similar procedure can be used while the electrogoniometer is attached to the joint of interest.

Angular Kinematic Data Presentation

The presentation of angular kinematic data is similar to that for linear kinematics, with the most common format being the graphing of θ, ω, or α as a function of time throughout the movement (a history or time series). This format is useful for comparison with other kinematic and kinetic signals that occur simultaneously, particularly when one is trying to relate the observed kinematics to the underlying forces and torques (see chapters 4 and 5). Another display format unique to angular kinematics is the *angle-angle diagram*, which illustrates the coordinated motion of two segments or joints by plotting one versus the other. The graphic must be carefully described and annotated because the sense of time during the movement is lost, but this weakness can be overcome by marking specific movement **events** on the angle-angle plot and placing small arrows that indicate the relative timing sequence.

Another unique format for presenting angular results is known as a *phase plot, phase portrait,* or *phase diagram.* Here, the relationship between θ and ω for a specific segment or joint is depicted, with θ on the horizontal axis and ω on the vertical axis. This presentation has become popular among biomechanists studying movement from a *dynamic systems* perspective (see chapter 13), in which the emphasis is on meaningfully representing the kinematic state of a system as a window to the underlying control of that system. The phase plot therefore is seen as an expression of the control of the segment or joint. Figure 1.28 shows examples of an angular kinematic time series, angle-angle diagram, and phase portrait. Note that in presenting angular kinematic data for groups of trials or subjects, one can use the ensemble averaging techniques that were discussed in the earlier section on the presentation of linear kinematic variables.

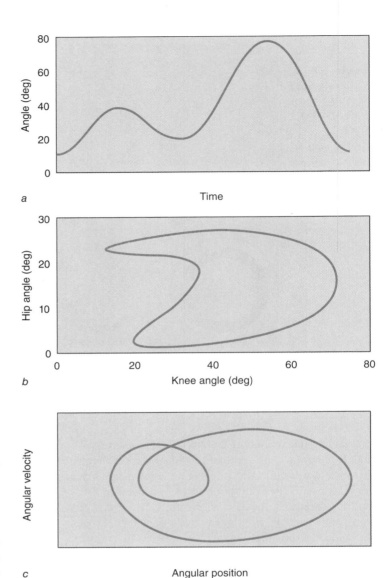

▲ Figure 1.28 *(a)* Time series of knee angle; *(b)* angle-angle plot of hip versus knee angle; and *(c)* phase plot of knee angular velocity versus angular position.

SUMMARY

This chapter outlined the major tools used by biomechanists to collect, process, and present data about the motion patterns of planar human movements. Data that describe motions are called *kinematics.* The next chapter expands on these concepts to handle motions in 3-D. Although many different tools are available, most kinematic studies begin with acquiring data from some sort of imaging or motion capture system, usually video. Equations for differentiating these data to obtain both linear and angular velocities and accelerations were presented. Additional technical information about processing and removing invalid information from these data can be found in chapter 12.

Although describing motion may be an end in itself, the most important reason for collecting kinematic data is to derive various kinetic quantities. Kinetics concerns the causes of motion and often can be indirectly computed from kinematic data and the inertial properties of the bodies (see chapter 3). Chapters 5, 6, and 7 outline how to derive such kinetic measures as mechanical work, energy, power, and momentum from kinematics.

SUGGESTED READINGS

Beer, F.P., E.R. Johnston Jr., D.F. Mazurek, P.J. Cornwell, and E.R. Eisenberg. 2010. *Vector Mechanics for Engineers; Statics and Dynamics.* 9th ed. Toronto: McGraw-Hill.

Chapman, A.E. 2008. *Biomechanical Analysis of Fundamental Human Movements.* Champaign, IL: Human Kinetics.

Hamill, J., and K.M. Knutzen. 2009. *Biomechanical Basis of Human Movement.* 3rd ed. Baltimore: Williams & Wilkins.

Nigg, B.M., and W. Herzog. 2007. *Biomechanics of the Musculo-Skeletal System.* 3rd ed. Toronto: Wiley.

Robertson, D.G.E. 2004. *Introduction to Biomechanics for Human Motion Analysis.* 2nd ed. Waterloo, ON: Waterloo Biomechanics.

Winter, D.A. 2009. *Biomechanics and Motor Control of Human Movement.* 4th ed. Toronto: Wiley.

Zatsiorsky, V.M. 1998. *Kinematics of Human Motion.* Champaign, IL: Human Kinetics.

Chapter 2

Three-Dimensional Kinematics

Joseph Hamill, W. Scott Selbie, and Thomas M. Kepple

3-D kinematics is the description of motion in 3-D space without regard to the forces that cause the motion. This chapter is a guide to the principles and computations of 3-D kinematics of the lower extremity. For more detailed and theoretical views of 3-D kinematics, readers should refer to works by Zatsiorsky (1998), Nigg and Herzog (1994), and Allard and colleagues (1998). Because 3-D kinematics relies heavily on vector operations and matrix algebra, we recommend a review of the principles of scalars, vectors, and matrices presented in appendixes D and E.

In this chapter, we

- ▶ present principles of 3-D data collection from optical sensors,

- ▶ define a link rigid segment model,

- ▶ introduce linear and rotational transformations between coordinate systems,

- ▶ mathematically define local coordinate systems for each segment of the lower extremity,

- ▶ present three methods for estimating the 3-D pose (position and orientation) of the model,

- ▶ present methods for representing 3-D angles, and

- ▶ present methods to calculate 3-D angular velocities and angular accelerations.

VISUAL3D EDUCATIONAL EDITION

The Visual3D Educational Edition software included with this text includes data sets that represent walking and running with full body sets of markers to help you further explore and understand the type of analysis presented in this chapter. Using the Visual3D Educational Edition software, you can experiment with all of the modeling capabilities of the professional Visual3D software by manipulating the model definitions, signal definitions, and basic signal processing in these sample data sets. To download the software, visit http://textbooks.c-motion .com/ResearchMethodsInBiomechanics2E.php.

COLLECTION OF THREE-DIMENSIONAL DATA

All 3-D motion-capture systems use multiple input sensors for the estimation of 3-D data. Typical sensors used in biomechanics include inertial sensors comprising accelerometers, gyroscopes, and sometimes magnetometers; electromagnetic sensors; linear sensors; and array sensors (all optical or camera-based systems). This chapter focuses on optical array sensors or cameras, but the principles of modeling and analysis can be applied to all sensors. The arrangement of cameras in a 3-D setup is not as rigorous as in a 2-D setup (figure 2.1), which may seem surprising. In a 2-D setup, because the motion is isolated to a single plane, the cameras must be placed precisely to capture the motion in this plane (see chapter 1 for more in-depth explanation of Planar Kinematics).

Each camera from a set of multiple cameras provides a unique view of the scene and records the 2-D location of specific markers in a camera coordinate system. From these sets of 2-D camera coordinates, 3-D global coordinates can be estimated. The most straightforward method for computing

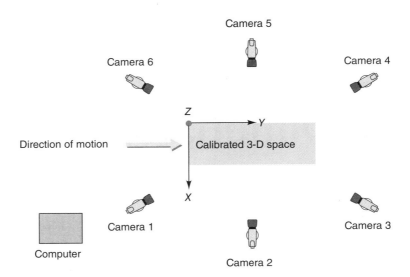

▲ **Figure 2.1** Typical multicamera setup for a 3-D kinematic analysis.

3-D coordinates from 2-D camera coordinates, called the *direct linear transformation* (DLT) method (Abdel-Azis and Karara 1971), assumes a linear relationship between the 2-D camera coordinates of a marker and the 3-D laboratory coordinates of the same marker. The DLT method is described in more detail in papers Marzan and Karrara (1975) and is not elaborated here.

COORDINATE SYSTEMS AND ASSUMPTION OF RIGID SEGMENTS

In this chapter, we define a number of Cartesian coordinate systems required for a 3-D analysis. These are referred to as a *global* or *laboratory coordinate system* (GCS), a *segment* or *local coordinate system* (LCS), and a *force platform coordinate system (FCS)*.

For this chapter, a biomechanical model is a collection of rigid segments. A segment's interaction with other segments is described by joint constraints permitting zero to six degrees of freedom, and subject-specific scaling is defined using palpable anatomical landmarks.

These rigid segments represent skeletal structures, which are not always represented ideally as rigid segments. For example, some segments, like the foot or the torso, often have one segment representing several bones. It is incorrect to assume that skeletal structures are rigid, but it makes the mathematics more palatable. The assumption of rigidity also aids the establishment of an LCS.

Global or Laboratory Coordinate System

The GCS refers to the capture volume in which we represent the 3-D space of the motion-capture system (also referred to as the *inertial reference system*). Recorded data are resolved into this fixed coordinate system. In this chapter, the GCS is designated using uppercase letters with the arbitrary designation of XYZ. The Y-axis is nominally directed anteriorly, the Z-axis is directly superiorly, and the X-axis is perpendicular to the other two axes. Because the subject may move anywhere in the data collection volume, only the vertical direction needs to be defined carefully, and that is only so we have a convenient representation of the gravity vector. The unit vectors for the GCS are $\hat{i}$, $\hat{j}$, $\hat{k}$ (see figure 2.2). In this chapter, and in the biomechanics literature, the GCS is a right-handed orthogonal system with an origin that is fixed in the laboratory. Note that a coordinate system is right-handed if and only if

$$\hat{k} = \hat{i} \times \hat{j} \text{ and } \hat{i} \cdot \hat{j} = 0 \tag{2.1}$$

Segment or Local Coordinate System

A mathematically convenient consequence of the assumption of rigidity is that in the context of kinematics, each segment is defined completely by an LCS fixed in the segment; as the segment moves, the LCS moves correspondingly. Like the GCS, the LCS is right-handed and orthogonal. In this chapter, the LCS is designated in lowercase letters x, y, z and unit vectors $\hat{i}'$, $\hat{j}'$, and $\hat{k}'$, respectively. In this chapter, the LCS is oriented such that the y-axis points anteriorly, the z-axis points axially (typically vertically), and the x-axis is perpendicular to the plane of the other two axes with its direction defined by the right-hand rule. Thus, on the right side the x-axis is directed from medial to lateral, whereas on the left side it is directed from lateral to medial. The orientation of the LCS with respect to the GCS defines the orientation of the body or segment in the GCS, and it changes as the body or segment moves through the 3-D space (see figure 2.2).

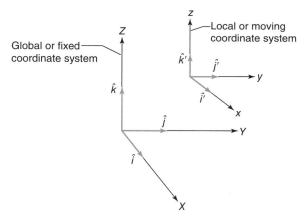

▲ **Figure 2.2** The global or fixed coordinate system, *XYZ*, with unit vectors, $\hat{i}'$, $\hat{j}'$, and $\hat{k}'$ and the local or moving coordinate system, *xyz*, and its unit vectors, $\hat{i}'$, $\hat{j}'$, and $\hat{k}'$.

TRANSFORMATIONS BETWEEN COORDINATE SYSTEMS

We have identified two types of coordinate systems (GCS and LCS) that exist in the same 3-D motion-capture volume. The descriptions of a rigid segment moving in space in different coordinate systems can be related by means of a *transformation* between the coordinate systems (see figure 2.3). A transformation allows one to convert coordinates expressed in one coordinate system to those expressed in another coordinate system. In other words, we can look at the same location in different ways based on which coordinate system we are using. At first glance this may seem redundant because we have not added new information by describing the same point in different ways. It is, however, convenient because objects move in the GCS but attributes of a segment, such as an anatomical landmark (e.g., segment endpoint), are constant in the LCS. We generally refer to transformations as linear or rotational.

Linear Transformation

In figure 2.4, a point is described by the vector $\vec{P}'$ in the LCS and by $\vec{P}$ in the GCS. The linear transformation between the LCS and the GCS can be defined by a vector $\vec{O}$, which specifies the origin of the LCS relative to the GCS. The components of $\vec{O}$ can be written as a column matrix in the form

$$\vec{O} = \begin{bmatrix} O_x \\ O_y \\ O_z \end{bmatrix} \qquad (2.2)$$

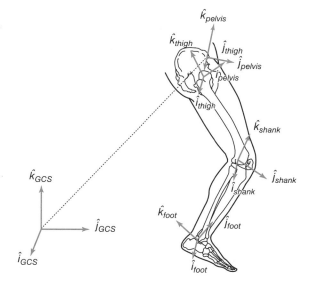

▲ **Figure 2.3** The global coordinate system and the local coordinate systems of the right-side lower extremity.

If we assume no rotation of the LCS relative to the GCS, converting the coordinates of a point $\vec{P}'$ in LCS to $\vec{P}$ in GCS can be expressed as

$$\vec{P} = \vec{P}' + \vec{O} \qquad (2.3)$$

$$\text{or} \quad \begin{bmatrix} P_x \\ P_y \\ P_z \end{bmatrix} = \begin{bmatrix} P_x' \\ P_y' \\ P_z' \end{bmatrix} + \begin{bmatrix} O_x \\ O_y \\ O_z \end{bmatrix} \qquad (2.4)$$

Conversely, conversion of the coordinates of a point $\vec{P}$ in GCS to $\vec{P}'$ in LCS is expressed as

$$\vec{P}' = \vec{P} - \vec{O} \qquad (2.5)$$

Rotational Transformation

If we assume no translation of the LCS relative to the GCS, converting the coordinates of a point $\vec{P}$ in GCS to $\vec{P}'$ in LCS can be expressed as

$$\vec{P}' = R\vec{P} \qquad (2.6)$$

where R is a matrix made up of orthogonal unit vectors (orthonormal matrix) that rotates the GCS about its origin, bringing it into alignment with the LCS. Conversely, converting the coordinates of a point in the LCS to a point in the GCS can be accomplished by

$$\vec{P} = R'\vec{P}' \qquad (2.7)$$

where R' is the inverse (and the transpose) of R. In this chapter we consistently use R as the transformation from GCS to LCS and R' as the transformation from LCS to GCS.

Consider the LCS unit vectors $\hat{i}'$, $\hat{j}'$, $\hat{k}'$ expressed in the GCS. The rotation matrix from GCS to LCS is as follows:

$$R = \begin{bmatrix} \hat{i}_x' & \hat{i}_y' & \hat{i}_z' \\ \hat{j}_x' & \hat{j}_y' & \hat{j}_z' \\ \hat{k}_x' & \hat{k}_y' & \hat{k}_z' \end{bmatrix} \qquad (2.8)$$

If we consider translation and rotation of the LCS relative to the GCS, converting the coordinates of a point $\vec{P}$ in GCS to $\vec{P}'$ in LCS can be expressed as

$$\vec{P}' = R(\vec{P} - \vec{O}) \qquad (2.9)$$

Conversely, converting the coordinates of a point $\vec{P}'$ in the LCS to point $\vec{P}$ in GCS can be accomplished by

$$\vec{P} = R'\vec{P}' + \vec{O} \qquad (2.10)$$

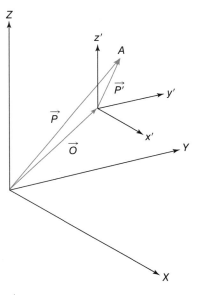

▲ **Figure 2.4** The point A is defined by vector $\vec{P}$ in *XYZ*, whereas the same point is defined by vector $\vec{P}'$ in *x'y'z'*. The linear transformation between *x'y'z'* and *XYZ* can be defined by a vector $\vec{O}$, which specifies the origin of the LCS relative to the GCS.

DEFINING THE SEGMENT LCS FOR THE LOWER EXTREMITY

In this chapter we use three noncollinear points to define a segment's LCS. *Noncollinear* means that the points are not aligned or in a straight line. The method presented is consistent with most models presented in the biomechanics literature. In this chapter, the LCS is described for the right-side segments of a lower-extremity model consisting of a pelvis, thigh, shank, and foot segments (the left side uses a similar derivation). The LCS of each segment is created based on a standing calibration trial and on palpable anatomical landmarks (see table 2.1). It is important to note that both the tracking and calibration markers are captured at the same time. However, in figure 2.5, only the calibration markers are shown.

Table 2.1 Abbreviations for Calibration Markers

Right	Description	Left	Description
$\vec{P}_{RASIS}$	Right anterior-superior iliac spine	$\vec{P}_{LASIS}$	Left anterior-superior iliac spine
$\vec{P}_{RPSIS}$	Right posterior-superior iliac spine	$\vec{P}_{LPSIS}$	Left posterior-superior iliac spine
$\vec{P}_{RLK}$	Right lateral femoral epicondyle	$\vec{P}_{LLK}$	Left lateral femoral epicondyle
$\vec{P}_{RMK}$	Right medial femoral epicondyle	$\vec{P}_{LMK}$	Left medial femoral epicondyle
$\vec{P}_{RLA}$	Right lateral malleolus	$\vec{P}_{LLA}$	Left lateral malleolus
$\vec{P}_{RMT5}$	Right fifth metatarsal head	$\vec{P}_{LMT5}$	Left fifth metatarsal head
$\vec{P}_{RMT1}$	Right first metatarsal head	$\vec{P}_{LMT1}$	Left first metatarsal head
$\vec{P}_{RHEEL}$	Right heel	$\vec{P}_{LHEEL}$	Left heel
$\vec{P}_{RTOE}$	Right toe	$\vec{P}_{LTOE}$	Left toe

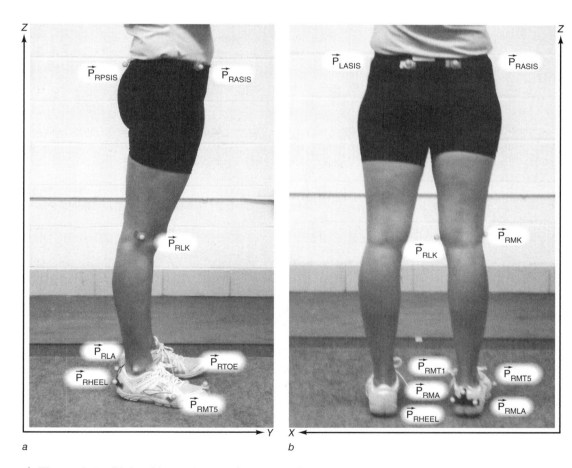

a   *b*

▲ **Figure 2.5** Right-side marker configuration of the calibration markers used in this chapter to compute the local coordinate system of each segment: *(a)* sagittal plane; *(b)* frontal plane. The $\vec{P}_{LPSIS}$ and $\vec{P}_{LASIS}$ are not seen in part *a* but are necessary for calibration. The $\vec{P}_{RPSIS}$, $\vec{P}_{LPSIS}$, and $\vec{P}_{RTOE}$ are not seen in part *b* but are necessary for calibration. Note that the tracking markers on the thigh, shank, and foot are necessary in the calibration trial to associate these with the LCS of each segment.

Pelvis Segment LCS

Markers are placed on the following palpable bony landmarks: right and left anterior-superior iliac spine ($\vec{P}_{RASIS}$, $\vec{P}_{LASIS}$) and right and left posterior-superior iliac spine ($\vec{P}_{RPSIS}$, $\vec{P}_{LPSIS}$) (figure 2.6). The origin of the LCS is midway between $\vec{P}_{RASIS}$ and $\vec{P}_{LASIS}$ and can be calculated as follows:

$$\vec{O}_{PELVIS} = 0.5 * (\vec{P}_{RASIS} + \vec{P}_{LASIS}) \qquad (2.11)$$

To create the *x*-component (or lateral direction) of the pelvis, a unit vector $\hat{i}'$ is defined by subtracting $\vec{O}_{PELVIS}$ from $\vec{P}_{RASIS}$ and dividing by the norm of the vector:

$$\hat{i}' = \frac{\vec{P}_{RASIS} - \vec{O}_{PELVIS}}{\left| \vec{P}_{RASIS} - \vec{O}_{PELVIS} \right|} \qquad (2.12)$$

Next we create a unit vector from the midpoint of $\vec{P}_{RPSIS}$ and $\vec{P}_{LPSIS}$ to $\vec{O}_{PELVIS}$:

$$\hat{v} = \frac{\vec{O}_{PELVIS} - 0.5 * (\vec{P}_{RPSIS} + \vec{P}_{LPSIS})}{\left| \vec{O}_{PELVIS} - 0.5 * (\vec{P}_{RPSIS} + \vec{P}_{LPSIS}) \right|} \qquad (2.13)$$

A unit vector normal to the plane (in the superior direction) containing $\hat{i}'$ and $\hat{v}$ is computed from a cross product:

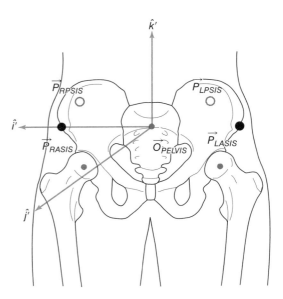

▲ **Figure 2.6** The origin of the pelvis LCS ($\vec{O}_{PELVIS}$) is midway between the right and left anterior-superior iliac spines. The right and left anterior-superior iliac spines ($\vec{P}_{RASIS}$ and $\vec{P}_{LASIS}$) and the posterior-superior iliac spines ($\vec{P}_{RPSIS}$ and $\vec{P}_{LPSIS}$) can be used to derive the pelvis LCS.

$$\hat{k}' = \hat{i}' \times \hat{v} \qquad (2.14)$$

Note that the order in which the vectors $\hat{i}'$ and $\hat{v}$ are crossed to produce a superiorly directed unit vector is determined by the right-hand rule. At this point we have defined the lateral direction and the superior direction. The anterior unit is created from the cross product

$$\hat{j}' = \hat{k}' \times \hat{i}' \qquad (2.15)$$

The rotation matrix describing the orientation of the pelvis, which will be used in later calculations, is constructed from the unit vectors as

$$R_{PELVIS} = \begin{bmatrix} \hat{i}'_x & \hat{i}'_y & \hat{i}'_z \\ \hat{j}'_x & \hat{j}'_y & \hat{j}'_z \\ \hat{k}'_x & \hat{k}'_y & \hat{k}'_z \end{bmatrix} \qquad (2.16)$$

Thigh Segment LCS

The thigh is defined by one virtual location and two marker locations (figure 2.7). The proximal end (and origin) of the thigh is coincident with the location of a virtual hip joint center. A number of studies have described regression equations for estimating the location of the hip joint center in the pelvis LCS (Andriacchi et al. 1980; Bell et al. 1990; Davis et al. 1991; Kirkwood et al. 1999). In this chapter we use equations derived from Bell and colleagues (1989) to compute a landmark that represents the hip joint center $\vec{P}_{RHIP}$ in the pelvis LCS:

$$\vec{P}'_{RHIP} = \begin{bmatrix} 0.36 * \left| \vec{P}_{RASIS} - \vec{P}_{LASIS} \right| \\ -0.19 * \left| \vec{P}_{RASIS} - \vec{P}_{LASIS} \right| \\ -0.30 * \left| \vec{P}_{RASIS} - \vec{P}_{LASIS} \right| \end{bmatrix} \qquad (2.17)$$

We can transform the location of the hip joint from the pelvis LCS to the GCS as follows:

$$\vec{O}_{RTHIGH} = \vec{P}_{RHIP} = R'_{PELVIS} * \vec{P}'_{RHIP} + \vec{O}_{PELVIS} \qquad (2.18)$$

To develop the thigh LCS, a superior unit vector is created along an axis passing from the distal end (midpoint between the femoral epicondyles $\vec{P}_{RLK}$ and $\vec{P}_{RMK}$) to the origin ($\vec{O}_{RTHIGH}$) as follows:

$$\hat{k}' = \frac{\vec{O}_{RTHIGH} - 0.5 * (\vec{P}_{RLK} + \vec{P}_{RMK})}{\left| \vec{O}_{RTHIGH} - 0.5 * (\vec{P}_{RLK} + \vec{P}_{RMK}) \right|} \qquad (2.19)$$

We then create a unit vector passing from the medial to the lateral femoral epicondyle:

$$\hat{v} = \frac{(\vec{P}_{RLK} - \vec{P}_{RMK})}{\left| \vec{P}_{RLK} - \vec{P}_{RMK} \right|} \qquad (2.20)$$

The anterior unit vector is determined from the cross product of the $\hat{k}'$ and $\hat{v}$ vectors as follows:

$$\hat{j}' = \hat{k}' \times \hat{v} \qquad (2.21)$$

Care should be taken in the placement of the knee markers. The lateral marker is placed at the most lateral aspect of the femoral epicondyle. The medial marker should be located so that the lateral and medial knee markers and the hip joint define the frontal plane of the thigh.

Last, the unit vector in the lateral direction is formed from the cross product:

$$\hat{j}' = \hat{k}' \times \hat{v} \qquad (2.22)$$

The rotation matrix describing the orientation of the thigh is constructed from the thigh unit vectors as

$$R_{RTHIGH} = \begin{bmatrix} \hat{i}'_x & \hat{i}'_y & \hat{i}'_z \\ \hat{j}'_x & \hat{j}'_y & \hat{j}'_z \\ \hat{k}'_x & \hat{k}'_y & \hat{k}'_z \end{bmatrix} \qquad (2.23)$$

Because the LCS is orthogonal and $\hat{k}'$ is defined explicitly to pass between the segment endpoints, the lateral unit vector $\hat{i}'$, which is perpendicular to $\hat{k}'$, is not necessarily parallel to an axis passing between the epicondyles. In other words, the placement of the medial and lateral knee markers does not define the flexion-extension axis.

Shank Segment LCS

For the shank or leg segment, the LCS is defined from four palpable landmarks: the lateral and medial malleoli $\vec{P}_{RLA}$ and $\vec{P}_{RMA}$ and the lateral and medial femoral epicondyles, $\vec{P}_{RLK}$ and $\vec{P}_{RMK}$ (figure 2.8). The origin of the LCS is at the midpoint between the femoral epicondyles and can be calculated as

$$\vec{O}_{RSHANK} = 0.5 * (\vec{P}_{RLK} + \vec{P}_{RMK}) \qquad (2.24)$$

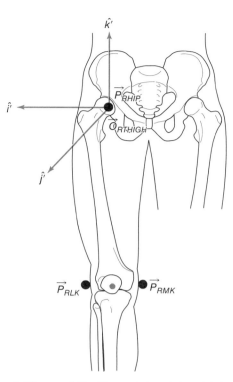

▲ **Figure 2.7** The origin of the thigh LCS ($\vec{O}_{RTHIGH}$) is at the hip joint center. The position of hip joint center ($\vec{P}_{RHIP}$) and the lateral and medial femoral epicondyles ($\vec{P}_{RLK}$ and $\vec{P}_{RMK}$) can be used to calculate the thigh LCS.

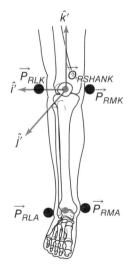

▲ **Figure 2.8** The origin of the shank LCS ($\vec{O}_{RSHANK}$) is located at the midpoint of the lateral and medial epicondyles ($\vec{P}_{RLK}$ and $\vec{P}_{RMK}$). The positions of the lateral and medial epicondyles ($\vec{P}_{RLK}$ and $\vec{P}_{RMK}$) and the midpoint of the lateral and medial malleoli ($\vec{P}_{RLA}$ and $\vec{P}_{RMA}$) can be used to derive the local coordinate system of a proximal biased shank.

To construct the shank LCS, we start by creating a superior unit vector based on an axis passing from the distal end (midpoint between the malleoli) to the segment origin as follows:

$$\hat{k}' = \frac{\vec{O}_{RSHANK} - 0.5 * (\vec{P}_{RLA} + \vec{P}_{RMA})}{\left| \vec{O}_{RSHANK} - 0.5 * (\vec{P}_{RLA} + \vec{P}_{RMA}) \right|} \tag{2.25}$$

Next, we create a unit vector passing from the medial epicondyle to the lateral epicondyle:

$$\hat{v} = \frac{\left(\vec{P}_{RLK} - \vec{P}_{RMK} \right)}{\left| \vec{P}_{RLK} - \vec{P}_{RMK} \right|} \tag{2.26}$$

We then create the anterior unit vector from the cross product of the $\hat{k}'$ and $\hat{v}$ unit vectors:

$$\hat{j}' = \hat{k}' \times \hat{v} \tag{2.27}$$

Last, we create the third lateral unit vector from the cross product:

$$\hat{i}' = \hat{j}' \times \hat{k}' \tag{2.28}$$

Finally, the orientation of the shank rotation matrix can be described using the unit vectors as

$$\mathrm{R}_{RSHANK} = \begin{bmatrix} \hat{i}'_x & \hat{i}'_y & \hat{i}'_z \\ \hat{j}'_x & \hat{j}'_y & \hat{j}'_z \\ \hat{k}'_x & \hat{k}'_y & \hat{k}'_z \end{bmatrix} \tag{2.29}$$

This definition of the shank LCS can be referred to as a proximally biased LCS because the frontal plane is defined by two proximal markers and one distal marker. A distally biased LCS could be defined by replacing equation 2.26 with the following:

$$\hat{v} = \frac{\left(\vec{P}_{RLA} - \vec{P}_{RMA} \right)}{\left| \vec{P}_{RLA} - \vec{P}_{RMA} \right|} \tag{2.30}$$

The proximal biased shank LCS and the distal biased LCS may differ because of tibial torsion. If substantial tibial torsion is evident, it is often useful to define two shank LCSs, the proximal biased shank for defining the knee angle and the distal biased shank for defining the ankle angle.

Foot Segment LCS

The foot LCS is defined from five markers, two placed on the lateral and medial malleoli, $\vec{P}_{RLA}$ and $\vec{P}_{RMA}$, two on the first and fifth metatarsal heads, $\vec{P}_{RMH1}$ and $\vec{P}_{MH5}$, and one on the calcaneus, $\vec{P}_{RHEEL}$ (figure 2.9). The origin of the LCS is at the midpoint between the malleoli and can be calculated as

$$\vec{O}_{RFOOT} = 0.5 * (\vec{P}_{RLA} + \vec{P}_{RMA}) \tag{2.31}$$

The axially directed unit vector is created by subtracting the midpoint of the metatarsal heads $\vec{P}_{RMH5}$ and $\vec{P}_{RMH1}$ from the origin $\vec{O}_{RFOOT}$:

$$\hat{k}' = \frac{\vec{O}_{RFOOT} - 0.5 * (\vec{P}_{RMH5} + \vec{P}_{RMH1})}{\left| \vec{O}_{RFOOT} - 0.5 * (\vec{P}_{RMH5} + \vec{P}_{RMH1}) \right|} \tag{2.32}$$

Next, we create a unit vector passing from the medial malleolus to the lateral malleolus as

$$\hat{v} = \frac{\left(\vec{P}_{RLA} - \vec{P}_{RMA} \right)}{\left| \vec{P}_{RLA} - \vec{P}_{RMA} \right|} \tag{2.33}$$

and create the anterior unit vector from the cross product of the $\hat{k}'$ and $\hat{v}$ unit vectors:

$$\hat{j}' = \hat{k}' \times \hat{v} \tag{2.34}$$

Last, we create the third unit vector that is laterally directed from the cross product:

$$\hat{i}' = \hat{j}' \times \hat{k}' \tag{2.35}$$

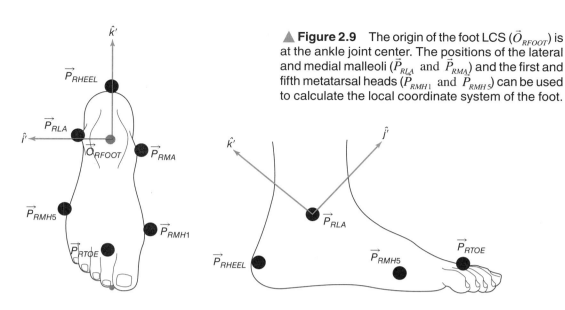

Finally, the orientation of the foot rotation matrix can be described using the unit vectors as

$$R_{RFOOT} = \begin{bmatrix} \hat{i}'_x & \hat{i}'_y & \hat{i}'_z \\ \hat{j}'_x & \hat{j}'_y & \hat{j}'_z \\ \hat{k}'_x & \hat{k}'_y & \hat{k}'_z \end{bmatrix} \tag{2.36}$$

Note that the orientation of the LCS of the foot is different than the LCS of the other segments in that the $\hat{k}'$ unit vector (i.e., the axis from the distal to the proximal end of the foot) is not oriented superiorly like the other segments. The $\hat{j}'$ unit vector points superiorly and the i' unit vector still points in the lateral direction. This is a convenient representation of the foot for kinetics because the proximal end of the foot and the distal end of the shank are coincident at the ankle, which is the point at which the segments are constrained.

This is not, however, a convenient representation of the foot for joint angles (as described later in this chapter), so often a second LCS will be created for the foot segment (figure 2.10) as follows. The origin of the LCS is at the heel marker:

$$\vec{O}_{RFOOT2} = \vec{P}_{RHEEL} \tag{2.37}$$

The axially directed unit vector is created by subtracting the marker on the toe ($\vec{P}_{RTOE}$) from the origin $\vec{O}_{RFOOT2}$:

$$\hat{j}' = \frac{\vec{O}_{RFOOT2} - \vec{P}_{RTOE}}{\left|\vec{O}_{RFOOT2} - \vec{P}_{RTOE}\right|} \tag{2.38}$$

Next, we create a unit vector passing from $\vec{O}_{RFOOT2}$ to the midpoint of $\vec{P}_{RLA}$ and $\vec{P}_{RMA}$:

$$\hat{v} = \frac{0.5 * \left(\vec{P}_{RLA} - \vec{P}_{RMA}\right) - \vec{O}_{RFOOT2}}{\left|0.5 * \left(\vec{P}_{RLA} - \vec{P}_{RMA}\right) - \vec{O}_{RFOOT2}\right|} \tag{2.39}$$

It is important to place the heel marker so that the desired sagittal plane is defined by the heel, the midpoint of the malleoli, and the toe. The sagittal plane defines the inversion-eversion of the foot during the standing trial. We can create the lateral unit vector from the cross-product of the $\hat{j}'$ and $\hat{v}$ unit vectors as

$$\hat{i}' = \hat{j}' \times \hat{v} \tag{2.40}$$

Last, we create the third unit vector that is superiorly directed from the cross product:

$$\hat{k}' = \hat{i}' \times \hat{j}' \tag{2.41}$$

This second foot LCS now has a superior unit vector consistent with the superior unit vectors of the other segments. The orientation of the second foot LCS can be described using the unit vectors as

$$R_{RFOOT2} = \begin{bmatrix} \hat{i}'_x & \hat{i}'_y & \hat{i}'_z \\ \hat{j}'_x & \hat{j}'_y & \hat{j}'_z \\ \hat{k}'_x & \hat{k}'_y & \hat{k}'_z \end{bmatrix} \quad (2.42)$$

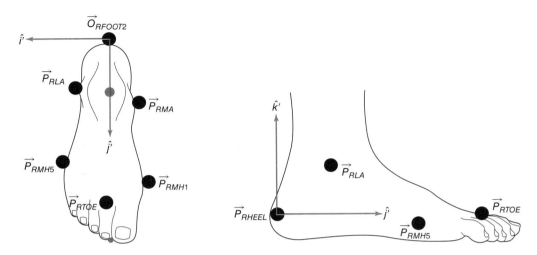

▲ **Figure 2.10** The origin of the foot LCS ($\vec{O}_{RFOOT2}$) is at the heel marker ($\vec{P}_{RHEEL}$). Using the positions of the lateral and medial malleoli ($\vec{P}_{RLA}$ and $\vec{P}_{RMA}$) and the first and fifth metatarsal heads ($\vec{P}_{RMH1}$ and $\vec{P}_{RMH5}$), we can derive the local coordinate system of the foot.

Marker Placement

If marker placement is used to establish the LCS, practical guidelines should be used to ensure appropriate coordinate systems. It is important that the collection of 3 markers repeatably define the LCS, and this rarely happens by placing markers cavalierly. Also, the computed LCS must make intuitive sense, so you should always survey the subject and model. If, for example, the placement doesn't look correct visually, such that the left lateral knee marker is much higher than the right lateral knee marker on a subject that otherwise appears symmetrical, or the computed coordinate systems don't reflect your intuition (e.g., the anterior direction of a segment that is pointed laterally), there is something wrong that should be addressed. It is highly recommended that you compute and display the local coordinate systems for a representative trial (usually a standing trial in the case of gait) before accepting the marker placement and collecting data. It is usually difficult to modify segment coordinate systems post hoc without "making up" data.

Right Side Versus Left Side

When we adopt the convention that the z-axis is directed superiorly and the y-axis is directed anteriorly, enforcing the right-hand rule causes the x-axis to be inconsistent anatomically between the left- and right-side LCSs of the body. To maintain a right-hand coordinate system, segments on the right side of the body have the x-axis in the lateral direction and segments on the left side have the x-axis in the medial direction. The axial segments, such as the torso, are treated with the x-axis pointing in the lateral direction (e.g., consistent with the right side). Note that this is not the only convention used in the biomechanics literature. Some authors prefer to specify the y-axis between the segment endpoints and the x-axis as the anterior direction. These are equivalent mathematically as both systems are right-handed and orthogonal, but the reader must be careful to identify the convention used by an author.

POSE ESTIMATION: TRACKING THE SEGMENT LCS

We will present three classes of algorithms for estimating the pose (position and orientation) of rigid bodies. In this chapter we use the nomenclature of Lu and O'Connor (1999) to describe a *direct* method, a *segment optimization* method, and a *global optimization* method. All three algorithms refer to the previously defined segment LCSs so that differences between the algorithms will be independent of the choice of the LCS.

The pose of an unconstrained rigid segment requires six independent variables (commonly referred to as degrees of freedom): three to specify the location of the origin and three to specify the orientation. The relationship between the number of degrees of freedom and the number of markers is as follows: one marker attached to a segment is sufficient to define three degrees of freedom, another marker adds two more degrees of freedom, and a third marker adds the last degree of freedom. This means that in order to fully describe the pose of a segment (six degrees of freedom), we must locate at least three noncollinear points on the segment.

Tracking a segment refers to the process of estimating the pose of the segment from motion data. *Tracking markers* refer to the markers attached to the segment during the movement trials that are used for tracking the pose of the segment (see figure 2.11). The principle assumption of the three pose estimation algorithms is that the markers move rigidly with the body segments to which they are attached; that is, marker coordinates in a segment LCS do not change during movement. It is accepted, however, that markers attached to the skin move relative to the underlying skeleton (Cappozzo et al. 1996b; Karlsson and Tranberg 1999), referred to as *soft tissue artifact*, and that motion-capture markers produce data that may be noisy, distorted, or missing. Marker noise and especially soft tissue artifact result in poor estimations of pose, and minimizing the effect of this noise through judicial marker placement or choosing an appropriate estimation algorithm improves the estimation of the pose. Although motion-capture systems provide a rough estimate of the sensor noise, it is challenging to quantify soft tissue artifact because it appears to be systematic yet varies case by case. The three pose estimation algorithms described in this text differ markedly in their ability to compensate for this soft tissue artifact.

Direct Pose Estimation

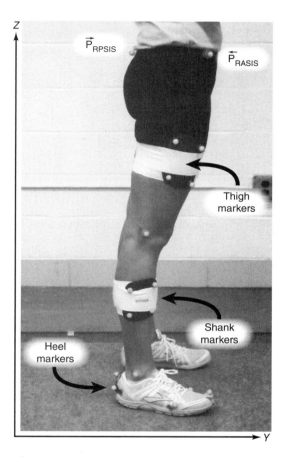

▲ **Figure 2.11** Right-side marker configuration used in this chapter for the tracking markers on each segment. Clusters of markers on rigid plates are used for the foot, shank, and thigh segments. For the pelvis, the calibration markers of the right and left anterior-superior iliac spines ($\vec{P}_{RASIS}$, $\vec{P}_{LASIS}$) and the right and left posterior-superior iliac spines ($\vec{P}_{RPSIS}$, $\vec{P}_{LPSIS}$) are used as tracking markers. The other calibration markers on the knee, ankle, and foot may be removed during the movement trials.

Direct pose estimation algorithms compute the LCS of each segment in a motion trial in the same fashion as they compute the LCS in the standing trial. An important limitation of this strict application of the direct method is that there is no redundancy in the markers and no leniency in marker placement; the markers must be placed precisely where the method expects them to be placed, and all markers must appear in every frame of the motion trial. If we consider the right thigh segment LCS described previously, a strict direct method would require the medial knee marker to appear in every frame of data, which is a practical problem because this marker is susceptible to being knocked off during the motion trials. To compensate for this unfortunate reality, a more convenient tracking marker can be added to the thigh noncollinearly with the other tracking markers. In some cases this is projected from the segment on a post, which is an attempt to improve the accuracy of tracking the axial rotation.

To estimate the pose during a frame of the motion trial in which the medial knee marker has been removed, a "virtual" representation of the marker can be computed that can be used to create the thigh LCS. This virtual marker is computed in a technical coordinate system (TCS) defined by the hip joint center, the lateral knee, and the additional thigh marker $\vec{P}_{RTH}$ (see figure 2.12). The origin of the TCS is located at the hip joint center:

$$\vec{O}_{TCS} = \vec{P}_{RHIP} \tag{2.43}$$

The first unit vector is created by subtracting $\vec{P}_{RLK}$ from $\vec{O}_{TCS}$:

$$\hat{k}' = \frac{\vec{O}_{TCS} - \vec{P}_{RLK}}{\left|\vec{O}_{TCS} - \vec{P}_{RLK}\right|} \tag{2.44}$$

Next, create a unit vector by subtracting of $\vec{P}_{RLK}$ from $\vec{P}_{RTH}$:

$$\hat{v} = \frac{\vec{P}_{RTH} - \vec{P}_{RLK}}{\left|\vec{P}_{RTH} - \vec{P}_{RLK}\right|} \tag{2.45}$$

A second unit vector is created from the cross product of the $\hat{k}'$ and $\hat{v}$ unit vectors:

$$\hat{j}' = \hat{k}' \times \hat{v} \tag{2.46}$$

Last, we create the third unit vector from the cross product.

$$\hat{i}' = \hat{j}' \times \hat{k}' \tag{2.47}$$

The orientation of the TCS can now be described as

$$R_{TCS} = \begin{bmatrix} \hat{i}'_x & \hat{i}'_y & \hat{i}'_z \\ \hat{j}'_x & \hat{j}'_y & \hat{j}'_z \\ \hat{k}'_x & \hat{k}'_y & \hat{k}'_z \end{bmatrix} \tag{2.48}$$

We can transform the medial knee marker from the GCS into the TCS by

$$\vec{P}'_{RMK} = R_{TCS}(\vec{P}_{RMK} - \vec{O}_{TCS}) \tag{2.49}$$

If we assume that $\vec{P}'_{RMK}$ remains constant in the TCS (e.g., markers are attached rigidly to the segment), we can compute a virtual location in the motion trial even though the physical marker has been removed. At each frame of the motion trial, R_{TCS} is created. We then transform $\vec{P}'_{RMK}$ into the GCS to produce a virtual marker, $\vec{P}_{VRMK}$, which can be used to compute the LCS as

$$\vec{P}_{VRMK} = R'_{TCS}\,\vec{P}'_{RMK} + \vec{O}_{TCS} \tag{2.50}$$

A first limitation of this direct algorithm is that only 3 markers are used to define the LCS, which means there is no redundancy in the representation of the segment; if a tracking marker is occluded, the LCS cannot be computed. A second limitation of the direct method is that it does not use the rigid body assumption (i.e., the expected distribution of the markers will not change during the movement) to minimize the effects of soft tissue artifact. Thus, if there is an error in the location of a marker, it will usually result in a direct error in the estimation of LCS. A third limitation of

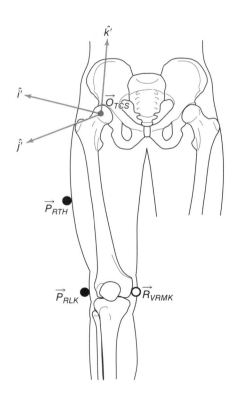

▲ **Figure 2.12** A technical coordinate system is defined by three markers: $\vec{O}_{TCS}$ and $\vec{P}_{RLK}$ and an additional marker on the right thigh ($\vec{P}_{RTH}$). The origin of the technical coordinate system ($\vec{O}_{TCS}$) is at the right hip ($\vec{P}_{RHIP}$). The local coordinate system is defined from these markers. A virtual marker at the right medial knee (R_{VRMK}) can be created.

the direct method is that all the segments below the pelvis use the virtual joint center created by the proximal segment when computing its pose. Therefore, as the segments' poses are computed, errors will be propagated from distal to proximal segments down the entire linkage. For example, errors in locating a target on the pelvis will result in errors in locating all of the other segments. An additional limitation of the direct method is that because the distal end of the segment on the proximal side of the joint and the proximal end of the segment on the distal side of the joint always share a point in common (a joint center location), actual joint translations cannot be measured. In summary, *direct pose estimation* is the least effective of the three pose estimation algorithms described in this chapter.

Pose Estimation Using Segment Optimization

The term *segment optimization* (coined by Lu and O'Connor 1999) is derived from the term *optimal tracking* coined by Cappozzo and colleagues (1995). Cappozzo and colleagues' use of the term *optimal tracking* refers to the fact that some mathematical optimization algorithm is used during the pose estimation of each segment. The word *optimal*, however, should not necessarily be construed to mean "best" tracking. Segment optimization is also commonly referred to as six degrees of freedom (six DOF) method because each segment has at least three tracking markers and because all six variables that describe its pose (three variables that describe the position of the origin and three variables that describe the rotation about each of the principal axes of the segment local coordinate system) are estimated. Tracking each segment independently means that there is no explicit linkage connecting segments (i.e., preconceived assumptions about joint properties); the endpoints of the proximal and distal segment move relative to each other based directly on the recorded motion-capture data (e.g., in other words, the joint may be considered to have six DOF).

In this section we use the same definition of the LCS described earlier for the direct method, and therefore the same anatomical markers must be placed for a standing trial. In addition to these anatomical markers, a set of three or more tracking markers is attached rigidly to each segment (Cappozzo at al. 1997 recommend four tracking markers). In general it is not necessary that tracking markers be independent of the anatomical markers, but it allows us to emphasize the distinction between identification of the LCS and pose estimation. In this scenario all anatomical and tracking markers must be placed on the subject for the standing trial, but the anatomical markers can be removed before the motion trials are recorded.

As with the direct method, the six DOF method assumes that the location of a tracking marker in the LCS is fixed. In principle, tracking markers can be placed anywhere on a rigid segment. In practice, the effects of artifacts in the data are reduced by distributing markers over the entire surface of a segment and placing markers in areas that exhibit minimal soft tissue artifact (Cappozzo, Catani et al. 1996). The anatomical markers are used only when the subject is stationary, so rigidity in their placement is not as important.

To understand the algorithm, consider a segment LCS defined in the standing trial by a rotation matrix, R_{SEG}, and an origin, $\vec{O}_{SEG}$. Consider 3 tracking markers located on the segment at position $\vec{P}_i^{'}$ in the LCS and $\vec{P}_i$ in the GCS (figure 2.13):

$$\vec{P}_i = R'_{SEG}\ \vec{P}_i^{'} + \vec{O}_{SEG} \tag{2.51}$$

If the segment undergoes motion, the new orientation matrix R'_{SEG} and translation vector $\vec{O}_{SEG}$ may be computed at any instant, provided that three noncollinear points $\vec{P}_i^{'}$ are predetermined from the standing trial and $\vec{P}_i$ are recorded at each frame of motion data. The matrix, R'_{SEG}, and the origin, $\vec{O}_{SEG}$, are estimated by minimizing the sum of squares error expression as follows:

$$E = \sum_{i=1}^{m}((\vec{P}_i - R'_{SEG}\ \vec{P}_i^{'}) - \vec{O}_{SEG})^2 \tag{2.52}$$

under the orthonormal constraint

$$R'_{SEG}\ R_{SEG} = I \tag{2.53}$$

where *m* is equal to the number of markers on the segment (*m* > 2).

The following section is included for completeness of this technique. Equation 2.52 represents a constrained maximum-minimum problem that can be solved using the method of Lagrangian multipliers (Spoor and Veldpaus 1980). There are an infinite number of solutions for R_{SEG} and $\vec{O}_{SEG}$ that will produce minima for equation 2.52. The method of Lagrangian multipliers uses the following boundary condition (derived from equation 2.53) to reduce this to a unique solution:

$$g\left(R_{SEG}\right) = R'_{SEG}\, R_{SEG} - I = 0 \qquad (2.54)$$

The method sets the gradient of equation 2.52 equal to the gradient of equation 2.54 times a set of Lagrangian multipliers:

$$\nabla f\left(R_{SEG}, \vec{O}_{SEG}\right) = \lambda \nabla g\left(R_{SEG}\right) \qquad (2.55)$$

This results in the equation

$$\nabla f\left(R_{SEG}, \vec{O}_{SEG}\right) - \lambda \nabla g\left(R_{SEG}\right) = 0 \qquad (2.56)$$

for which there exists an exact solution for R_{SEG} and $\vec{O}_{SEG}$ as long as $m > 2$.

In other words, the mapping of motion-capture markers to six DOF segments is a matter of tracking a set of markers that are linked rigidly to the segment. This least squares solution

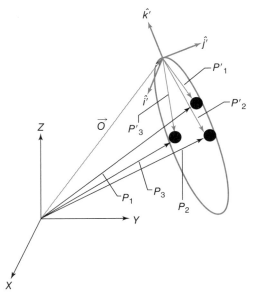

▲ **Figure 2.13** Relationship between tracking markers in the GCS and the LCS. The markers P_1, P_2, and P_3 are represented in the GCS (XYZ) and also as P'_1, P'_2, and P'_3 in the LCS ($\hat{i}'$, $\hat{j}'$, $\hat{k}'$) of the segment.

may be considered a straightforward pattern recognition algorithm; the pattern (configuration) of the tracking markers in each LCS is specified in a standing trial, and this pattern is fit to the homologous marker configuration in each frame of motion-capture data. There is considerable benefit to the segment optimization approach to pose estimation because it is straightforward to use, is easy to understand and requires no subjective guidance, and the solution has no local minima.

Segment optimization methods assume that segments are linked implicitly by the motion-capture data (e.g., segments did not come apart because the subject did not come apart when the motion was captured) and the joints were modeled with six degrees of freedom (e.g., all segments were treated as if they were independent). Tracking the pose of a segment using segment optimization does not, however, constrain the endpoints of the proximal and distal segment to remain fixed relative to each other. This movement may be real (e.g., the knee joint is not a fixed axis) or may be caused by errors due to noise or skin movement artifact. Excessive movement of the endpoints between two segments may indicate a serious problem in the data collection that should be addressed.

Segment optimization places no restrictions on where the tracking markers may be placed on a segment, so there is tremendous leeway in placing markers. Segment optimization estimation allows us to explore marker placements that might reduce soft tissue artifact. This method also exploits the fact that a least squares fit can be done on an over-specified system $m > 3$. This allows an unlimited number of markers to track each segment. If noise or artifact in the data is uncorrelated, the computed pose will minimize the effects of the noise, and if some targets are missing, the over-specification allows you to calculate segment pose as long as $m > 2$.

Pose Estimation Using Global Optimization

Lu and O'Connor (1999) described a global optimization process with which physically realistic joint constraints can be added to the model to minimize the effect of the soft tissue and measurement error. Global optimization is often referred to as inverse kinematics (IK), and the solution depends explicitly on the choice of a hierarchical model because the task is to identify an articulated figure consisting of a set of rigid segments connected with joints.

Global optimization computes the pose of a model that best matches the motion-capture data in terms of a global criterion. In other words, global optimization is the search, in each data frame, for an optimal pose of a multilink model that minimizes the differences (in a least squares sense) between the measured and model-determined marker coordinates across all the body segments. It considers measurement error distributions in the system and provides an error compensation mechanism between body segments that can be regarded as optimal at the system level. Mathematically, van den Bogert and Su (2007) described the configuration of the total body using a set of generalized coordinates $\vec{q}$. Generalized coordinates are the minimum set of independent variables that describe the pose of the model. Global optimization is an extension to the segment optimization pose estimation because if all joints are ascribed six degrees of freedom, the solutions are equivalent.

We can extend equation 2.51 to a function of the generalized coordinates $\vec{q}$ as follows:

$$\vec{P}_i = R'(\vec{q})\,\vec{P}'_i + \vec{O}(\vec{q}) \tag{2.57}$$

and the expression (equation 2.52) that is minimized now becomes

$$E(\vec{q}) = \sum_{i=1}^{m} \{\vec{P}_i - R'(\vec{q})\vec{P}'_i - \vec{O}(\vec{q})\}^2 \tag{2.58}$$

where m is the total number of targets on all the segments in the IK chain.

In the general case, there is no analytic solution for the global optimization problem. It is beyond the scope of this chapter to elaborate on the solution to this minimization problem; common minimization algorithms in order of computational efficiency include Levenberg-Marquardt, quasi-Newton, and simulated annealing algorithms.

To estimate the pose of the segments, we record the position, orientation, or both of sensors attached to the segments. The number of markers required and the number of segments to which markers are attached depend on the hierarchical structure of the model and the pose estimation algorithm used. The most important concept is observability. A system is observable if the data are sufficient to describe uniquely the pose of the model.

For segment optimization, a segment is observable if there are three or more noncollinear tracking markers, so this is straightforward to identify intuitively. The complexity of the global optimization hierarchical model precludes a simple "counting" approach to determine the number of markers required for observability. For example, consider an overly simplistic example of a two-segment IK model in which six markers are attached to one segment and no markers are attached to the other segment. Even though there are 2 × 3 markers in the model, this model is clearly not observable. If each segment of the global optimization model were required to have at least three tracking markers, observability would be ensured. One of the advantages of the IK approach, however, is that a model may be observable with fewer than three tracking markers per segment because of the joint constraints. For example, if a segment has a parent joint that permits only one degree of freedom, only one tracking marker is required on that segment.

In many circumstances the IK solution is preferred to the six DOF solution because the inclusion of the joint constraints is a means to minimize artifact, but one must determine the appropriateness of the joint constraints. For example, an experiment focused on understanding the kinematics of a damaged knee (e.g., anterior cruciate ligament injury) might not benefit from an IK solution because the prescribed motion of the knee joint may *hide* the damage. IK is an extension to the segment optimization pose estimation because if a joint is ascribed six degrees of freedom, the solutions are equivalent.

It is well known that residual errors $E(\vec{q})$ (see equation 2.58) computed by optimization algorithms reflect marker noise and soft tissue artifact. For segment optimization methods, placing many distributed markers on a segment will reduce the effect of uncorrelated noise (Cappozzo et al. 1997), and joint constraints used by the global optimization method reduce additionally the effect of errors normal to the constraints. What we really want to do, however, is to minimize the effect of systematic error on the estimated pose. As reported by Cereatti and colleagues (2006), there have been several attempts to modify optimization methods to minimize soft tissue artifact (Andriacchi et al. 1998; Cappozzo et al. 1997) but none of the approaches were satisfactory because discriminative models have no mechanism to compute a compensation for systematic errors even when the presence of this soft tissue artifact can be modeled. It has been proposed that pose estimation from noisy motion-capture data is better tackled by assuming uncertainty in the data and using well-established probabilistic algorithms based on Bayesian inference (Todorov 2007). Bayesian statistics is particularly well suited for dealing with uncertain data because it provides a framework for making inferences based on uncertain information. It is well beyond the scope of this chapter to elaborate on probabilistic pose estimation, but we would be remiss in leaving the reader with the impression that there is no potential solution to the issue of soft tissue artifact.

JOINT ANGLES

A joint angle is the relative orientation of one local coordinate system with another local coordinate system and is independent of the position of the origin of these coordinate systems. Joint angles represent a conceptual challenge to many biomechanists, so we have attempted to explain them carefully. It is important that the reader be aware of the following challenges:

1. Joint angles are rarely represented by an orientation matrix but instead are represented by a parameterized representation of this orientation matrix, and often the resulting angle is not a vector. This means that the angles cannot be added or subtracted, which makes the specification of a reference angle awkward.

2. There are joints, such as the shoulder, for which there is no single definition of a joint angle that is anatomically meaningful for the full range of motion of the joint.

3. A number of clinical and sport-related conventions specify an angle relative to a motion-based coordinate system (e.g., a swing plane in golf, or the direction of travel of a thrown object) and not to a structural or anatomically based coordinate system. This requires the creation of *virtual coordinate systems* that vary from trial to trial rather than by the definition of the segments.

4. 2-D projected angles in the sagittal, frontal, and transverse planes, although conceptually straightforward to understand, do not constitute a 3-D angle. In fact, reporting multiple planar views of angles as if they were a 3-D angle is incorrect.

Several methods can be used to parameterize the relative orientation of two coordinate systems (e.g., Grood and Suntay 1983; Spoor and Veldpaus 1980; Woltring 1991). Three of the most commonly used methods, the Cardan-Euler method (e.g., Davis et al. 1991; Engsberg & Andrews 1987), the joint coordinate system method (e.g., Grood and Suntay 1983; Soutas-Little et al. 1987), and the helical angle method (e.g., Woltring 1991), are presented in this chapter.

To illustrate the notion of a rotating coordinate system, we can look at a 2-D rotation about a single axis (in this case, the *x*-axis). In figure 2.14, the coordinate systems designated as *xyz* and *x'y'z'* initially are coincident with each other. That is, the origins are the same and the *y*- and *y'*-axes are parallel. The LCS *x'y'z'* is then rotated from the right horizontal of LCS *xyz* by an angle φ (e.g., positive rotation about the *x*-axis). The rotation matrix R_x describing the transformation from $\vec{P}$ to $\vec{P}'$ in 2-D is

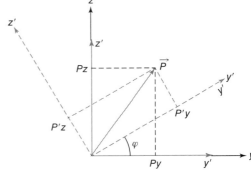

▲ **Figure 2.14** The coordinate systems here are designated as *xyz* and *x'y'z'*. The rotation occurs about the *x* and *x'* axes, which are not shown in this figure. The vector $\vec{P}$ can be represented both in the coordinate system *yz* as (P_y, P_z) and the coordinate system *'z'* as (P_y, P_z). If the coordinate system *yz* is rotated to *y'z'* through the angle φ, a rotation matrix can be determined to transform the vector $\vec{P}$ into $\vec{P}'$.

$$R_x = \begin{bmatrix} \cos\varphi & \sin\varphi \\ -\sin\varphi & \cos\varphi \end{bmatrix} \tag{2.59}$$

A vector $\vec{P}$ can be represented in either LCS, either $\vec{P} = (\vec{P}_y, \vec{P}_z)$ or $\vec{P}' = (\vec{P}'_y, \vec{P}'_z)$ using the rotation matrix (equation 2.59). The vector $\vec{P}$ can be transformed into $\vec{P}'$ as follows:

$$\vec{P}' = R_x\vec{P} \quad \text{or} \quad \begin{bmatrix} y' \\ z' \end{bmatrix} = R_x \begin{bmatrix} y \\ z \end{bmatrix} \tag{2.60}$$

Similarly, the vector $\vec{P}'$ can be transformed into $\vec{P}$ as follows:

$$\vec{P} = R'_x \vec{P}' \quad \text{or} \quad \begin{bmatrix} y \\ z \end{bmatrix} = R'_x \begin{bmatrix} y' \\ z' \end{bmatrix} \tag{2.61}$$

In this example, vectors can be presented in different coordinate systems. That is, given a vector in one coordinate system, the same vector can be represented in another coordinate system by means of a rotation matrix.

Cardan-Euler Angles

A 3-D rotation matrix (in other words, the orientation of one LCS with respect to another LCS) can be represented by three successive rotations about unique axes. This means that three elements (angles) fully specify the nine components of a 3×3 rotation matrix. The order of the rotations matters greatly, and for clarity we describe one particular sequence fully and comment on other sequences.

The Cardan rotation sequence XYZ is often used in biomechanics (Cole et al. 1993). This sequence involves three steps: first, rotation about the laterally directed axis (X); second, rotation about the anteriorly directed axis (Y); and third, rotation about the vertical axis (Z). Remember that the X-axis is directed laterally for the right-side segments and medially for the left-side segments. This sequence of rotations is diagrammed in figure 2.15.

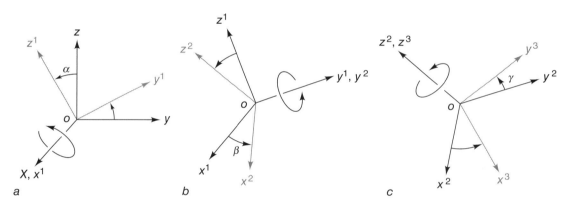

▲**Figure 2.15** Cardan sequence XYZ of rotations first about (a) the X-axis of the stationary coordinate system (α); then about (b) the new y^1-axis (β); and finally about (c) the z^2-axis (γ).

Using figure 2.15, we see that the first rotation (α) takes place about the X-axis and leads to new orientations of the y- and z-axes (y^1 and z^1), with the X-axis remaining in the same orientation and now labeled x^1. The second rotation (β) about the y^1-axis leads to new positions of the x^1- and z^1-axes (x^2 and z^2). For the third rotation (γ) about the z^2-axis, the x^2- and y^2-axes assume the new orientation of x^3 and y^3.

For illustrative purposes, we will first compute a 3-D *segment angle*, which is the rotation of a segment LCS relative to the laboratory GCS (see figure 2.16). Segment angles only have anatomical meaning when the coordinate systems are defined consistently (e.g., in this chapter Z-axial, Y-anterior, X-lateral). This means that the direction of walking in a gait trial must be in the anterior direction of the laboratory GCS,

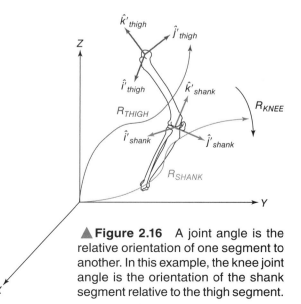

▲**Figure 2.16** A joint angle is the relative orientation of one segment to another. In this example, the knee joint angle is the orientation of the shank segment relative to the thigh segment.

which is naturally restrictive, so many laboratories define a virtual GCS whose anterior direction is the direction of walking.

In an earlier section of this chapter we demonstrated how to compute the LCS of a segment at each frame of data (e.g., equation 2.16 for the pelvis). The resulting orientation matrix is exactly what we need for computing a segment angle. By elaborating the rotation matrix for the XYZ sequence, we will demonstrate how to extract the three Cardan angles. The angles for the XYZ sequence are designated α (alpha) for the first rotation, β (beta) for the second rotation, and γ (gamma) for the third rotation. The rotation matrix R for an XYZ rotation sequence is

$$R = R_z R_y R_x \tag{2.62}$$

where

$$R_x = \begin{bmatrix} 1 & 0 & 0 \\ 0 & \cos\alpha & \sin\alpha \\ 0 & -\sin\alpha & \cos\alpha \end{bmatrix} R_y = \begin{bmatrix} \cos\beta & 0 & -\sin\beta \\ 0 & 1 & 0 \\ \sin\beta & 0 & \cos\beta \end{bmatrix} R_z = \begin{bmatrix} \cos\gamma & \sin\gamma & 0 \\ -\sin\gamma & \cos\gamma & 0 \\ 0 & 0 & 1 \end{bmatrix} \tag{2.63}$$

The rotation matrix is computed by multiplying the three matrices of equation 2.63 to generate

$$R = \begin{bmatrix} \cos\gamma\cos\beta & \cos\gamma\sin\beta\sin\alpha + \sin\gamma\cos\alpha & \sin\gamma\sin\alpha - \cos\gamma\sin\beta\cos\alpha \\ -\sin\gamma\cos\beta & \cos\alpha\cos\gamma - \sin\gamma\sin\beta\sin\alpha & \sin\gamma\sin\beta\cos\alpha + \cos\gamma\sin\alpha \\ \sin\beta & -\cos\beta\sin\alpha & \cos\beta\cos\alpha \end{bmatrix} \tag{2.64}$$

The elements in the combined matrix represent the relative orientation of an LCS relative to the GCS. This matrix is often called the *decomposition matrix*. The Cardan angle calculations are derived directly from the matrix R in equation 2.64. The angle α is computed from elements (3, 2) and (3, 3) as follows:

$$\alpha = \tan^{-1}\left(\frac{-R_{32}}{R_{33}}\right) \tag{2.65}$$

The angle β is computed from elements (1, 1), (2, 2) and (3, 1) as follows:

$$\beta = \tan^{-1}\left(\frac{R_{31}}{\sqrt{R_{11}^2 + R_{21}^2}}\right) \tag{2.66}$$

The angle γ is computed from elements (2, 1) and (1, 1) as follows:

$$\gamma = \tan^{-1}\left(\frac{-R_{21}}{R_{11}}\right) \tag{2.67}$$

Computationally, if you use $\tan^2$ instead of $\tan^{-1}$, the range for all three components is in the interval $[-\pi, \pi]$ radians.

The derivation of a segment angle can be extended to the derivation of a joint angle, which is the rotation of one segment LCS to another segment LCS. The movement at a joint is often defined as the orientation of a distal segment relative to a proximal segment (Woltring 1991). For example, the knee joint angles can be computed as the orientation of the shank relative to the thigh. Computing a joint angle requires an additional step to the computation of a segment angle because the two segments that we will use both have rotation matrices that describe their orientation relative to the laboratory GCS.

As an illustration, consider the knee joint angle (see figure 2.16). The transformation of the thigh LCS to the laboratory GCS was presented as R_{RTHIGH} in equation 2.23. The transformation of the shank LCS to the laboratory GCS was presented as R_{RSHANK} in equation 2.29. The transformation of the shank LCS to the thigh LCS can be expressed as follows

$$R_{RKNEE} = R_{RSHANK} R'_{RTHIGH} \tag{2.68}$$

where R'_{RTHIGH} is the transpose of the thigh LCS matrix.

The Cardan angles can be derived from the rotation matrix R_{RKNEE} using equations 2.65, 2.66, and 2.67. For the sequence XYZ and the above proximally referenced knee joint angle, the three Cardan-Euler angles can be interpreted to represent (1) flexion-extension—the laterally directed x-axis of the thigh, (2) abduction-adduction—a nodal axis (i.e., an axis perpendicular to both the flexion-extension and axial rotation axes), and (3) axial rotation—the longitudinal z-axis of the shank. It is important to note that this interpretation of α, β, and γ is only valid if β is less than approximately 40°, and it cannot be computed at all if β equals 90°. If β is greater than 40°, as it often is at the shoulder, a different Cardan-Euler sequence is required to generate an angle with anatomical meaning. For the lower extremities, the XYZ sequence is used and this interpretation is usually valid.

We have presented only 1 of the 12 Euler sequences. Six of the 12 sequences, including the one presented here, involve rotations about three different axes (e.g., the previous sequence of XYZ) and are referred to as Cardan angles. The first rotation is about an axis fixed in LCS^1 (x^1, y^1, or z^1), the second is about a floating axis (an axis that changes according to the orientations of the first and third axes), and the third is about an axis fixed in LCS^2 (x^2, y^2, or z^2). The remaining sequences have a terminal rotation axis identical to the first rotation axis. These six sequences are *Euler rotations* (e.g., *XYX, ZYZ*, and so on) and the angles that define these rotations are *Euler angles*.

Both Euler and Cardan angles describe the orientation of one coordinate system relative to another coordinate system as a sequence of ordered rotations from the initial position of one coordinate system. In biomechanics, a Cardan sequence of rotations is used more often for the lower-extremity joint angles than is an Euler sequence. For the lower-extremity joint angles, and the segment coordinate system definitions used in this chapter, only the Cardan XYZ sequence has the anatomical meaning of flexion-extension, abduction-adduction, axial rotation, so it is very important to be aware of the specification of the segment LCS and the Euler-Cardan sequence used. If you elect to change your coordinate system convention from the one presented in this chapter (e.g., Z-axial, Y-anterior), make sure you understand the consequences of that choice. For a coordinate system with Y-axial and X-anterior, the lower extremity joint angles often use the sequence ZXY.

From Figure 2.17, we can see that the flexion-extension graph is similar to a comparable 2-D graph. It is also clear that the range of motion of the axial and abduction-adduction rotation angles is small relative to the flexion angle. Davis and colleagues (1991) stated that abduction-adduction and axial rotation angles are not used in their clinical laboratory because of the poor signal-to-noise ratio associated with these data. However, although these angles are small, they may contain information that is pertinent to understanding human locomotion.

The hip angles are calculated with the same formulas that were used for the knee angles but use the thigh LCS relative to the pelvis LCS. The same formulas are also solved for the ankle angles,

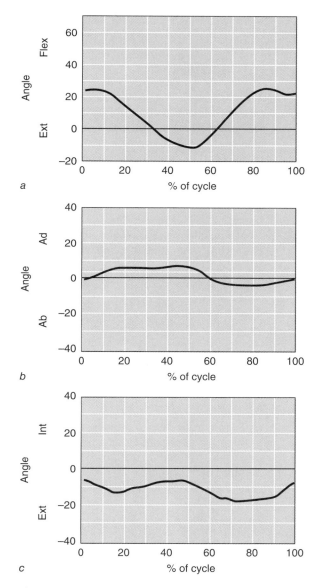

▲ **Figure 2.17** Walking 3-D hip angles of a healthy individual over a stride cycle: top, sagittal plane (flexion-extension); middle, frontal plane (adduction-abduction); and bottom, transverse plane (internal-external rotation).

using the foot LCS relative (note that this is typically R_{FOOT2} described by equation 2.42), but the terms concerning the actions are different. For the ankle, the flexion-extension angle, α, refers to ankle dorsiflexion-plantar flexion; the second angle, β, is referred to as ankle inversion-eversion; and the axial rotation angle, γ, is referred to as ankle abduction-adduction.

For the left lower extremity, the LCS of each segment is computed in exactly the same manner as for the right limb. However, because of the strict use of the right-hand rule, the i' direction is medially directed, whereas it was laterally directed in the right limb. To maintain the same anatomical polarity between right and left joint angles, the left limb abduction-adduction and axial rotation angles are multiplied by -1.

FROM THE SCIENTIFIC LITERATURE

Siegel, K.L., T.M. Kepple, P.G. O'Connell, L.H. Gerber, and S.J. Stanhope. 1995. A technique to evaluate foot function during the stance phase of gait. *Foot & Ankle International*. 16(12):764-770.

The purpose of this study was to develop and describe a biomechanical 3-D evaluation of foot function during gait. This study included one participant without a foot abnormality, four participants with rheumatoid arthritis, and one participant with an excessively pronated foot. The investigators evaluated foot function using a 3-D motion-capture system. Kinematic data were expressed relative to one of three coordinate systems: in the local coordinate system the z-axis was vertical, the y-axis corresponded to the path of progression, and the x-axis was perpendicular to the other two. Kinematic data selected for analysis included the absolute segment and relative joint angular displacements. The results indicated that foot function clearly changed with increasing severity of forefoot disease (see figure 2.18). This measurement technique was able to discriminate patients with rheumatoid arthritis from a healthy population and to discriminate different levels of the disease. Future studies will use this technique to study the impact of disease and treatment on foot function.

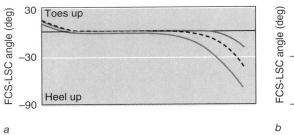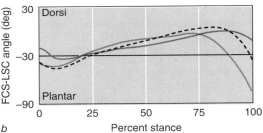

▲ **Figure 2.18** Plots showing the angles of the foot (FCS) relative to the leg (LCS) as a percent of the support phase. Healthy subject, blue line; female rheumatoid arthritis patient, dashed line; and male rheumatoid arthritis patient, gray line. *(a)* Foot-floor contact angle; *(b)* foot-shank dorsi–plantar flexion angle.

Adapted from K.L. Siegel et al., 1995, "A technique to evaluate foot function during the stance phase of gait," *Foot & Ankle International* 16(12): 764-770.

Joint Coordinate System

The joint coordinate system (JCS) was first proposed to describe the motion of the knee joint (Grood and Suntay 1983) and has since been applied to the other lower-extremity joints. The method was developed so that all three rotations between body segments had a functional, anatomical meaning. The JCS approach uses one coordinate axis from each LCS of the two segments that constitute the joint. In the example of the knee (see figure 2.19), the longitudinal axis of the JCS is the z-axis of the LCS of the distal segment $(\hat{k}'_{shank})$ and the laterally directed axis is the x-axis from the LCS of the proximal segment $(\hat{i}\,'_{thigh})$. The third axis is a floating axis that is the cross product of the longitudinal and lateral axes and thus is perpendicular to the plane formed by these directed axes $(\hat{k}\,'_{shank} \times \hat{i}\,'_{thigh})$. It should be clear that the vertical and lateral axes of this system are not necessarily perpendicular. The JCS is, therefore, not an orthogonal system. The JCS for an instant in time for the knee joint is schematically presented in figure 2.19.

Angles of the JCS are designated as α for flexion-extension, β for abduction-adduction, and γ for external-internal rotation with precisely same meaning as in the Cardan angles presented earlier. Flexion-extension is assumed to be a rotation about the laterally directed axis of the proximal segment $(\hat{i}\,'_{thigh})$, abduction-adduction is a rotation about the floating axis $(\hat{k}\,'_{shank} \times \hat{i}\,'_{thigh})$, and axial rotation is a rotation about the vertical axis of the distal segment $(\hat{k}\,'_{shank})$. It has been shown that the JCS is equivalent to an *XYZ* Cardan rotation sequence described earlier but only when the proximal segment is the reference segment (Cole et al. 1993).

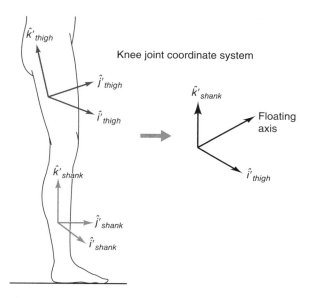

▲ **Figure 2.19** Representation of the JCS of the knee. The vertical axis is the $\hat{k}$-axis of the shank LCS, the mediolateral axis is the $\hat{i}$-axis of the thigh LCS, and the floating axis is calculated by the cross product $\hat{k} \times \hat{i}$.

FROM THE SCIENTIFIC LITERATURE

McClay, I., and K. Manal. 1997. Coupling parameters in runners with normal and excessive pronation. *Journal of Applied Biomechanics* 13:109-24.

The purpose of this study was to investigate differences in the coupling behavior of foot and knee motions during the support phase of running in normal pronators (NL) and in those with excessive pronation (PR). The authors collected three-dimensional data and then computed ankle and knee angles using the JCS (figure 2.20).

Excursion ratios between rear-foot eversion and tibial internal rotation were compared between the two groups and found to be significantly lower in the pronator subjects. Timing between peak eversion, knee flexion, and knee internal rotation was also assessed. The timing between peak knee and rear-foot angles was not significantly different between the groups, although times were more closely matched in the normal subjects. Results of this study suggested that increased motion of the rear foot can lead to excessive movement at the knee.

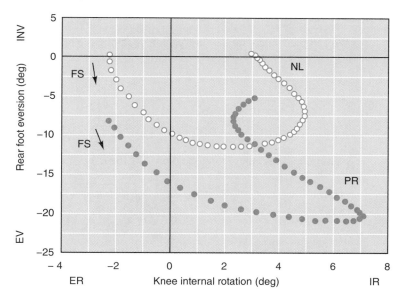

▲ **Figure 2.20** Ensemble angle-angle diagram of ankle eversion and knee internal rotation for normal and excessive pronation subjects.

Reprinted, by permission, from I. McClay and K. Manal, 1997, "Coupling parameters in runner with normal and excessive pronation," *Journal of Applied Biomechanics* 13: 109-124.

Helical Angles

Another method of parameterizing the orientation of one LCS to another LCS is based on the *finite helical axis* or *screw axis* (Woltring et al. 1985; Woltring 1991). In this technique, a position vector and an orientation vector are defined. Any finite movement from a reference position can be described in terms of a rotation about and translation along a single directed line or axis (i.e., the helical or screw axis) in space with unit direction ñ (figure 2.21). Note that in many cases this axis will not coincide with any of the defined axes of the LCS of either segment.

The orientation vector is defined from the rotation matrix R' calculated previously (see equation 2.61). The orientation vector components are calculated using the relationships outlined by Spoor and Veldpaus (1980). The orientation components can be determined as

$$\sin\theta\,\vec{n} = \frac{1}{2}\begin{bmatrix} R_{23} - R_{32} \\ R_{31} - R_{13} \\ R_{12} - R_{21} \end{bmatrix} \qquad (2.69)$$

If $\vec{n}^T\vec{n} = 1$ and $\sin\theta \leq \frac{1}{2}\sqrt{2}$, we can use the following equation to solve for $\sin\theta$:

$$\sin\theta = \frac{1}{2}\sqrt{\left(R_{23} - R_{32}\right)^2 + \left(R_{31} - R_{13}\right)^2 + \left(R_{12} - R_{21}\right)^2} \qquad (2.70)$$

However, if $\sin\theta > \frac{1}{2}\sqrt{2}$, we can use the following equation to solve for $\cos\theta$:

$$\cos\theta = \frac{1}{2}\left(R_{11} + R_{22} + R_{33} - 1\right) \qquad (2.71)$$

We can then calculate the unit vector, $\vec{n}$, along the helical axis as

$$\vec{n} = \frac{\frac{1}{2}\begin{bmatrix} R_{23} - R_{32} \\ R_{31} - R_{13} \\ R_{12} - R_{21} \end{bmatrix}}{2\sin\theta} \qquad (2.72)$$

Putting it all together from equation 2.71 the magnitude of θ applied to equation 2.72 as:

$$\theta = \sin^{-1}\left(\frac{1}{2}\sqrt{\left(R_{23} - R_{32}\right)^2 + \left(R_{31} - R_{13}\right)^2 + \left(R_{12} - R_{21}\right)^2}\right)$$
$$\text{or } \theta = \cos^{-1}\left(\frac{R_{11} + R_{22} + R_{33} - 1}{2}\right) \qquad (2.73)$$

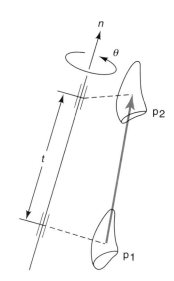

▲ **Figure 2.21** A finite helical axis is defined from the translation *(t)* along and a rotation *(θ)* about the helical axis *(n)* from point p₁ to point p₂.

Adapted from *Human Movement Science*, Vol. 10, "Representation and calculation of 3-D joint movement," H.J. Woltring, pg. 603-616, copyright 1991, with permission of Elsevier.

Cardan-Euler System Versus Joint Coordinate System Versus Helical Angles

Of the methods presented here, the Cardan-Euler and JCS techniques are the most widely used for calculating 3-D joint angles. Neither of these two approaches appears to have any obvious advantages or disadvantages over the other. In fact, the principle of the two approaches is the same, and, for the

XYZ sequence, the angles calculated by each should give exactly the same results. There are advantages and drawbacks, however, relative to using helical angles.

The major advantage for Cardan-Euler angles over helical angles is that Cardan-Euler angles are widely used in biomechanics and provide a well-understood anatomical representation of joint angles of the lower extremities. The *XYZ* sequence fails to have anatomical meaning if the *Y* rotation is greater than approximately 40° of abduction and, in the case of *gimbal lock* (when the second rotation equals ±90°), has no meaning at all. However, this is generally a much greater problem in the upper extremity than in the lower extremity.

Helical angles are especially appropriate when the rotation is very small. In addition, they eliminate the problem of gimbal lock. However, the representation of joint motion provided by helical angles does not always correspond with an anatomical representation that is clinically meaningful. In addition, helical angles are very sensitive to noisy coordinate data. Therefore, the coordinate data must be substantially smoothed before the helical angles are calculated.

Joint Angle Normalization

Joint angles parameterize the rotation matrix representing the transformation of one segment coordinate system into another segment coordinate system. Joint angles are therefore directly dependent on the orientation of the segment LCS. Many investigators, however, prefer to declare that the standing posture should be considered a reference posture from which all joint angles are computed. In other words, all joint angles in the standing posture should be considered zero. Referencing a joint angle to a reference posture is referred to as *angle normalization*.

The process of normalization is much more complicated than many people suspect because joint angles are not vectors. This means that they cannot be added or subtracted. Therefore, a normalized joint angle is not computed simply by subtracting the joint angle in the reference posture from the computed joint angle at a given frame of data.

The rotation matrix that defines the relative orientation of a segment, R_{Seg}, with a reference segment, R_{Ref}, can be expressed as

$$R = R_{Seg} R'_{Ref} \tag{2.74}$$

For a normalized joint angle, the rotation matrix must include the orientation of the segments in the calibration pose R_{CalSeg} and R_{CalRef} and is expressed as follows:

$$R = \left(R'_{CalSeg} R_{Seg} \right) \left(R'_{CalRef} R_{Ref} \right)' \tag{2.75}$$

The normalized joint angles can then be extracted from *R* as was done previously. One of the risks of this form of normalization is that the reference poses may result in gimbal lock relative to the movement trials. This can occur if the subject in the standing trial is orientated perpendicular to the movement trial.

An explicit (and equivalent) way to force the reference posture to represent a zero angle is to define all segment coordinate systems in the standing trial to have precisely the same orientation but have the anterior direction consistent with anterior direction of the subject. One way to accomplish this is to ignore the anatomical landmarks and force all segments to be aligned with the laboratory GCS. The effect is the same as the computation of the normalized joint angle previously described. There are many more subtle tricks for aligning segment coordinate systems, but these are beyond the scope of this chapter.

Some authors use normalization with the intention to clean up errors in the determination of the segment coordinate systems caused by the misplacement of markers. In effect, these authors are assuming that making the reference posture a zero angle will result in uniformity across subjects and data collection sessions (e.g., zero angle will have a consistent meaning). This assumption is in error because the coordinate systems may still be incorrect. The reference posture may indeed have a common angle of zero for the reference posture across sessions, but if the segment coordinate systems are not aligned consistently across sessions, no posture other than the reference posture will be guaranteed to be equivalent (note: a stopped clock is correct twice a day).

JOINT ANGULAR VELOCITY AND ANGULAR ACCELERATION OF CARDAN JOINT ANGLES

In a 2-D analysis, to obtain the joint angular velocity, the joint angle is simply differentiated with respect to time, and for angular acceleration, the joint angle can be differentiated twice with respect to time. However, in a 3-D analysis, the derivative of the joint angles $(\dot{\alpha}, \dot{\beta}, \dot{\gamma})$ is not equivalent to the joint angular velocity nor is the double derivative equal to angular acceleration because Cardan angles are not vectors.

We can compute the angular velocity of a segment relative to the laboratory GCS by differentiating the rotation matrix using finite differences. The angular velocity ω_i at time t_i is computed from the transformation between the rotation matrix $R_{t_{i-1}}$ at time t_{i-1} and $R_{t_{i+1}}$ at time t_{i+1} as follows:

$$R_\Delta = R_{t_{i+1}} R_{t_{i-1}}$$ (2.76)

$$|\omega_i| = \frac{\delta}{t_{i+1} - t_{i-1}}$$ (2.77)

where $\delta = \cos^{-1}\left(\dfrac{R_{\Delta 11} + R_{\Delta 22} + R_{\Delta 33} - 1}{2}\right)$.

We can compute the unit vector as:

$$\vec{v} = \frac{\begin{bmatrix} R_{\Delta 23} - R_{\Delta 32} \\ R_{\Delta 31} - R_{\Delta 13} \\ R_{\Delta 12} - R_{\Delta 21} \end{bmatrix}}{2\sin\delta}$$ (2.78)

and transform the unit vector into the GCS from the coordinate system from which the relative rotation began:

$$\hat{u} = R'_{t_{i-1}} \hat{v}$$ (2.79)

The angular velocity vector is therefore

$$\omega_i = |\omega_i| \hat{u}$$ (2.80)

Joint angular velocity is a vector that describes the relative angular velocity of one segment to another. In this case the rotation matrix R can be replaced by the rotation matrix determined using equation 2.61.

The joint angular velocity, $\omega = \begin{bmatrix} \omega_X \\ \omega_Y \\ \omega_Z \end{bmatrix}$, can be expressed in terms of the derivatives of the Cardan angles, $\dot{\theta} = \begin{bmatrix} \dot{\alpha} \\ \dot{\beta} \\ \dot{\gamma} \end{bmatrix}$, by transforming the second and third rotations back into the first rotation coordinate system as follows:

$$\begin{bmatrix} \omega_X \\ \omega_Y \\ \omega_Z \end{bmatrix} = \begin{bmatrix} \dot{\alpha} \\ 0 \\ 0 \end{bmatrix} + R'_x \begin{bmatrix} 0 \\ \dot{\beta} \\ 0 \end{bmatrix} + R'_x R'_y \begin{bmatrix} 0 \\ 0 \\ \dot{\gamma} \end{bmatrix}$$ (2.81)

Because angular velocity is a vector, we can compute angular acceleration, $\begin{bmatrix} \alpha_X \\ \alpha_Y \\ \alpha_Z \end{bmatrix}$, as the first derivative of angular velocity:

$$\begin{bmatrix} \alpha_X \\ \alpha_Y \\ \alpha_Z \end{bmatrix} = \begin{bmatrix} \dot{\omega}_X \\ \dot{\omega}_Y \\ \dot{\omega}_Z \end{bmatrix} \tag{2.82}$$

SUMMARY

3-D kinematics are straightforward mathematically but conceptually challenging for everyone, and the reader should focus carefully on the details of the transformations between coordinate systems. It should be stressed that 3-D joint angles are not equivalent to projected planar 2-D angles.

Various marker setups, all of which have pros and cons, can be used. Minimally, however, three noncollinear markers must define a segment LCS and three noncollinear markers must track the pose of the LCS. The three pose estimation algorithms presented—direct pose estimation, segment optimization, and global optimization—vary in mathematical complexity and in their ability to cope with artifacts in the data.

Several methods are used to calculate joint angles, including (1) the Cardan-Euler approach, (2) the JCS approach, and (3) the finite helical axis approach. The most widely used in the biomechanics literature for the lower extremity are the Cardan-Euler and JCS approaches.

SUGGESTED READINGS

Allard, P., A. Capozzo, A. Lundberg, and C. Vaughan. 1998. *Three-Dimensional Analysis of Human Locomotion*. Chichester, UK: Wiley.

Berme, N., and A. Cappozzo (eds.). 1990. *Biomechanics of Human Movement: Applications in Rehabilitation, Sports and Ergonomics*. Worthington, OH: Bertec Corporation.

Nigg, B.M., and W. Herzog. 1994. *Biomechanics of the Musculo-Skeletal System*. New York: Wiley.

Vaughan, C.L., B.L. Davis, and J.C. O'Connor. 1992. *Dynamics of Human Gait*. Champaign, IL: Human Kinetics.

Zatsiorsky, V.M. 1998. *Kinematics of Human Motion*. Champaign, IL: Human Kinetics.

PART II

KINETICS

Kinetics is the study of the causes of motion. In essence, it is the recording and study of forces and how they affect motion. Forces can change the linear or angular motion of a body by changing the body's linear or angular momentum. Forces can also work by changing the mechanical energy of a body either by increasing it (called positive work) or by reducing it (called negative work). In analyzing human motion, it is often necessary to quantify the inertial properties of various parts of the body. Chapter 3 introduces commonly used methods to determine segmental inertial properties such as mass, center of mass (also called center of gravity), and moment of inertia. In chapter 4, the basic laws concerning forces are outlined along with the concepts of their vector nature, their ability to create rotation (moment of force), and the principle of impulse and momentum. Finally, methods for directly recording forces are described. Chapter 5 continues the study of kinetics by explaining how two-dimensional kinematics coupled with inertial properties can indirectly quantify the net forces and net moments of force that must be present at human joints to produce the observed motions. This process is called inverse dynamics. Next, chapter 6 introduces the concepts of mechanical energy, work, and power, two-dimensionally. This chapter includes how to compute the work done by the moments of force computed in chapter 5. Chapter 7 provides the mathematical processes necessary for performing inverse dynamics and computing the powers produced by the resulting moments of force in three dimensions instead of just two. Appendixes D and E provide background mathematics for chapter 7.

Chapter 3

Body Segment Parameters

D. Gordon E. Robertson

Anthropometry is the discipline concerned with the measurement of the physical characteristics of humans. Biomechanists are mainly interested in the inertial properties of the body and its segments, a subfield of anthropometry called **body segment parameters**. Before a kinetic analysis of human movement is possible, each segment's physical characteristics and inertial properties must be determined. The relevant characteristics are the segmental mass, locations of the segmental centers of gravity, and segmental mass moments of inertia.

This chapter is not meant to be a detailed account of the study of anthropometry or body segment parameters; they are described elsewhere (e.g., Contini 1972; Drillis et al. 1964; Nigg and Herzog 1994; Zatsiorsky 2002). In this chapter, we

▶ present background material on measuring and estimating body segment parameters,

▶ outline computational methods for quantifying total body and segmental inertial characteristics for planar (2-D) motion analyses (essential for understanding chapters 5 and 6), and

▶ outline computational methods for quantifying total body and segmental inertial characteristics for spatial (3-D) motion analyses (essential for understanding chapter 7).

To learn about planar motion analysis only, you can omit the section titled Three-Dimensional (Spatial) Computational Methods from the course of study. Equivalently, if spatial (i.e., 3-D) analysis is undertaken, the section titled Mass Moment of Inertia (part of the Two-Dimensional (Planar) Computational Methods section) can be omitted.

METHODS FOR MEASURING AND ESTIMATING BODY SEGMENT PARAMETERS

A major concern for biomechanists is the assumption that body segments behave as rigid bodies during movements. This assumption obviously is not valid, because bones bend, blood flows, ligaments stretch, and muscles contract. It is also common to model some body parts as single rigid bodies despite the fact that they consist of several segments. For instance, the foot is commonly considered to be one segment even though it clearly can bend at the metatarsal-phalangeal joints. Similarly, the trunk often is treated as one single rigid body or sometimes as two or three. In reality, however, it is a series of interconnected rigid bodies including the many interconnected vertebrae, pelvis, and scapulae. These assumptions simplify the otherwise complex musculoskeletal system and eliminate the necessity of quantifying the changes in mass distribution caused by tissue deformation and movement of bodily fluids. By also assuming that segmental mass distribution is similar among members of a particular population, the researcher can estimate an individual's segmental parameters by applying equations based on the averages obtained from samples taken from the population. Several sources offer these averaged parameters, but it is best to select from a population that closely matches the subject (Hay 1973).

Attempts to quantify body segment parameters generally fall into four categories: cadaver studies (e.g., Braune and Fischer 1889, 1895-1904; Dempster 1955; Fischer 1906; Harless 1860), mathematical modeling (e.g., Hanavan

1964; Hatze 1980; Yeadon 1990a, 1990b; Yeadon and Morlock 1989), scanning and imaging techniques (e.g., Durkin and Dowling 2003; Durkin et al. 2002; Mungiole and Martin 1990; Zatsiorsky and Seluyanov 1983), and kinematic measurements (e.g., Dainis 1980; Hatze 1975; Vaughan et al. 1992). Each of these techniques has both advantages and disadvantages. We next review some of the literature concerning studies that used one or more of these techniques.

Cadaver Studies

Inertial properties (mass, center of mass, moment of inertia) are difficult to determine for a particular living person. If you were to quantify these properties for a robot, you would separate each segment and analyze it individually by performing specific tests. Because this is not possible for living persons, indirect methods must be used. For example, the *coefficient method* uses tables of proportions that predict the body segment parameters from simple, noninvasive measures such as total body mass, height, and segment lengths. The earliest attempts at defining these proportions date to the works of Harless (1860), Braune and Fischer (1889), and Fischer (1906), but the most significant advancement was the work done by W.T. Dempster, published in 1955 as the monograph *Space Requirements of the Seated Operator*. This document, which Dempster compiled while working for the U.S. Air Force, not only outlined the procedures for measuring body segment parameters from cadaveric

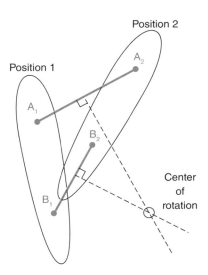

▲ **Figure 3.1** Reuleaux's (1876) method for calculating the center of rotation of a body. Points A and B are fixed on the rigid body. After movement from position 1 to position 2, dotted lines are drawn as right bisectors of the lines A1-A2 and B1-B2. The center of rotation is at the intersection of the dotted lines.

materials but also included tables for proportionally determining the body segment parameters needed to biomechanically analyze human motion.

Dempster collected data from living persons, from anatomical specimens, and, most important (for biomechanists), from eight complete cadavers. First, he used

Table 3.1 Dempster's Segment Endpoint Definitions

Segment	Proximal end	Distal end
Clavicular link	Sternoclavicular joint center: midpoint of palpable junction between proximal end of clavicle and upper border of sternum	Claviscapular joint center: midpoint of line between coracoid tuberosity of clavicle and acromioclavicular articulation at lateral end of clavicle
Scapular link	Claviscapular joint center: see Clavicular link, Distal end	Glenohumeral joint center: midpoint of palpable bony mass of head and tuberosities of humerus
Humeral link	Glenohumeral joint center: see Scapular link, Distal end	Elbow joint center: midpoint between lowest palpable point of medial epicondyle of humerus and a point 8 mm above the radiale (radiohumeral junction)
Radial link	Elbow joint center: see Humeral link, Distal end	Wrist joint center: distal wrist crease at palmaris longus tendon or palpable groove between lunate and capitate bone in line with metacarpal bone III
Hand link	Wrist joint center: see Radial link, Distal end	Center of mass of hand: midpoint between proximal transverse palmar crease and radial longitudinal crease in line with third digit
Femoral link	Hip joint center: tip of femoral trochanter, 1 cm anterior to most laterally projecting part of greater trochanter	Knee joint center: midpoint between centers of posterior convexities of femoral condyles
Leg link	Knee joint center: see Femoral link, Distal end	Ankle joint center: level of a line between tip of lateral malleolus of fibula and a point 5 mm distal of the tibial malleolus
Foot link	Ankle joint center: see Leg link, Distal end	Center of gravity of foot: midway between ankle joint center and ball of foot at head of metatarsal II

the method of Reuleaux (1876) to determine the average center of rotation at each joint (see figure 3.1). At some joints, particularly the shoulder, the center of rotation was difficult to identify, and these locations became the endpoints for the various body segments. Table 3.1 defines these endpoints.

The cadavers were then segmented according to Dempster's own techniques, and their lengths, masses, and volumes were carefully recorded. Dempster then calculated the location of the center of gravity (using a balancing technique) and the moment of inertia (using a pendulum technique) for each segment. Finally, Demp-

ster created tables showing the segmental masses as proportions of the total body mass and the locations of the centers of gravity and lengths of the radii of gyration as proportions of the segments' lengths. (The radius of gyration was used as an indirect means of calculating rotational inertia. Its meaning and relationship with moment of inertia are presented in the next section.) These data in modified form (Miller and Nelson 1973; Plagenhoef 1971; Winter 1990) appear in table 3.2. Later, Barter (1957), working with Dempster's data, performed stepwise regression analysis to derive regression equations that more accurately compute segmental masses.

Table 3.2 Dempster's Body Segment Parameters

Segment	Endpoints[a] (proximal to distal)	SEGMENTAL MASS/TOTAL MASS (P)[b]	CENTER OF MASS/SEGMENT LENGTH $(R_{proximal})$[c]	(R_{distal})[c]	RADIUS OF GYRATION/ SEGMENT LENGTH (K_{cg})[d]	$(K_{proximal})$[d]	(K_{distal})[d]
Hand	Wrist center to knuckle II of third finger	0.0060	0.506	0.494	0.298	0.587	0.577
Forearm	Elbow to wrist center	0.0160	0.430	0.570	0.303	0.526	0.647
Upper arm	Glenohumeral joint to elbow center	0.0280	0.436	0.564	0.322	0.542	0.645
Forearm and hand	Elbow to wrist center	0.0220	0.682	0.318	0.468	0.827	0.565
Upper extremity	Glenohumeral joint to wrist center	0.0500	0.530	0.470	0.368	0.645	0.596
Foot	Ankle to ball of foot	0.0145	0.500	0.500	0.475	0.690	0.690
Leg	Knee to ankle center	0.0465	0.433	0.567	0.302	0.528	0.643
Thigh	Hip to knee center	0.1000	0.433	0.567	0.323	0.540	0.653
Lower extremity	Hip to ankle center	0.1610	0.447	0.553	0.326	0.560	0.650
Head	C7-T1 to ear canal	0.0810	1.000	0.000	0.495	1.116	0.495
Shoulder	Sternoclavicular joint to glenohumeral joint center	0.0158	0.712	0.288			
Thorax	C7-T1 to T12-L1	0.2160	0.820	0.180			
Abdomen	T12-L1 to L4-L5	0.1390	0.440	0.560			
Pelvis	L4-L5 to trochanter	0.1420	0.105	0.895			
Thorax and abdomen	C7-T1 to L4-L5	0.3550	0.630	0.370			
Abdomen and pelvis	T12-L1 to greater trochanter	0.2810	0.270	0.730			
Trunk	Greater trochanter to glenohumeral joint	0.4970	0.495	0.505	0.406	0.640	0.648
Head, arms, and trunk	Greater trochanter to glenohumeral joint	0.6780	0.626	0.374	0.496	0.798	0.621
Head, arms, and trunk	Greater trochanter to mid-rib	0.6780	1.142	−0.142	0.903	1.456	0.914

[a]Endpoints are defined in table 3.1.

[b]A segment's mass as a proportion of the total body mass.

[c]The distances from the proximal and distal ends of the segment to the segment's center of gravity as proportions of the segment's length.

[d]The radii of gyration about the center of gravity, proximal and distal ends of the segment to the segment's center of gravity as proportions of the segment's length.

Adapted from D.A. Winter 1990.

Many other body segment parameter studies have been conducted since Dempster's groundbreaking work. Two in particular, by Clauser and colleagues (1969) and Chandler and colleagues (1975), are noteworthy because they defined body segments using palpable bony landmarks instead of estimated, averaged joint centers of rotation. Clauser and colleagues also segmented an additional 13 cadavers (only 6 were used by Chandler et al.) and presented the results in a table similar to Dempster's. The segment definitions appear in table 3.3 and the values are presented in table 3.4.

Table 3.3 Clauser and Colleagues' (1969) Definitions of Endpoints of Body Segment Parameters

Endpoint	Definition
Acromion	Point at the superior and external border of the acromion process of the scapula. This point is easier to find if the subject bends laterally at the trunk to relax the deltoid muscle.
Glabella	Bony ridge under the eyebrows.
Metacarpale III	Distal head of the third metacarpal (proximal knuckle of the middle finger).
Occiput	Occipital process.
Radiale	Point at the proximal and lateral border of the head of the radius. Palpate downward in the lateral dimple at the elbow. Have the subject pronate and supinate the forearm slowly so that the radius can be felt rotating beneath the skin.
Sphyrion	Most distal point of the medial malleolus of the tibia. Place the marker on the fibula at the level of the sphyrion. Do not use the sphyrion fibulare, as it is more distal than the sphyrion.
Stylion	Styloid process of the radius. The styloid process of the ulna may be used.

Table 3.4 Clauser and Colleagues' Body Segment Parameters

Segment	Endpoints[a] (proximal to distal)	SEGMENTAL MASS/TOTAL MASS (P)[b]	CENTER OF MASS/SEGMENT LENGTH $(R_{proximal})$[c]	(R_{distal})[c]	RADIUS OF GYRATION/ SEGMENT LENGTH (K_{cg})[d,e]	$(K_{proximal})$[d]
Hand	Stylion to metacarpale III	0.0065	0.1802	0.8198	0.6019	0.6283
Forearm	Radiale to stylion	0.0161	0.3896	0.6104	0.3182	0.5030
Upper arm	Acromion to radiale	0.0263	0.5130	0.4870	0.3012	0.5949
Forearm and hand	Radiale to stylion	0.0227	0.6258	0.3742		
Upper extremity	Regression equation[f]	0.0490	0.4126	0.5874		
Foot	Heel to toe II	0.0147	0.4485	0.5515	0.4265	0.6189
Foot	Sphyrion to floor	0.0147	0.4622	0.5378		
Leg	Tibiale to sphyrion	0.0435	0.3705	0.6295	0.3567	0.5143
Thigh	Trochanter to tibiale	0.1027	0.3719	0.6281	0.3475	0.5090
Leg and foot	Tibiale to floor	0.0582	0.4747	0.5253		
Lower extremity	Trochanter to floor (sole)	0.1610	0.3821	0.6179		
Trunk	Chin-neck intersection to trochanter[g]	0.5070	0.3803	0.6197	0.4297	0.5738
Head	Top of head to chin-neck intersection	0.0728	0.4642	0.5358	0.6330	0.7850
Trunk and head	Chin-neck intersection to trochanter	0.5801	0.5921	0.4079		
Total body		1.0000	0.4119	0.5881	0.7430	0.8495

[a]Endpoints are defined in table 3.3.

[b]A segment's mass as a proportion of the total body mass.

[c]The distances from the proximal and distal ends of the segment to the segment's center of gravity as proportions of the segment's length.

[d]The radii of gyration about the center of gravity and the proximal end of the segment to the segment's center of gravity as proportions of the segment's length.

[e]Computed from $K_{cg} = \sqrt{K^2_{proximal} - R^2_{proximal}}$

[f]Regression equation for arm length = 1.126 (acromion to radiale distance) + 1.057 (radiale to stylion distance) + 12.52 (all lengths in cm).

[g]Chin-neck intersection. The point superior to the cartilage, at the level of the hyoid bone. Marker should be placed level with the intersection but at the lateral aspect of the neck.

From Clauser et al. 1969.

For their 3-D kinematic and kinetic investigation of the lower extremity during human gait, Vaughan and colleagues (1992) enhanced the methods used by Chandler and colleagues (1975). Vaughan and colleagues created regression equations for the masses of the lower extremities that included various segmental anthropometrics, such as calf and midthigh circumference, in addition to segment length and body mass (table 3.5). Hinrichs (1985) also extended Chandler's work by applying regression equations for the calculation of segmental moments of inertia. For in-depth reviews of the historical and scientific development of body segment parameters, refer to Contini (1972), Drillis and colleagues (1964), and Hay (1973, 1974).

Mathematical Modeling

Mathematical modeling of the inertial properties of human body segments was pioneered by the work of Hanavan (1964) when it became necessary to model the body for 3-D analyses. Hanavan made the assumption that mass was uniformly distributed within each segment and that the segments were rigid bodies that could be represented by geometric shapes. Most of the

segments were modeled as frusta of right-circular cones, as shown in figure 3.2. (A *frustum* of a cone—*frusta* in the plural—is the base of a cone that has had its vertex removed.) The hands were modeled as spheres, the head as an ellipsoid, and the trunk segments as elliptical cylinders. Hanavan's methods were later enhanced to include more segments and more directly measured anthropometric measures. For example, Hatze (1980) developed a system that uses 242 direct anthropometric measurements to define the segmental properties of a 42-DOF, 17-segment model of the body.

Scanning and Imaging Techniques

Another approach to determining inertial properties and then estimating body segment parameters involves scanning the living body with various radiation techniques. For instance, Zatsiorsky and Seluyanov (1983, 1985) presented data from an extensive study on the body segment parameters of both male ($n = 100$) and female ($n = 15$) living subjects. To compute mass distribution, the investigators used gamma mass scanning to quantify

Table 3.5 Body Segment Parameters for Equations Developed by Vaughan and Colleagues

Segment	Axis	Regression equation[a]
Foot	Flexion-extension	$I_{f/e} = 0.00023 m_{total}\left[4\left(h_m^2\right) + 3\left(l_{foot}^2\right)\right] + 0.00022$ $h_m = malleolus\ height\ (m)$
	Abduction-adduction	$I_{a/a} = 0.00021 m_{total}\left[4\left(h_m^2\right) + 3\left(l_{foot}^2\right)\right] + 0.00067$
	Internal-external rotation	$I_{i/e} = 0.00041 m_{total}\left[h_m^2 + w_{foot}^2\right] - 0.00008$ $w_{foot} = foot\ width\ (m)$
Leg	Flexion-extension	$I_{f/e} = 0.00347 m_{total}\left[l_{leg}^2 + 0.076 c_{leg}^2\right] + 0.00511$ $c_{leg} = leg\ circumference\ (m)$
	Abduction-adduction	$I_{a/a} = 0.00387 m_{total}\left[l_{leg}^2 + 0.076 c_{leg}^2\right] + 0.00138$
	Internal-external rotation	$I_{i/e} = 0.00041 m_{total} c_{leg}^2 + 0.00012$
Thigh	Flexion-extension	$I_{f/e} = 0.00762 m_{total}\left[l_{thigh}^2 + 0.076 c_{thigh}^2\right] + 0.01153$ $c_{thigh} = thigh\ circumference\ (m)$
	Abduction-adduction	$I_{a/a} = 0.00762 m_{total}\left[l_{thigh}^2 + 0.076 c_{thigh}^2\right] + 0.01186$
	Internal-external rotation	$I_{i/e} = 0.00151 m_{total} c_{thigh}^2 + 0.00305$

[a]Moments of inertia are in kilogram meters squared (kg·m²), m_{total} is the total body mass in kilograms, and l_{thigh}, l_{leg}, and l_{foot} are the segment lengths in meters.

Reprinted, by permission, from C.L. Vaughan, B.L. Davis, and J.C. O'Connor, 1992, *Dynamics of human gait* (Champaign, IL: Human Kinetics). By permission of C.L. Vaughan.

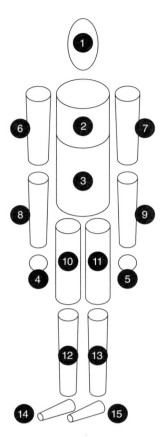

▲ **Figure 3.2** Hanavan's geometric model of the body.

the density of incremental slices of each segment. This method enabled estimations to be made of the mass, center of mass, and principal moments of inertia in 3-D for a total of 15 segments. Their subjects included indi-

viduals younger than the ages of the cadavers used in previous studies; furthermore, the investigators applied regression equations to customize body segment parameters. These values and methods were more recently documented and republished by Zatsiorsky (2002, pp. 583-616).

In addition to gamma mass scanning, other techniques that have been developed to quantify human body segment parameters include photogrammetry (Jensen 1976, 1978), magnetic resonance imaging (Mungiole and Martin 1990), and dual-energy X-ray absorptiometry (DEXA; Durkin and Dowling 2003; Durkin et al. 2002).

Some researchers have provided data from populations that are not well represented by the previously mentioned studies, which were based primarily on middle-aged male or young adult Caucasians. For example, Jensen and colleagues provided data for children (1986, 1989) and pregnant women (1996), Schneider and Zernicke (1992) quantified regression equations for the body segment parameters of infants, and Pavol and colleagues (2002) estimated inertial properties for the elderly.

Kinematic Techniques

Kinematic techniques are methods that measure kinematic characteristics to indirectly determine the inertial properties of segments. Hatze (1975) developed an oscillation technique that derives the mass, center of mass, and moment of inertia of the segments of the extremities and the damping coefficients of joints. The technique, which cannot be used for the trunk segments, requires that a body part be set into oscillation with an instrumented spring. The muscles must be relaxed so

FROM THE SCIENTIFIC LITERATURE

Schneider, K., and R.F. Zernicke. 1992. Mass, center of mass and moment of inertia estimates for infant limb segments. *Journal of Biomechanics* 25:145-8.

The purpose of this study was to develop linear regression equations that could quantify the body segment parameters for infants (0.4-1.5 years). The authors recognized that before they could analyze the mechanics of infant motion, they needed to develop a new set of anthropometric proportions applicable to infants. Clearly, the mass distribution of infants is different from that of adults. This is the reason that adult seat belts cannot be used by infants or young children, whose total body center of gravity is proportionally higher than that of an adult.

The authors collected data from the upper limbs of 44 infants and the lower limbs of 70 infants. They adapted a 17-segment mathematical model developed by Hatze (1979, 1980) to data from 18 infants. Hatze's model

requires 242 anthropometric measurements to compute the volumes of the 17 segments. These data were adjusted so that there was a close agreement between the model's estimates of total body mass and the infants' measured masses. Regression analysis was then used to derive mass, center of mass, and moment of inertia for the three segments of the upper limb and three segments of the lower limb. The researchers found that infants' segment masses and moments of inertia changed remarkably during the first 1 1/2 years of growth but that the center of mass was not greatly influenced by age. They also concluded that because of the high correlations achieved by their linear regression equations (64%-98% variance accounted), nonlinear regression was unnecessary.

that they do not influence the damped oscillations of the spring-limb system. Mathematically derived equations based on small oscillation theory are then used to estimate the properties of the joint and segment based on a passively damped reduction of the oscillations of the spring-limb system.

Another technique for estimating mass moment of inertia is called the *quick-release method* (Drillis et al. 1964). This technique also assumes that the muscles are relaxed and that the acceleration of a rapidly accelerated segment is affected only by the segment's rotational inertia. Thus, by measuring the angular acceleration of a segment after release of a known force *(F)* or moment *(M)*, we can derive the moment of inertia *(I)* of the segment from the following relationship:

$$I = M/\alpha = (Fd)/\alpha \qquad (3.1)$$

where M (or Fd) is the moment of force immediately before release, and α is the angular acceleration after release. Obviously, this method can be used only for terminal segments because applying it to other body parts would be difficult.

TWO-DIMENSIONAL (PLANAR) COMPUTATIONAL METHODS

In the following sections, we present methods for computing body segment parameters for 2-D analyses based on the traditional proportional methods developed by Dempster (1955). These are the simplest types of equations that yield reasonable results for people of average dimensions. Researchers who need more accurate measures should consult the research literature for methods that are suitable for their population of subjects and that use procedures and equipment that are within the financial and practical constraints of their project. Many excellent reviews of such literature are presented in the Suggested Readings for this chapter.

Segment Mass

The standard method for computing segment mass is to weigh the subject and then multiply the total body mass by the proportion that each segment contributes to the total. These proportions (*P* values) may be taken from tables 3.2 or 3.4, but these values were derived from middle-aged male cadavers and may not be appropriate for all subjects, particularly children and women. For young adult subjects, the values presented by Zatsiorsky (2002) can be used. Segment mass (m_S) is defined as

$$m_S = P_s\, m_{total} \qquad (3.2)$$

where m_{total} is the total body mass and P_s is the segment's mass proportion. Note that the sum of all the P values should equal 1; otherwise, the calculated body weight will be incorrect. That is,

$$\sum_{s=1}^{S} P_s = 1.000 \qquad (3.3)$$

where S is the number of body segments and s is the segment number. The values in table 3.2 were derived from the work of Dempster (1955), but they have been adjusted so that they total unity (i.e., 100%). Dempster's original proportions totaled less than 1 as a result of fluid loss during dissection and other sources of error. Miller and Nelson (1973) worked out adjustments to compensate for these losses.

EXAMPLE 3.1

Use the proportions in table 3.2 to calculate the thigh mass of a person who weighs 80.0 kg.

See answer 3.1 on page 377.

Center of Gravity

Center of gravity and center of mass are in essentially the same locations. In biomechanics, the terms are used interchangeably. A difference occurs only when a body is far from the earth's surface or otherwise affected by a large gravitational source. The **center of gravity** is the point where a motionless body, if supported at that point, will remain balanced—the balance point of a body. It is the point on a rigid body where the mass of the body can be considered to be concentrated for translational motion analyses. In other words, the translational motion of a rigid body concerns only the motion of the body's center of gravity. This lessens the amount of information that needs to be recorded about the body. The body's shape and structure can be ignored and only its center of gravity needs to be quantified.

Segment Center of Gravity

To simplify and generalize the process of calculating segment centers of gravity, Dempster (1955) developed the technique of representing the distances from each endpoint of a segment to that segment's center of gravity as proportions (*R* values) of the segment's length *(L)*. Given that $r_{proximal}$ and r_{distal} are the distances from the proximal and distal ends to the segment's center of gravity, respectively, these proportions are defined as

$$R_{proximal} = r_{proximal}/L \qquad (3.4)$$

$$R_{distal} = r_{distal}/L \qquad (3.5)$$

To compute a segment's center of gravity, the coordinates of the segment's endpoint must be quantified. Next, the *R* value for the particular segment must be selected from a table of proportions suitable for the subject. In general, table 3.2 can be used for adult males because these proportions are averages derived from eight male cadavers. Other tables may be used for other populations. One can select either $R_{proximal}$ or R_{distal}, but by consensus, segment centers are usually defined from their proximal ends. The center is then computed as a proportion of the distance from the proximal end toward the distal end, as depicted in figure 3.3. These equations can then be used to determine the *X-Y* coordinates of the segment's center *(x_cg, y_cg)*:

$$x_{cg} = x_{proximal} + R_{proximal} (x_{distal} - x_{proximal}) \quad (3.6)$$

$$y_{cg} = y_{proximal} + R_{proximal} (y_{distal} - y_{proximal}) \quad (3.7)$$

where *(x_proximal, y_proximal)* and *(x_distal, y_distal)* are the coordinates of the proximal and distal ends, respectively. Note that $R_{proximal} + R_{distal} = 1.000$, because they represent the total length *(L)* of the segment. Also, the actual distance from the proximal end to the center of gravity *(r_proximal)* can be computed from

$$r_{proximal} = R_{proximal} L \quad (3.8)$$

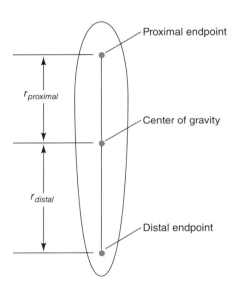

▲ Figure 3.3 Center of gravity in relation to segment endpoints.

Limb and Total Body Center of Gravity

To compute the center of gravity of a limb or combination of segments, a "weighted" average of the segments that make up the limb is computed. These equations define this process:

EXAMPLE 3.2

Calculate the center of gravity of the thigh using the proportions from table 3.2, given that the proximal end (the hip) of the thigh has the coordinates (−12.80, 83.3) cm and the distal end (the knee) has the coordinates (7.30, 46.8) cm.

See answer 3.2 on page 377.

$$x_{limb} = \frac{\sum_{s=1}^{L} P_s x_{cg_s}}{\sum_{s=1}^{L} P_s} \quad (3.9)$$

$$y_{limb} = \frac{\sum_{s=1}^{L} P_s y_{cg_s}}{\sum_{s=1}^{L} P_s} \quad (3.10)$$

where *L* is the number of segments in the limb, *(x_limb, y_limb)* represents the limb's center of gravity, *(x_cg, y_cg)* represents each segment's center of gravity, and the P_s are each segment's mass proportion. The total body's center of gravity is computed in a similar fashion. It is the weighted average of all the segments of the body. That is,

$$x_{total} = \sum_{s=1}^{L} P_s x_{cg_s} \quad (3.11)$$

$$y_{total} = \sum_{s=1}^{S} P_s y_{cg_s} \quad (3.12)$$

where *(x_total, y_total)* is the total body's center of gravity coordinates, *(x_cg, y_cg)* are the coordinates of the segments' centers of gravity, and *S* is the number of body segments. Each segment is weighted according to its mass proportion, P_s. In other words, each segment's center of gravity contributes proportionally to its *P* value, which is its mass as a proportion of the total body mass (figure 3.4). The heavier the segment, the more it affects the location of the center of gravity. Lighter segments have less influence on the center of gravity. Notice that there is no divisor in these equations because, as shown previously, the sum of all the *P* values is 1, namely, $\Sigma P_s = 1.0$.

Mass Moment of Inertia

The **mass moment of inertia**, also called *rotational inertia*, is the resistance of a body to change in its rotational motion. It is the angular or rotational equivalent of mass. In the following pages, the term *moment of inertia* is used instead of *mass moment of inertia*. There is another

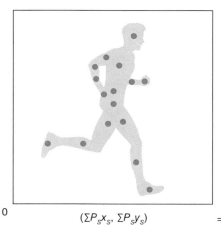 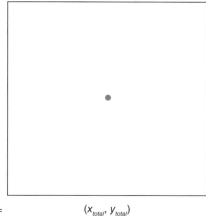

0      $(\Sigma P_s x_s, \Sigma P_s y_s)$ = (x_{total}, y_{total})

▲ **Figure 3.4** The total body center of gravity is a weighted average of the segment centers of gravity (x_s, y_s).

moment of inertia, one rarely used in biomechanics, that is called the **area moment of inertia**. It is a geometric property that does not concern itself with how mass is distributed within a body. Because it is not used in this textbook, any reference to the term *moment of inertia* is to the mass moment of inertia.

Moments of inertia are needed whenever the rotational motion of a body is investigated. Moment of inertia is not as important a measure as mass because, relatively speaking, moment of inertia has less influence on the motion of human bodies than mass does. This is because humans are more often concerned with translational motions and because human movements rarely rotate at high rates. Of course, such measures become important when acrobatic motions are examined, such as diving, trampolining, figure skating, gymnastics, skateboarding, and aerial skiing. In these sporting activities, 3-D measurements of moments of inertia are required. A simple method for computing the moment of inertia for planar (2-D) analyses of motion is presented next.

Segment Moment of Inertia

Computing the moment of inertia of a segment is not as straightforward as computing the center of gravity. This is because the moment of inertia is not linearly related to segment length. Classically, the **moment of inertia** is defined as the "second moment of mass"; it is the sum (integral) of mass particles times their squared distances (moments) from an axis. That is,

$$I_{axis} = \int r^2 \, dm \qquad (3.13)$$

where r is the distance of each mass particle *(dm)* from a point or axis of rotation.

Because moments of inertia differ nonlinearly with the axis about which they are computed, they cannot be computed directly using proportions, as are the other

body segment parameters. To simplify the computation of moment of inertia, an indirect method is used that involves calculating the **radius of gyration**. The radius of gyration is the distance that represents how far the mass of a rigid body would be from an axis of rotation if its mass were concentrated at a point. This concept is illustrated in figure 3.5. Thus, a rigid body can be considered *rotationally* equivalent to a point mass that is located at this certain distance—called the radius of gyration, k—from the axis of rotation. This distance differs from the **center of mass**, which is the point at which a rigid body's mass can be considered to be concentrated for linear motions. Note that whereas the center of gravity of a rigid body is always at the same location on the body, the radius of gyration varies depending on the axis about which the body is rotating. The value of determining radii of gyration is that they can be made proportional to a segment's length, thereby simplifying the task of determining segmental moments of inertia for any axis.

Thus, to compute a segment's moment of inertia using proportions of segment length, first compute the radius of gyration. The radius of gyration (k_{axis}) of a rigid body or segment is defined as the length (radius) that satisfies this relationship:

$$k_{axis} = \sqrt{\frac{I_{axis}}{m}} \qquad (3.14)$$

where I_{axis} is the moment of inertia about the axis *(axis)* and m is the mass of the rigid body or segment.

Radius of gyration changes depending on the axis of rotation selected. The minimum radius of a rigid body occurs when the body rotates about its own center of gravity, called its **centroidal moment of inertia**. Using table 3.2 or 3.3, we can compute the radius of gyration of a segment rotating about its center of gravity from

$$k_{cg} = K_{cg} L \qquad (3.15)$$

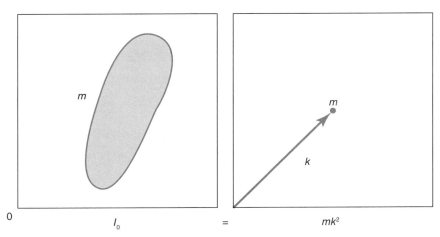

▲ **Figure 3.5** Relationship between the centroidal moment of inertia (I_o) of a rigid body and the radius of gyration (k) of a rotationally equivalent point mass (m).

where k_{cg} is the radius of gyration in meters and K_{cg} is the length of the radius of gyration as a proportion of the segment's length, L, in meters.

The segment's centroidal moment of inertia can then be computed from

$$I_{cg} = mk_{cg}^2 \qquad (3.16)$$

where I_{cg} is the segment's moment of inertia for rotations about the segment's center of gravity and m is the segment's mass in kilograms. Note that the moment of inertia of a segment about any arbitrary axis is equal to the segment's centroidal moment of inertia, I_{cg}, plus a term equal to the segment's mass times the square of the distance between the segment's center of gravity and the arbitrary axis. This is called the *parallel axis theorem*. That is,

$$I_{axis} = I_{cg} + mr^2 \qquad (3.17)$$

where I_{axis} is the moment of inertia about an arbitrary axis and r is the distance from the axis to the center of gravity. Because segments tend to rotate about either their proximal or distal ends, it is often desirable to compute the segment's moment of inertia about them. Using the proportions $K_{proximal}$ and K_{distal} and knowing the moment of inertia about the segment's center of gravity (I_{cg}), we can apply the parallel axis theorem thus:

$$I_{proximal} = I_{cg} + m(R_{proximal} \times L)^2 \qquad (3.18)$$

$$I_{distal} = I_{cg} + m(R_{distal} \times L)^2 \qquad (3.19)$$

where $I_{proximal}$ and I_{distal} are the moments of inertia at the proximal and distal ends, respectively; m is the segment mass; and L is the segment length.

A direct method (Plagenhoef 1971, pp. 43-44) for determining the moment of inertia of a rigid body involves measuring the period of oscillation (the duration

EXAMPLE 3.3

Calculate the thigh's moment of inertia about its center of gravity for a 80.0 kg person using the proportions from table 3.2 given that the proximal end (the hip) of the thigh has the coordinates (−12.80, 83.3) cm and the distal end (the knee) has the coordinates (7.30, 46.8) cm (the same as in example 3.2).

See answer 3.3 on page 377.

of a cycle) of the object as it oscillates as a pendulum. Typically, 10 to 20 periods are timed and then the duration is computed for a single cycle. The equation for the moment of inertia of the object about the suspension point is computed from

$$I_{axis} = \frac{mgrt^2}{4\pi^2} \qquad (3.20)$$

where I_{axis} is the moment of inertia of the object about an axis through the pivot point of the pendulum in kilograms meters squared (kg·m²), m is the mass of the object in kilograms, g is 9.81 m/s², r is the distance in meters from the suspension point to the center of gravity of the object, and t is a single period of oscillation in seconds. Note that for this equation to be valid, the pendulum should not swing more than 5° to either side of the vertical.

Total Body Moment of Inertia

Although the total body moment of inertia is rarely used in biomechanics, occasionally a researcher needs to compute this value. It is tempting to simply add all the segments' centroidal moments of inertia (I_{cg}), but this ignores the fact that each segment has a different location for its center of gravity. Therefore, the parallel axis theorem must be applied to each segment where the distance between each segment's center of gravity and the total body's center of gravity needs to be calculated (r_s in equation 3.21).

Consequently, the total body's centroidal moment of inertia (I_{total}) is the sum of each segment's centroidal moment of inertia (I_{cg}) plus transfer terms based on the parallel axis theorem. That is,

$$I_{total} = \sum_{s=1}^{S} I_{cg_s} + \sum_{s=1}^{S} m_s r_s^2 \qquad (3.21)$$

where S is the number of segments and r_s is the distance between the total body's center of gravity and each segment's center of gravity.

If it then becomes necessary to compute the total body's moment of inertia about a secondary axis (I_{axis}), the parallel axis theorem can be applied once again. Thus,

$$I_{axis} = I_{total} + mr^2 \qquad (3.22)$$

where r is the distance between the total body's center of gravity and the secondary axis.

Note that this moment of inertia can also be calculated more directly as

$$I_{axis} = \sum_{s=1}^{S} I_{cg_s} + \sum_{s=1}^{S} mr_s^2 \qquad (3.23)$$

where r_s is the distance between a segment center and the secondary axis *(axis)*.

Center of Percussion

The **center of percussion** is not strictly speaking a body segment parameter. It is usually associated with sporting implements, such as baseball bats, rackets, and golf clubs. It is the point on an implement, sometimes called the *sweet spot*, where the implement when struck experi-

ences no pressure at its grip or point of suspension. Thus, a bat suspended about its grip end and struck at its center of percussion will rotate only about the suspension point.

In contrast, if the bat is struck at its center of gravity, it moves in pure translation with no rotational motion. Any other striking point produces some combination of translation and rotation (figure 3.6). Note that striking at the center of percussion does not produce pure rotation of the body, only pure rotation of the body about its suspension point. Most implements are not held at their center of gravity, so contacting at the center of percussion produces minimal forces to the hands holding the implement.

The center of percussion can be determined either empirically by striking the implement in various places and observing which location causes pure rotation about the suspension point or computationally (Plagenhoef 1971):

$$q_{axis} = \frac{k_{axis}^2}{r_{axis}} \qquad (3.24)$$

where q_{axis} is the center of percussion, k_{axis} is the radius of gyration, and r_{axis} is the distance to the center of gravity from the axis of rotation at the point of suspension.

Note that there is a sequential relationship between r, k, and q. The radius of gyration (k) is always farther than the center of gravity (r), and the center of percussion (q) is always farther than the radius of gyration (k) from the axis of rotation (figure 3.7). The one exception is for a uniformly dense spherical body in which all three locations are coincident.

The center of percussion is not the only sweet spot. There are other locations on an implement that produce special effects. For example, all rackets, depending on their construction, their shape, and how they are strung, have a location that provides maximum rebound. This location can be determined experimentally by dropping

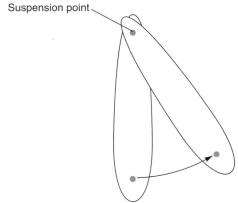
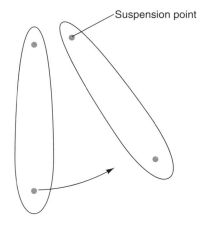

▲ Figure 3.6 Effects of striking a bat at the center of percussion and away from the center of percussion.

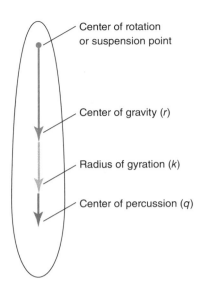

▲ Figure 3.7 Relative locations of the center of gravity *(r)*, radius of gyration *(k)*, and center of percussion *(q)*.

balls onto the motionless racket and observing which location produces the highest bounce (figure 3.8). The square root of the ratio of the bounce height to the drop height is called the **coefficient of restitution**:

$$c_{restitution} = \sqrt{h_{bounce} / h_{drop}} \qquad (3.25)$$

The larger the coefficient of restitution, the greater the ball's velocity after impact. The coefficient of restitution generally is calculated by measuring the velocities of the racket (or other implement) and the ball before and after an impact. Note that the ball's velocity should be measured only after the ball is undeformed after the impact. The coefficient is then defined by the ratio of the

relative velocities after and before an impact (see Hatze 1993; Plagenhoef 1971):

$$c_{restitution} = -\left(\frac{v_{bat\ after} - v_{ball\ after}}{v_{bat\ before} - v_{ball\ before}} \right) \qquad (3.26)$$

THREE-DIMENSIONAL (SPATIAL) COMPUTATIONAL METHODS

In the following sections, methods for computationally determining body segment parameters for 3-D analyses are presented. The main differences between 2- and 3-D analyses lie in the computation of mass moments of inertia. The methods outlined for calculating mass and center of mass are not necessarily different from the equations used for two dimensions that were presented previously, but for the purposes of this text and to use a consistent system for 3-D analysis with Visual3D software, the methods used by Hanavan (1964) are emphasized.

Segment Mass and Center of Gravity

There are few differences between the equations used for two and three dimensions when one is computing segment mass or center of gravity. In chapter 7, segment mass is calculated using the proportions listed in table 3.2. An alternate method using regression equations developed by Vaughan and colleagues (1992) is useful for estimating the masses of the lower extremity. These equations require the anthropometric measures listed in table 3.6 and illustrated in figure 3.9. The segment

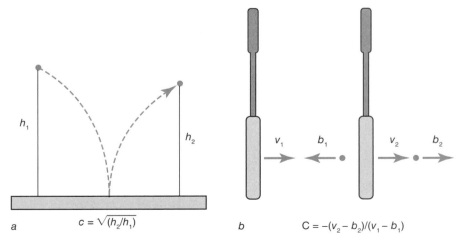

▲ Figure 3.8 Methods of calculating the coefficient of restitution *(c)*.

masses of the three segments of the lower extremity are computed from these equations:

$$m_{foot} = 0.0083m_{total}$$
$$= 254.5(L_{foot}\, h_{malleolus}\, w_{malleolus}) - 0.065 \quad (3.27)$$

$$m_{leg} = 0.0226m_{total} + 31.33(L_{leg}\, c_{leg}^2) + 0.016 \quad (3.28)$$

$$m_{thigh} = 0.1032m_{total} + 12.76(L_{thigh}\, c_{midthigh}^2) - 1.023 \quad (3.29)$$

where m_{total} is the total body mass; the Ls represent the segment lengths; $h_{malleolus}$ and $w_{malleolus}$ are the malleolus height and width, respectively, c_{leg} is the leg (calf) circumference; and $c_{midthigh}$ is the midthigh circumference. Note that all masses are in kilograms and all other measurements are in meters.

To compute segment center of gravity, an additional equation, identical to those presented previously for planar mechanics, may be applied. That is, the center of gravity in the Z direction (z_{cg}) is

$$z_{cg} = z_{proximal} + R_{proximal}\, (z_{distal} - z_{proximal}) \quad (3.30)$$

where $z_{proximal}$ and z_{distal} are the Z coordinates of the proximal and distal ends, respectively, and $R_{proximal}$ is the distance from the proximal end of the segment to the center as a proportion of the segment's length.

Table 3.6 Anthropometric Data Required for Body Segment Parameter Equations of Vaughan, Davis, and O'Connor

Number	Anthropometric measure	Units
1	Total body mass	kg
2	Anterior-superior iliac spine	m
3	Thigh length	m
4	Midthigh circumference	m
5	Leg (calf) length	m
6	Leg circumference	m
7	Knee diameter	m
8	Foot length (heel-toe)	m
9	Malleolus height	m
10	Malleolus width	m
11	Foot breadth	m

Reprinted, by permission, from C.L. Vaughan, B.L. Davis, and J.C. O'Connor, 1992, *Dynamics of human gait* (Champaign, IL: Human Kinetics). By permission of C.L. Vaughan.

Vaughan and colleagues (1992) used the proportions (R values) reported by Chandler and colleagues (1975). These are listed in table 3.4. That is, $R_{proximal}$(foot) = 0.4485, $R_{proximal}$(leg) = 0.3705, and $R_{proximal}$(thigh) = 0.3719.

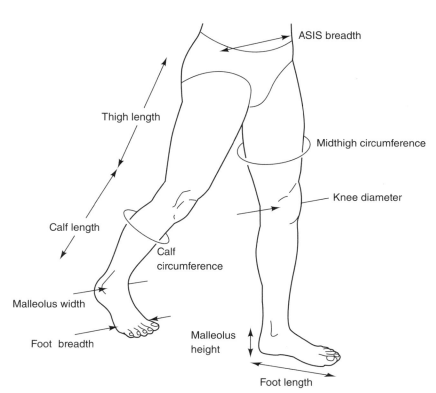

▲ **Figure 3.9** Anthropometric measurements needed for the body segment parameter equations of Vaughan and colleagues (1992).

Note that these researchers provided body segment parameter information only for the lower extremity.

Alternatively, chapter 7 presents equations based on the geometric properties of uniformly dense solids of revolution as was outlined by Hanavan (1964). This approach assumes that most segments can be modeled as the frusta (cones with the peaks removed) of right circular cones. Other geometric shapes commonly used to model body parts include an ovoid for the head, spheres for the hands, and elliptical cylinders for the torso and pelvis. Software such as Visual3D uses equations from calculus to determine the mass centers of such geometric bodies.

Segment Moment of Inertia

Computing segmental moments of inertia in 3-D requires the same basic approach that is used for planar (2-D) mechanics, but instead of using a single scalar quantity for moment of inertia *(I)*, we use a **moment of inertia tensor** ($\overline{I}$). This **tensor** is a 3 × 3 matrix:

$$\begin{bmatrix} I_x & P_{xy} & P_{xz} \\ P_{yx} & I_y & P_{yz} \\ P_{zx} & P_{zy} & I_z \end{bmatrix} \quad (3.31)$$

where the diagonal elements *(I$_x$, I$_y$, and I$_z$)* are called the **principal mass moments of inertia** and the off-diagonal elements are called the mass **products of inertia** (the *P*s).

In general, all nine elements of this tensor must be measured, computed, or estimated to define, for example, the resultant moment of force *($\vec{M}_R$)* with respect to a spe-

cific frame of reference or axis. That is, $\vec{M}_R = \overline{I}\vec{\alpha}$, where $\vec{\alpha}$ is the angular acceleration vector of the rigid body or segment and $\overline{I}$ is the centroidal (i.e., at the center of gravity) moment of inertia tensor. This situation is simplified by using an axis system that places one axis along the longitudinal axis of the segment reducing the inertia tensor to a diagonal matrix in which all the products of inertia are zero. That is, the inertia tensor is reduced to

$$\begin{bmatrix} I_x & 0 & 0 \\ 0 & I_y & 0 \\ 0 & 0 & I_z \end{bmatrix} \quad (3.32)$$

In this way, only the elements along the principal diagonal need to be computed. Details about 3-D mechanics are presented in chapter 7.

As mentioned previously, Hanavan in 1964 (see also Miller and Morrison 1975) laid down the most commonly used methods for modeling human body segments in three dimensions. Hanavan used the empirically derived data provided by Dempster (1955) and then measured various anthropometric measures, such as midthigh circumference, malleolus height, knee diameter, and biacromial breadth. These were then used as parameters to mathematically derive the principal mass moments of inertia based on equations derived from integral calculus. Figure 3.2 illustrates Hanavan's mathematical model of the body. Table 3.7 shows equations for computing the principal mass moments of inertia of various geometric, uniformly dense solids.

Later, other researchers developed additional methods of estimating the principal moments of inertia as well

Table 3.7 Principal Mass Moments of Inertia of Solid Geometric Shapes

	I_x	I_y	I_z	
Slender rod	0	1/12ml^2	1/12ml^2	

m = mass; *l* = length of rod

Rectangular plate	1/12$m(b^2 + c^2)$	1/12mc^2	1/12mb^2	

m = mass; *b* = height of plate; *c* = width of plate

	I_x	I_y	I_z	
Thin disk	$1/2mr^2$	$1/4mr^2$	$1/4mr^2$	
m = mass; r = radius of disk				
Rectangular prism	$1/12m(b^2 + c^2)$	$1/2m(a^2 + c^2)$	$1/2m(a^2 + b^2)$	
m = mass; a = depth *(x)*; b = height *(y)*; c = width *(z)*				
Circular prism	$1/2mr^2$	$1/12m(3r^2 + l^2)$	$1/12m(3r^2 + l^2)$	
m = mass; l = length of cylinder; r = radius				
Elliptical cylinder	$1/12m(3c^2 + l^2)$	$1/12m(3b^2 + l^2)$	$1/4m(b^2 + c^2)$	
m = mass; l = length of cylinder *(x)*; b = height/2 *(y)*; c = width/2 *(z)*				
Circular cone	$3/10mr^2$	$3/5m(1/4r^2 + l^2)$	$3/5m(1/4r^2 + l^2)$	
m = mass; l = length of cone; r = radius at base				
Sphere	$2/5mr^2$	$2/5mr^2$	$2/5mr^2$	
m = mass; r = radius				
Ellipsoid	$1/5m(b^2 + c^2)$	$1/5m(a^2 + c^2)$	$1/5m(a^2 + b^2)$	
m = mass; a = depth *(x)*; b = height *(y)*; c = width *(z)*				

as mass proportions and segmental centers of gravity. As stated earlier, Zatsiorsky and Seluyanov (1983, 1985) used gamma mass scanning to acquire tables of body segment parameters for young males and females. These authors divided the body into 15 segments, 3 for the trunk and 3 for each extremity. They also used bony landmarks, as Clauser and colleagues (1969) and Chandler and colleagues (1975) had, to define segment endpoints, in contrast to Dempster (1955) who used joint centers of rotation. Paolo de Leva (1996) provided methods that converted Zatsiorsky and Seluyanov's data to the form used by Dempster and Hanavan (i.e., segment endpoints defined by their joint centers of rotation).

In chapter 7, additional equations for computing the 3-D kinetics of the lower extremities are presented. The equations presented in chapter 7 for the principal moments of inertia of the frustum are based on the geometric properties of uniformly dense solids of revolution. These values are automatically calculated by software such as Visual3D based on the segment properties defined when each segment is created during the modeling process. The radii of the proximal and distal ends are either based on markers placed on either side of the joints or on joint-width measurement, using that value as the radius about the location of the joint center.

SUMMARY

Three methods for computationally obtaining body segment parameters were outlined in this chapter. The first method, based primarily on the techniques and data derived from Dempster (1955) or Clauser and colleagues

(1969) and Chandler and colleagues (1975), was outlined for 2-D analyses. This method, based on proportions and measured anthropometrics, can estimate the body segment parameters for linked-segment models of the body. The researcher decides on the number of segments according to the complexity of the motion being investigated. Proportions can also be obtained from various databases, depending on the population (male or female, young or old, and so on) being investigated.

The second method is based primarily on techniques defined by Hatze (1980), Vaughan and colleagues (1992), and Zatsiorsky and Seluyanov (1983, 1985). This method requires the measurement of specific anthropometric dimensions so that 2-D or 3-D body segment parameters can be estimated. Again, the equations presented can be extended to other segments by resorting to the research literature.

The third method, which is mainly used for 3-D analyses, uses Dempster's data for segment masses and the properties of uniformly dense geometric solids of revolution. This method is used in chapter 7 to conduct 3-D kinetic analyses and is used by Visual3D software.

Although accurate body segment parameters are desirable, errors in these parameters may have little effect on kinetic measurements, especially when the body is in contact with the environment. In such cases, large errors in the body segment parameters have little influence on the computation of, for example, joint moments of force or moment powers. This is because the relative magnitudes of the inertial forces $(-ma)$ and especially the moments of force $(-I\alpha)$ are small compared with the moments caused by the ground reaction forces.

SUGGESTED READINGS

De Leva, P. 1996. Adjustments to Zatsiorsky-Seluyanov's segment inertia parameters. *Journal of Biomechanics* 29:1223-30.

Jensen, R.K. 1993. Human morphology: Its role in the mechanics of motion. *Journal of Biomechanics* 26(Suppl. No. 1):81-94.

Krogman, W.M., and F.E. Johnston. 1963. *Human Mechanics: Four Monographs Abridged.* AMRL Technical Document Report 63-123. Wright-Patterson Air Force Base, OH.

Nigg, B.M., and W. Herzog. 2007. *Biomechanics of the Musculo-Skeletal System.* 3rd ed. Toronto: Wiley.

Vaughan, C.L., B.L. Davis, and J.C. O'Connor. 1992. *Dynamics of Human Gait.* Champaign, IL: Human Kinetics.

Winter, D.A. 2009. *Biomechanics and Motor Control of Human Movement.* 4th ed. Toronto: Wiley.

Zatsiorsky, V.M. 2002. *Kinetics of Human Motion.* Champaign, IL: Human Kinetics.

Chapter 4

Forces and Their Measurement

Graham E. Caldwell, D. Gordon E. Robertson, and Saunders N. Whittlesey

The field of mechanics is partitioned into the study of motion **(kinematics)** and the study of the causes of motion **(kinetics)**. Chapters 1 and 2 covered important aspects of the kinematics of human movement, and we now turn our attention to the underlying kinetics. In this chapter, we

▶ introduce the concepts of force, which causes linear motion, and torque, also called moment of force, or simply moment, which causes angular motion;

▶ discuss the effect of applied forces and moments of force through the consideration of laws and equations set forth by Newton and Euler;

▶ explain how to create and use free-body diagrams;

▶ identify various forces encountered in biomechanical investigations;

▶ define the mechanical concepts of impulse and momentum, which dictate the effect of changing levels of force and moment applied over a duration; and

▶ describe how to measure force and moment for human biomechanics research.

FORCE

The term *force* is common in everyday language. Curiously, in its use in physics, force can only be defined by the effect that it has on one or more objects. More specifically, **force** represents the action of one body on another. In our study of human biomechanics, we are interested in two specific effects related to force. The first is the effect that a force has on a particle or a perfectly rigid body, and the second is the effect that a force has on a deformable body or material. The effects of forces on particles or rigid bodies can be assessed using Newton's laws and are critical to understanding the causes of the kinematic data described in earlier chapters. A **rigid body** is one in which the constituent particles have fixed positions relative to each other. The effects of forces on a **deformable body** or material are important when we consider **internal forces** within biological tissues and seek to understand how it is possible to measure **external forces** applied to or by humans.

Force is a **vector** quantity defined by its magnitude, direction, and point of application. For the linear motion of particles or rigid bodies, only the force magnitude and direction are important. However, the point of application is critical if angular motion is also under consideration. Furthermore, although the concept of a force vector leads one to consider a "point" of application, in reality most external forces are generated by direct contact spread over a finite area rather than at a single point. This introduces the concept of **pressure**, that is, force distributed over an area of contact. Within the domain of human biomechanics, kinetic analyses usually concern either external forces and pressures applied through direct contact with the ground or an object (e.g., tool, machine, ball, bicycle, or keyboard) or internal forces within muscles, ligaments, bones, and joints. Although kinetic parameters sometimes can be measured directly, often they must be calculated or estimated based on measurement of the observed kinematics.

NEWTON'S LAWS

In 1687, Isaac Newton published a book on "rational mechanics" titled *Philosophiae Naturalis Principia Mathematica,* or *Mathematical Principles of Natural Philosophy,* with two revised editions appearing before his death in 1727. Although our knowledge of the physics of motion has evolved substantially since its publication, the importance of the *Principia* can be appreciated by the fact that its tenets are still taught in high schools and universities throughout the world more than 300 years after its first publication.

Although Einstein's 1905 theory of relativity demonstrated that Newton's laws of mechanics were incomplete, Newton's laws are still valid for all but the movement of subatomic particles and situations in which velocity approaches the speed of light, and they remain the principal tools of biomechanics and engineering. The laws that Newton set forth in the *Principia* define the relationship between force and the linear motion of a particle or rigid body (referred to henceforth as a *body* or *object*) to which it is applied. Three such relationships are described here. These laws may be described in slightly different forms in other texts as a result of differing translations from the original Latin, changes in common English over the past 300 years, and varying levels of background preparation in the target student audience.

Newton's first law, the **law of inertia**, describes how a body moves in the absence of external forces, stating that "a body will remain in its current state of motion unless acted upon by an external force." The "state of motion" is described by the body's momentum *(p)*, defined as the product of its mass and linear velocity *(p = mv)*. Simply put, if no net external force is applied to a body, its momentum will remain constant. This field of study is called **statics**. For many situations in biomechanics, mass will be constant, so the absence of force means that the body will remain at constant linear velocity. Juxtaposed with this situation is the **law of acceleration** (Newton's second law), which describes how a rigid body moves when an external force is applied—the field of study called **dynamics**. This law states, "An external force will cause the body to accelerate in direct proportion to the magnitude of the force and in the same direction as the force." This proportionality can be stated as an equality with the introduction of the body's mass, resulting in the famous Newtonian equation $F = ma$. The International System unit for force is the newton (N), with 1 N being the force needed to cause a mass of 1 kg to accelerate by 1 m/s². Simply put, an applied force causes a rigid body to accelerate by changing its velocity, direction of motion, or both.

Newton's third law describes how two masses interact with each other. The **law of reaction** states that "when one body applies a force to another body, the second body applies an equal and opposite reaction force on the first body." A common example in biomechanics involves a human in contact with the surface of the earth, as when running (figure 4.1). When the runner's foot strikes the ground, the runner applies a force to the earth. This force can be represented as a vector having a certain magnitude and direction. At the same time, the earth applies to the runner a reaction force of equal magnitude but opposite direction. Other examples include a person

▲ **Figure 4.1** The physical interaction of two bodies results in the application of *action* and *reaction* forces. In this example, the runner is pushing against the ground. The figure on the left depicts the force acting on the earth resulting from the runner's muscular efforts. On the right, the equal but opposite reaction force that acts on the runner is shown, giving rise to the term *ground reaction force*.

making contact with a ball, bat, tool, or other handheld implement. Because of the opposition in direction between the so-called **action** and **reaction** forces, be careful in these situations to correctly recognize which force acts on which body when you are assessing the effect caused by the force.

In many situations, more than one force acts on a body at a given point in time. This situation is easily handled within Newton's laws through the concept of a **resultant** force vector. Because each force is a vector quantity, a set of forces acting on a body can be combined through vector summation into a single resultant force vector. Newton's laws can then be considered using this single **resultant force**. A useful tool when dealing with kinetics problems is the **free-body diagram** (FBD), which is a simple sketch of the body that includes all of the forces acting on it. Drawing an FBD reminds researchers of the existence of each force and helps them to visualize the direction of reaction forces that are acting on the body in question.

FREE-BODY DIAGRAMS

Diagrams usually are helpful in visualizing mechanics problems. The FBD is a formal means of presenting the forces, moments of force, and geometry of mechanical systems. This diagram is all-important; the first step in the solution of a mechanical problem is to draw the FBD. The second step is to use the FBD to derive the equations of motion of the object. Finally, known numerical values are substituted and the equations are solved for the unknown terms. There are several formalities to FBD construction:

▶ Draw the object of interest in minimalist form (either an outline or even just a single line), free of the environment and other bodies.

▶ Write out the coordinates of the object to completely specify its position.

▶ Indicate the object's center of mass with a marker; it is from here that the accelerations are drawn.

▶ Draw and label all external reaction forces and moments of force. Base the directions for these forces and moments on how the object experiences them. For example, direct vertical ground reaction forces upward and frictional forces opposite to the direction of motion of the contacting surfaces.

▶ Draw all unknown forces and moments with positive coordinate system directions. Unknown forces must be applied wherever the body is in contact with the environment or other bodies (or body segments).

▶ It is also desirable to draw and label global coordinate system (GCS) axes off to the side of the diagram indicating the positive directions.

Figure 4.2a shows a runner crossing a force plate. In this case, because of the relative complexity of the body shape, we chose to draw a stick figure of the runner. At the mass center we drew the force of gravity (mg) and the force of the wind at the center of the frontal area. The ground reaction forces are included at the subject's stance foot. Note that these are the reaction forces acting on the runner, rather than the forces that the runner is applying to the earth. Figure 4.2b is an FBD of a bicycle crank showing the known pedal forces. The crank is represented as a single line and its mass center is indicated with a dot. At the distal end of the crank are the horizontal and vertical forces of the pedal; we measured these forces and labeled their magnitudes. Because the pedal has an axle with smooth bearings, we assumed that there was no moment of force exerted about the pedal axle. At the proximal end of the crank, we drew the reaction forces on the crank's axle (R_{Ax}, R_{Ay}). These are unknown, so we gave them names and drew them in the positive GCS directions. We also drew a resultant moment (M_A) about the crank axle, which we gave a positive counterclockwise direction because its magnitude is unknown. This moment is nonzero because of the resistance of the chain driving the bicycle.

EXAMPLE 4.1

1. Draw the FBD for a rowing oar.
2. Draw the FBD for a running human, including wind resistance. What is the problem with drawing the wind resistance?

See answer 4.1 on page 377.

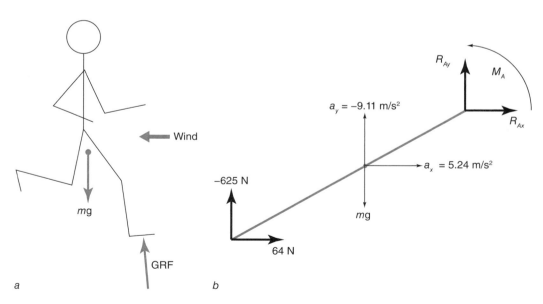

▲ **Figure 4.2** Free-body diagrams of a runner *(a)* and a bicycle crank *(b)*. It is assumed that there is no moment of force at the pedal but there is a moment at the crank axle *(M$_A$)* due to the chain ring.

TYPES OF FORCES

At present, physicists contend that there are four fundamental forces in nature: the "strong" and "weak" nuclear forces, electromagnetic force, and gravitational force. Of these forces, only gravitational and electromagnetic forces concern the biomechanist. All forces experienced by the body are some combination of these two.

Another of Newton's substantial contributions to our understanding of motion is his description of how masses interact even when they are not in contact. His **universal law of gravitation** states that "two bodies attract each other with a force that is proportional to the product of their masses and inversely proportional to the square of the distance between them." With the introduction of the universal constant of gravitation (G), we can quantify the magnitude of the gravitational force: $F_g = G(m_1 \times m_2)/r^2$, where m_1 and m_2 are the masses of the two bodies and r is the distance between their mass centers. Whereas Newton considered this concept of gravity in his quest to understand planetary motion, in the realm of biomechanics its most important application is for the effect that the earth's gravity has on bodies near its surface. In this case, m_1 is the mass of the Earth, m_2 is the mass of an object on the surface of the Earth, and r is the specific distance, R, from the body's center of mass to the center of mass of the Earth. If we consider the gravitational force acting on mass m_2, the effect of this force is described by the law of acceleration $(F = m_2a)$. By substitution into the gravitational equation, we see that $m_2a = G(m_1 \times m_2)/R^2$. The body's mass, m_2, appears on both sides of the equation and therefore drops out, leaving an expression for the acceleration of the body caused by the earth's gravitational force, $a = Gm_1/R^2$. Because all of the terms on the right-hand side are constants, the acceleration, a, is also a constant. This is the acceleration resulting from Earth's gravity, commonly called g and approximately equal to 9.81 m/s². Furthermore, the gravitational force is given the specific name **weight**, with the equation $W = mg$ being a specific form of the more general $F = ma$. (Note that body weight is a force and therefore a vector. In contrast, body mass is a scalar quantity.) Therefore, on an FBD for an object on or near the earth's surface, the weight force should be drawn acting in the direction toward the earth's center, usually designated as vertically down.

If a person stands quietly while waiting for a bus, his weight force vector is acting to accelerate him downward at 9.81 m/s². However, this acceleration does not occur, because a person's vertical velocity remains constant at zero. The reason for this lack of acceleration is explained by Newton's third law, which governs the forces associated with the contact between the person's feet and the pavement. The person is applying a force to the earth equal to his weight, while the earth is applying an equal and opposite reaction force on the person. On an FBD of the person (figure 4.3a), we would draw a force vector pointing upward (away from the Earth's center) called the **ground reaction force** (GRF or F_g). Therefore, there are two forces acting on the person in the vertical direction: the weight force downward (negative) and the GRF upward (positive). In this quiet-stance situation, these two forces have the same magnitude. Vector summation results in a resultant vertical force of zero ($W + F_g = 0$), which is the reason the person's vertical velocity remains constant.

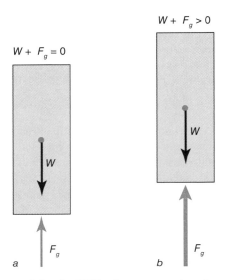

▲ **Figure 4.3** An FBD of a person standing on the earth must show two forces. The first is the weight force vector *(W)* acting downward (negative) because of the gravitational attraction of the Earth. The second force is the GRF *(F_g)* acting upward (positive) because of the physical contact between the person and the Earth. For a person in a quiet stance *(a)*, these two opposing forces are approximately equal in magnitude, and therefore the person's center of mass undergoes no acceleration ($\Sigma F = 0$). If the person uses muscular effort to push down on the ground *(b)*, the resulting increase in F_g will cause an upward acceleration because F_g will now be larger than W (i.e., $\Sigma F > 0$).

Although equal to the person's weight in this example, in general the magnitude of the GRF can vary and therefore can be greater than or less than the magnitude of the person's weight, which is a constant *($W = mg$)*. If a person standing on the ground activates his leg extensor muscles (i.e., pushes down on the ground), the GRF will rise above the magnitude of his weight (figure 4.3b). The resultant force vector will therefore be nonzero and directed upward ($F_g > W$, therefore $W + F_g > 0$). The law of acceleration dictates that this positive resultant force causes acceleration in the upward direction, the magnitude of which depends on the person's mass *(F = ma,* rearranged as *a = F/m)*. If the person had been standing quietly before activating his muscles, the initial vertical velocity would have been zero. The positive resultant force and acceleration result in a change in the velocity from zero to an upwardly directed positive velocity. In contrast, if the person had just jumped off a step, his initial velocity at ground contact would have been downward (negative). The positive resultant force and acceleration act to reduce this negative velocity (slow the downward motion). If large enough and applied long enough, the positive resultant force reduces the velocity to zero, stopping the downward movement, or even reversing the movement from downward to upward.

Although this example refers to the vertical direction, the GRF is a 3-D vector that can have components in the horizontal plane that are often referenced to body position, with anterior-posterior (A/P) and medial-lateral (M/L) components (figure 4.4). The 3-D direction of the GRF vector depends on how the person applies the force to the ground, and it dictates the relative size of the vertical, A/P, and M/L components. For example, a soccer player trying to initiate forward motion pushes downward and backward on the ground. The GRF therefore is directed upward and forward, resulting in positive vertical and anterior GRF components. A similar push-off to initiate a forward and lateral movement results in a lateral GRF component as well. The ability to generate these horizontal components depends on the nature of the foot-ground contact. Recall that in the vertical direction, the GRF was the result of the resistance to the gravitational attraction provided by the earth's solid surface. In the horizontal directions, the foot-ground interface must provide a similar resistance; the person must be able to push on the earth to generate a GRF. A force known as **friction** can provide this resistance. If a person stands on a truly frictionless surface, it is impossible to generate A/P or M/L GRF components.

Friction is a specific force that acts whenever two contacting surfaces slide over each other. Frictional forces are directed **parallel** to the surfaces and always oppose the relative motion of the two surfaces. In some cases, the frictional force is large enough to prevent movement; this is known as **static friction**. In other cases, applied forces are great enough to cause movement, and then **kinetic friction** acts to resist the motion. Imagine that a block sitting on a level surface is subjected to a slowly increasing applied horizontal force, $F_{applied}$ (figure 4.5). When $F_{applied}$ is small, the frictional force ($F_{friction}$) is able to resist movement, and so the block remains stationary ($F_{applied} = -F_{friction}$; $F_{applied} + F_{friction} = 0$). As $F_{applied}$ increases, $F_{friction}$ continually increases to match it in

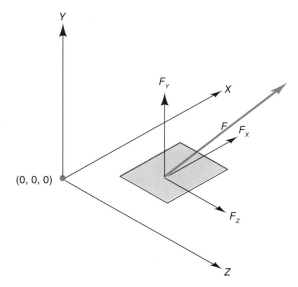

▲ **Figure 4.4** The GRF vector *F* can be resolved into its three components, F_X, F_Y, and F_Z.

magnitude to keep the block stationary. However, at some point, $F_{friction}$ will reach its maximal static value, and further increases in $F_{applied}$ will not be met by equal increases in $F_{friction}$. It is at this point that movement of the block commences. The maximal value of $F_{friction}$ in static conditions is known as the **limiting static friction force**, and it is calculated using the equation $F_{maximum} = \mu_{static}N$, where μ_{static} is the **coefficient of static friction** and N is the normal force acting across the two surfaces (figure 4.5). As the magnitude of $F_{applied}$ exceeds that of $F_{maximum}$, the block is set in motion, with $F_{applied}$ tending to accelerate it, whereas the kinetic frictional force $F_{kinetic}$ opposes the motion (i.e., tends to slow it down). The magnitude of $F_{kinetic}$ is somewhat less than that of $F_{maximum}$ and is approximately constant despite the magnitude of $F_{applied}$ or the velocity attained by the block. The kinetic friction force can be calculated with the formula $F_{kinetic} = \mu_{kinetic}N$, where $\mu_{kinetic}$ is the **coefficient of kinetic friction** and N is the normal force. The **normal force** is the force perpendicular to the surfaces that keeps the surfaces in contact. Note that μ_{static} and $\mu_{kinetic}$ both depend on the nature of the two surfaces involved and that μ_{static} is always slightly greater than $\mu_{kinetic}$.

Readers will encounter other classes of forces throughout the biomechanics literature. Internal forces usually refer to those generated or borne by tissues within the body, such as muscles, ligaments, tendons, cartilage, or bone. In contrast, external forces are those imposed on the body by contact with other objects, such as the reaction forces described earlier. *Inertia forces* are those associated with accelerating bodies, and they arise from a slightly different expression of Newton's law of

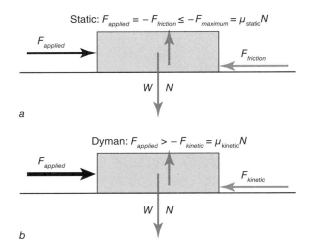

Static: $F_{applied} = -F_{friction} \leq -F_{maximum} = \mu_{static}N$

$F_{applied}$ $F_{friction}$

W | N

a

Dyman: $F_{applied} > -F_{kinetic} = \mu_{kinetic}N$

$F_{applied}$ $F_{kinetic}$

W | N

b

▲ **Figure 4.5** When a force, $F_{applied}$, is applied in a direction that would tend to cause an object to slide across a surface, a frictional force vector, $F_{friction}$, opposes the applied force. A static situation is shown in the top panel, in which $F_{applied}$ and $F_{friction}$ are less than the limiting static friction force $F_{maximum} = \mu_{static}N$, where μ_{static} is the coefficient of static friction and N is the normal force acting across the two surfaces. In this static case, $F_{applied} = -F_{friction}$. If $F_{applied}$ becomes greater than $F_{maximum}$, the object will slide, resisted by the dynamic frictional force $F_{kinetic} = \mu_{kinetic}N$, where $\mu_{kinetic}$ is the coefficient of dynamic friction (bottom panel).

acceleration $(F = ma)$. If the right-hand side is subtracted from both sides of the equation, the expression becomes $F - ma = 0$; this formulation is known as d'Alembert's principle. The term $-ma$ is called the **inertial force**; it is dimensionally equivalent to other forces that constitute the resultant force F on the left-hand side of the equation. This so-called **pseudo-force** is felt when an elevator rapidly slows as it approaches its destination floor. The body's inertia wishes to continue upward, and the inertial force results in a decrease in the reaction force between the feet and the floor. Another example is the **g-force** experienced during rapid acceleration in a car or plane; in these cases, the body wants to remain stationary while the vehicle moves rapidly forward. In these situations, a person feels like she is being pushed into the seat but the real force is the seat pushing her forward.

Another type of pseudo-force arises when we consider objects rotating about an axis and is associated with the ever-changing linear direction of particles within the rotating rigid body. In this circumstance, an outwardly directed **centrifugal force** represents the inertial tendency for the particles to continue moving away from the axis of rotation, while the inwardly directed **radial** or **centripetal force** acts to prevent such an occurrence. A third pseudo-force, the **Coriolis force**, occurs in rela-

tion to the rotations of reference systems within a given system. Readers are directed to physics or engineering texts (e.g., Beer et al. 2010) for a more complete description and the computation of these forces.

MOMENT OF FORCE, OR TORQUE

Earlier we noted that the point of application of a force vector is important only if angular motion of a rigid body is under consideration. By definition, angular motion takes place around an **axis of rotation** (see chapters 1 and 3). If a force vector $\vec{F}$ is applied to a rigid body so that its line of action passes directly through the axis of rotation, no angular motion is induced (figure 4.6). However, if force vector $\vec{F}$ is moved to a parallel location so that its line of action falls some distance from the axis, the force tends to cause rotation. Forces that do not pass through the axis of rotation are known as **eccentric** *(off-center)* **forces**. If a displacement vector, $\vec{r}$, is defined from the axis of rotation to the point of force application, the vector cross product of the force and displacement vector is known as the **moment of force** *(M)* so that $\vec{M} = \vec{r} \times \vec{F}$. The perpendicular distance from the axis of rotation to the force line of action is known as the **moment arm** *(d)* of the force. Figure 4.6 illustrates that $d = r \sin \theta$, where θ is the angle formed by the lines of action of the displacement vector $\vec{r}$ and force vector $\vec{F}$. Using the perpendicular distance, d, we can compute the magnitude of the moment of force as $M = Fd$. The unit for a moment of force (or simply "moment") is the newton meter (N·m). Clearly, the point of application of a force vector dictates the magnitude of the moment, because altering the application point changes the force moment arm.

A single eccentric force produces both linear and rotational effects. The force itself causes the object to accelerate according to Newton's law of acceleration regardless of whether the force is directed through the axis of rotation. A purely rotational motion, with no linear acceleration, can be produced by two forces acting as a force couple. A **force couple** consists of two noncollinear but parallel forces of equal magnitude acting in opposite directions. For example, in figure 4.6, the equal parallel forces F and $-F$ are separated by a perpendicular distance, d, and thereby form a force couple that applies a moment equal to Fd. Because F and $-F$ are in opposite directions, they sum to zero, and thus the resultant force applied to the object is zero. This results in an absence of linear acceleration.

Another term commonly used instead of *moment of force* is **torque**. Some physics and engineering texts distinguish between the two terms, associating torque

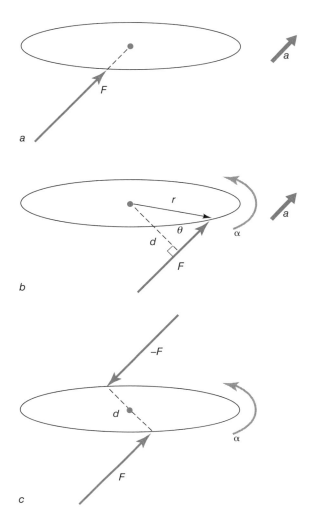

▲ **Figure 4.6** *(a)* The force *F* acts through the body's center of mass, which also is its axis of rotation. Therefore, the force causes only translation. *(b)* If *F* is applied so that it does not act through the axis of rotation, a moment of force, *M*, is created *(M = Fd)* and the body undergoes both linear and angular acceleration. *(c)* A force couple consisting of parallel forces *F* and *−F*, separated by the moment arm *d*. The force couple causes angular acceleration only.

with either force couples or "twisting" movements where it is difficult to identify a single force vector and point of application. Within the biomechanics community, the terms are used interchangeably by most, and we use both terms throughout this text.

The exact rotational kinematic effect of an applied torque is dictated by the angular version of Newton's laws of motion, the first two of which deal with angular momentum in the absence or presence of torque. In the absence of an applied torque, a rotating body continues to rotate with constant angular momentum, analogous to the linear law of inertia. Angular momentum, *L*, is

defined as the product of the body's mass moment of inertia, *I*, and its angular velocity, *ω*, so that $L = I\omega$.[1] The units for angular momentum are therefore kg·m²/s. When a moment of force is applied to a rigid body, the angular momentum changes so that $M = dL/dt$, where *dL/dt* is the time derivative of angular momentum. This is known as Euler's equation, after the famous 18th-century Swiss mathematician Leonhard Euler. It is the angular equivalent of $F = ma$, which Newton originally expressed as $F = dp/dt$, where *dp/dt* is the time derivative of the linear momentum *(p)* of a particle or rigid body. In the linear case, a change in velocity (acceleration) is the only possibility because the mass of a rigid body is a constant.[2] In the angular case, if the mass moment of inertia, *I*, is a constant, Euler's equation becomes $M = I\alpha$, where *α* is the rotating body's angular acceleration. However, when one is measuring the human body, changes in configuration are possible, and thus the moment of inertia can change. Therefore, the more general form of Euler's equation dictating changes in angular momentum (rather than acceleration) is useful, even though the body's mass does not change. Note that Euler's full equations for 3-D motion are more complex and will not be dealt with here. Consult an engineering mechanics text such as Beer and colleagues (2010) for a complete description of the 3-D case.

[1] *L* is the accepted SI abbreviation for angular momentum. Many textbooks use *H*.

[2] Note that scientists in the aerospace industry must use Newton's original formula because much of a rocket's mass is the fuel used for propulsion; therefore, the mass of the rocket continually changes.

LINEAR IMPULSE AND MOMENTUM

Newton's second law, $F = ma$, can be applied instantaneously or when an average force is considered. When a researcher wants to know the influence of a force that varies over its duration of application, the impulse-momentum relationship becomes useful. This relationship is directly derivable from Newton's second law; as noted previously, it was originally written as a relationship between force and momentum. At the time, the term *momentum* was not used; Newton referred instead to the quantity of motion. A mathematical derivation of the impulse-momentum relationship for forces begins with Newton's law of acceleration:

$$F = ma = m\frac{dv}{dt} \qquad (4.1)$$

Next, rearrange the equation by multiplying both sides by *dt*.

$$F\,dt = m\,dv \qquad (4.2)$$

Finally, integrating both sides of the equation yields the impulse-momentum relationship

$$\int F\,dt = mv_{final} - mv_{initial} \quad (4.3)$$

where the left-hand side is the linear impulse of the resultant force, F, and the right-hand side represents the change in linear momentum of mass, m. The terms mv_{final} and $mv_{initial}$ are the final and initial linear momenta of the body, respectively. The units of linear impulse are newton seconds (N·s), which are dimensionally equivalent to the units for linear momentum (kg·m/s).

Thus, the linear **impulse** of a force is defined as the integral of the force over its period of application, and this impulse changes the body's momentum. Graphically, linear impulse is the area under a force history. Figure 4.7 illustrates that *(a)* increasing the **amplitude** of the force, *(b)* increasing the duration of the force, *(c)* increasing both amplitude and duration, and *(d)* increasing the number of impulses (i.e., the frequency of impulses) increase the impulse on a body.

Measured Forces, Linear Impulse, and Momentum

The impulse-momentum relationship can be used to evaluate the effectiveness of a force in altering the **momentum** or velocity of a body. For example, a person performing a start in sprinting (Lemaire and Robertson 1990b) or swimming (Robertson and Stewart 1997) tries to apply horizontal reaction forces to initiate horizontal motion, and sensors capable of recording these forces can quantify the effectiveness of the start. Figure 4.8 shows the horizontal impulses of a start from instrumented track starting blocks (Lemaire and Robertson 1990b), a force platform mounted on swimmers' starting blocks (Robertson and Stewart 1997), and a force platform imbedded in an ice surface (Roy 1978). Integrating the areas shown in figure 4.8 and dividing by the mass of the athlete permit us to determine the change in the athlete's horizontal velocity. It is assumed with these types of skills that the initial velocity is zero, which is required for these skills; otherwise, the start is considered *false*, for which the athlete is penalized. This requirement is not applicable to relay starts, where the athlete is allowed to have a *running start*. In such a situation, the researcher has to measure the athlete's velocity before the start of force application. Thus, for a given impulse, the velocity of the person after the impulse is

$$v_{final_x} = \frac{\displaystyle\int_{t_{initial}}^{t_{final}} F_x\,dt}{m} + v_{initial_x} \quad (4.4)$$

where m is the person's mass, $v_{initial_x}$ is the initial velocity (zero if the motion starts from rest), and the numerator is the impulse of the horizontal force—that is, the area under the horizontal reaction force history—from time $t_{initial}$ to t_{final}.

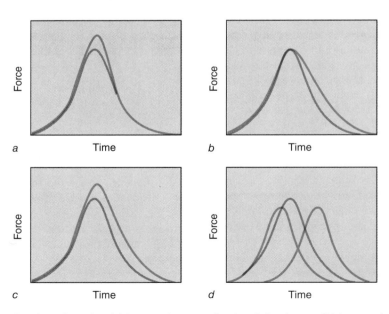

▲**Figure 4.7** Increasing impulses by *(a)* increasing amplitude of the force, *(b)* increasing the duration of the force, *(c)* increasing both the amplitude and duration, and *(d)* increasing the frequency or number of impulses. Note that the gray curve is identical in all frames.

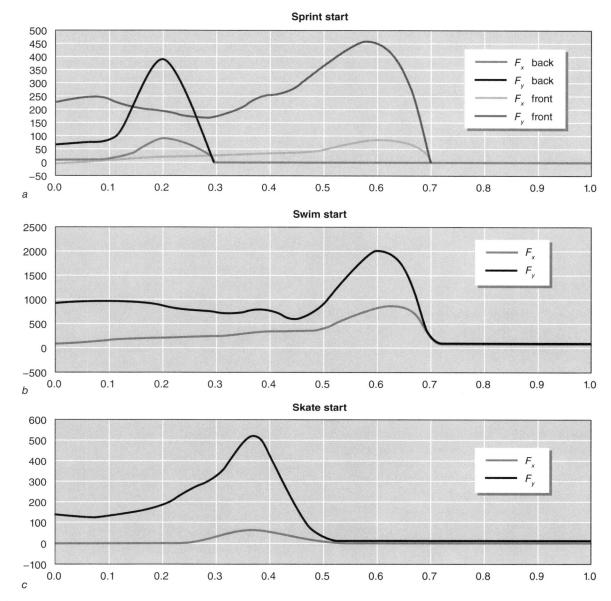

▲ Figure 4.8 Horizontal (F_x) and vertical (F_y) impulses of a start from the front and back blocks of instrumented track starting blocks *(a)*, from a force platform mounted on a swimmer's starting platform *(b)*, and from a force platform imbedded in an ice surface *(c)*. Abscissa is time in seconds, and ordinate is force in newtons.

A similar application concerns research on vertical jumping or landing from a jump, in which vertical force platform signals are integrated over time to obtain the changes in the jumper's vertical momentum. In the case of standing jumps, the athlete's initial velocity is zero; therefore, the takeoff velocity of the jumper can be computed directly from the force platform signals (i.e., the change in velocity is equivalent to the takeoff velocity). There is a slight difference in the equation for the vertical impulse because gravity must be excluded to obtain the vertical velocity. The equation for the vertical velocity is therefore

$$v_{final_y} = \frac{\int_{t_{initial}}^{t_{final}} (F_y - W)\, dt}{m} + v_{initial_y} \qquad (4.5)$$

where W is the person's weight in newtons, $v_{initial_y}$ is the initial vertical velocity (zero if the motion starts from rest), and F_y is the vertical GRF. Of course, this equation applies only if all forces act against the force platform. For example, if one of the person's feet is off the platform, the vertical velocity will be underestimated. Similarly, it is assumed that no other body part acts against the ground or the environment. If so, additional force-measuring instruments must be used.

Other applications for the impulse-momentum relationship include activities such as rowing, canoeing, golf, batting, and cycling, in which the forces applied by the hands or feet can be measured by force-sensing elements to quantify their impulses. Again, the effectiveness of the force can be directly quantified by integrating the applied force over time. The simplest method for computing these integrals is to use **Riemann integration**. With this method, the force signals collected from an analog-to-digital (A/D) converter of a computer are added and then the sum is multiplied by the sampling interval (Δt). If the sampling rate of the force signal is 100 Hz, the sampling interval is 0.01 s. The equation for a Riemann integral is

$$\text{Impulse} = \Delta t \sum_{i=1}^{n} F_i \qquad (4.6)$$

where F_i is the sampled forces and n is the number of force samples. Another, more accurate integral uses **trapezoidal integration**:

$$\text{Impulse} = \Delta t \left(\frac{F_1 + F_n}{2} + \sum_{i=2}^{n-1} F_i \right) \qquad (4.7)$$

Essentially, this is half the sum of the first and last forces plus the sum of the remaining forces times the sampling duration. Other, more sophisticated integrals are possible, such as **Simpson's rule integration**, but with a sufficiently high sampling rate there is little difference in the resulting integrals. Readers are directed to college calculus or numerical analysis textbooks for more information on these integration techniques. Also important in this process is the selection of an appropriate sampling rate and smoothing function used during data collection and reduction (see chapter 12 for data-smoothing functions). The sampling rate should be selected so that it is neither too low, in which the true peaks and valleys in the force history are clipped, nor too high, which increases errors due to the integration process. A suitable sampling rate for many jumping and starting situations is 100 Hz. After calculating the impulse, compute the change in velocity by dividing the impulse by the body's mass.

A slightly different approach can be used to observe the instantaneous changes in velocity resulting from force application. We begin by converting the GRFs into acceleration histories by dividing by the person's mass at each instant in time. The vertical GRF must be reduced by subtracting the person's weight (W):

$$a_x = F_x / m$$
$$a_y = (F_y - W) / m \qquad (4.8)$$

Note that W must be very accurate; otherwise, there will be an ever-increasing error during the integration process. The best solution is to record a brief period immediately before the impulse starts from which the person's weight can be determined. It happens that a person's weight as registered by a force platform varies slightly depending on where the feet are placed.

These acceleration patterns have the same shape as the force profiles because they have merely been scaled by the constant mass. The acceleration histories are then integrated to obtain the velocity histories by iteratively adding successive changes in velocity. The velocity history is computed by repeatedly applying this integration equation:

$$v_i = v_{i-1} + a_i(\Delta t) \qquad (4.9)$$

where v_i is the velocity at time i, v_{i-1} is the previous time interval's velocity, a_i is the acceleration, and Δt is the sampling time interval. The first initial velocity, called a constant of integration in calculus, must be known. If the activity starts statically, the first initial velocity is zero; if not, then the researcher must compute or measure the initial velocity with another system, such as videography. In a similar manner, the second integral of force theoretically can yield the displacement of the body. The integration process is repeated using the computed velocity signal to obtain the displacement history, which introduces another constant representing the initial position of the person. For simplicity, we can set the initial position to zero and then determine displacement (s_i) that occurs after starting the integration. That is, $s_i = s_{i-1} + v_i(\Delta t)$, where v_i is the velocity computed from the previous iteration of the equation.

Note that this integration process can become unstable if instrumentation problems arise. If the force signal drifts (i.e., low frequency changes in the baseline) as the subject stands motionless on the force plate, the computed displacement signal rapidly becomes unrealistic. This drift is a characteristic of piezoelectric force plates. It is advisable, therefore, to conduct this type of measurement by minimizing the integration time and calculating body weight from the force recording at the instant immediately before the integration starts, when the person is standing motionless. Small errors in weight determination occur when a person stands on a force platform in response to exact foot placement and other environmental factors. These slight inaccuracies in the subject's weight cause large errors in the displacement record because of the double integration of the force signal (Hatze 1998). This is the inverse of the situation that occurs when one derives acceleration from displacement, in which small, high-frequency displacement errors create large acceleration errors (see chapters 1 and 12).

Segmental and Total Body Linear Momentum

When we analyze human motion, it is not always possible to directly measure external forces that act on the body. Instead, momentum can be computed indirectly from the kinematics of body markers and computed segment centers of gravity (chapter 3). Once the segment centers are known, it is a relatively simple matter of multiplying the velocity vectors by the segment masses (see chapter 3). That is,

$$\vec{p} = m\vec{v} \quad \text{or}$$
$$p_x = mv_x, \, p_y = mv_y, \, p_z = mv_z \quad (4.10)$$

Total body momentum is also easy to compute with the necessary input data, because momenta can be summed vectorially. The total body linear momentum is therefore the sum of its segmental momenta. That is,

$$\vec{p}_{total} = \sum_{s=1}^{S} m_s \vec{v}_s \quad (4.11)$$

where m_s is the segment masses, $\vec{v}$ is the velocity vectors of the segment centers, and S is the total number of segments. The scalar versions are

$$P_{total\,x} = \sum_{s=1}^{S} m_s v_{sx}$$
$$P_{total\,y} = \sum_{s=1}^{S} m_s v_{sy} \quad (4.12)$$
$$P_{total\,z} = \sum_{s=1}^{S} m_s v_{sz}$$

This measurement is not often used in biomechanics because it requires recording the kinematics of all the body's segments, which typically is difficult, especially in three dimensions. However, this technique has been used for the study of airborne **dynamics**, such as long jumping (Ramey 1973a, 1973b), high jumping (Dapena 1978), diving (Miller 1970, 1973; Miller and Sprigings 2001), and trampolining (Yeadon 1990a, 1990b). In these situations, conservation of linear momentum occurs in the horizontal direction and momentum decreases predictably in the vertical direction because of gravity.

ANGULAR IMPULSE AND MOMENTUM

Angular impulse and **angular momentum** are the rotational equivalents of linear impulse and linear momentum. They are derived from Euler's equation *(M = Iα)* in a fashion similar to the linear *F = ma*. Recall that Euler expressed his equation as *M = dL/dt*, where

dL/dt is the time derivative of angular momentum *(L = Iω)*. Rearranging Euler's equation results in *MΔt = ΔIω*. Given a resultant moment of force, M_R, acting on a body, integrating over time yields the angular impulse applied to the body. That is,

$$\text{Angular Impulse} = \int_{t_{initial}}^{t_{final}} M_R \, dt \quad (4.13)$$

where $t_{initial}$ and t_{final} define the duration of the impulse in seconds. Given that the angular momentum of a system of rigid segments is L, the angular impulse-momentum relationship can be written

$$L_{final} = L_{initial} + \text{Angular Impulse} \quad (4.14)$$

That is, the angular momentum of the system after an angular impulse is equal to the angular momentum of the system before the impulse plus the angular impulse. Note that the system's moment of inertia may have a different value before and after the duration of the impulse, depending upon segmental configuration. Unlike the constant mass assumption of linear motion, the rotational moment of inertia can be quite different from one instant to another. For example, a diver or gymnast in the layout position can have a tenfold greater moment of inertia than when in the tuck position. Thus, the effects of an angular impulse vary with changes in the body's moment of inertia. This factor is taken into consideration by segmenting the body into a linked system of rigid bodies and computing each segment's contribution to the total body's angular momentum. Each segment contributes two terms to the angular momentum of the whole body: one term sometimes called the **local angular momentum** and another called the **moment of momentum** or **remote angular momentum**. The first term describes rotation of the segment about its own center of gravity, whereas the second, the moment of momentum, corresponds to the angular momentum created by the segment's center of gravity rotating about the total body's center of gravity. These terms are defined next.

Segmental Angular Momentum

Whereas the linear momentum of a segment is the product of its mass and linear velocity, the angular momentum of a segment (L_s) rotating about its center of gravity is the product of its moment of inertia and its angular velocity:

$$L_s = I_s \omega_s \quad (4.15)$$

where I_s is the moment of inertia (in kg·m²) of the segment about its center of gravity and ω_s is the angular velocity of the segment (in rad/s). Of course, segments rarely rotate solely about their own center of gravity. To determine the angular momentum of a segment about another axis (e.g., the total body center of gravity or the proximal end of the segment), the moment of momentum (L_{mofm}) is needed. This term, based on the parallel axis theorem, is defined as

$$L_{mofm} = \left[\vec{r}_s \times m_s \vec{v}_s\right]_z = m_s\left(r_x v_y - r_y v_x\right) \quad (4.16)$$

where (r_x, r_y) is the position vector that goes from the axis of rotation to the segment's center of gravity, m_s is the segment's mass, and (v_x, v_y) is the linear velocity vector of the segment. Note that the symbol $\times$ means that the two vectors are multiplied as a cross or vector product and that the symbols $[\vec{r}_s \times m_s \vec{v}_s]_z$ mean that only the scalar component about the Z-axis is to be considered.

Total Body Angular Momentum

To obtain the angular momentum of a whole body, several different approaches may be taken. If the body is made up of a series of interconnected segments (as is the human body), then the total body angular momentum is the sum of all the segment angular momenta plus their associated moments of momentum. For example, to calculate the total body angular momentum (L_{total}) about the total body center of gravity for planar analyses, this equation applies:

$$L_{total} = \sum_{s=1}^{S} I_s\,\omega + \sum_{s=1}^{S}\left[\vec{r}_s \times m_s \vec{v}_s\right]_z \quad (4.17)$$

where $\vec{r}_s$ represents the position vector connecting the total body center of gravity to the center of gravity of the segment, that is, $(x_s - x_{total}, y_s - y_{total})$; $\vec{v}_s$ is the velocity of the segment's center; and S represents the number of body segments.

For most human movements, moment of momentum terms are larger than the local angular momentum terms because a segment's moment of inertia is usually less than 1 (in kg·m²) whereas the mass of a segment is greater than 1 (in kg). Furthermore, the position vectors for the least massive segments can be quite large, and consequently their cross products with velocity are relatively large compared with the segment's rotational velocity.

Angular Impulse

An alternative way of determining total body angular momentum makes use of the external forces and moments of force acting on the body and the angular impulses they produce. Figure 4.9 shows four examples of external forces that produce angular impulses and consequently affect the angular momentum of the bodies. Notice that in all cases the lines of action of the external forces do not pass through the bodies' centers of gravity. To determine how much angular impulse is produced, one must measure the forces over time and, simultaneously, record the body's center of gravity trajectory. In addition, all other external forces must be quantified to determine their magnitude, direction, and points of application on the body. The only external force that

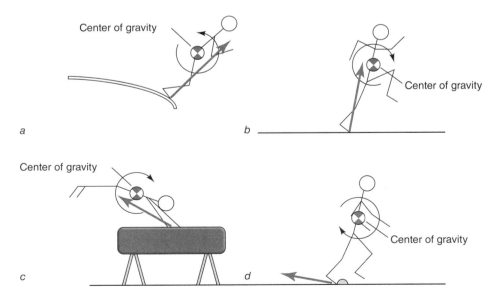

▲ **Figure 4.9** Examples of forces that cause angular impulses and angular momenta. Curved lines show the direction of the applied angular impulse. Example *a* is a diver, *b* is a long jumper at takeoff, *c* is a gymnast, and *d* is someone tripping while walking.

does not need to be measured is gravity, because it is a **central force** that passes through the body's center of gravity and therefore causes no rotational momentum.

Recall that angular impulse is the time integral of the resultant moment of force or eccentric force acting on the body (figure 4.10), whereas angular momentum is the quantity of rotational motion of the body. In mathematical form,

$$\text{Angular Impulse} = \int_{t_i}^{t_f} M_R \, dt \qquad (4.18)$$

or, if the moment of force, M_R, is constant, then

$$\text{Angular Impulse} = M_R \Delta t \qquad (4.19)$$

where Δt is the duration of the impulse.

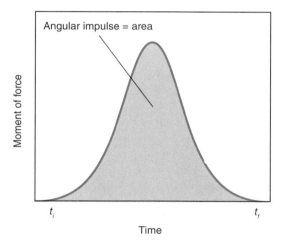

Figure 4.10 Angular impulse is defined as the area under the moment of force versus time curve.

Alternatively, if there is a single external force acting on the body (figure 4.11), the angular impulse can be quantified:

$$\text{Angular Impulse} = \int \left(\vec{r} \times \vec{F} \right) dt \qquad (4.20)$$

where $\vec{F}$ is the applied force and $\vec{r}$ is the position vector from the total body center of gravity to the force's point of application.

Note that $\vec{r} \times \vec{F}$ is the cross product of the force and position vectors. In two dimensions, the magnitude of this product is $r_x F_y - r_y F_x$ (see appendix D for vector products in three dimensions). If the applied force is a central force such as gravity, it causes no angular impulse and therefore causes no change in the angular momentum of the body. This important principle, called the *law of conservation of angular momentum*, applies when the body is airborne and the only external force is gravity (i.e., ignoring air resistance).

Other situations also follow this law if any rotational friction can be neglected, such as standing on a frictionless turntable or spinning on an icy surface. Thus, the person is assumed to spin with constant angular momentum but not necessarily constant angular velocity. The person's spin rate (angular velocity) may be increased or decreased, respectively, by decreasing or increasing the total body's moment of inertia through movements of body segments. The law of conservation of angular momentum says that the total body angular momentum (L_{total}) stays constant about any axis whenever the applied moments are zero and the resultant force is a central force:

$$L_{total} = \text{Constant} \qquad (4.21)$$

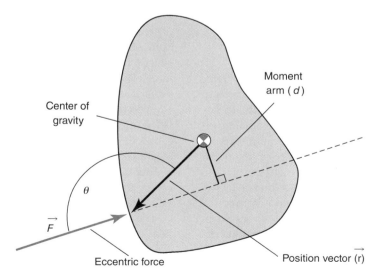

Figure 4.11 Eccentric forces are capable of producing angular impulses.

FROM THE SCIENTIFIC LITERATURE

Yu, B., and J.G. Hay. 1995. Angular momentum and performance in the triple jump: A cross-sectional analysis. *Journal of Applied Biomechanics* 11:81-102.

This project quantified the segmental and total body angular momenta for four airborne phases of the triple jump—the last stride, the hop, the step, and the jump. Elite-level triple jumpers ($n = 13$) were videotaped during a competition, and only their longest effort was analyzed. Three-dimensional coordinates of the segments and segmental and total body angular momenta were computed for the airborne portions of the four phases. The average values of the angular momentum for each phase and each axis of rotation were then normalized by mass and height squared and correlated to actual jump performance.

The authors found that there was a statistically significant nonlinear correlation between the mediolateral angular momentum of the support phase of the step and the jump distance achieved ($r = 0.86$). They concluded that to achieve the required angular momentum at the step, momentum must be obtained during the support phase of the hop. Furthermore, the step phase should minimally change the angular momentum created by the hop.

This was the first paper to report the angular momenta in 3-D for such a complicated motion as the triple jump. Not only was it a breakthrough for quantifying such a dynamic motion, but it was done on four different skills (the run, the hop, the step, and the jump) performed by elite athletes in a highly competitive situation (U.S. Olympic trials).

The majority of researchers who have attempted to quantify angular momentum have been concerned with the study of athletic airborne motions, for example, diving (Miller 1970, 1973; Murtaugh and Miller 2001), figure skating (Albert and Miller 1996), long jumping (Lemaire and Robertson 1990a; Ramey 1973a, 1973b, 1974), gymnastics (Gervais and Tally 1993; Kwon 1996), and trampolining (Yeadon 1990a, 1990b).

MEASUREMENT OF FORCE

There are many tools available to the biomechanist for the measurement of force and moment of force. Although all of these tools can be called transducers, we separate them into force platforms, pressure distribution sensors, internally applied force sensors, and isokinetic devices.

Force Transducers

Much of our discussion has dealt with the effects caused by forces applied to a particle or rigid body. In fact, modeling any object or body segment as "rigid" in the presence of an applied force is only an approximation of reality. Because all objects deform to a certain extent, our definition of a rigid body (all particles having fixed positions relative to each other) is not strictly correct. In many cases, the deformation and errors associated with "nonrigidity" are small. Furthermore, this deformation can be useful to biomechanists because it allows them to measure the applied forces using force **transducers**.

Sensing elements of different types can be adhered to or built into deformable materials. When a force is applied, the sensing elements register the amount the material deforms. Typically, the sensing elements have electrical properties and constitute part of an electronic circuit. For example, *resistive* or *piezoresistive* elements can serve as resistors within a circuit such as a Wheatstone bridge. Deformation causes structural and geometric changes in resistors that alter their electrical resistance (e.g., a thin piece of metal becomes thinner when stretched, altering its ability to transmit electric charge). This change in resistance alters the voltage decrease in an appropriate electronic circuit. Piezoresistive elements are based on semiconductor materials such as silicon and are more sensitive than ordinary resistive materials. Because force, deformation, resistance, and voltage are directly related, knowledge of these relationships allows one to calculate the force applied by measuring the voltage change in such a circuit. A different type of sensing element is made from **piezoelectric** crystal, a naturally occurring mineral that produces electric charge in response to deformation from an applied force. The piezoelectric crystals must be connected to a *charge amplifier*, and the associated electronic circuit is much different from that in the piezoresistive case. However, the concept is the same: An applied force causes deformation that leads to a measurable electrical response, and the magnitude of the electrical response is directly related to the magnitude of the force. Readers are directed to appendix C for more information on electronic circuits.

The quality of a force transducer depends on the relationships between the applied force, the deformation, the electrical characteristics of the sensing element, and the measured electrical output. If we ignore the underlying electrical details, the relationship that is of interest is the one between the applied force input and the registered voltage output (figure 4.12). The responses shown in figure 4.12 are derived from static measurements, for which a constant force is applied to a transducer and the resulting (and, we hope, constant) voltage is registered. Each data point represents a given force level and the voltage response. When the magnitude of the constant applied force is changed, voltage responses across a range of forces are recorded. If the resulting relationship is linear (figure 4.12*a*), the slope of the relationship (Δ*F*/Δ*V*) represents a *calibration coefficient*, called the **sensitivity**, which can be used to convert from measured voltage into force in newtons (e.g., 6.5 volts [V] × 63 N/V = 409.5 N). **Linearity** is determined by computing the Pearson product moment correlation coefficient, called simply the correlation coefficient (*r*). A sufficiently high *r* value implies a linear relationship between the actual force and the measured voltage. If the relationship is nonlinear (figure 4.12*b*), the data can be fitted to a *calibration equation* (e.g., second- or third-order polynomial) that can be used for the conversion of voltage to force.

Another consideration is the range of forces a transducer will measure before its response changes markedly or it is damaged. Transducers are rated for a particular force range, over which their response is linear; if higher forces are applied, the voltage output may saturate at a given level. A related issue is the sensitivity of the transducer. A force transducer should be matched to the range of forces one wishes to measure; it should be sensitive enough to detect small changes in applied force yet still have enough range. Another point of concern is **hysteresis** (figure 4.12*c*), in which a different force-to-voltage relation is found when the force is incrementally increased compared with when it is incrementally decreased. This is undesirable because, in theory, different calibration coefficients or equations should be used in loading and unloading situations.

Static characteristics are important and easy to assess, but in most biomechanics applications, the applied forces continually change in magnitude. Therefore, the **dynamic response** capabilities of the transducer are equally important. The **frequency response** characteristics of the transducer should be matched to those of the applied force. The physical characteristics of the transducer's construction permit it to respond to a limited range of input force frequencies. If the input force changes too rapidly, the transducer may be unable to respond quickly enough to faithfully register the true time history of the force. This is analogous to a low-pass filter, which attenuates or eliminates the higher-frequency components of the input force signal (see chapter 12, Signal Processing). However, the transducer's construction may cause the unwanted amplification of some frequencies within the input force. Any physical

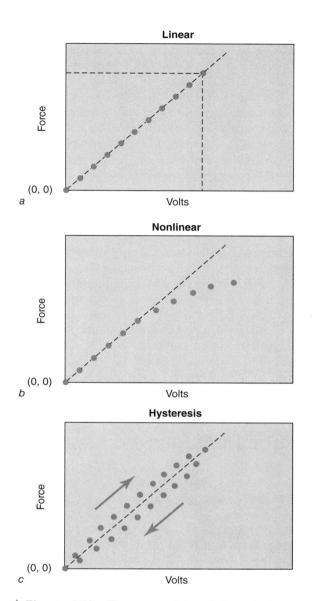

▲ **Figure 4.12** Three possible relations between an input force and the output response voltage for a force transducer: *(a)* a linear response, in which the slope of the force-voltage relation can be used as a calibration coefficient; *(b)* a nonlinear relation; *(c)* the concept *of hysteresis.*

structure or system will respond to a forced vibration in a characteristic way based on its internal mass, elasticity, and damping. The mass and elasticity dictate the **natural frequency** of a structure, and the structure resonates when an external vibration at or above the natural frequency is imposed. This concept is useful in the construction of tuning forks, which oscillate at their natural frequency when struck with an applied force. In biomechanics, this response is undesirable, because an applied force that contains significant energy at the natural frequency of a force transducer causes the transducer to vibrate on its own, therefore exaggerating the force response at that frequency.

Transducers can be purchased based on their inherent response properties, such as range, linearity, sensitivity, frequency response, and natural frequency. However, transducers can have very different characteristics when they are mounted or installed as one component in a typical measurement system. For example, a transducer with a high natural frequency (e.g., 2000 Hz) vibrates excessively at lower frequencies if it is attached to a wood frame with springs. One area of particular concern involves the construction of measurement devices with multidirectional transducers that measure force in three orthogonal directions. The transducers may have excellent isolation of forces applied in the three directions, but the measurement device in which they are installed may have considerable elasticity that transmits forces across these orthogonal axes, leading to **crosstalk**. For example, the application of a force purely in the F_x direction may result in the transducer responding in the F_y and F_z directions as well. Such a response can be dealt with by using a calibration matrix that relates the response of the transducer in all three directions to given input forces.

Force Platforms

The most commonly used type of force transducer in biomechanics is the **force platform** (or **force plate**), which is an instrumented plate installed flush with the ground for the registration of GRF. Early models used springs (Elftman 1934) and inked rubber pyramids to display pressure patterns. Currently, two types of sensors are used in commercially available force platforms: strain gauges and piezoelectric crystals. The strain-gauge models are less expensive and have good static capabilities but do not have the range and **sensitivity** of the piezoelectric models. The piezoelectric platforms have high-frequency response but must have special electronics to enable measurement of static force.

Early force platforms were designed with a single instrumented central column or pedestal, whereas recent designs usually have four instrumented pedestals located near the corners of the plate. Researchers have abandoned the single-pedestal design because forces tend to become more inaccurate the farther an applied force moves away from the pedestal. With the four-pedestal models, forces are accurate whenever the contact force is applied within the area bounded by the pedestals. The pedestals are instrumented to measure forces and bending moments that result from all forces and moments applied to the platform's top plate. Most commercial platforms are instrumented to measure in three dimensions—vertical

(Z), along the length of the plate (Y), and across the width of the plate (X). Force platforms operate on the principle that no matter how many objects apply force to different locations on the top plate, there is one resultant force vector—the GRF—that is numerically and physically equivalent to all the applied forces.

Any single 3-D force vector applied to a force plate can be described by nine quantities. The three orthogonal components of the force vector are designated as F_x, F_y, and F_z, whereas three spatial coordinates, x, y, and z, designate the force vector location with respect to the *plate reference system* (PRS) origin. Because the force vector usually results from a distribution of forces to an area of contact on the surface of the plate, its location is often called the *center of pressure* (COP). The final three quantities are orthogonal moments M_x, M_y, and M_z, taken with respect to the PRS origin. The exact location of the PRS origin depends on the specific location of the force-sensing pedestals, but in general the origin is located in the middle of the plate slightly beneath the level of the top surface (figure 4.13). Thus, the COP coordinate z is a constant equal to the depth of the PRS origin beneath the top of the plate.

Although these nine parameters represent the force applied to the plate, we generally are interested in only six quantities with respect to the reaction force vector applied by the force plate to the human performer. This change in emphasis comes from our interest in the motion of the human, not the force plate itself. These six quantities are the three GRF components $(R_x, R_y,$ and $R_z)$, the COP location of the reaction force vector in a GCS (x, y), and a moment known as the **free moment**, M_z'. Note that in the GCS, the Z-axis is in the vertical direction, the Y-axis is horizontal in the direction of progression, and the horizontal X-axis is aligned roughly in the medial/lateral direction. Only the x and y COP coordinates need to be computed, because the vertical coordinate is on the force platform's top plate, usually designated as z = 0 within the GCS. The free moment M_z' represents the reaction to a twisting moment applied by the subject about a vertical axis located at the COP coordinates. The free moments about the X- and Y-axes are assumed to be zero, because these can occur only if there is a direct connection between the shoe and plate (as with glue). Note that these six quantities are related to, but not equal to, the nine force platform parameters described in the previous paragraph, for three reasons:

▶ the directions of the PRS and GCS axes may not coincide,

▶ our interest is in the reaction forces $(R_x, R_y,$ and $R_z)$ rather than the applied forces $(F_x, F_y,$ and $F_z)$, and

▶ we need to know the COP location in the GCS rather than in the PRS.

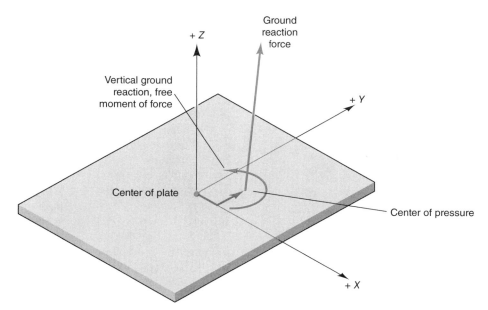

▲ **Figure 4.13** Force platform with its reaction to an applied force and vertical moment of force.

In the next section, we investigate how these six quantities are computed from the nine force platform measures.

Processing Force Platform Signals

Each type of force platform has its own unique set of equations from which the six measures of the ground reaction are computed. Note that each force platform manufacturer may use a unique PRS and that these reference systems do not necessarily coincide with the GCS established for the kinematic data-capture system (see chapters 1 and 2). For consistency, we will use the PRS system established in the previous section (figure 4.13). Example equations are given for two commercial brands, the strain-gauge AMTI plate (AMTI, Watertown, MA) and the piezoelectric Kistler platform (Kistler AG, Winterthur, Switzerland). Other manufacturers may have slightly different configurations, but we should still be able to compute the six measures. The equations for an AMTI force platform are derived from output signals labeled F_x, F_y, F_z, M_x, M_y, and M_z (figure 4.14). The six AMTI equations are

$$F_x' = F_x f_x \quad F_y' = F_y f_y \quad F_z' = F_z f_z$$

$$x = -(M_y g_y + F_x'z)/F_z'$$

$$y = (M_x g_x - F_y'z)/F_z'$$

$$M_z' = M_z g_z + F_x'y - F_y'x$$

(4.22)

where $(F_x'$, F_y', and $F_z')$ are the components of the GRF; $(x, y, 0)$ are the coordinates of the COP; M_z' is the free moment of force; f_x, f_y, and f_z are scale factors that convert

the forces from voltages to newtons; and g_x, g_y, and g_z are scale factors that convert the bending moments from voltages to newton meters. The six output signals from an AMTI plate are fed to an amplifier unit with selectable gain levels that determine the exact values of the f_i and g_i scale factors. Each AMTI plate has a unique, factory-calibrated value for z, the distance from the top of the plate to the PRS origin. Note that the applied moments $(M_x$, M_y, and $M_z)$ about the PRS origin are used to only calculate the COP location.

Kistler force platforms have 12 piezoelectric force sensors, packaged as three orthogonal cylinders in each of the four pedestals (figure 4.14). Horizontal, but not vertical, sensors are summed in pairs so that the following eight signals are output: F_{x12}, F_{x34}, F_{y14}, F_{y23}, F_{z1}, F_{z2}, F_{z3}, and F_{z4}. The equations for computing the six quantities of the GRF are

$$F_x' = (F_{x12} + F_{x34})f_{xy}$$

$$F_y' = (F_{y14} + F_{y23})f_{xy}$$

$$F_z' = (F_{z1} + F_{z2} + F_{z3} + F_{z4})f_z$$

(4.23)

$$x = -[a(-F_{z1} + F_{z2} + F_{z3} - F_{z4})f_z - F_x'z]$$

$$y = [b(F_{z1} + F_{z2} - F_{z3} - F_{z4})f_z + F_y'z]F_z'$$

$$M_z' = b(-F_{x12} - F_{x23})f_{xy} + a(F_{y14} - F_{y23})f_{xy} - xF_y' + yF_x'$$

where $(F_x'$, F_y', $F_z')$ are the components of the GRF, $(x, y, 0)$ are the coordinates of the COP, M_z' is the free moment of force, f_{xy} and f_z are scale factors that convert the forces from voltages to newtons, and a and b are the distances

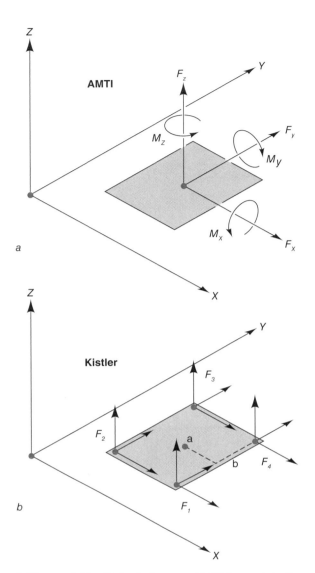

▲ Figure 4.14 Outputs from an AMTI force plate (top panel) include three forces, F_x, F_y, and F_z, and three moments, M_x, M_y, and M_z. The bottom panel shows the four piezoelectric sensors in a Kistler force plate, spaced equidistant from the plate center (lengths a and b). The three force components from each of the four sensors are used either alone (F_z) or in summed pairs (F_x, F_y) in the calibration equations (see text).

between the sensors and the plate center in the X and Y directions, respectively. The eight output signals are fed to a charge amplifier unit that includes selectable gain levels for each channel that determines the exact values of the f_{xy} and f_z scale factors. As with the AMTI plates, each Kistler platform has its own factory-calibrated specific value for z, the distance from the top of the plate to the PRS origin.

The equations for both the AMTI and Kistler force plates result in the computation of six quantities, but they are not yet in the form needed for analysis of their effects on the human subject. The three force components $(F_x', F_y',$ and $F_z')$ are in reference to the plate and must be multiplied by −1 to provide the reaction forces (R_x, R_y, R_z) applied to the subject. The same reversal in sign applies to the free moment M_z'. The next potential adjustment occurs because the three axes of the PRS and the GCS may be labeled differently or misaligned. The GCS origin and axes' directions are set when calibrating the imaging system used in kinematic data capture. For example, in locomotor studies, the subject usually progresses in such a way that her foot will land on the force plate along its greatest length, Y, in our PRS. In the GCS defined above, this may also correspond to the Y direction. However, the Y directions of the GCS and PRS may be opposite in polarity (figure 4.15). In addition, for 2-D studies, the direction of progression may be designated as the X-axis (see chapter 1). If the force plate data are to be combined with the kinematic data in an inverse dynamics analysis (chapter 5), the researcher must ensure that force data from individual PRS directions are correctly transformed into the GCS directions.

The same PRS/GCS alignment issue also applies to the COP coordinate data, which is a matter of **spatial synchronization**. Recall that the calculated x and y coordinates are in the PRS, but we want to express them in relation to the subject's kinematics in the GCS. Thus, we need to know the location of the PRS origin with respect to the GCS origin. An easy solution is to locate the GCS origin at the very center of the plate so that the origin x and y coordinates coincide (the vertical origins will be separated by the distance z). Another solution is to place a reflective marker at a known coordinate location in the PRS (e.g., on one corner of the plate). Knowledge of that marker's location in both the GCS and PRS will allow the coordinates x and y to be transformed into the GCS by a simple linear transformation.

A more complex problem occurs when the PRS and GCS axes are misaligned so that the two horizontal PRS axes are rotated by an angle, ϕ, with respect to the two GCS axes (figure 4.15). If the angle ϕ is known, a rotational transformation as described in chapter 2 can be performed to mathematically align the two reference systems. This situation can usually be avoided by carefully selecting the GCS axes' directions in relation to the force platform at the time of calibration.

Another issue relevant to the combination of force and kinematic data is the **temporal synchronization** of the two data sets. If the kinematic data-capture system is separate from the analog-to-digital converter used to collect the force plate data, the same event must be recorded on both systems to ensure proper time synchronization of the two systems. Synchronization units that use the vertical channel from the force plate to trigger a light-

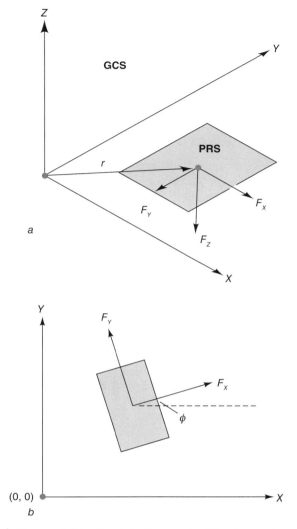

▲ Figure 4.15 Spatial alignment of the force plate PRS and GCS is critical for inverse dynamics analyses in which kinematic and force data are combined. The top panel illustrates an example of differences in the origin location and axes' orientation of the PRS and GCS. The position vector r describes the location of the PRS origin within the GCS. The bottom panel shows an overhead view of angular misalignment between the PRS and GCS in the horizontal plane. If the angle ϕ is known, this misalignment can be corrected using rotational transformation techniques.

emitting diode (LED) in the view of the imaging system are commonly used. When a small threshold vertical force is applied (such as when the subject first touches the plate), the LED is turned on; this identifies the instant of first foot contact in both data collection systems. If the motion studied begins with the subject on the plate (e.g., jumping), the point of takeoff can be established as the time the LED turns off. Some commercial products (e.g.,

Motion Analysis, Vicon, Qualisys, and Optotrak) ensure correct time synchronization with analog-to-digital conversion modules that are operated within the kinematic capture process. These systems capture the force and kinematic data simultaneously and allow the kinematic data to be sampled at a rate lower than for the analog data. One caveat is that the analog sampling rate must be an even multiple of the kinematic data-capture rate (e.g., 1000 Hz force to 200 Hz kinematics, or multiples of 60 Hz for video capture systems).

Center of Pressure on the Foot

Once the COP has been located within the GCS through the calculation of its *(x, y)* coordinate pair, it is important to examine how this location relates to the position of the subject. During the stance phase of walking or running, the portion of either foot in contact with the ground continually changes, and the COP constantly moves along the surface of the plate. When only one foot is on the plate, the COP lies within the part of the foot or shoe that contacts the platform, and it moves beneath the foot in a characteristic manner typically from the heel to the toes (Cavanagh 1978). When more than one foot is in contact with the force platform (a situation that normally should be avoided), the COP lies within the areas formed by either foot or the area between the two feet. If a kinematic analysis is done simultaneously with the force measurement, both COP and foot-location data will be in the GCS, and their relative positions can be computed. Note that foot position is usually designated based on the locations of several markers on the foot (see chapters 1 and 2).

Several difficulties arise when comparing the COP and foot locations. The location of the COP is highly variable, especially during periods when the foot is initially contacting or leaving the force plate (i.e., the landing and liftoff phases). This variability comes from the equations used to calculate the COP coordinates *(x, y)*, which are unstable when the vertical forces are small, near the beginning and end of the stance phase. Also, slippage may occur during loading and unloading, causing further errors in the COP computation. Often, unrealistic COPs must be discarded, such as any positions that are outside the area of the foot that is in contact with the platform. However, the foot-marker data do not define the outline of this contact area, so it is difficult to implement such a checking algorithm. Frequently, the researcher must visually inspect the individual marker coordinate data (e.g., the heel marker, ankle marker, fifth metatarsal head marker) to determine which portion of the foot is on the ground at a particular time and check that the COP coordinates are realistic. Finally, averaging the COPs is difficult because the subject will land at a different place on the force platform each time. However, Cavanagh (1978) provided a means of averaging COP paths in relation to the outside edges of the shoe during gait (figure 4.16).

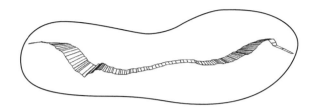

▲ **Figure 4.16** Center of pressure path relative to a footprint. Line segments represent directions of force vectors in the transverse plane. Line closest to medial side of foot is path of center of pressure.

Combining Force Platform Data

Occasionally, the GRFs from two or more force platforms need to be combined to yield a single force. Gerber and Stuessi (1987) provided the equations necessary to perform this operation.

Figure 4.17 shows the forces from a single footfall in which the rear foot landed on one force platform and the forefoot on another, a setup used in a two-segment foot model to improve accuracy in modeling the foot dynamics during gait (Cronin and Robertson 2000). Also shown is the combined single GRF. This process enables the computation of the moments of force across the metatarsophalangeal joint, which is usually modeled as a rigid body.

Presentation of Force Platform Data

The six quantities of the GRF can be displayed in many different ways. As with kinematic data presentation (chapter 1), a common method is to present the GRF, COP coordinates, and free moment as time-series or history plots (figure 4.18). To facilitate the comparison of the COP and the foot location, it is sometimes helpful to plot their location data together (figure 4.16). Another useful graph is a trajectory plot of the COP *(x, y)* coordinates; again, inclusion of the foot-marker data can be helpful. Sample trajectory plots are shown in figure 4.19 for a person walking (tracing *a*) and standing (tracing *b*) and are frequently used for posture and balance research.

A *force signature*, vector, or *butterfly* graph combines the three force components and COP location into one 2-D image that can be rotated to view in three dimensions

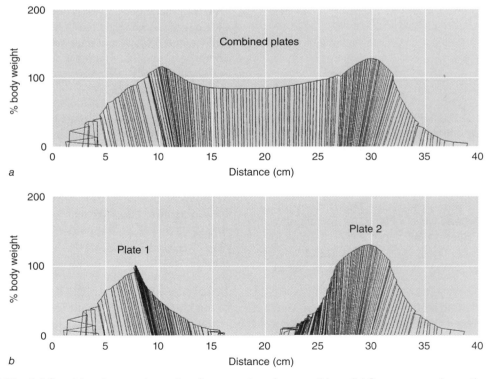

▲ **Figure 4.17** *(a)* Combined ground reaction force vectors from walking. *(b)* Same ground reaction forces from force platforms under the rear foot and forefoot. Line segments indicate direction and relative magnitudes of GRFs. The lower ends of the segments correspond to the location of the center of pressure in the A/P direction.

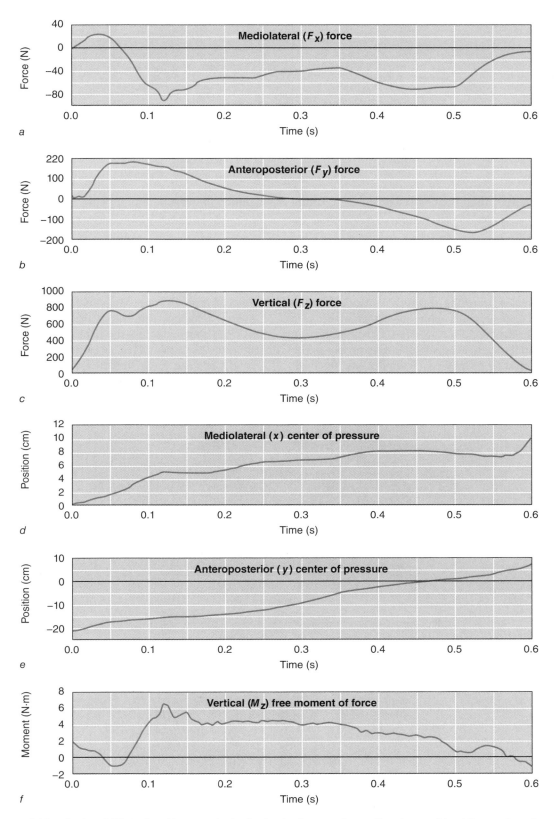

▲ **Figure 4.18** *(a, b, c)* Histories (time series) of a typical ground reaction force, *(d, e)* its center of pressure location, and *(f)* its vertical (free) moment of force during walking.

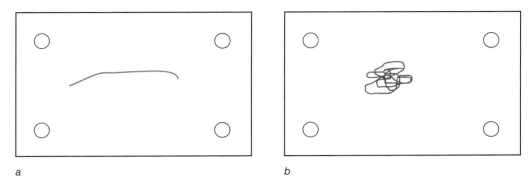

a   *b*

▲**Figure 4.19**   Trajectory plot of center of pressure paths during *(a)* walking and *(b)* standing.

(Cavanagh 1978). Figure 4.20 shows force signatures for walking and running. A typical walking signature has two humps and often a spike immediately after initial foot contact. This spike has been attributed to the passive characteristics of the subject's footwear and the anatomical characteristics of the subject's foot, because its characteristics exhibit too rapid a change to have been caused by the relatively slow-acting muscles. The double hump disappears during running, and the peak vertical force increases to approximately three times the body weight. During walking, the peak vertical forces fluctuate around the body-weight line by approximately 30% of body weight.

The force signature is an especially useful tool if an inverse dynamics study (chapters 5 and 7) is conducted. Inverse dynamics determines the internal forces and moments that act across the joints in response to such external forces as GRFs. By displaying the motion pattern and the force signature on the same display, the researcher can ensure that the two data-collection

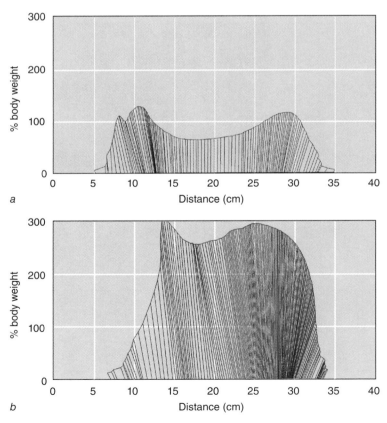

a

b

▲**Figure 4.20**   Force signature of *(a)* walking and *(b)* running. Note the initial spike at the beginning of each signature and notice that the walking trial has two major peaks (excluding the initial spike).

systems (motion and force) and their reference axes are synchronized temporally and spatially. Figure 4.21 shows four different sagittal views of the force signatures for walking from left to right. Only tracing *a* has the correct shape of the force signatures. In the other three, the directions of the horizontal GRF, the COP, or both are reversed. By walking across a force platform and examining the resulting force signature, the researcher can check that the axes of the force platform match the axes of the motion-collection system. Displaying stick figures of the subject's motion along with the synchronized force signature can also be used as a validation check. The force signature's vectors should appear to be synchronized in time and space beneath the subject's foot. Figure 4.22 illustrates correct and incorrect spatial synchronization of the forces during walking.

Another way of combining GRFs is to average a series of signals from the same subject or across several subjects. Averaging the forces is not difficult, although each force time series is of a different duration. By standardizing the time base to percentages of the footfall duration, often called the *stance phase*, we can average the various trials (see chapter 1 for details on ensemble averaging). Of course, the resulting averages are no longer time based; they are instead percentages of the stance phase.

Pressure Distribution Sensors

We have discussed force as a vector quantity characterized by magnitude, direction, and point of application. This viewpoint is a mathematical construct useful for the description and application of Newton's laws of motion. Usually, a force is distributed over an area of contact rather than concentrated at one specific point of application. For example, a GRF vector has a COP point location, but in reality an ever-changing area of the shoe's sole is in contact with the ground at any given time during a stance. The force vector is in fact distributed over this contact area, and its distribution can be analyzed using the concept of **pressure**, defined as the force per unit area and expressed in newtons per square meter (N/m^2), also called pascals (Pa, although kilopascals [kPa] are more common). This can be conceptualized as an array of force vectors distributed evenly across the area of contact, each one applied to a unit surface area (e.g., 1 mm^2). Some forces within this array are larger than others; the overall pattern of these force vectors constitutes the force distribution across the contact area. The summation of these distributed forces equals the magnitude of the overall force vector measured with a force plate.

Measurement of pressure distribution over an area is possible with the use of *capacitance* or *conductance*

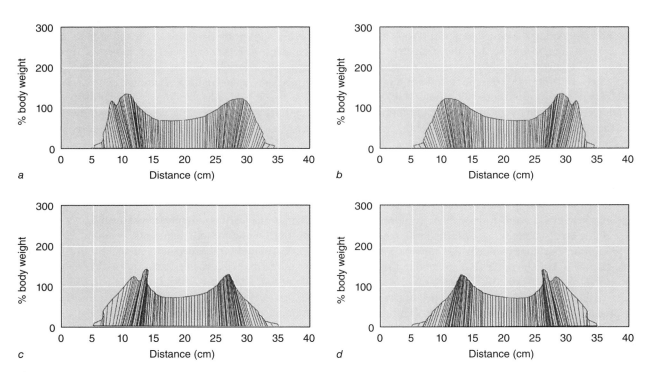

▲ Figure 4.21 Four different representations of a force signature for different combinations of directions of the horizontal force and center of pressure. The correct pattern is *(a)*; both the horizontal forces and centers of pressure are reversed in *(b)*; only the horizontal forces are reversed in *(c)*; and in *(d)* only the horizontal centers of pressure are reversed.

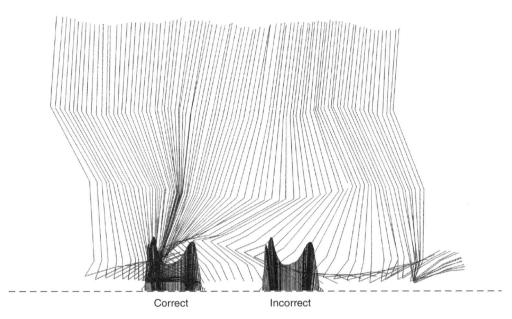

▲ **Figure 4.22** Correct and incorrect spatial synchronization of ground reaction force vectors with a stick figure representation of the lower-extremity and trunk motion patterns.

sensors. These sensors consist of multilayered materials that can constitute part of an electronic circuit, in much the same manner as was described earlier for force transducers. Capacitance sensors consist of two electrically conductive sheets separated by a thin layer of a nonconducting dielectric material. When a normal force is applied to the sensor, the dielectric is compressed, reducing the distance between the two conducting sheets and changing the magnitude of a measured electric charge. Appropriate calibration establishes the relationship between the magnitudes of the applied force and the subsequent measured current. Conductance sensors are similar in construction, but a conductive rather than dielectric material separates the two conductive sheets. The separating material has electrical properties different from those of the two sheets and offers some resistance to the flow of current between the two sheets. Application of a normal force causes compression of the middle layer, changing its electrical resistance. This alteration in resistance can be used to invoke a measurable change in voltage that will be related to the applied force magnitude. Appendix C gives more details on electronic circuits. Note that both types of sensors are responsive to normal forces, and therefore horizontal shear forces cannot be detected.

Capacitance and conductance materials are manufactured so that independent cells of equal area are formed, and the circuit is designed to measure the pressure within each cell. The resolution of the pressure distribution is dictated by the size of the individual cells, typically on the order of 0.5 cm². Sheets of these materials can be formed into pressure mats or insoles that are placed inside shoes. Approximately 400 individual pressure cells may line the bottom of the foot. To extract the data from such a large number of cells, individual wires would be impractical. Therefore, the insoles are constructed like flexible circuit boards by using thin conductive strips within the insole to carry the signals to a small connecting box worn near the subject's ankle. Other systems use individual *piezoceramic* sensors that are glued onto specific sites of interest (e.g., directly under the lateral heel or head of the first metatarsal during gait analysis of the foot). Figure 4.24 illustrates pressure-distribution patterns measured with pressure insoles that were taken from two specific points within the stance phase of a runner.

Internal Force Measurement

In many cases, it is of interest to know the internal forces acting on individual ligaments, tendons, and joints. Unfortunately, measurement of these internal forces requires highly invasive techniques, although they are commonly used in animal studies (e.g., Herzog and Leonard 1991) and have been used in a limited number of human studies (e.g., Gregor et al. 1991). Although it is doubtful that these invasive protocols will ever be widely adopted for human research, a few options are mentioned here. A less-invasive technique is to use musculoskeletal modeling to estimate these forces during human movements (see chapters 9 and 11).

The most common technique for measuring force in tendons (or ligaments) is the buckle transducer, which consists of a small rectangular frame that mounts on the

Caldwell, G.E., L. Li, S.D. McCole, and J.M. Hagberg. 1998. Pedal and crank kinetics in uphill cycling. *Journal of Applied Biomechanics* 14:245-59.

Research on the biomechanics and neural control of cycling has benefited greatly from the advent of instrumented pedals that measure the force vector applied during various cycling conditions. This study used a pedal based on piezoelectric crystal technology (Broker and Gregor 1990) to examine pedal and crank kinetics during both seated and standing uphill cycling. These authors measured both **normal forces** and **tangential forces** acting on the surface of the pedal as elite cyclists climbed a simulated 8% grade in a laboratory. A unique aspect of measuring pedal forces is that the transducer orientation changes as the subject manipulates his foot position, unlike a force platform rigidly embedded in a laboratory floor. By measuring the angle of the pedal with kinematic markers, the normal and tangential forces were

transformed into a GCS (i.e., vertical and horizontal force components). The forces were further transformed into components that acted perpendicularly and parallel to the crank arm as it rotated through a complete 360° revolution. The perpendicular force components were considered effective because they contributed to crank torque ($T = Fd$, where d represents the length from the pedal to the crank axle) to propel the bicycle. The magnitude and pattern of crank torque were altered substantially when the subjects changed from a seated to a standing posture (figure 4.23). This was attributed to more effective use of gravitational force in parts of the crank cycle associated with upward and forward movement of the subject's center of mass as he rose from the saddle.

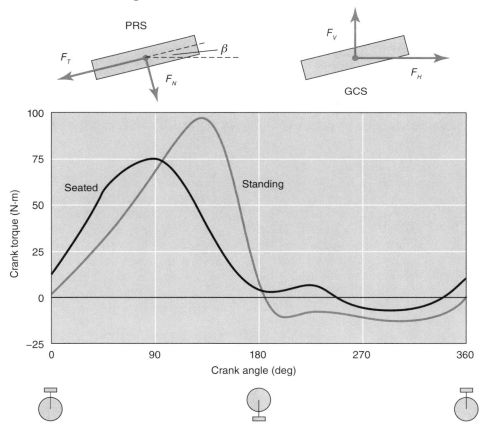

▲ **Figure 4.23** The top panel illustrates that instrumented pedals measure forces relative to a pedal reference system. In a sagittal view, the forces are measured in a manner tangential (F_T) and normal (F_N) to the pedal surface. By accounting for the pedal angle, β, the investigator can transform these forces into the GCS for registration of vertical (F_V) and horizontal (F_H) force components. The pedal forces can be used to calculate the torque exerted about the crank, which is the propelling force for the bicycle. The crank torque pattern over one complete crank arm revolution (360°) is shown in the bottom panel for a subject climbing a hill while seated and while standing.

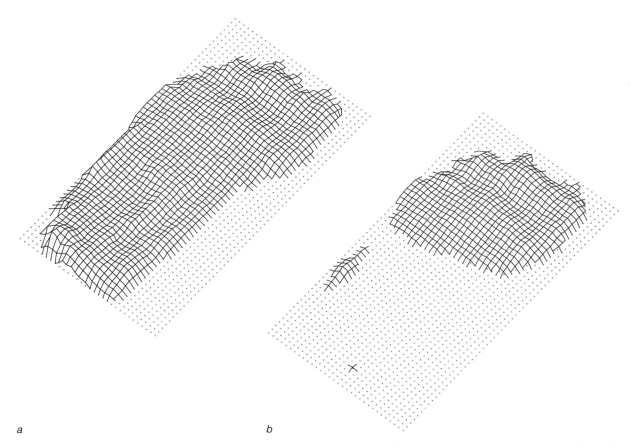

▲**Figure 4.24**   Two in-shoe pressure patterns from the left foot at different times in the stance phase of running.

tendon and a transverse steel beam that rests against the tendon (figure 4.25). Strain gauges change their electrical resistance when the tendon transmits force and deforms the transverse beam. For animal preparations, the transducer can be surgically inserted and the wound allowed to heal, with the thin electrical wires used to carry the signals through the skin to associated external circuitry still exposed. A similar approach has been used in the few human studies, although little healing time was permitted; the transducer insertion, experimental data collection, and transducer removal were performed all in 1 day (Gregor et al. 1991). Other transducer designs include foil strain gauges or liquid metal gauge transducers, which are small tubes filled with a conductive liquid metal that can be inserted within a ligament. Axial strain

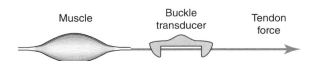

▲**Figure 4.25**   Buckle-style muscle force transducer. Note that the muscle and transducer are not drawn to scale; muscle is much larger.

in the ligament causes the tube to elongate and become thinner, thereby altering the electrical resistance afforded by the liquid metal (Lamontagne et al. 1985). Another device is the **Hall-effect transducer**, which consists of a small, magnetized rod and a tube containing a Hall generator. As a force is imposed on a tendon, the rod moves with respect to the tube, causing the Hall generator to produce a voltage. Finally, the use of fiber-optic transducers has enabled measurement of forces in large, superficial tendons (Komi et al. 1996).

Isovelocity Dynamometers

One of the most commonly used strength-testing devices is the *isokinetic dynamometer*, which tests the muscular strength capabilities of patients and athletes in different stages of recovery, training, and rehabilitation. These machines were designed to measure the torque a subject produces at a single joint under such controlled kinematic conditions as **isometric** (a fixed joint angular position), **isotonic** (constant force or moment of force), and **isovelocity** (a predetermined constant angular velocity). In general, the torque produced during these muscular efforts changes continuously, although the term *isokinetic* means "same force" or "same torque." The term *isovelocity* ("same velocity") is therefore preferred (Bobbert and van Ingen Schenau 1990; Chapman 1985). The dynamic isovelocity conditions are further categorized into **concentric**, when the muscles crossing the joint shorten, and **eccentric**, when the muscles lengthen.

Although there are several commercial dynamometers, some research laboratories have chosen to build their own to give them better control over the machine's specifications (e.g., its maximal rotational velocity). In either case, the machines permit a subject to apply torque to the dynamometer using a single, isolated joint. The subject is usually seated with the more distal segment of the joint attached to a solid metal bar that rotates about a fixed axis of rotation (figure 4.26). The subject either grasps the bar by its handle or applies force to a padded plate attached to it. A series of straps firmly affix the distal segment to the padded plate and the proximal segment and trunk to the seat. This prevents extraneous movement and ensures that only the muscles surrounding the tested joint generate the force applied to the bar. It is important to position the subject with the joint and machine arm axes of rotation aligned, because misalignment can cause errors in the measurement of the joint torque (Herzog 1988). In some cases, the torque is measured with a torque transducer element built into the physical axis of rotation of the bar. In other designs, a force sensor is placed between the bar and the padded plate, and torque is calculated from the recorded force

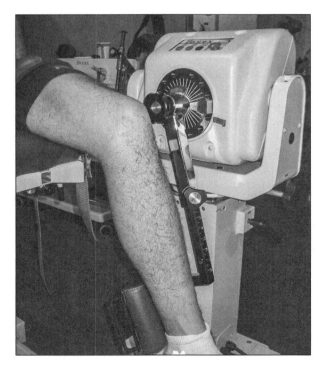

▲ **Figure 4.26** Subject in an isokinetic dynamometer.

and the perpendicular distance from the plate to the axis of rotation. A potentiometer built into the axis of rotation measures the angle, and in some machines a tachometer independently measures angular velocity.

Within the research community, dynamometers are used most often to study the *torque-angle (T-θ)* and *torque–angular velocity (T-ω)* relationships for individual joints. The isometric *T-θ* relation is generated from a series of muscular efforts performed with the dynamometer bar fixed (i.e., at velocity equal to zero) in various angular positions. Beginning from rest in each position, the subject is encouraged to produce maximal joint torque against the padded plate of the machine bar so that the applied torque will increase from zero to a more or less steady plateau level (figure 4.27*a*). Averaging a section of the torque plateau generates a single point in the isometric *T-θ* relation (i.e., the maximal isometric torque value at that specific angle). Repeating this procedure at different angular positions generates a series of points that define the joint's maximal isometric capabilities across the range of motion (figure 4.27*b*). The reasons that the maximal isometric torque varies with joint angle include the muscular force-length relationship (see chapter 9), the change in muscle moment arm with angle, and the ability to completely activate muscles in all joint positions. Also, passive structures such as ligaments and bony contact points may produce torque at the extremes of joint motion, adding to the active contributions of the muscles themselves.

A similar protocol is used to determine the dynamic torque–angular velocity relationship at a joint. In this case, the subject performs a series of maximal muscular efforts while the dynamometer bar is constrained to move at or below a predetermined angular velocity (e.g., 30°/s). Beginning at a set angle from rest, the subject exerts maximal effort against the bar. The applied torque causes the bar to rotate, and if the subject exerts enough torque, the bar attains the predetermined velocity. The torque profile here (figure 4.28*a*) is quite different from in the isometric case, because it rises to a peak value and then falls. Recording the peak torque and the angular velocity gives one point for the *T-ω* relation, and the complete joint *T-ω* capability can be ascertained by performing repeated trials across a wide range of angular velocities (figure 4.28*b*). As with the isometric *T-θ* relation, several reasons can account for the variation in torque potential with angular velocity, the most important of which is the force-velocity characteristics of the muscles crossing the tested joint (see chapter 9).

A number of issues must be considered when collecting data with an isovelocity dynamometer (see Herzog 1988). If the movement takes place within a sagittal plane, the distal segment and dynamometer arm will change their orientations with respect to gravity as the movement unfolds. The body segment and dynamometer arm both

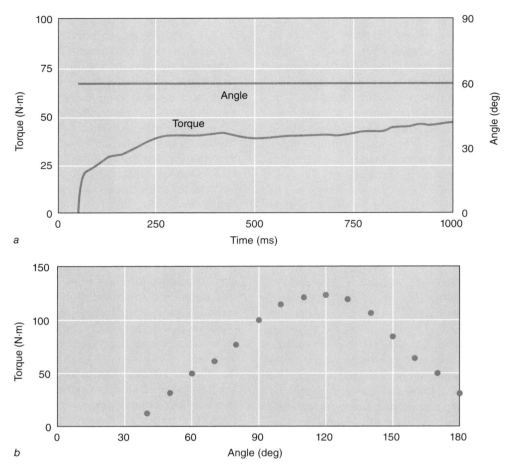

▲ Figure 4.27 *(a)* A typical isometric maximal effort contraction at a fixed angular position (*θ* = 60°). *(b)* A typical *T-θ* relation for the extensor muscles of the knee joint. Each data point *(T)* was taken from a maximal effort isometric contraction at a fixed angular position *(θ)*.

have a finite weight and will affect the measured joint torque (Winter et al. 1981). Most commercial dynamometers include a gravity correction option for handling this situation. Another consideration is the speed at which the dynamometer bar is moving. Although moving the segment and bar at a constant angular velocity is the goal, the movement by necessity begins and ends with a velocity of zero. This means that there are periods of acceleration and deceleration within each experimental trial and only a finite length of time and movement at the desired constant velocity. Acceleration periods can produce undesirable inertial loads and an overshoot phenomenon. Furthermore, because the bar is constrained to move at or below a set velocity, in some trials the subject may be unable to achieve the desired target velocity. The experimental data records should be examined carefully to ensure that the conditions under which the torque is attained are correct for the research questions being posed.

A second group of concerns involves the meaning of the data extracted from the experimental records. The

T-θ and *T-ω* relationships clearly illustrate that torque potential varies isometrically as a function of angle and dynamically as a function of velocity. Although these relationships are found using two different test protocols (i.e., one static and one dynamic), it has been noted that underlying muscular properties are "in effect" during all experimental trials. The torque-angle relationship should be taken into account when assessing the dynamic torque-velocity relation to ensure that variation in peak torque at different velocities is not partly the result of peaks occurring at different angular positions. Indeed, in plantar flexion, peak torque occurs at progressively greater angles as concentric velocity increases (Bobbert and van Ingen Schenau 1990). One method used to prevent this problem is to measure the torque at one specific angle, regardless of whether a peak occurred in the torque time record (e.g., Froese and Houston 1985). Although this seems sensible, it is flawed because of the passive series elasticity that resides in muscle (Hill 1938). This series elastic component changes length if the torque either rises or falls when the

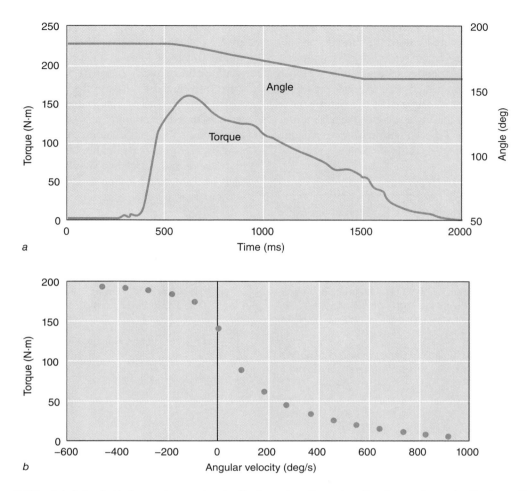

▲**Figure 4.28** *(a)* A typical dynamic maximal effort contraction at a fixed angular velocity ($\omega = 60°/s$). *(b)* A typical dynamic T-ω relation for the knee extensors. Each data point *(T)* represents the peak torque from a maximal effort isovelocity contraction at a fixed angular velocity (ω).

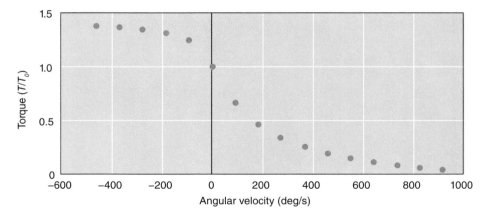

▲**Figure 4.29** Standardized dynamic T-ω relation for the knee extensors (from figure 4.28). Each data point *(T)* represents the peak torque from a maximal effort isovelocity contraction at a fixed angular velocity *(ω)*. However, to account for the occurrence of each peak torque at a different angular position, each "raw" torque has been divided by the isometric maximum *(T_0)* for the angle at which the peak occurred.

angle-specific measurement is taken, meaning that the force-generating structures within the muscles are not at the same velocity as registered by the dynamometer (Caldwell et al. 1993). A better approach is to establish the isometric T-θ relation and then express each peak torque found during the dynamic trials as a percentage of the isometric value for the angle at which the peak torque occurred. Such a scaled T-ω relationship is shown in figure 4.29 for the "raw" T-ω relationship shown in figure 4.28.

SUMMARY

In this chapter, we introduced the conceptual basics for understanding Newton's laws of motion and how these laws can be used to study human motion. Of importance are the concepts of force applied to particles and forces and moments of force applied to rigid bodies and the resulting linear and angular effects. Free-body diagrams were introduced as a way of visualizing systems in which multiple forces act. The measurement of forces was discussed, including the types of force transducers most commonly used in biomechanics. Readers should now be able to understand how forces are measured in biomechanics research and should be able to undertake their own force and torque measurements if the proper equipment is available.

SUGGESTED READINGS

Beer, F.P., E.R. Johnston Jr., D.F. Mazurek, P.J. Cornwell, and E.R. Eisenberg. 2010. *Vector Mechanics for Engineers; Statics and Dynamics*. 9th ed. Toronto: McGraw-Hill.

Hamill, J., and K.M. Knutzen. 2009. *Biomechanical Basis of Human Movement*. 3rd ed. Baltimore: Williams & Wilkins.

Nigg, B.M., and W. Herzog. 2007. *Biomechanics of the Musculo-Skeletal System*. 3rd ed. Toronto: Wiley.

Zatsiorsky, V.M. 2002. *Kinetics of Human Motion*. Champaign, IL: Human Kinetics.

Chapter 5

Two-Dimensional Inverse Dynamics

Saunders N. Whittlesey and D. Gordon E. Robertson

Inverse dynamics is the specialized branch of mechanics that bridges the areas of kinematics and kinetics. It is the process by which forces and moments of force are indirectly determined from the kinematics and inertial properties of moving bodies. **Direct dynamics**, in contrast, determines the motion of bodies under the influence of applied forces. In principle, inverse dynamics also applies to stationary bodies, but usually it is applied to bodies in motion. It derives from Newton's second law, where the resultant force is partitioned into known and unknown forces. The unknown forces are combined to form a single **net force** that can then be solved. A similar process is done for the moments of force so that a single **net moment of force** is computed. This chapter

- defines the process of inverse dynamics for planar motion analysis,

- presents the standard method for numerically computing the internal kinetics of planar human movements,

- describes the concept of general plane motion,

- outlines the method of sections for individually analyzing components of a system or segments of a human body,

- outlines how inverse dynamics aids research of joint mechanics, and

- examines applications of inverse dynamics in biomechanics research.

Inverse dynamics of human movement dates to the seminal work of Wilhelm Braune and Otto Fischer between 1895 and 1904. These works were later revisited by Herbert Elftman for his research on walking (1939a, 1939b) and running (1940). Little follow-up research was conducted until Bresler and Frankel (1950) conducted further studies of gait in three dimensions (3-D) and Bresler and Berry (1951) expanded the approach to include the powers produced by the ankle, knee, and hip moments during normal, level walking. Because Bresler and Frankel's 3-D approach measured the moments of force against a Newtonian or absolute frame of reference, they could not determine the contributions made by the flexors or extensors of a joint versus the abductors and adductors (a problem since solved by the 3-D methods described in chapter 7).

Few inverse dynamics studies of human motion were conducted until the 1970s, when new research was spurred by the advent of commercial force platforms to measure the ground reaction forces (GRFs) during gait and inexpensive computers to provide the necessary processing power. Another important development has been the recent propagation of automated and semiautomated motion-analysis systems based on video or infrared camera technologies, which greatly decrease the time required to process the motion data.

Inverse dynamics studies have since been carried out on such diverse movements as lifting (McGill and Norman 1985), skating (Koning et al. 1991), jogging (Winter 1983a), race walking (White and Winter 1985), sprinting (Lemaire and Robertson 1989), jumping (Stefanyshyn and Nigg 1998), rowing (Robertson and Fortin 1994; Smith 1996), and kicking (Robertson and Mosher 1985), to name a few. Yet inverse dynamics has not been applied to many fundamental movements: swimming and skiing, because of the unknown external forces of water and snow; or batting, puck shooting, and golfing, because of the indeterminacy caused when the two arms and the implement (the bat, stick, or club) form a closed kinematic chain. Future research may be able to overcome these difficulties.

PLANAR MOTION ANALYSIS

One of the primary goals of biomechanics research is to quantify the patterns of force produced by the muscles, ligaments, and bones. Unfortunately, recording these forces directly (a process called **dynamometry**) requires invasive and potentially hazardous instruments that inevitably disturb the observed motion. Some technologies that measure internal forces include a surgical staple for forces in bones (Rolf et al. 1997) and mercury strain gauges (Brown et al. 1986; Lamontagne et al. 1985) or buckle force transducers (Komi 1990) for forces in muscle tendons or ligaments. Although these

devices enable the direct measurement of internal forces, they have been used only to measure forces in single tissues and are not suitable for analyzing the complex interaction of muscle contractions across several joints simultaneously. Figure 5.1 (Seireg and Arvikar 1975) shows the complexity of forces that a biomechanist must consider when trying to analyze the mechanics of the lower extremity. In figure 5.2, the lines of action of only the major muscles of the lower extremity have been graphically presented by Pierrynowski (1982). It is easy to imagine the difficulty of and the risks associated with attempting to attach a gauge to each of these tendons.

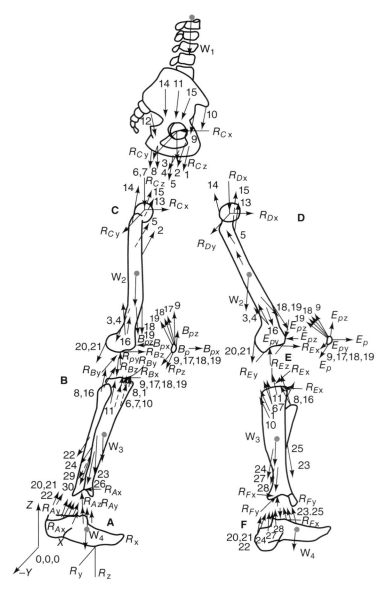

▲ **Figure 5.1** Free-body diagrams of the segments of the lower extremity during walking.

Reprinted from *Journal of Biomechanics,* Vol. 18, A. Seireg and R.J. Arvikar, "The prediction of muscular load sharing and joint forces in the lower extremities during walking," pgs. 89-102, copyright, with permission of Elsevier.

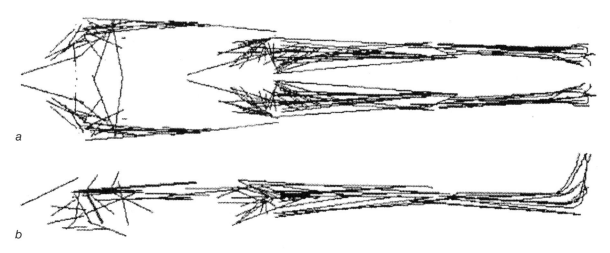

▲ Figure 5.2 Lines of action of the muscle forces in the lower extremity and trunk: *(a)* front view, *(b)* side view.

Adapted from data, by permission, from M.R. Pierrynowski, 1982, *A physiological model for the solution of individual muscle forces during normal human walking* (Simon Fraser University).

Inverse dynamics, although incapable of quantifying the forces in specific anatomical structures, is able to measure the *net* effect of all of the internal forces and moments of force acting across several joints. In this way, a researcher can infer what total forces and moments are necessary to create the motion and quantify both the internal and the external work done at each joint. The steps set out next clarify the process for reducing complex anatomical structures to a solvable series of equations that indirectly quantify the kinetics of human or animal movements.

Figure 5.3 shows the space and free-body diagrams of one lower extremity during the push-off phase of running. Three equations of motion can be written for each segment in a two-dimensional (2-D) analysis, so for the foot, three unknowns can be solved. Unfortunately, because there are many more than three unknowns (figure 5.4), the situation is called *indeterminate*. Indeterminacy occurs when there are more unknowns than there are independent equations. To reduce the number of unknowns, each force can be resolved to its equivalent force and moment of force at the segment's endpoint. The process starts at a terminal segment, such as the foot or hand, where the forces at one end of the segment are known or zero. They are zero when the segment is not in contact with the environment or another object. For example, the foot during the swing phase of gait experiences no forces at its distal end; when it contacts the ground, however, the GRF must be measured by, for example, a force platform.

A detailed free-body diagram (FBD) of the foot in contact with the ground is illustrated in figure 5.4. Notice the many types of forces crossing the ankle joint, including muscle and ligament forces and bone-on-bone forces; many others have been left out (e.g., forces from

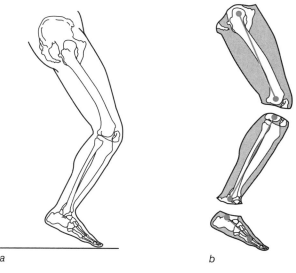

▲ Figure 5.3 *(a)* Space and *(b)* free-body diagrams of the foot during the push-off phase of running.

skin, bursa, and joint capsule). Furthermore, the foot is assumed to be a "rigid body," although some researchers have modeled it as having two segments (Cronin and Robertson 2000; Stefanyshyn and Nigg 1998). A rigid body is an object that has no moving parts and cannot be deformed. This state implies that its inertial properties are fixed values (i.e., that its mass, center of gravity, and mass distribution are constant).

Figure 5.5 shows how to replace a single muscle force with an equivalent force and moment of force about a common axis. In this example, the muscle force exerted by the tibialis anterior muscle on the foot segment is replaced by an equivalent force and moment of force at the ankle center of rotation. Assuming that the foot

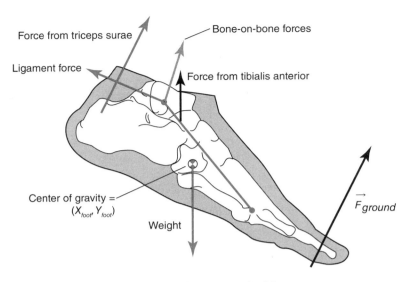

▲ **Figure 5.4**   Free-body diagram of the foot showing anatomical forces.

is a "rigid body," a force ($\vec{F}^*$) equal in magnitude and direction to the muscle force ($\vec{F}$) is placed at the ankle. Because this would unbalance the free body, a second force ($-\vec{F}^*$) is added to maintain equilibrium (figure 5.5b). Next, the force couple ($\vec{F}$ and $-\vec{F}^*$) is replaced by the moment of force ($M_F\vec{k}$). The resulting force and moment of force in figure 5.5c have the same mechanical effects as the single muscle force in figure 5.5a, assuming that the foot is a rigid body.

The first step to simplifying the complex situation shown in figure 5.4 is to use the process illustrated in figure 5.5 to replace every force that acts across the ankle with its equivalent force and moment of force about a common axis. Figure 5.6 shows this situation. Note that forces with lines of action that pass through the ankle

joint center produce no moment of force around the joint. Thus, the major structures that contribute to the net moments of force are the muscle forces. The ligament and bone-on-bone forces contribute mainly to the net force experienced by the ankle and only affect the ankle moment of force when the ankle is at the ends of its range of motion.

Muscles attach in such a way that their turning effects about a joint are enhanced, and most have third-class leverage to promote speed of movement. Thus, muscles rarely attach so that they cross directly over a joint axis of rotation because that would eliminate their ability to create a moment about the joint. Ligaments, in contrast, often cross joint axes, because their primary role is to hold joints together rather than to create rotations of the

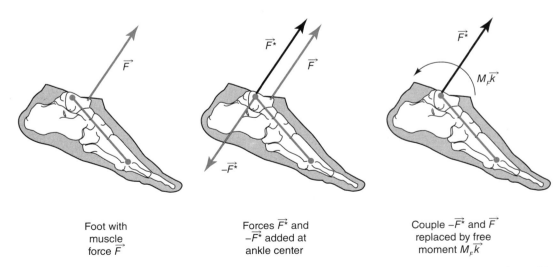

| Foot with muscle force $\vec{F}$ | Forces $\vec{F}^*$ and $-\vec{F}^*$ added at ankle center | Couple $-\vec{F}^*$ and $\vec{F}$ replaced by free moment $M_F\vec{k}$ |

▲ **Figure 5.5**   Replacement of a muscle force by its equivalent force and moment of force at the ankle axis of rotation.

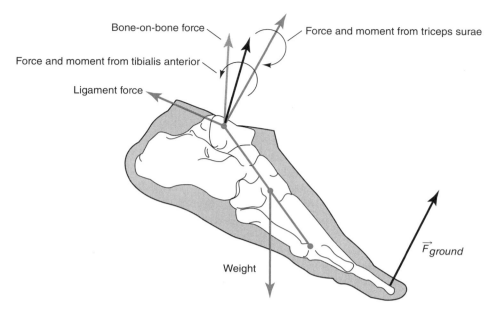

▲ **Figure 5.6** Free-body diagram of the foot showing the muscle forces replaced by their equivalent force and moment about the ankle.

segments that they connect. They do, however, have a role in producing moments of force when the joint nears or reaches its range-of-motion limits. For example, at the knee, the collateral ligaments prevent varus and valgus rotations and the cruciate ligaments restrict hyperextension. Often, the ligaments and bony prominences produce force couples that prevent excessive rotation, such as when the olecranon process and the ligaments of the elbow prevent hyperextension of the elbow.

To complete the inverse dynamics process for the foot, every anatomical force, including ligament and bone-on-bone (actually, cartilaginous) forces, must be transferred to the common axis at the ankle. Note that

only forces that act across the ankle are included in this process. Internal forces that originate and terminate within the foot are excluded, as are external forces in contact with the sole of the foot. Figure 5.7 represents the situation after all of the ankle forces have been resolved. In this figure, the ankle forces and moments of force are summed to produce a single force and a moment of force, called the *net force* and *net moment of force*, respectively. They are also sometimes called the *joint force* and *joint moment of force*, but these are confusing because many different joint forces are included in this sum, such as those caused by the joint capsule, the ligaments, and the articular surfaces (cartilage). Other confusing terms

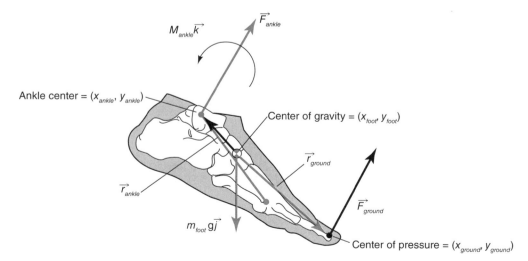

▲ **Figure 5.7** Reduced free-body diagram showing net force and moment of force.

are *resultant joint force* and *resultant moment of force*, because these terms may be confused with the resultant force and moment of force of the foot segment itself. Recall that the resultant force and moment of force of a rigid body are the sums of all forces and moments acting on the body. These sums are *not* the same as the net force and moment of force just defined. The resultant force and moment of force concern Newton's first and second laws.

The moment of force is often called *torque* in the scientific literature. In engineering, torque is usually considered a moment of force that causes rotation about the longitudinal axis of an object. For example, a torque wrench measures the axial moment of force when one is tightening nuts or bolts, and a torque motor generates spin about an engine's spin axis. In the biomechanics literature, however, as stated in chapter 4, torque and moment of force are used interchangeably.

Another term related to moment of force is the *force couple*. A force couple occurs when two parallel, noncollinear forces of equal magnitude but opposite direction act on a body. The effect of a force couple is special because the forces, being equal but opposite in direction, effect no translation on the body when they act. They do, however, attempt to produce a pure rotation, or torque, of the body. For example, a wrench (figure 5.8*a*) causes two parallel forces when applied to the head of a nut. The nut translates because of the threads of the screw but turns around the bolt because of the rotational forces (i.e., moment of force, force couple, or torque) created by the wrench.

Another interesting characteristic of a force couple, or *couple* for short, is that when the couple is applied to a rigid body, the effect of the couple is independent of its point of application. This makes it a *free moment*, which means that the body experiencing the couple will react in the same way wherever the couple is applied as long

as the lines of axis of the force are parallel. For example, a piece of wood that is being drilled will react the same no matter where the drill contacts the wood as long as the drill bit enters the wood from the same parallel direction. Of course, how the wood actually reacts will depend on friction, clamping, and other forces, but the drill will inflict the same rotational motion on the wood no matter where it enters.

The work done by the net moments of force quantifies the mechanical work done by only the various tissues that act across and contribute a turning effect at a particular joint. All other forces, including gravity, are excluded from contributing to the net force and moment of force. More details about how the work of the moment of force is calculated are delineated in chapter 6.

Net forces and moments are not real entities; they are mathematical concepts and therefore can never be measured directly. They do, however, represent the summed or net effect of all the structures that produce forces or moments of force across a joint. Some researchers (e.g., Miller and Nelson 1973) have called the source of the net moment of force a "single equivalent muscle." They contend that each joint can be thought of as having two single, equivalent muscles that produce the net moments of force about each joint—for example, one for flexion and the other for extension—depending on the joint's anatomy. Others have called the net moments of force "muscle moments," but this nomenclature should be avoided, because even though muscles are the main contributors to the net moment, other structures also contribute, especially at the ends of the range of motion. An illustration of this situation is when the knee reaches maximal flexion during the swing phase of sprinting. Lemaire and Robertson (1989) and others showed that although a very large moment of force occurs, the likely cause is not an eccentric contraction of the extensors;

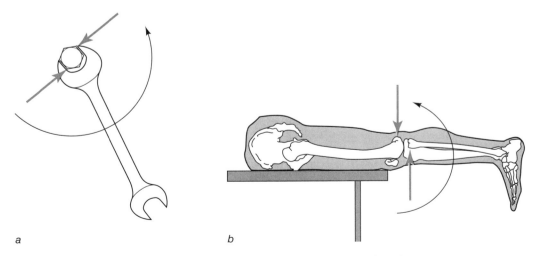

a　　　　　　　　　　　　　　　*b*

▲ **Figure 5.8**　Force couples produced by a wrench and the ligaments of the knee.

instead it is the result of the calf and thigh bumping together. However, the same cannot be said to occur for the negative work done by the knee extensors during the swing phase of walking, because the joint does not fully flex and therefore muscles must be recruited to limit knee flexion (Winter and Robertson 1979).

NUMERICAL FORMULATION

This section presents the standard method in biomechanics for numerically computing the internal kinetics of planar human movements. In this process, we use body kinematics and anthropometric parameters to calculate the net forces and moments at the joints. This process uses three important principles: Newton's second law ($\Sigma \vec{F} = m\vec{a}$), the principle of *superposition*, and an engineering technique known as the *method of sections*. The principle of superposition holds that in a system with multiple factors (i.e., forces and moments), given certain conditions, we can either sum the effects of multiple factors or treat them independently. In the method of sections, the basic idea is to imagine cutting a mechanical system into components and determining the interactions between them. For example, we usually section the human lower extremity into a thigh, leg, and foot. Then, via Newton's second law, we can determine the forces acting at the joints by using measured values for the GRFs and the acceleration and mass of each segment. This process, called the *linked-segment* or *iterative Newton-Euler method*, is diagrammed in figure 5.9. The majority of this chapter explains how this method works. We will begin with kinetic analysis of single objects in 2-D, then demonstrate how to analyze the kinetics of a joint via the method of sections, and finally explain the general procedure diagrammed in figure 5.9 for the entire lower extremity.

Note the diagram conventions used in this chapter: Linear parameters are drawn with straight arrows, and angular parameters are drawn with curved arrows. Known kinematic data (linear and angular accelerations) are drawn with green lines. Known forces and moments are drawn with black arrows. Unknown forces and moments are drawn with gray arrows. These conventions will assist you in visualizing the solution process.

General Plane Motion

General plane motion is an engineering term for 2-D movement. In this case, an object has three degrees of freedom (DOF): two linear positions and an angular position. Often, we draw these as translations along the *x*- and *y*-axes and a rotation about the *z*-axis. As discussed in chapter 2, many lower-extremity movements can be analyzed using this simplified representation, including

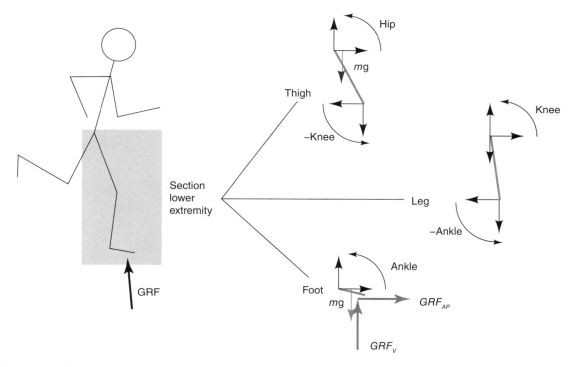

▲ **Figure 5.9** Space diagram of a runner's lower extremity in the stance phase, with three free-body diagrams of the segments.

walking, running, cycling, rowing, jumping, kicking, and lifting. However, despite the simplification to 2-D analysis, the resulting mechanics can still be complicated. For example, a football punter has three lower-extremity segments that swing forward much like a whip, kick the ball, and elevate. Even the ball has a somewhat complicated movement, translating both horizontally and vertically and rotating. To determine the kinetics of such situations, we treat the three DOF independently. That is, we exploit the fact that an object accelerates in the vertical direction only when acted on by a vertical force and accelerates in the horizontal direction only when acted on by a horizontal force. Similarly, the body does not rotate unless a moment (torque) is applied to it. The principle of superposition states that when one or more of these actions occur, we can analyze them separately. We therefore separate all forces and moments into three coordinates and solve them separately.

To illustrate, let us consider the football example. In figure 5.10*d*, a football being kicked is subjected to the force of the punter's foot. The ball moves horizontally and vertically and also rotates. Our goal is to determine the force with which the ball was kicked. We cannot measure the force directly with an instrumented ball

or shoe. However, we can film the ball's movement and measure its mass and moment of inertia. Given these data, we are left with an apparently confusing situation to analyze: The single force of the foot has caused all three coordinates to change. However, the situation becomes simpler when we use superposition. The horizontal and vertical accelerations of the ball's mass center must be proportional to their respective forces, and the angular acceleration must be proportional to the applied moment. Consider the examples in figure 5.10. Figure 5.10*a* and *b* are rather obvious but are presented for the sake of demonstration. In figure 5.10*a*, a horizontal force is applied to the ball through its center of mass, and the ball will accelerate horizontally; it will not accelerate vertically because there is no vertical force. Similarly, in figure 5.10*b*, the ball will accelerate only vertically because there is no horizontal force. In figure 5.10*c*, the force acts at a 45° angle, and therefore the ball accelerates at a 45° angle. This is just a superposition of the situations in figure 5.10*a* and *b*. We do not deal with the force at this angle; rather, we measure the accelerations in the horizontal and vertical directions, and therefore we can determine the forces in the horizontal and vertical directions. In Figure 5.10*d*, the applied force is not

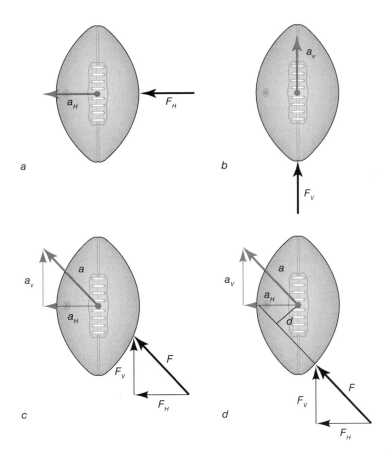

▲ **Figure 5.10** Four free-body diagrams of a football experiencing four different external forces.

directed through the center of mass. In this case, the force is the same as it is in figure 5.10c, so the ball's center of mass has the same acceleration. However, there is also an angular acceleration proportional to the product of the force, F, and the distance, d, between its line of action and the center of mass. The acceleration, a, in this case is the same as in figure 5.10c. However, the ball will also rotate.

To reiterate, a force causes a body's center of mass to accelerate in the same direction as that force. A force does not cause a body to rotate; only the moment of a force causes a body to rotate. These principles derive from Newton's first and second laws. If a ball is kicked, a resulting force is imparted on the ball. This force is the reaction of the ball being accelerated. Whether the ball was kicked through its center of mass, causing it to rotate, is irrelevant. Rotation is affected in proportion to the distance by which the applied force and mass center are out of line.

Let us explore moments further. Referring to figure 5.11, a moment can be defined as the effect of a force-couple system, that is, two forces of equal magnitude and opposite direction that are not collinear (figure 5.11a). In this system, the sum of the forces is zero. However, because the two forces are noncollinear, they cause the body to rotate. This is drawn diagrammatically as a curved arrow (figure 5.11b).

To return to the football in figure 5.10, there is a force couple in all four diagrams, the applied force, F, and the reaction, ma. In figure 5.10, a through c, this force couple is collinear, so there is no moment. However, in figure 5.10d, the force and reaction are noncollinear. There is a perpendicular distance between the line of action of the force and the reaction of the mass center, and the result is a moment that rotates the ball.

Let us formalize the present discussion. Given a 2-D FBD, the process is to use Newton's second law in the horizontal, vertical, and rotational directions:

$$\sum F_x = ma_x \quad \sum F_y = ma_y \quad \sum M = I\alpha \quad (5.1)$$

The mass, m, and moment of inertia, I, for the object in question are determined beforehand. The linear and angular accelerations are determined from camera data. The sum of the forces or moments on the left side of each equation (i.e., each Σ term) can combine many forces or moments, but it should contain only one unknown—a net force or moment—to solve for. Because of this, we usually must solve for the two forces before we solve for the unknown moment.

When putting together the sums of forces and moments, we must adhere to the sign conventions established by the FBD. Inverse dynamics problems often require careful bookkeeping of positive and negative signs. As the examples in this chapter show, the FBD takes care of this as long as we honor the sign conventions that have been drawn. For example, in many cases we solve for a force or moment even though we are uncertain of its direction. This is not a problem with FBDs: We merely draw the force or moment with an assumed direction. If, in fact, the force points in the opposite direction, our calculations simply return a negative numerical value.

A sum of moments must be calculated about a single point on the object in question. There is no right or wrong point about which to calculate; however, some points are simpler than others. If we sum moments about a point where one or more forces act, then the moments of those forces will be zero because their moment arms are zero. Therefore, sometimes in human movement, it is convenient to calculate about a joint center. However, in most cases we calculate moments about the mass center; there, we can neglect the reaction force (ma) and gravity terms because their moment arms are zero.

This text uses the convention that a counterclockwise moment is positive, also called the *right-hand rule*. Coordinate system axes on FBDs establish their positive force directions. When solving a problem from an FBD, the proper technique is to first write the equations in algebraic form, honoring the sign conventions, and then to substitute known numerical values with their signs once

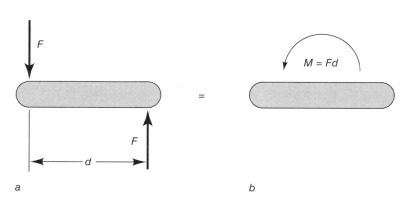

a *b*

▲ **Figure 5.11** *(a)* A force couple and *(b)* its equivalent free moment.

all of the algebraic manipulations have been carried out. These procedures are only learned by example, so we now present a few of them.

EXAMPLE 5.1

5.1a—Suppose that in our football example, the ball was kicked and its mass center moved horizontally. The horizontal acceleration was −64 m/s², the angular acceleration was −28 rad/s², the mass of the ball was 0.25 kg, and its moment of inertia was 0.05 kg·m². What was the kicking force? How far from the center of the ball was the force's line of action?

See answer 5.1a on page 377.

5.1b—Now solve the same problem assuming that there was a force plate under the football and that the tee on which the football rested resisted the kicking force. The force of the tee was 4 N in the horizontal direction; its **center of pressure** (COP) was 15 cm below the mass center of the ball.

See answer 5.1b on page 378.

EXAMPLE 5.2

A commuter is standing inside a subway car as it accelerates away from the station at 3 m/s². She maintains a perfectly rigid, upright posture. Her body mass of 60 kg is centered 1.2 m off the floor, and the moment of inertia about her ankles is 130 kg·m². What floor forces and ankle moment must the commuter exert to maintain her stance?

See answer 5.2 on page 378.

EXAMPLE 5.3

A tennis racket is swung in the horizontal plane (X and Y axis are both horizontal). The racket's mass center is accelerating at 32 m/s² in the Y direction and its angular acceleration is 10 rad/s². Its mass is 0.5 kg and its moment of inertia is 0.1 kg·m². From the base of the racket, the locations of the hand and racket mass center are 7 and 35 cm, respectively. If we ignore the force of gravity, what are the net force and moment exerted on the racket? Given that the hand is about 6 cm wide, interpret the meaning of the net moment (i.e., what is the actual force couple at the hand?).

See answer 5.3 on page 379.

In these examples, we provided distances between various points on the FBDs. These distances are not the data we measure with camera systems and force plates. Rather, these instruments measure the locations of points, specifically the locations of joint centers and the GRF (the COP). We need these points to calculate the moment arms of the various forces in the FBD. This is simply a matter of subtracting corresponding positions in the global coordinate system (GCS) directions. However, a commonly made error is that the moment arm of a force in the *x* direction is a distance in the perpendicular *y* direction, and vice versa. This is a very important point that we again illustrate with examples.

EXAMPLE 5.4

Given the FBD of the arbitrary object following, calculate the reactions R_x, R_y, M_z at the unknown end. Its mass is 8.0 kg and its moment of inertia is 0.2 kg·m² about the mass center.

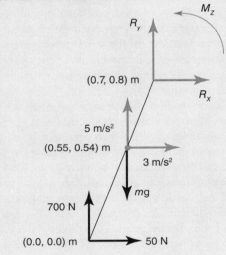

See answer 5.4 on page 379.

EXAMPLE 5.5

Given the following data for a bicycle crank, draw the FBD and calculate the forces and moments at the crank axle: pedal force x = −200 N; pedal force y = −800 N; pedal axle at (0.625, 0.310) m; crank axle at (0.473, 0.398) m; crank mass center at (0.542, 0.358) m. The crank mass is 0.1 kg and its moment of inertia is 0.003 kg·m². Its accelerations are −0.4 m/s² in the *x* direction and −0.7 m/s² in the *y* direction; its angular acceleration is 10 rad/s².

See answer 5.5 on page 380.

Having discussed general plane motion for a single object, we now turn to the solution technique for multiobject systems like arms and legs. The procedure for each body segment is exactly the same as for single-body systems. The only difference is that the method of sections is applied to manage the interactive forces between segments. In fact, this technique was applied in the previous example of the bicycle crank: We had to "section" the crank, that is, we imagined cutting it off at the axle, to determine the forces and moment at the axle. Let us explore this technique in detail.

Method of Sections

Engineering analysis of a mechanical device usually focuses on a limited number of key points of the structure. For example, on a railroad bridge truss, we normally study the points where its various pieces are riveted together. The same is true when we analyze the kinetics of human movement. We generally do not concern ourselves with a complete map of the forces and moments within the body. Rather, we study specific points of the body—most commonly the joints. We therefore "section" the body at the joints and calculate the reactions between adjacent segments that keep them from flying apart. These forces and moments at a sectioned joint are unknown. Therefore, when constructing our FBD, we must draw a reaction for each DOF—that is, a horizontal force, a vertical force, and a moment. In some calculations, it is possible for one or two of these reactions to be zero, but the method of sections requires that each be drawn and solved for because they are unknowns.

The method of sections is straightforward:

▶ Imagine cutting the body at the joint of interest.
▶ Draw FBDs of the sectioned pieces.
▶ At the sectioned point of one piece, draw the unknown horizontal and vertical reactions and the net moment, honoring the positive directions of the GCS.
▶ At the sectioned point of the other piece, draw the unknown forces and moment in the negative directions of the GCS. This is Newton's third law.
▶ Solve the three equations of motion for one of the sections.

Single Segment Analysis

In many cases, we may be interested in only one of the sectioned pieces. Let us start with three such examples and then progress to multisegmented analysis.

Multisegmented Analysis

Complete analysis of a human limb follows the procedure used in the previous examples. We simply have to for-

EXAMPLE 5.6

Consider an arm being held horizontally. What are the shoulder reaction forces and joint moment? Assume that the arm is stationary and rigid. The weights of the upper arm, forearm, and hand are 4, 3, and 1 kg, respectively, and their mass centers are, respectively, 10, 30, and 42 cm from the shoulder.

See answer 5.6 on page 381.

EXAMPLE 5.7

Suppose the hand is holding a 2 kg weight. What are the shoulder reaction forces and joint moment now?

See answer 5.7 on page 381.

EXAMPLE 5.8

The elbow is 22 cm from the shoulder. What are the reactions at the elbow on the forearm? On the upper arm? Solve again for the reactions at the shoulder using the FBD of the upper arm.

See answer 5.8 on page 382.

malize our solution process. This process has a specific order: We start at the most distal segment and continue proximally. The reason for this is that we have only three equations to apply to each segment, which means that we can have only three unknowns for each segment—one horizontal force, one vertical force, and one moment. However, we can see (return to figure 5.9, if necessary) that if we were to section and analyze either the thigh or the leg, we would have six unknowns—two forces and one moment at each joint. The solution is to start with a segment that has only one joint (i.e., the most distal) and from there proceed to the adjacent segment. For this we use Newton's third law, applying the negative of the reactions of the segment that we just solved, as shown for the lower extremity in figure 5.12. Of the lower-extremity segments, only the foot has the requisite three unknowns, so we solve this segment first. Note how the actions on the ankle of the foot have corresponding equal and opposite reactions on the ankle of the leg. Then we can calculate the unknown reactions at the knee. These are drawn in reverse in the FBD for the thigh, and then we solve for the hip reactions.

One final, but important, detail about this process is that the sign of each numerical value does not change from one segment to the next. From Newton's third

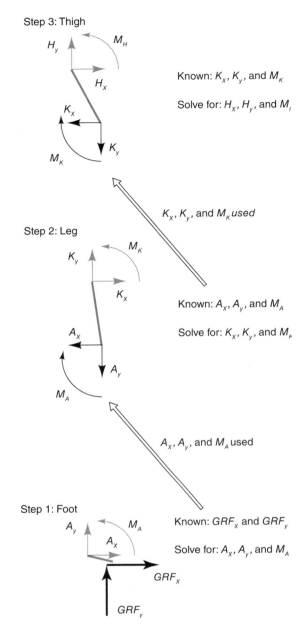

Step 3: Thigh

Known: K_x, K_y, and M_K

Solve for: H_x, H_y, and M_I

K_x, K_y, and M_K used

Step 2: Leg

Known: A_x, A_y, and M_A

Solve for: K_x, K_y, and M_k

A_x, A_y, and M_A used

Step 1: Foot

Known: GRF_x and GRF_y

Solve for: A_x, A_y, and M_A

▲ **Figure 5.12** Free-body diagrams of the segments of the lower extremity.

law, every action has an equal but opposite reaction. At the joints, therefore, the forces on the distal end of one segment must be equal to but opposite of those on the proximal end of the adjacent segment. However, we never change the signs of numerical values. The FBDs take care of this. Note that in figure 5.12, the knee forces and moments are drawn in opposing directions. Following the procedures shown previously, first construct the equations for a segment from the FBD without considering the numeric values. Once the equations are constructed, the numeric values with their signs are

substituted into these equations. Note how this process is carried out in data tables for examples 5.9 and 5.10.

We provide example calculations there for each of the two distinct phases of human locomotion, swing and stance. When we are solving the kinetics of a swing limb, the process is almost exactly the same as for stance. The only difference is that the GRFs are zero, and thus we can neglect their terms in the equations of motion. The procedure is virtually identical to the calculations that would be carried out in computer code for one frame of data.

HUMAN JOINT KINETICS

What exactly are these forces and moments that we have just calculated? The answer, which was given earlier in this chapter, deserves revisiting: The net forces and moments represent the sum of the actions of all the joint structures. A common error is thinking of a joint reaction force as the force on the articular surfaces of the bone and a joint moment as the effect of a particular muscle. These interpretations are incorrect because joint reaction forces and moments are more abstract than that; they are sums, net effects. In examples 5.9 and 5.10, we calculate relative measures to compare the efforts of the three lower-extremity joints in producing the movement and forces that were measured. We do not even have estimates of the activities of the quadriceps, triceps surae, or any other muscle. We do not have estimates of the forces on the articular surfaces of the joints or any other anatomical structure. Let us explain this further.

As depicted in figure 5.2, many muscle-tendon complexes, ligaments, and other joint structures bridge each joint. Each of these structures exerts a particular force depending on the specific movement. In figure 5.2, we neglected all friction between the articular surfaces as well as between all other adjacent structures. It cannot be overemphasized that joint reaction forces and moments should be discussed without reference to specific anatomical structures. There are several reasons for this. Equal joint reactions may be carried by entirely different structures. Consider, for example, the elbow joint of a gymnast: When the gymnast is hanging from the rings, the elbow is subjected to a tensile force that must be borne by the various tendons, ligaments, and other structures crossing the elbow. In contrast, when the gymnast performs a handstand, many of these same structures can be lax, because much of the load is shifted to the articular cartilage. From the sectional analysis presented in this chapter, we would calculate equal but opposite joint reaction forces in each case; both would equal one-half of body weight minus the weight of the forearms. However, the distribution of the force among the joint structures is completely different in these two cases.

EXAMPLE 5.9

Determine the joint reaction forces and moments at the ankle, knee, and hip given the following data. These occurred during the swing phase of walking, so the GRFs are zero.

The ankle is at (0.303, 0.189) m, the knee is at (0.539, 0.420) m, and the hip is at (0.600, 0.765) m.

See answer 5.9 on page 383.

	Mass (kg)	I (kg·m^2)	a_x (m/s^2)	a_y (m/s^2)	α (rad/s^2)	Center of mass (m)
Foot	1.2	0.011	−4.39	6.77	5.12	(0.373, 0.117)
Leg	2.4	0.064	−4.01	2.75	−3.08	(0.437, 0.320)
Thigh	6.0	0.130	6.58	−1.21	8.62	(0.573, 0.616)

EXAMPLE 5.10

Determine the joint reaction forces and moments at the ankle, knee, and hip given the following data. These occurred during the stance phase of walking, so the ground reaction forces are nonzero. The solution process is almost identical.

The ankle is at (0.637, 0.063) m, the knee at (0.541, 0.379) m, and the hip at (0.421, 0.708) m. The horizontal GRF is −110 N, the vertical GRF is 720 N, and its center of pressure is (0.677, 0.0) m.

See answer 5.10 on page 384.

	Mass (kg)	I (kg·m^2)	a_x (m/s^2)	a_y (m/s^2)	α (rad/s^2)	Center of mass (m)
Foot	1.2	0.011	−5.33	−1.71	−20.2	(0.734, 0.089)
Leg	2.4	0.064	−1.82	−0.56	−22.4	(0.583, 0.242)
Thigh	6.0	0.130	1.01	0.37	8.6	(0.473, 0.566)

Even if we are not analyzing an agile gymnast, the fact remains that we do not know from net joint forces and moments how loading is shared between various structures. This situation, in which there are more forces than equations, is said to be *statically indeterminate*. In plain language, it is a case in which we know the total load on a system, but we are not able to determine the distribution of the load without considering the specific properties of the load-bearing structures. This is analogous to a group of people moving a heavy object such as a piano: We know that they are carrying the total weight of the piano, but without a force plate under each individual, we cannot know how much weight each person bears.

Consider another specific example, a human standing quietly with straight lower extremities. We could calculate a joint reaction force at the knee. Assuming that the limbs share body weight evenly, the knee reaction force would equal one-half of the body weight minus the weight of the leg and foot. If we asked the subject to clench his muscles as much as possible, the joint reaction forces would not change. However, the tensions in the tendons would increase, as would the compressive

forces on the bones. The changes are equal but opposite and thus do not express themselves as an external, joint reaction force change.

Before discussing patterns of joint moments during human movements, we detail the limitations of these measures. As stated earlier, they are somewhat abstract. However, the purpose of the previous discussion was simply to delineate the limits of joint moments. This being done, these data can be discussed appropriately.

Limitations

Aside from the intrinsic limitations of 2-D kinematics discussed in chapter 1, there are several other important limitations in our foregoing analysis of 2-D kinetics.

Effects of friction and joint structures are not considered. The tensions of various ligaments become high near the limits of joints, and thus moments can occur when muscles are inactive. Also, although the frictional forces in joints are very small in young subjects, this is often not the case for individuals with diseased joints. Interested readers should refer to Mansour and Audu (1986),

Audu and Davy (1988), or McFaull and Lamontagne (1993, 1998). Segments are assumed to be rigid. When a segment is not rigid, it attenuates the forces that are applied to it. This is the basis for car suspension systems: The forces felt by the occupant are less than those felt by the tires on the road. Human body segments with at least one full-length bone such as the thigh and leg are reasonably rigid and transmit forces well. However, the foot and trunk are flexible, and it is well documented that the present joint moment calculations are slightly inaccurate for these structures (see, for example, Robertson and Winter 1980). In the foot, for example, various ligaments stretch to attenuate the GRFs. It is for this reason that individuals walking barefoot on a hard floor tend to walk on their toes: The calcaneus-talus bone structures are much more rigid than the forefoot.

The present model is sensitive to its input data. Errors in GRFs, COPs, marker locations, segment inertial properties, joint center estimates, and segment accelerations all affect the joint moment data. Some of these problems are less significant than others. For example, GRFs during locomotion tend to dominate stance-phase kinetics; measuring them accurately prevents the majority of accuracy problems. However, during the swing phase of locomotion, data-treatment techniques and anthropometric estimates are critical. Interested readers can refer to Pezzack and colleagues (1977), Wood (1982), or Whittlesey and Hamill (1996) for more information. Exact comparisons of moments calculated in different studies are not appropriate; we recommend allowing at least a 10% margin of error.

Individual muscle activity cannot be determined from the present model. We do not know the tension in any given muscle simply because muscle actions are represented by moments. Moreover, we do not even know the moment of a single muscle because multiple muscles, ligaments, and other structures cross each joint. Muscle forces are estimated using the musculoskeletal techniques discussed in chapter 9. An important aspect of this is that *cocontraction* of muscles occurs in essentially all human movements. Thus, for example, if knee extensor moments decrease under a certain condition, we do not know whether the decrease occurs because of decreased quadriceps activity or increased hamstring activity. As another example, a subject asked to stand straight and clench her lower-extremity muscles would have joint moments close to zero even though her muscles were fully activated; their actions would work against each other.

Two-joint muscles are not well represented by the present model. Although the moments of two-joint muscles are included in joint moment calculations, the segment calculations effectively assume that muscles only cross one joint. Again, this is a problem that is addressed by musculoskeletal models (see chapter 9).

People with amputations require different interpretations than other individuals. For example, an ankle moment can be measured for prostheses even when the leg and foot are a single, semirigid piece. Prosthetic knees have stops to prevent hyperextension and have other components, such as springs and frictional elements, to control their movement. These cause knee moments that require completely different interpretations from those seen in intact subjects. Similar considerations also apply to subjects with braces such as ankle-foot orthoses.

It was a conscious choice on the authors' part to present the interpretation of joint moments after this discussion of their limitations. Limitations do not invalidate model data, but they do limit the extent of interpretation. Note in the following discussion that no reference is made to specific muscles or muscle groups and that the magnitudes of different peaks are not referenced to the nearest 0.1 of a newton meter.

Relative Motion Method Versus Absolute Motion Method

The approach presented so far is just one way of computing the net forces and moments at the joints. Plagenhoef (1968, 1971) called this approach the *absolute motion method* of inverse dynamics because the segmental kinematic data are computed based on an absolute or fixed frame of reference. An alternative approach outlined by Plagenhoef is the *relative motion method*. This method quantifies the motion of the first segment in a kinematic chain from an absolute frame of reference, but all other segments are referenced to moving axes that rotate with the segment. Thus, each segment's axis, except that of the first, moves relative to the preceding segment. This method has the advantage of showing how one joint's moment of force contributes to the moments of force of the other joints in the kinematic chain. The drawback is that the level of complexity of the analysis increases with every link added to the kinematic chain. Furthermore, the method requires the inclusion of Coriolis forces, which are forces that appear when an object rotates within a rotating frame of reference. These fictitious forces—sometimes called pseudoforces—only exist because of their rotating frames of reference. From an inertial (i.e., fixed or absolute) frame of reference, they do not exist.

Rarely have researchers tried to compare the two methods. However, Pezzack (1976) did compare both methods using the same coordinate data and found that the relative motion method was less accurate, especially as the kinematic chain got longer (had more segments). For short kinematic chains (two or three segments), both methods yielded similar results. Most researchers have adopted the absolute motion method, because most data-collection systems measure segmental kinematics with respect to axes affixed to the ground or laboratory floor.

APPLICATIONS

There are many uses for the results of an inverse dynamics analysis. One application sums the extensor moments of a lower extremity during the stance phase of walking and jogging to find characteristic patterns and predict whether people with artificial joints or pathological conditions have a sufficient support moment to prevent collapse (Winter 1980, 1983a; see From the Scientific Literature). Others have used the net forces and moments in musculoskeletal models to compute the compressive loading of the base of the spine for research on lifting and low back pain (e.g., McGill and Norman 1985). An extension of this approach is to calculate the compression and shear force at a joint. To do this, the researcher must know the insertion point of the active muscle acting

FROM THE SCIENTIFIC LITERATURE

Winter, D.A. 1980. Overall principle of lower limb support during stance phase of gait. *Journal of Biomechanics* 13:923-7.

This paper presents a special way of combining the moments of force of the lower extremity during the stance phase of gait. During stance, the three moments of force—the ankle, knee, and hip—combine to support the body and prevent collapse. The author found that by adding the three moments in such a way that the extensor moments had a positive value, the resulting "support moment" followed a particular shape. The support moment $(M_{support})$ was defined mathematically as

$$M_{support} = -M_{hip} + M_{knee} - M_{ankle} \qquad (5.2)$$

Notice that the negative signs for the hip and ankle moments change their directions so that an extensor moment from these joints makes a positive contribution to the support moment. A flexor moment at any joint reduces the amplitude of the support moment. Figure 5.13 shows the average support moment of normal subjects and the support moment and hip, knee, and ankle moments of a 73-year-old male with a hip replacement.

This useful tool allows a clinical researcher to monitor a patient's progress during gait rehabilitation. As the patient becomes stronger or coordinates the three joints more effectively, the support moment gets larger. People with one or even two limb joints that cannot adequately contribute to the support moment will still be supported by that limb if the remaining joints' moments are large enough to produce a positive support moment.

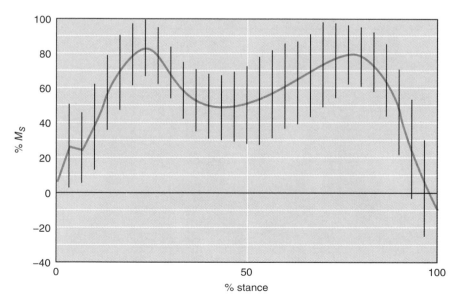

▲ **Figure 5.13a** Averaged support moment of subjects with intact joints.

Reprinted from *Journal of Biomechanics,* Vol. 13, D.A. Winter, "Overall principle of lower limb support during stance phase of gait," pgs. 923-927, copyright 1980, with permission of Elsevier.

(continued)

(continued)

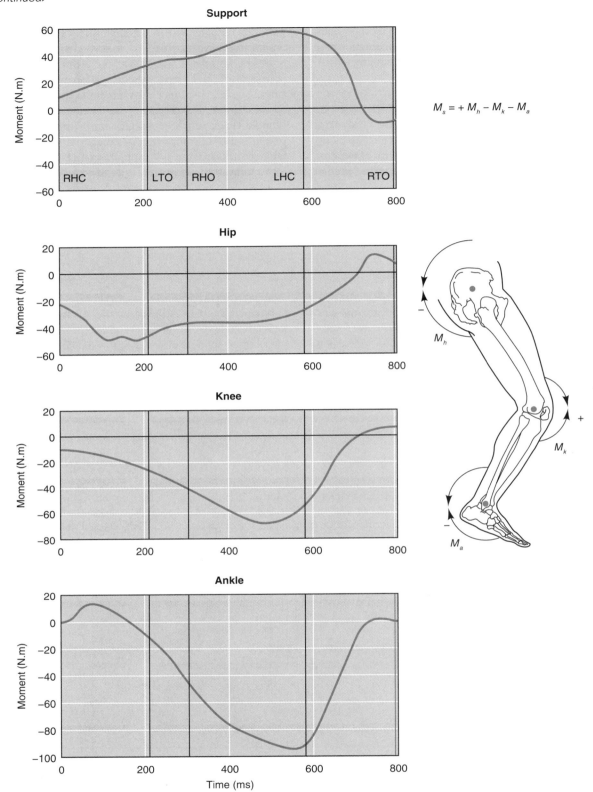

$$M_s = + M_h - M_k - M_a$$

▲ **Figure 5.13b** The support moment and hip, knee, and ankle moments of a 73-year-old male with a hip replacement during the stance phase of walking.

Reprinted from *Journal of Biomechanics,* Vol. 13, D.A. Winter, "Overall principle of lower limb support during stance phase of gait," pgs. 923-927, copyright 1980, with permission of Elsevier.

across a joint and assume that there are no other active muscles.

Computing the force in a muscle requires several assumptions to prevent indeterminacy (i.e., too few equations and too many unknowns). For example, if a single muscle can be assumed to act across a joint and no other structure contributes to the net moment of force, then, if the muscle's insertion point and line of action are known (from radiographs or estimation), the muscle force (F_{muscle}) is defined:

$$F_{muscle} = M/(r \sin \theta) \qquad (5.3)$$

where M is the net moment of force at the joint, r is the distance from the joint center to the insertion point of the muscle, and θ is the angle of the muscle's line of action and the position vector between the joint center and the muscle's insertion point. Of course, such a situation rarely occurs because most joints have multiple synergistic muscles with different insertions and lines of action as well as antagonistic muscles that often act in cocontraction. By monitoring the electrical activity (with an electromyograph) of both the agonists and antagonists, one can reduce these problems, but even an inactive muscle can create forces, especially if it is stretched beyond its resting length. The contributions of other tissues to the net moment of force can be minimized also as long as the motion being analyzed does not include the ends of the range of motion, when these structures become significant.

Once the researcher has estimated the force in the muscle, the muscle stress can be computed by determining the cross-sectional area. The cross-sectional areas of particular muscles can be derived from published reports or measured directly from MRI scans or radiographs. Axial stress (σ) is defined as axial force divided by cross-sectional area. For pennate types of muscles in which the force is assumed to act along the line of the muscle, the stress is defined as $\sigma = F_{muscle}/A$, where F_{muscle} is the muscle force in newtons and A is the cross-sectional area in square meters. The units of stress are called pascals (Pa), but because of the large magnitudes, kilopascals (kPa) are usually used. Of course, true stress on the muscle cannot be quantified because of the difficulty of directly measuring the actual muscle force.

The following sections present and discuss the patterns of planar lower-extremity joint moments during walking and running. These movements and others such as stair ascent and descent often can be analyzed two-dimensionally because their principal motions occur primarily in the sagittal plane. The convention

FROM THE SCIENTIFIC LITERATURE

McGill, S., and R.W. Norman. 1985. Dynamically and statically determined low-back moments during lifting. *Journal of Biomechanics* 18:877-85.

This study presents a method for computing the compressive load at the L4-L5 joint from data collected on the motion of the upper body during a manual lifting task. Three different methods for computing the net moments of force at L4-L5 were compared. A conventional dynamic analysis was performed using planar inverse dynamics to compute the net forces and moments at the shoulder and neck, from which the net forces and moments were calculated for the lumbar end of the trunk (L4-L5). Second, a static analysis was done by zeroing the accelerations of the segments. A third, quasi-dynamic approach assumed a static model but used dynamic information about the load. Once the lumbar net forces and moments were calculated, it was assumed that a "single equivalent muscle" was responsible for producing the moments of force and that the effective moment arm of this muscle was 5 cm. The magnitude of the compressive force $(F_{compress})$ on the L4-L5 joint was then computed by measuring the angle of the lumbar spine (actually, the trunk), θ. The equation used was

$$F_{compress} = \frac{M}{r} + F_y \cos\theta + F_x \sin\theta \qquad (5.4)$$

where M is the net moment of force at the L4-L5 joint, r is the moment arm of the single equivalent extensor muscle across L4-L5 (5 cm), (F_x, F_y) is the net force at the L4/L5 joint, and θ is the angle of the trunk (the line between L4-L5 and C7-T1).

The researchers showed that there were "statistically significant and appreciable difference(s)" among the results of the three methods for the pattern and peak values of the net moment of force at L4-L5. In general, the dynamic method yielded larger peak moments than the static approach but smaller values than the quasi-dynamic method. Comparisons of a single subject's lumbar moment histories and the averaged histories of all subjects showed that each approach produced very different patterns of activity. These comparisons also showed that although the static approach produced compressive loads that were less than the 1981 National Institute for Occupational Health and Safety (NIOSH) lifting standard, the more accurate dynamic model produced loads greater than the "maximum permissible load." The quasi-dynamic approach produced even higher compressive loads. Clearly, then, one should apply the most accurate method to obtain realistic conclusions.

for presenting these figures is that extensor moments are presented as positive and flexor moments as negative. This is in agreement with the engineering standard that mechanical actions that lengthen a system are positive (positive strain) and actions that shorten the system are negative (negative strain). In figure 5.9, hip flexor moments and dorsiflexor moments were calculated as positive. Therefore, these two moments are usually presented as the negative of what is calculated.

Walking

Joint moments during walking have many typical features. Sample data are presented in figure 5.14 for the ankle, knee, and hip joint moments. These data are presented as percentages of the gait cycle; the vertical line at 60% of the cycle represents toe-off, and the 0% and 100% points of the cycle reflect heel-contact. On the vertical axes, the joint moments are presented in newton meters.

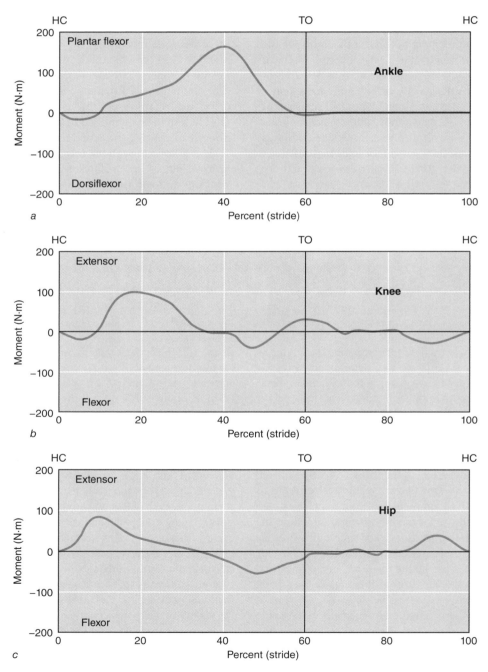

▲ **Figure 5.14** Moments of force at the *(a)* ankle, *(b)* knee, and *(c)* hip during normal level walking. HC = heel-contact; TO = toe-off.

Sometimes these values are scaled to the percentage of body mass or percentage of body mass times leg length to assist in comparing different subjects. We keep these data in newton meters for continuity with the preceding examples. In general, joint moments do scale up and down with body size. They also change magnitude with the speed of movement.

The ankle moment depicted in figure 5.14a shows that there was a dorsiflexor moment after heel-contact that peaked at about 15 N·m. This moment prevents the foot from rotating too quickly from heel-contact to foot-flat, a condition known clinically as foot-slap. Although 15 N·m is a relatively small moment on the scale of this graph, this peak is nonetheless a very common feature of normal walking. Thereafter, we see a large plantar flexor moment that peaked at about 160 N·m near 40% of the stride duration. This reflects the effort necessary to effect push-off. As this peak diminishes, the limb becomes unloaded. As toe-off occurs, we again see a small dorsiflexor moment of about 10 N·m. This action, although small, is important because it lifts the toes out of the way of the ground. Individuals with dorsiflexor dysfunction have a problem with their toes catching the ground at this part of the gait cycle. For the rest of the swing phase, the ankle moment is near zero.

In figure 5.14b, there are four distinct peaks of the knee moment. The largest peak during stance is extensor, typically peaking around 100 N·m. During this peak, the limb is loaded and thus the extensor moment acts to prevent collapse of the limb. Note that this peak occurs slightly earlier than the peak of the ankle moment. Often we see a smaller knee flexor peak before toe-off as the leg is pulled through the remainder of the stance. During swing, the first peak is extensor; it limits the flexion of the knee that occurs because the lower extremity is being swung forward from the hip. Without this peak, the knee would reach a highly flexed position, especially at faster walking velocities. The second swing-phase peak is a flexor peak of about 30 N·m; it slows the leg before the knee reaches full extension.

Figure 5.14c shows that the hip moment tends to have an 80 N·m extensor peak during the first half of stance. At toe-off, there is a flexor moment that is needed to swing the lower extremity forward. Then, similar to the knee moment, the hip moment has an extensor peak of about 40 N·m at the end swing to slow the lower extremity before heel-contact.

The hip moment is the most variable of the three lower-extremity joint moments. The foot moment tends to be the least variable because the segment is constrained by the ground. The hip, in contrast, not only is responsible for lower-extremity control but also has to control the balance of the torso. This is a significant task, given that the torso accounts for at least two-thirds of the body weight of an average individual. Winter and Sienko (1988) showed that the movement of the torso reflects a majority of the variability in hip moment. In this regard, the hip moment becomes somewhat hard to interpret during stance.

Running

Lower-extremity joint moments during running are shown in figure 5.15. These moments have patterns similar to their analogues in walking. The most prominent differences are their magnitudes. Running is a more forceful activity than walking; thus, just as GRFs are larger during running, so, too, are the joint moments. Furthermore, the stance phase (HC to TO) is a smaller percentage of the running cycle than during walking. During stance, the ankle moment (figure 5.15a) has a larger plantar flexor peak of about 200 N·m. Thereafter, the swing-phase (TO to second HC) ankle moment is near zero. The ankle moment at heel-contact varies slightly depending on running style. Runners with a heel-toe footfall pattern exhibit a small dorsiflexor peak at heel-contact, much like in walking (as shown in figure 5.14). Individuals who run foot-flat or on their toes do not have this peak because there is no need to control foot-slap.

The stance-phase knee moment (Figure 5.15b), like the ankle moment, consists primarily of a single extensor peak of about 250 N·m. Prior to toe-off, there is usually a flexor moment of about 30 N·m. This action flexes the knee rapidly before the lower extremity is swung forward. During the swing, there is an extensor phase in which the leg is swung forward. Finally, there is a flexor phase before heel-contact to slow the leg.

The hip moment (figure 5.15c) is extensor for much of stance, peaking at more than 200 N·m. We then see a shift to net flexor activity around toe-off to swing the lower extremity forward. Then there is an extensor action to slow the thigh before heel-contact.

SUMMARY

This chapter focused heavily on proper technique for inverse dynamics problems. This focus is necessary simply because the technique clearly has many steps and potential pitfalls (Hatze 2002). Students are encouraged to practice such problems until they can solve them without referring to this book. Students should be able to draw an FBD for any segment, construct the three equations of motion, and solve them.

Students are also encouraged to be mindful of the limitations of joint moments. Joint moments are only a summary representation of human effort, one step more advanced than the information obtained from a force plate. Joint moments are not an end-all statement of human kinetics; rather, they are convenient standards

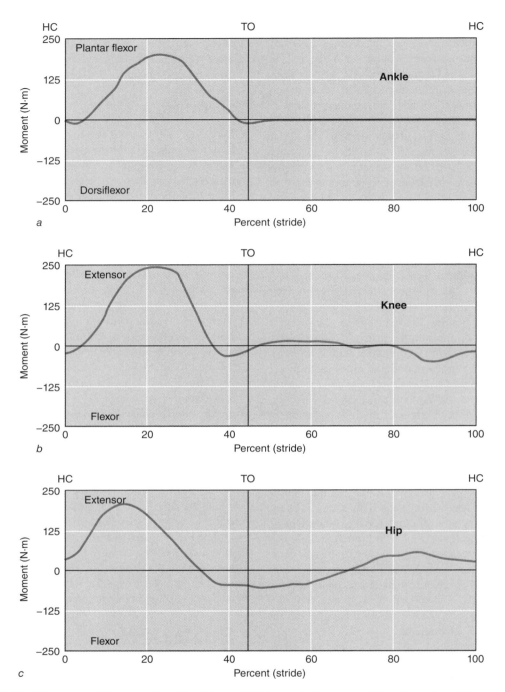

▲ Figure 5.15 Moments of force at the *(a)* ankle and *(b)* knee during normal level running. HC = heel-contact; TO = toe-off.

for evaluating the relative efforts of different joints and movements. Researchers interested in specific muscle actions must use either electromyography (chapter 8), musculoskeletal models (chapter 9), or both. It is noteworthy that the great Russian scientist Nikolai Bernstein (see Bernstein 1967) refers to moments but in fact preferred to use segment accelerations as overall representations of segmental efforts. It is hard to argue that joint moments offer much more information. A 200 or 20 N·m joint moment is in itself fairly meaningless. It is only by comparing specific cases like the walking and running examples in figures 5.14 and 5.15 that we can develop a relative basis for the magnitudes of joint moments.

The problem with analyzing limbs segmentally is that it can promote the misconception that each joint is controlled independently. We stated that two-joint muscles are poorly treated with this method. In addition, the individual segments of an extremity interact. For example, we discussed the fact that thigh moments affect the movement of the leg. Therefore, in terms of joint moments, we know that a hip moment will affect the thigh; however, we do not establish the resulting effect on the leg. Many studies have noted the importance of intersegmental coordination (see, for example, Putnam 1991); in fact, Bernstein cited the use of segment interactions as the final step in learning to coordinate a movement. Clinicians are also beginning to recognize such effects in various populations, particularly in people with amputation. Systemic analyses such as the Lagrangian approach, outlined in chapter 10, may be preferable for these situations.

SUGGESTED READINGS

Chapman, A.E. 2008. *Biomechanical Analysis of Fundamental Human Movements*. Champaign, IL: Human Kinetics.

Nigg, B.M., and W. Herzog. 1999. *Biomechanics of the Musculo-Skeletal System*. 2nd ed. Toronto: Wiley.

Özkaya, N., M. Nordin, D. Goldsheyder, and D. Leger. 2012. *Fundamentals of Biomechanics*. 3rd ed. New York: Van Nostrand Reinhold.

van den Bogert, A.J., and A. Su. 2007. A weighted least squares method for inverse dynamic analysis. *Computer Methods in Biomechanics and Biomedical Engineering* 11:1, 3-9.

Winter, D.A. 2009. *Biomechanics and Motor Control of Human Movement*. 4th ed. Toronto: Wiley.

Zatsiorsky, V.M. 2001. *Kinetics of Human Motion*. Champaign, IL: Human Kinetics.

Chapter 6

Energy, Work, and Power

D. Gordon E. Robertson

Energy is a well-known physical quantity that, despite its notoriety, is not well understood. For instance, physicists have yet to identify any atomic or subatomic particle that corresponds to a basic unit, or quantum, of energy. One of the difficulties with understanding energy is that it takes many forms. Matter itself is one form of energy, which Einstein was able to quantify with his most famous equation, $E = mc^2$, but this energy is only manifest when the matter itself is torn apart. Other forms of energy include nuclear, electrical, thermal (heat), solar, light, chemical, and, the one of greatest interest to biomechanists, mechanical.

In this chapter, we are mainly concerned with mechanical energy, and so we

- ▶ examine the concepts of energy, work, and the laws of thermodynamics;
- ▶ introduce the concepts of conservation of mechanical energy and a conservative force;
- ▶ outline methods for directly measuring mechanical work, called *direct ergometry*;
- ▶ outline methods for indirectly quantifying mechanical work, including the use of results from inverse dynamics; and
- ▶ examine the relationship between work and the cost of doing work, called *mechanical efficiency*.

ENERGY, WORK, AND THE LAWS OF THERMODYNAMICS

Energy can be defined as the ability to perform work: in other words, the ability to affect the state of matter. In a sense, energy is the motion of particles or the potential to create motion. For instance, heat, a ubiquitous form of energy, is the rate of vibration of molecules—the greater the vibration or agitation, the greater the heat or thermal energy. The quantity of heat in a substance is measured by its temperature. All matter vibrates to a certain extent; as this vibration is reduced, the matter's temperature is reduced and we say it is *cooler*. The lowest temperature, called *absolute zero*, corresponds to the complete absence of vibration, which, according to the third law of thermodynamics, can never be achieved; that is, no substance can be lowered to absolute zero (0 kelvins).

Thermodynamics is the field of study concerned with energy and its quantification, transmission, and transduction (change) from one form to another. It was once thought that energy was a fluid (caloric) that flowed from one object to another. We now understand that energy is a property of matter. According to the first law of thermodynamics, also called the law of conservation of energy, the quantity of energy in the universe is a constant. In simpler terms, we can say that in a closed system, the quantity of energy in the system is a fixed amount. A closed system is a volume that no energy can enter or leave. The energy within the system can change form, but as long as no energy enters or escapes the volume, the total amount inside is a fixed amount. Of course, it is not a simple matter to quantify all the energy sources within a system. If we can measure some of the sources and assume that the others do not change, then by monitoring changes in the known sources, we can evaluate any transformations of energy to determine the work done and the mechanical efficiencies of any machines within the volume.

The second law of thermodynamics, first elucidated by Rudolf Clausius in 1865, states that when energy is transformed from one form to another—for example, when electricity produces light, water power produces electricity, or biochemical energy produces a muscle contraction—some of the energy is wasted and can no longer be transformed into another usable form of energy. Clausius named this unusable energy **entropy**, to sound like energy. Entropy can be considered energy that can no longer perform useful work.

Work can be defined as the changing of energy to another useful form of energy, also called *transduction*. For example, when a fire heats a pot of water, or steam creates motion in a piston, or electricity passing through a tungsten wire creates light, energy is transduced from one form to another. In every such transformation, some energy is produced that is not directly associated with the intended work. Much of this lost energy takes the form of heat. The heating of gears, springs, surfaces, and air is usually wasted energy. In some cases, this heat can be collected and reused, but some energy always escapes into the surrounding atmosphere and cannot be recovered. This heat becomes entropy. Entropy is forever increasing as the universe ages; in other words, the universe is gradually burning down or becoming increasingly chaotic. We will leave it to the physicists and philosophers to predict the eventual outcome. For the biomechanist, it is enough to understand that there are costs associated with the transformation of energy. This cost is manifested in the heating of machines, muscles, and the surrounding environment.

Figure 6.1 shows a simplified schematic of the flow of energy through the human body and identifies several areas where entropy occurs. For example, in the conversion of chemical energy to mechanical energy, some energy is lost as heat that is dissipated by the skin or gases expired into the environment. Two other means of loss include mechanical friction and viscosity. Fric-

tional losses occur whenever the body rubs against the environment or tissues within the body are subjected to internal friction. Viscous losses occur because of drag forces induced by moving through fluid media, such as water or air, or as a result of the viscoelastic properties of various tissues within the body.

Mechanical work is the work done when the total mechanical energy of a body changes. This principle, called the *work-energy relationship*, is based on Newton's second law. With the appropriate mathematical operations, Newton's second law can be transformed into the following relationship:

Mechanical Work = Change in Mechanical Energy

$$W = \Delta E = E_0 - E_N \qquad (6.1)$$

where the total mechanical energy of a body *(E)* is defined as the sum of the potential and kinetic energies of the body (Winter 1976). These types of energies are defined later in this chapter. Figure 6.2 illustrates the relationship between work and energy. Notice that between points A and B, the mechanical energy of the body changes. This change in energy represents the work that was done on the body. If the duration of the work is known, the *average power* can be computed from

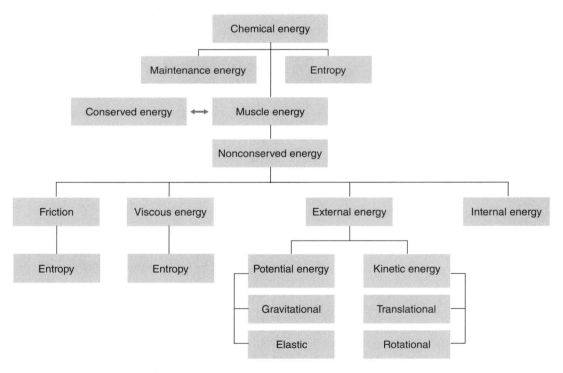

▲ **Figure 6.1** Energy flow through the human body to the environment. Maintenance energy includes energy to all tissues, excluding the skeletal muscles. Entropy includes all energy that can no longer be recovered to perform useful work as a result of heating the environment or creating turbulence in fluids in the environment (air or water). Conservative energy is the energy that recycles by changing form within segments or exchanges between and among segments or other bodies.

$$\overline{P} = \Delta E/\Delta T = \text{Work/Duration} = W/\Delta t \qquad (6.2)$$

The *instantaneous power* at any particular instant in time can also be computed by taking the time derivative of the energy history; that is,

$$P = dE/dt \qquad (6.3)$$

Note that the units of measure of work and energy must be equivalent. By international agreement, the units of work and energy are called joules. The *joule* is the work done by a 1-newton (N) force that moves an object through 1 m in the direction of the force. The joule is dimensionally equivalent to the units used for moments of force, newton meters (N·m). Physicists therefore gave work and energy units a different name—joule (J)—in honor of James Prescott Joule. Joule established numerous relationships between the work and heat energy of electrical motors. The joule is the International System of Units (SI) unit of work for energy (mechanical, electrical, solar, and so on) and the newton meter is the SI unit for moment of force or torque. The SI unit for **power** is the watt (abbreviated W), which is defined as the rate of doing work at 1 J per second. The watt was named to honor James Watt, who designed efficient steam engines that were fundamental to the development of the Industrial Revolution.

When work is done by muscles, some of the energy produced may be used to move internal structures and some may be used to do work on the environment. The former is called *internal energy*, the latter *external energy* (see figure 6.1). Some of the work done may also be recycled by conservative forces, such as elastic storage and recoil of muscle tendons or pendulum actions of swinging limbs. This pathway is also included in figure 6.1. Energy recycled in this way reduces the amount of muscle work and chemical energy required by the system. Work done on the environment can be done against friction or viscous forces (drag) or can be manifested in changes in the potential or kinetic energies of objects in the environment. For example, when an object is lifted to a height, its *gravitational potential energy* is increased. *Elastic potential energy* is increased by depressing a springlike object, such as a diving board, springboard, or pole vault pole. When a ball is thrown, the external energy done by the body appears in the form of *translational* or *rotational kinetic energy* of the ball. The translational kinetic energy appears as the ball's linear velocity, whereas the rotational kinetic energy manifests itself in the ball's spin.

The internal energy referred to in figure 6.1 appears as potential and kinetic energy, but instead of being used to perform work on external objects, it is used to move internal structures. These movements appear as movements of the upper and lower extremities and are considered to be a cost of performing the task. Some tasks, such as level walking, running, and cycling, require small amounts of external work to get the person up to speed but then require mainly internal work to maintain the speed. Other tasks, such as lifting and vertical jumping, require small amounts of internal work but relatively large amounts of external work. The following sections outline methods for computing these various quantities for a wide variety of human movements.

CONSERVATION OF MECHANICAL ENERGY

To conserve mechanical energy, special circumstances must occur. *Conservation of mechanical energy* occurs when all the forces and moments of force acting on a body or a single segment are conservative, that is, the resultant force acting on the body is a conservative force. Forces or moments of force are considered conservative when the work they do in moving a body from one point to another is independent of the path taken, that is, when the work done depends only on the location of the two points. Conservative forces include gravitational forces, the force of an ideal spring, elastic collisions, the tensile force of an ideal pendulum, and the normal force of a frictionless surface. Illustrations of these conservative forces appear in figure 6.3.

Another corollary that follows from the definition of conservative forces is that they do no work after acting through a closed path. For example, if a weight is attached to a stretched spring and released and then the spring returns to its original stretched length without assistance from other forces, then no work is done and the spring force is considered to be conservative.

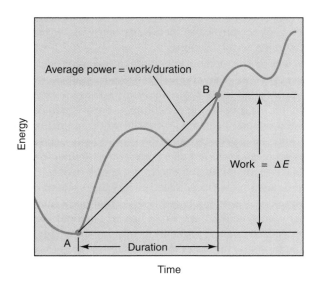

▲ **Figure 6.2** Relationship between energy, work, and average power.

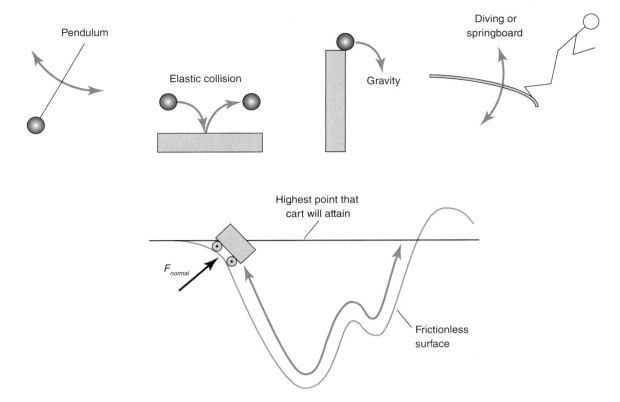

▲ **Figure 6.3** Examples of conservative forces.

The work done by a nonconservative force is affected by the path that is taken when an object is moved from one point to another. Nonconservative forces include frictional, viscous (e.g., fluid friction), and viscoelastic (e.g., muscle and ligament) forces and plastic deformations. Frictional forces are nonconservative because the amount of work they do increases with the length of the path taken from one point to another. The longer the path, the more work is required. The same is true for viscous forces. Viscous forces also increase the work done depending on how fast the path is taken—the faster the motion through a viscous medium, the greater the work. In general, the resistance offered by viscous forces increases with the square of the velocity. This is why the most efficiently run races are those in which a constant speed is maintained throughout the race.

Most conservative forces are ideals that cannot be realized physically. For example, an ideal spring returns all the energy that is imparted to it by compression or elongation. In reality, all springs obey the second law of thermodynamics and lose some energy in the form of heat. Similarly, "pure" elastic collisions do not occur in nature. An *elastic collision* occurs when a body is dropped from a certain height and rebounds to the same height. In reality, all such collisions result in lower rebound heights and some heating or deformation of the surfaces. A purely *plastic collision* occurs when the

body does not rebound at all, such as when putty or a soft snowball hits the ground.

Gravitational forces are truly conservative because, in the absence of nonconservative forces, any change in potential energy causes a corresponding change in kinetic energy. For example, if an object is allowed to fall, its kinetic energy will increase by the same amount that its potential energy decreases. Thus, the total energy of the body remains constant (i.e., its energy is conserved, assuming no air friction).

The tensile forces of pendulums are also considered to be conservative and typically work in conjunction with gravitational force. Figure 6.4 illustrates a simple pendulum. At time A, the pendulum or body is motionless at a particular height. All of its energy is in the form of gravitational potential energy, but as the body is released, its height and potential energy decrease. Simultaneously, its translational kinetic energy increases in exactly the same amount that its potential energy decreases until the bottom of the swing is reached. At that point (midway between A and B), the body has no potential energy; instead, all its energy is in the form of kinetic energy. The kinetic energy provides the necessary impetus to swing the body back to its original height at time B. From time B to C, a nonconservative force acts to raise the body to a new height and a new energy level—that is, work is done. If the body is then allowed to swing under the

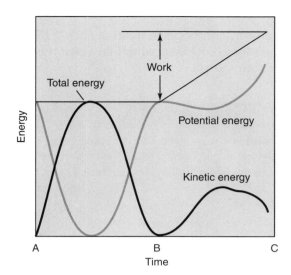

▲ **Figure 6.4** Energy of a pendulum. The pendulum is conservative from time A to time B. Between times B and C, work is done to raise the pendulum to a higher energy level. Amounts of translational and potential energies are also shown.

influence of only conservative forces (gravity and the tensile force of the pendulum string), the body will cycle back and forth at the new energy level, interchanging potential and kinetic energy at no cost to the system.

In humans, particularly during walking, some energy is conserved by permitting parts of the body to act like simple pendulums. There is, however, another way conservation can take place. It is possible to permit energy to transfer between adjacent segments, as is the case with compound pendulums. For this to occur, the segments must be connected with nearly frictionless joints, and the muscles that act across the joints must be passive or act to facilitate the transfer of energy. Elftman (1939a, 1939b), Winter and Robertson (1979), and others have shown that such situations occur during the swing phase of gait. They showed that energy originally in the leg and foot transfers to assist in the thigh's upward swing-through. Furthermore, this same energy transfers back to the leg and foot immediately before heel-strike, as the thigh reaches maximum flexion. Such mechanisms reduce the chemical energy costs needed for these movements. If these conservative mechanisms were not used, more chemical and mechanical energy would have to be produced to supply the necessary work to propel the segments. For example, during sprinting (Lemaire and Robertson 1989; Vardaxis and Hoshizaki 1989), there is insufficient time for the action of simple or compound pendulum motions to swing the leg through. These researchers showed that in some athletes, energy had to be supplied at the rate of 4000 J/s (watts) to swing the thigh upward and approximately 3600 W to drive it back

down prior to touchdown on the track. Athletes cannot wait for conservative forces to supply the energy necessary to cycle the lower extremity during running and sprinting; instead, nonconservative forces are required. The measurement of the work done by nonconservative forces is called *ergometry*.

ERGOMETRY: DIRECT METHODS

Ergometry is literally the measurement of work. An **ergometer** is any system or technology that quantifies the work done during an activity. Many commercial ergometers are used by kinesiologists and physical educators, but the most common one is probably the bicycle ergometer. Other devices commonly found in research laboratories and fitness rooms include rowing machines, running treadmills, isokinetic dynameters, and exercise machines; the latter, however, are so poorly calibrated that they are unsuitable for accurate scientific measurements.

To qualify as an ergometer, a device must measure at least two things—force and displacement or moment of force and angular displacement. This is because work is defined as the product of force times the displacement or distance traveled. More accurately, work is defined as the **scalar** or dot product of force and displacement. Thus, the work done by the resultant force acting on a body is

$$W = \vec{F} \cdot \vec{s} = F_x s_x + F_y s_y = Fs \cos \phi \quad (6.4)$$

where W is work, $\vec{F}$ and (F_x, F_y) are the resultant forces, $\vec{s}$ and (s_x, s_y) are the displacement, and ϕ is the angle between the force and the displacement vectors. When a moment of force is used, the work done is quantified by

$$W = M\theta \quad (6.5)$$

where M is the resultant moment of force and θ is the angular displacement of the object. If a body has both a nonzero resultant force and a nonzero moment of force, then the work done is the sum of the two equations just set out. Note that to add the work done by the resultant force and resultant moment of force, the units of measure must be the same. The units for the work of a force appear to be newton meters, but these units are dimensionally equivalent to the joule. The work done by the resultant moment of force appears to be in units of newton meters radians, but because newton meters are equivalent to joules and radians are dimensionless, these units are dimensionally equivalent to the units of the work produced by a force, namely joules.

In many cases, the work done can be measured using forces or moments of force. For example, on a Monark bicycle ergometer, the work done can be measured by the moment of force produced by the flywheel braking system times the number of rotations of the flywheel (times 2π to convert flywheel rotations to radians). The moment of force must be in units of newton meters and is dependent upon the position of the load arm of the ergometer. Most ergometers are calibrated using the obsolete unit of a **kilopond**, which is equivalent to the weight (force) of a mass of 1 kg. To convert a kilopond load to its equivalent force in newtons, multiply the load by the gravitational acceleration value of 9.81 m/s². That is,

$$F \text{ (newtons)} = \text{Load (kiloponds)}$$
$$\times\ 9.81 \text{ (newtons/kilopond)} \quad (6.6)$$

To obtain the moment of force created by the load, multiply the force times the radius of the flywheel. This distance is 25.46 cm (the circumference is 160.0 cm) for a Monark bicycle ergometer. Therefore, the work done *(W)* using the moment of force approach is

$$W = M\theta = Fr\theta \quad (6.7)$$

where F is the load in newtons, r is the radius of the flywheel, and θ is the angular displacement of the flywheel. The force or load is actually a braking belt that creates a rotational friction equivalent to a linear frictional force acting tangentially to the flywheel. Thus,

$$W = F\text{ (N)} \times \text{Radius (m)} \times \text{Rotations (revolutions)}$$
$$\times\ 2\pi \text{ (radians/revolution)} \quad (6.8)$$

To measure work for a Monark bicycle ergometer, the number of crank rotations usually is used instead of the number of flywheel revolutions. In a standard ergometer, every *crank* (not flywheel) revolution is equivalent to 6 m displacement of a point on the rim of the flywheel. Therefore,

$$W = \text{Load (kiloponds)} \times 9.81 \times \text{Crank Rotations}$$
$$\times\ 6 \text{ (m/rotation)} \quad (6.9)$$

In each case, the units of work are expressed in joules.

A treadmill can also be used as an ergometer. A person running at a constant speed up an incline is considered to have done an amount of work equivalent to raising the body weight up to a certain height. The height is computed by multiplying the speed of the treadmill times the length of the run in seconds times the sine of the angle of the incline. That is,

$$W = \text{Body Weight (N)} \times \text{Speed (m/s)} \times \text{Duration (s)}$$
$$\times\ \sin \gamma \quad (6.10)$$

where γ is the angle of incline of the treadmill. Notice that consideration is given not to the actual distance traveled but rather only to the height the person would be raised if lifted to an equivalent height. It is also assumed that no work is done against the treadmill belt as a result of stretching or friction and that the speed is constant throughout the run.

EXAMPLE 6.1

Calculate the work done on a bicycle ergometer if the person pedals against a 2.50 kilopond workload for 20 s at the rate of 60 revolutions per second. Assume that each revolution is equivalent to 6 m of linear motion of the flywheel.

See answer 6.1 on page 386.

ERGOMETRY: INDIRECT METHODS

In a sense, all methods of determining the mechanical energy of a system are indirect because there is no direct way to measure the flow of energy into the system like there is, for example, at an electrically metered house. When we refer to the indirect measurement of mechanical energy, we are referring to methods that quantify the motion of bodies, called their *kinetic energies*, and the positions of the bodies with respect to a fixed frame of reference, called their *potential energies*. The sum of the kinetic and potential energies of a body yields the body's total mechanical energy.

To simplify the analysis of some systems, the motion and position of the system can be reduced by considering only the system's center of gravity as a point mass. When we analyze a point mass, only two types of mechanical energy are present, translational kinetic energy and gravitational potential energy. A better model of the human body assumes that the body is a system of interconnected rigid bodies or segments. In this case, each segment includes an additional type of energy called *rotational kinetic energy*. It is also possible, although to date it has rarely been attempted, to model each segment as a deformable body, in which case elastic potential energy is also included in the computation of total mechanical energy.

Each of these energies can be quantified by knowing the kinematics of the body or its segments. In the following two sections, methods for determining the energy of point masses and systems of rigid bodies are presented. An alternative approach uses the work-energy relationship. This relationship equates the work done on a body to the change in mechanical energy of the body. Therefore, the changes in mechanical energy can be computed by determining the work done on the body by both internal and external forces. This approach is described in the section Inverse Dynamics Methods.

Point Mass Methods

If a body can be assumed to behave as a single point mass, its energy can be quantified from the linear kinematics of its center of gravity. The most common means of obtaining linear kinematics is to digitize the coordinates of markers attached to the body and then perform finite difference calculus to obtain the linear velocities (as outlined in chapter 1). A simpler but less accurate approach is to estimate the location of the center of mass and determine its trajectory. Either way, the total energy of a point mass is the sum of its gravitational potential energy plus its translational kinetic energy. That is,

Gravitational Potential Energy:

$$E_{gpe} = mgy \qquad (6.11)$$

Translational Kinetic Energy:

$$E_{tke} = 1/2 \; mv^2 = 1/2 \; m(v_x^2 + v_y^2) \qquad (6.12)$$

Total Mechanical Energy:

$$E_{tme} = E_{gpe} + E_{tke} \qquad (6.13)$$

where m is the mass of the body, g is 9.81 m/s², y is the height of the body above the horizontal reference axis, and (v_x, v_y) is the velocity of the point with respect to the stationary reference axes. Several researchers have used this approach to quantify the work done during walking, running, and sprinting (Cavagna et al. 1963, 1964, 1971).

EXAMPLE 6.2

Calculate the work done and the power produced when a 80.0 kg person starts from rest and achieves a speed of 6.00 m/s in 4.00 s. Assume that the person runs on a level surface and that the rotational energy is negligible.

See answer 6.2 on page 386.

Segmental Methods

The point mass method has been criticized for being inappropriate for human body analyses (Williams and Cavanagh 1983; Winter 1978) because it underestimates the true mechanical energy of the system. The more accurate method is to divide the body into segments and then sum the energies of each segment to determine the total body's energy. This method was first used by Wallace Fenn to determine the energetics of running (1929) and sprinting (1930).

To compute the total mechanical energy of a segment, usually only the gravitational potential energy and the translational and rotational kinetic energies are calculated. This is the correct formula for a rigid body, but if the body deforms, the elastic potential energy should also be included. In most biomechanical studies, it is not possible to compute the elastic potential energy because the amount of deformation is too small to measure and the deformation-force relationship is too expensive or difficult to obtain. Even if it were possible to determine these factors, the amount of energy stored in this way does not warrant the expense.

These equations define how the total mechanical energy (E_{tme}) of a segment is computed:

Gravitational Potential Energy:

$$E_{gpe} = mgy \qquad (6.14)$$

Translational Kinetic Energy:

$$E_{tke} = 1/2 \; mv^2 = 1/2 \; m(v_x^2 + v_y^2) \qquad (6.15)$$

Rotational Kinetic Energy:

$$E_{rke} = 1/2 \; I\omega^2 \qquad (6.16)$$

Total Mechanical Energy:

$$E_{tme} = E_{gpe} + E_{tke} + E_{rke} \qquad (6.17)$$

where m is the mass of the segment, g is 9.81 m/s², y is the height of the segment above the reference axis, (v_x, v_y) is the velocity of the segment center of gravity, I is the segment's mass moment of inertia about its center of gravity, and ω is the angular velocity of the segment.

To obtain the total body's mechanical energy (E_{tb}), the individual segments' total mechanical energies are summed. That is,

Total Body Mechanical Energy:

$$E_{tb} = \sum_{s=1}^{S} E_{tme,s} \qquad (6.18)$$

where S is the number of segments in the human body model and $E_{tme,s}$ is the total mechanical energy of segments. The number of segments varies from study to study depending on the motion and whether bilateral symmetry is assumed. For example, Winter and colleagues (1976) used three segments to investigate the work of walking by assuming that the head, arms, and trunk could be modeled as a single segment. Martindale and Robertson (1984) used six segments by assuming bilateral symmetry for rowing (on water and ergometer rowing). Twelve segments were used by Williams and Cavanagh (1983) for the analysis of running and by Norman and colleagues (1985) for cross-country skiing. Sometimes a single extremity is analyzed, in which case only two or three segments are quantified (Caldwell and Forrester 1992).

EXAMPLE 6.3

Calculate the mechanical energy of an 18.0 kg thigh that has a linear velocity of 8.00 m/s and a rotational velocity of 20.0 rad/s and whose center of gravity is 1.20 m high. The thigh's moment of inertia is 0.50 kg·m.

See answer 6.3 on page 386.

Although this method of calculating energy is relatively simple because only kinematic data and body segment parameters are needed, the method has several drawbacks for calculating the work done. In principle, calculating the external work done requires only a measurement of the changes in energy throughout the motion. In other words, the external work $(W_{external})$ is defined as

$$W_{external} = \sum_{n=1}^{N} \Delta E_{tb,n} = E_{tb,N} - E_{tb,1} \qquad (6.19)$$

where N is the number of frames of the time-sampled motion and $E_{tb,n}$ is the total mechanical energy of the body (the sum of segmental total mechanical energies) at time frame n. Notice that you only need to measure the energy before and after the duration of the motion to obtain the total external work done on the body because all of the intermediate measures cancel out.

This measure is useful whenever the body increases its height or its linear or angular speed, but in situations in which speed is constant or locomotion is along a level surface, the external work done is zero. As long as the work is measured at the same point in the motion's cycle, there will be no change in height and no change in linear or angular velocity and, therefore, no changes in any of the potential or kinetic mechanical energies. This is called the *zero-work paradox* (Aleshinsky 1986a) because although no external work is done (excluding work done against dry and fluid frictional forces), internal work is done to cycle the extremities throughout the motion. (More about this paradox is described later in the section on Mechanical Efficiency.) It is also obvious that one can locomote across a particular distance in an efficient way or in any number of inefficient ways that clearly consume more chemical and mechanical energy to achieve. For example, one could walk normally over a 10 m distance or hop from side to side over the same distance. The hopping motion obviously wastes energy, but the external work done in both cases is the same if the person arrives at the same point with the same velocity.

Norman and colleagues (1976) proposed a method of calculating pseudo-work that was later modified by Winter (1979a, b) to quantify the internal work during locomotion. They assumed that any change in total body mechanical energy required that mechanical work be done and that monitoring these changes in mechanical energy would permit calculation of the total mechanical work and internal work done. In effect, the method calculated the internal work by computing the total work done and then subtracting the external work done. The total work done (W_{total}) was equal to the sum of the absolute values of the changes in total mechanical energy. That is,

$$W_{total} = \sum_{n=1}^{N} \left| \Delta E_{tb,n} \right| \qquad (6.20)$$

The internal work was then computed by subtracting the external work (defined previously) from the total work (above):

$$W_{internal} = W_{total} - W_{external} \qquad (6.21)$$

Several researchers used these methods to investigate various locomotor movements, including overground walking (Winter 1979a), treadmill walking (Pierrynowski et al. 1980), load carriage (Pierrynowski et al. 1981), rowing (Martindale and Robertson 1984), and cross-country skiing (Norman et al. 1985; Norman and Komi 1987), to name a few.

Aleshinsky (1986b) has criticized this approach because it inaccurately measures the total mechanical work and therefore the internal work. This approach, called the *absolute energy approach* or the *mechanical energy approach*, assumes that any simultaneous energy increase and decrease of the segments reduce the total mechanical energy expenditure. Williams and Cavanagh (1983) tried to correct this flaw by permitting transfers of energy only between adjacent segments, but Winter and Robertson (1979) had already disproved this concept by showing that some of the energy generated at the ankle transfers from the foot to the leg, to the thigh, and even to the trunk. Wells (1988) demonstrated that the work done and energy transferred within the body can be estimated using an algorithm that predicts the recruitment patterns of mono- and biarticular muscles. The algorithm partitions the net moments of force at each joint into either mono- or biarticular muscles based on whether activating a two-joint muscle could reduce the levels of muscle force.

In summary, the total work and internal work done, as calculated from mechanical energy changes, can be used to estimate the mechanical work. This method can be used when other, more accurate methods cannot be applied. A better method, supported by Elftman (1939a), Robertson and Winter (1980), Aleshinsky (1986a), and van Ingen Schenau and Cavanagh (1990), uses inverse dynamics to compute the net moments of force at each joint and then measures the work done by these sources.

Inverse Dynamics Methods

Because the work on a body can be measured by the changes in mechanical energy or the work done by the resultant forces and moments of force acting on the body, another approach to computing the external, internal, and total work done is possible. Inverse dynamics permits the calculation of the net forces and moments of force acting on individual segments of an *n*-link system of rigid bodies. This section details how the results from an inverse dynamics analysis can be used to compute total body work measures.

Procedures for computing net forces and moments of force were presented in chapter 5. If the body is assumed to be deformable and to have friction in its joints, then the net forces at these joints may do work against friction, and energy may be stored and released because of the elasticity of the tissues. These sinks and sources of energy are difficult to quantify and therefore are usually assumed to be negligible or are attributed to other sources, such as the net moments of force. A *sink* is a structure that can dissipate or store energy and may or may not return the energy. In the body, muscles are sinks when they contract eccentrically to reduce the mechanical energy of the system. Ligaments, bones, bursae, and cartilage also act as energy sinks. Energy *sources* are structures that supply or return stored energy. Muscles are the primary biological structures that supply energy to the body. Elastic structures are both sources and sinks. They receive energy during deformation (elongation or compression) and then return some of the energy after the deforming force is released (although some stored elastic energy is lost as heat).

Most researchers model the human body as a linked system of rigid bodies and assume that the joints are frictionless. These assumptions simplify the computation of mechanical work by eliminating the possibility that the net forces will affect the total work done by the body or its internal or external work. The only sources of work become the net moments of force acting at each joint and any external forces acting on the body. Aleshinsky (1986a) presented a thorough explanation for this phenomenon. In the sections that follow, the equations for computing work based on the net forces and moments of

FROM THE SCIENTIFIC LITERATURE

Winter, D.A. 1976. Analysis of instantaneous energy of normal gait. *Journal of Biomechanics* 9:253-7.

This paper illustrates how quantifying the instantaneous energy patterns of linked-segment movements can identify conservation of mechanical energy. In this study, the segments of the lower extremity were studied in 2-D (the sagittal plane) during walking along a level surface. The potential, translational kinetic, rotational kinetic, and total energy patterns of the leg (shank), thigh, and torso were calculated, as were the total body energies. The total body energy was estimated by considering the trunk, head, and upper extremities as one segment and assuming bilateral symmetry of the lower extremities so that the patterns of one lower extremity could be phase-shifted to estimate the other side.

The study reached several important conclusions about walking. For example, the results showed that at normal walking speeds, rotational kinetic energies can be ignored because of their small magnitudes. More important, Winter identified the relative importance of translational kinetic energy changes in comparison with the other forms of mechanical energy changes for the thigh (figure 6.5*b*) and particularly the leg segments (figure 6.5*a*). In contrast, the torso's energy variations for translational kinetic energy and potential energy were similar in magnitude.

Winter explained that the leg segment could not conserve energy because the kinetic and potential energies increased and decreased synchronously. Thus, no passive energy exchange like the one that occurs with pendular motion could take place between these two energy sources. Conversely, the thigh (figure 6.5*b*) and torso (figure 6.5*c*) segments exhibited periods during which their potential and translational kinetic energy patterns were asynchronous—one source increased while the other decreased and vice versa. This phenomenon reduces the muscular work costs and shows that energy conservation occurred between the two segments. Winter also showed (figure 6.5*d*) that conservation of energy may occur between one lower extremity and the other; there were overlapping periods during which one extremity's total energy pattern decreased while the other's increased and vice versa. Although these data do not prove that conservation occurs from one side to the other, they do support the possibility. The mechanical energy approach used in this study cannot distinguish whether conservation occurs between and among segments, nor can it distinguish whether muscles acting across each segment act concentrically or eccentrically to cause the respective increases or decreases in energy. Later, we present a more sophisticated method that uses inverse dynamics to determine the work done across adjacent segments.

(continued)

(continued)

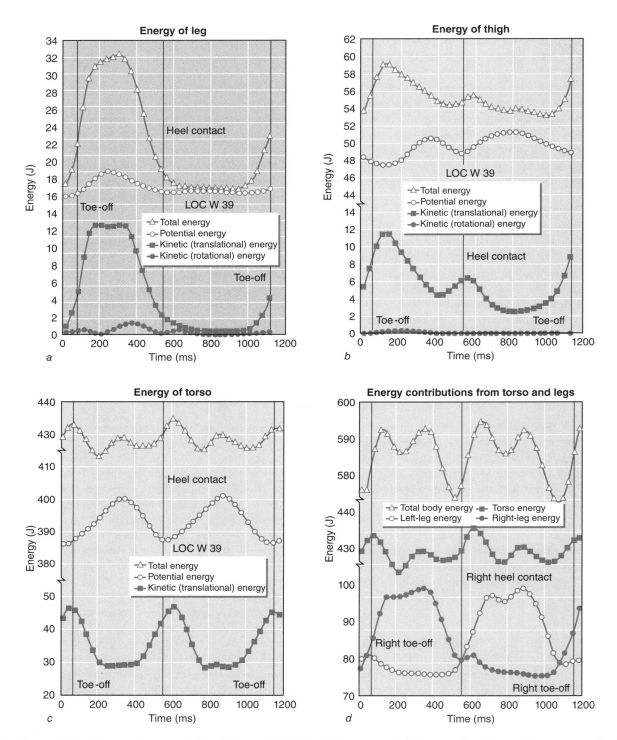

▲ Figure 6.5 Energy changes in the *(a)* leg, *(b)* thigh, *(c)* torso, and *(d)* lower extremities during one cycle of walking.

Reprinted from *Journal of Biomechanics,* Vol. 9, D.A. Winter, "Analysis of instantaneous energy of normal gait," pgs. 253-257, copyright 1976, with permission of Elsevier.

force are presented, along with a discussion of how the two methods (mechanical energy and inverse dynamics) can be applied together to better understand how energy is supplied, transferred within the body, and dissipated. First, a segmental approach is taken to investigate the transmission and use of mechanical energy by the segments. Second, a joint analysis is described, and the role of the moments of force at each joint and the transfers of energy by the net forces at each joint are defined. Third, methods for computing the total body power requirements are described. Fourth, the relationships between the two methods and the differences that are often observed are investigated.

Segmental Power Analysis

Segmental energy analysis requires only knowledge of segmental kinematics and inertial properties. Using inverse dynamics methods to obtain the work done by the moments of force at each joint requires additional information and a more complex analysis. The additional information includes the history of any external force that is in direct contact with the segment of interest. For example, to compute the work done at the ankle during stance, the GRF acting against the foot must be measured (Robertson and Winter 1980; Williams and Cavanagh 1983), so force platforms are necessary in gait laboratories. During the swing phase, however, a force platform is not needed because no external force (excluding gravity) acts on the foot. A number of studies of sprinting and running have been conducted this way (Caldwell and Forrester 1992; Chapman et al. 1987).

The following equations describe how to compute the power delivered to a segment from the net forces and moments of force at the segment's connections with the rest of the body.

Force Power:
$$P_F = \vec{F} \cdot \vec{v} = F_x v_x + F_y v_y \qquad (6.22)$$

Moment Power:
$$P_M = M_j \omega_j \qquad (6.23)$$

Total Power Delivered to Segment:
$$P_S = \sum_{j=1}^{J} \left(P_{F_j} + P_{M_j} \right) \qquad (6.24)$$

where J is the number of joints or other structures that are directly connected to the segment. For example, a foot segment has one attached segment (the leg) and the forearm has two (wrist and elbow), but the trunk could have two (both thighs), four (both thighs and both shoulders), or five (thighs, shoulders, and neck) connections, depending upon how it is modeled. Note that a foot in contact with a surface can gain or lose energy to the surface if there is relative motion (e.g., slippage, deformation, or motion such as with an elevator, escalator, diving board, or bicycle pedal). In such cases, the foot is considered to have two sources of power. For most locomotor studies, a single connection is the norm.

Another advantage of the inverse dynamics approach is that the powers delivered to each segment can be measured from the segment's rate of change of mechanical energy. This equality, derived from the work-energy relationship and sometimes called the *energy balance* or *power balance*, has been used to check the validity of assumptions about modeling segments as rigid bodies and synchronizing external forces with cinematographic or videographic data (Robertson and Winter 1980). It has also been used to indirectly determine the deformation energy of the foot during stance (Robertson et al. 1997; Winter 1996). The following equations express these relationships:

Rate of Change of Mechanical Energy:
$$P_E = \frac{dE_{tme}}{dt} \approx \frac{\Delta E_{tme}}{\Delta t} \qquad (6.25)$$

Total Power Delivered to Segment:
$$P_S = \sum_{j=1}^{J} \left(P_{F_j} + P_{M_j} \right) \qquad (6.26)$$

Power Balance:
$$P_E = P_S \qquad (6.27)$$

where E_{tme} is the total mechanical energy of a segment, P_F and P_M are the powers delivered by the forces and moments of force, respectively, and J is the number of joints connected to the segment.

Joint Power Analysis

Joint power analysis is a simplified version of segmental power analysis. It refers to examining the flow of energy (power) across a joint that results from the net force and moment of force at the joint. The powers from net forces are the same as those used in segmental power analyses, but the moment of force powers require the relative angular velocities *(ω)* of the joints rather than the segmental angular velocities. In other words, the power provided by the net moment of force, called the moment power, is the product of the net moment of force times the difference between the angular velocities of the two segments that make up the joint *($\omega_p - \omega_d$)*:

Force Power:
$$P_F = \vec{F} \cdot \vec{v} = F_x v_x + F_y v_y \qquad (6.28)$$

Moment Power:
$$P_M = M_j \omega_j = M_j \left(\omega_p - \omega_d \right) \qquad (6.29)$$

Robertson, D.G.E., and D.A. Winter. 1980. Mechanical energy generation, absorption and transfer amongst segments during walking. *Journal of Biomechanics* 13:845-54.

This paper had two purposes—to partially validate the calculation of inverse dynamics and to show how segmental power analyses can be used to investigate the delivery of power to the segments of the lower extremity during walking. The first purpose was realized by applying the work-energy relationship—more specifically, its derivative. That is, the instantaneous power of a rigid body is equal to the power delivered to the body by forces and moments of force. Computing the instantaneous power of a body requires only knowledge of the body's kinematics, namely its linear velocity, angular velocity, and height above a reference (usually the ground). From these (presented earlier in this chapter), the translational kinetic energy, rotational kinetic energy, and gravitational potential energy are computed. Summing these to obtain the total mechanical energy and taking the time derivative yield the instantaneous power of the body. The same can be done for segments of the body. These powers are not affected by GRF measurements or synchronizing motion data with external force data and therefore are relatively more accurate than the powers of the forces and moments of force determined by inverse dynamics methods (see chapter 5).

In theory, the instantaneous power of a body must be equal to the sum of the powers delivered to the body from all forces and moments of forces that act on the body. Because the force and moment powers can be contaminated by errors resulting from improper measurement or synchronization of external forces (e.g., GRFs), the authors believed that agreement between the two measures of a segment's power requirements provided evidence that the inverse dynamics calculations were accurate.

The authors tested this principle for the three segments of the lower extremity during walking and found excellent agreement for all but the foot segment during the periods of early weight acceptance and late push-off. They concluded that the poor agreement probably resulted from poor spatial or temporal synchronization of the GRFs with the motion kinematics. Later, Siegel and colleagues (1996) suggested that the cause of these discrepancies may be the assumption that the foot is a rigid body. They showed that the errors between the instantaneous powers and the sum of the powers from the forces and moments were significantly reduced by modeling the foot as a deformable body and using 3-D equations of motion. Winter (1996) also supported this concept, whereas Robertson and colleagues (1997) proposed that modeling the foot as two segments by dividing it at the metatarsal-phalangeal joints would improve the power discrepancies.

The second purpose of the Robertson and Winter (1980) paper was to show how these two measurements of power can be used to understand how a segment's energy is supplied by the net forces and moments of forces that act on the segment. Table 6.1 shows the various ways that a moment of force acting across a joint can transfer energy between, supply energy to, or dissipate energy from adjacent segments. The authors showed that each moment of force can simultaneously transfer and generate or dissipate (absorb) energy, depending on the angular velocities of the adjacent segments and joints. The authors also demonstrated the importance of energy transfers between segments as a result of the moments of force acting across the joints. These transfers had magnitudes comparable to the magnitudes of the powers generated during concentric contractions or dissipated during eccentric contractions. These latter roles of the muscle (doing positive and negative work) are usually considered to be the only functions of the moments of force and the muscles (Donelan et al. 2002). This research showed that an equally important role of muscles and moments of force is to transfer energy from segment to segment and thereby to conserve energy and reduce the physiological costs of motion.

Table 6.1 Possible Transfers, Generation, and Absorption of Energy by the Moment of Force Acting Across a Joint

Description of movement	Type of contraction	Directions of segmental original velocities	Muscle function	Amount, type, and direction of power
TWO SEGMENTS ROTATING IN OPPOSITE DIRECTIONS				
a. Joint angle decreasing	Concentric	ω_1 M ω_2	Mechanical energy generation	$M\omega_1$ generated to segment 1. $M\omega_2$ generated to segment 2.

Description of movement	Type of contraction	Directions of segmental original velocities	Muscle function	Amount, type, and direction of power
TWO SEGMENTS ROTATING IN OPPOSITE DIRECTIONS *(continued)*				
b. Joint angle increasing	Eccentric	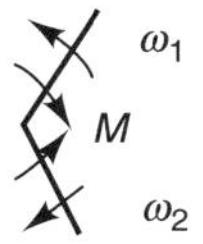 ω_1 M ω_2	Mechanical energy absorption	$M\omega_1$ absorbed from segment 1. $M\omega_2$ absorbed from segment 2.
BOTH SEGMENTS ROTATING IN SAME DIRECTION				
a. Joint angle decreasing (e.g., $\omega_1 > \omega_2$)	Concentric	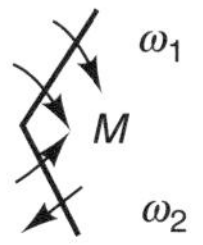 ω_1 M ω_2	Mechanical energy generation and transfer	$M(\omega_1 - \omega_2)$ generated to segment 1. $M\omega_2$ transferred to segment 1 from 2.
b. Joint angle increasing (e.g., $\omega_2 > \omega_1$)	Eccentric	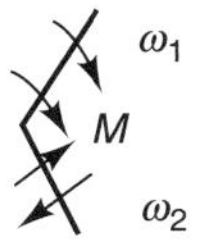 ω_1 M ω_2	Mechanical energy absorption and transfer	$M(\omega_2 - \omega_1)$ absorbed from segment 2. $M\omega_1$ transferred to segment 1 from 2.
c. Joint angle constant ($\omega_1 = \omega_2$)	Isometric (dynamic)	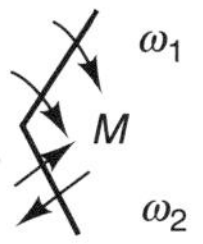 ω_1 M ω_2	Mechanical energy transfer	$M\omega_2$ transferred from segment 2 to 1.
ONE SEGMENT FIXED (e.g., SEGMENT 1)				
a. Joint angle decreasing ($\omega_1 = 0$, $\omega_2 > 0$)	Concentric	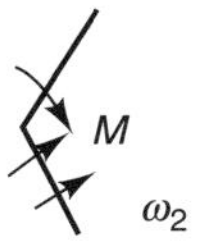 M ω_2	Mechanical energy generation	$M\omega_2$ generated to segment 2.
b. Joint angle increasing ($\omega_1 = 0$, $\omega_2 < 0$)	Eccentric	 M ω_2	Mechanical energy absorption	$M\omega_2$ absorbed from segment 2.
c. Joint angle constant ($\omega_1 = \omega_2 = 0$)	Isometric (static)	 M	No mechanical energy function	Zero

Reprinted from *Journal of Biomechanics*, Vol. 13, D.G.E. Robertson and D.A. Winter, "Mechanical energy generation, absorption and transfer amongst segments during walking," pgs. 845-854, copyright 1980, with permission of Elsevier.

This method is the one most often used for the analysis of a wide variety of human movements, including walking (Winter 1991), jogging (Winter and White 1983), race walking (White and Winter 1985), running (Elftman 1940), sprinting (Lemaire and Robertson 1989), jumping (Robertson and Fleming 1987), kicking (Robertson and Mosher 1985), skating (de Boer et al. 1987), and cycling (van Ingen Schenau et al. 1990).

In general, power histories are presented simultaneously with the net moment histories and the joint angular velocities or displacements. Figure 6.6 shows how these data are typically presented. Figure 6.6a shows the powers produced by the hip moment of force of an elite sprinter. Notice that the angular velocity curve at the top indicates when the joint is flexing or extending. That is, when the angular velocity is positive, the hip is flexing, and when it is negative, the hip is extending. Notice that the peak flexion velocity is nearly 15 rad/s, or approximately 860°/s, a speed significantly greater than what can be tested on isokinetic dynamometers such as the KinCom, Biodex, or Cybex, which are limited to velocities less than 360°/s.

The net moment curve indicates the direction of the net moment of force. For example, a positive hip moment of force indicates a flexor moment, whereas a negative moment indicates an extensor moment. The sense of the moment of force (flexor or extensor) depends on which joint is being analyzed and which way the subject is facing. If the subject is facing sideways and to the right,

a positive moment of force about the mediolateral axis is flexor for the hip, extensor for the knee, and dorsiflexor for the ankle. If the moment of force is negative, the hip moment is extensor, the knee moment is flexor, and the ankle moment is plantar flexor.

The power curve, which is the product of the other two curves, shows when positive work and negative work are being done. Positive work, also called concentric work, is done when the power history is positive, for example, between 0.01 and 0.16 s of the hip power. By looking at the moment curve above it, one can readily determine which moment of force (extensor or flexor) created the power. The area under the power curve determines

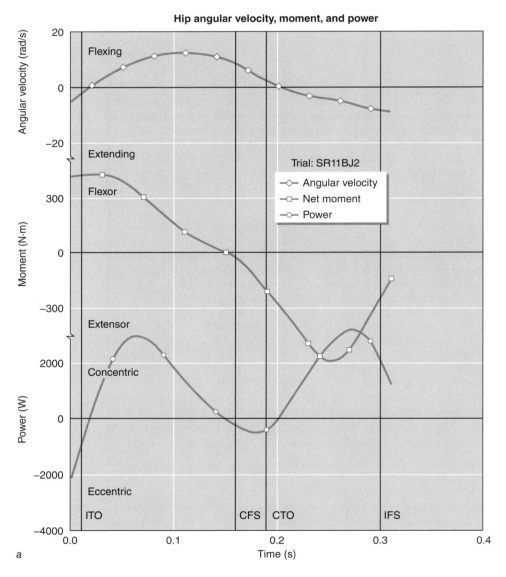

▲ Figure 6.6a Angular velocities, moments of force, and powers at the *(a)* hip and *(b)* knee during the swing phase (ITO to IFS) of an elite male sprinter running at 12 m/s. The data displayed are from the ipsilateral side. ITO = ipsilateral, toe-off; IFS = ipsilateral, foot-strike; CTO = contralateral, toe-off; CFS = contralateral, foot-strike.

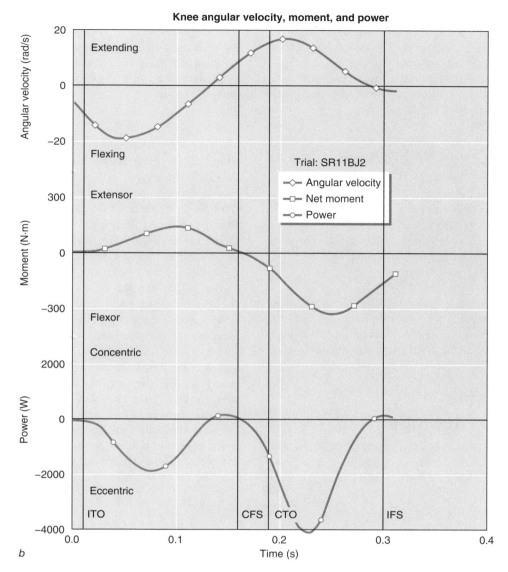

Knee angular velocity, moment, and power

▲ Figure 6.6b

the amount of work done. Conversely, when the power history is negative, for example, between 0.01 and 0.13 of the knee power, eccentric or negative work is being done by the associated moment of force, in this case extensor. Many authors use power histories to determine the sequence of events during a movement cycle (see Robertson and Winter 1980; Winter 1983c).

Total Body Work and Power

Computing total body work—equivalent to the external work done by or on the body—requires the summation of the work done by all of the body's moments of force. The work done by a moment of force, as outlined earlier, is the time integral of the product of the moment of force times the angular velocity of the joint it crosses. In equation form,

$$W_{tb} = \int_0^T \sum_{j=1}^J P_j \, dt \approx \sum_{n=1}^N \sum_{j=1}^J P_{n,j} \Delta t \qquad (6.30)$$

where J is the number of joints in the body, N is the number of time intervals, and $P_{n,j}$ is the moment power produced ($= M_{n,j}\omega_{n,j}$) by the jth net moment of force at time n. This relationship assumes that no external structure (elevator, escalator, diving board, car, and bicycle) does work on the body and that there are no losses of energy at the joints from friction or compression or from deformation of the skeletal system. These latter situations create forces that consume power and reduce the total body's mechanical energy. If the skeletal system is assumed to be rigid and the joints are assumed to be frictionless, only the moments of force can increase or decrease the total mechanical energy.

To determine the total body's instantaneous power, you simply add up the moment powers of all the joints at any instant in time (n). That is,
Total Body Power:

$$P_{tb} = \sum_{j=1}^{J} \int_{0}^{T} M_j \; \omega_j \, dt \approx \sum_{j=1}^{J} \sum_{n=1}^{N} M_{j,n} \; \omega_{j,n} \Delta t \quad (6.31)$$

Cappozzo and colleagues (1975, 1976) were among the first researchers to attempt to quantify the total body's mechanical energy and power in this way. They also validated their measure against the change in energy as measured by segmental energy changes. That is,

$$W_{tb} = \Delta E_{tme} \quad (6.32)$$

Using relatively crude data-smoothing techniques, the investigators showed that the two measures agreed. Differences were attributed to errors in the data collection procedures and the assumption that segments were rigid bodies.

Relationship Between Methods

The segmental and inverse dynamics methods are different ways of obtaining the work and power produced and distributed within the body. This relationship exists because the work-energy relationship demonstrates that the change in energy of a body or segment results from the work done on the body or segment. That is, $W = \Delta E$. Equivalently, the power of a body or segment is equal to the rate of change of the body's or segment's total energy. That is,

$$P = \Delta E / \Delta t \quad (6.33)$$

The two sides of these equations can be determined using the two methods just outlined. Table 6.2 shows an expanded version of these equations for a simple four-segment model of the body. Note that the powers (P) have been replaced by each segment's associated force ($P_F = \vec{F} \cdot \vec{v}$) and moment ($P_M = M\omega$) powers. This model assumes that the subject performs a bilaterally symmetric motion and that the head, arms, and trunk are combined into a single segment called *trunk*.

Notice that all of the $\vec{F} \cdot \vec{v}$ terms cancel out and only the $M\omega$ terms are preserved when all the segmental powers are summed. This is because it is assumed that no energy is dissipated across the joints from frictional losses or compression of the joint articulating surfaces, and thus the joint forces do no work on the segments. Joint forces transfer energy from segment to segment but do not influence the body's overall mechanical energy. Only the moments of force across the joints increase or decrease the body's energy levels. The final equation defines the rates (i.e., the powers) at which the moments affect the body, and it corresponds to the method of calculating the total body power outlined earlier.

Table 6.2   Segment and Total Powers of a Four-Segment Model of the Body

Joint	Ankle	Knee	Hip	Total
Foot	$+F_{ankle} \cdot v_{ankle} + M_{ankle} \omega_{foot}$			$= \dfrac{\Delta E_{foot}}{\Delta t}$
Leg	$-F_{ankle} \cdot v_{ankle} - M_{ankle} \omega_{leg}$	$+F_{knee} \cdot v_{knee} + M_{knee} \omega_{leg}$		$= \dfrac{\Delta E_{leg}}{\Delta t}$
Thigh		$-F_{knee} \cdot v_{knee} - M_{knee} \omega_{thigh}$	$+F_{hip} \cdot v_{hip} + M_{hip} \omega_{thigh}$	$= \dfrac{\Delta E_{thigh}}{\Delta t}$
Trunk			$-F_{hip} \cdot v_{hip} - M_{hip} \omega_{trunk}$	$= \dfrac{\Delta E_{trunk}}{\Delta t}$
Total	$M_{ankle} (\omega_{foot} - \omega_{leg})$ or $M_{ankle} \omega_{ankle}$	$M_{knee} (\omega_{leg} - \omega_{thigh})$ or $+ M_{knee} \omega_{knee}$	$M_{hip} (\omega_{thigh} - \omega_{trunk})$ or $+ M_{hip} \omega_{hip}$	$= \dfrac{\Delta E_{Total}}{\Delta t}$

F_{ankle} = net ankle force as applied to foot; F_{knee} = net knee force as applied to leg; F_{hip} = net hip force as applied to thigh; v_{ankle} = velocity of ankle; v_{knee} = velocity of knee; v_{hip} = velocity of hip; M_{ankle} = net ankle moment as applied to foot; M_{knee} = net knee moment as applied to leg; M_{hip} = net hip moment as applied to thigh; ω_{foot} = angular velocity of foot; ω_{leg} = angular velocity of leg; ω_{thigh} = angular velocity of thigh; ω_{trunk} = angular velocity of trunk; ΔE_{foot} = change in total mechanical energy of foot; ΔE_{leg} = change in total mechanical energy of leg; ΔE_{thigh} = change in total mechanical energy of thigh; ΔE_{trunk} = change in total mechanical energy of trunk; ΔE_{Total} = change in total mechanical energy of total body; Δt = duration of motion; ω_{ankle} = angular velocity of ankle; ω_{knee} = angular velocity of knee; and ω_{hip} = angular velocity of hip.

MECHANICAL EFFICIENCY

Traditionally, mechanical efficiency (ME) is defined as either the work done by a system divided by the energy cost of running the system (times 100%) or the power output over the power input. That is,

$$ME = 100\% \times W_{output}/W_{input}$$
$$= 100\% \times P_{output}/P_{input} \qquad (6.34)$$

For mechanical or electrical systems, measuring the input and output work or power is relatively easy. For example, an engine's input cost is measured by the amount of fuel consumed or electricity used. The output work or power is more difficult to determine, but, depending on what the engine is used for, it is measured by the useful work done. Of course, no machine can achieve 100% efficiency. Some energy is wasted as frictional heat or viscous damping or mechanical wear on the system parts. Typical mechanical systems rarely achieve MEs of greater than 30% and electrical systems rarely of greater than 40%.

For a biological system, ME is defined as the mechanical work done over the physiological cost times 100% (Cavagna and Kaneko 1977; Williams 1985; Zarrugh 1981). In general, the mechanical work done is considered to be the external work done by the body on its environment. This quantity was defined earlier. More recently, however, the work done has been defined as the total mechanical work done by the body—in other words, the external work plus the internal work done. These terms were also defined earlier in this chapter. The reason that this redefinition has occurred is the *zero-work paradox* (Aleshinsky 1986a).

The zero-work paradox occurs when a person or machine ambulates (walks, runs, paddles, crawls, or otherwise moves) at a constant average speed along a level surface. The input cost can be measured by calculating the energy cost of the activity. In the case of humans, the energy cost is measured from the utilization of biological fuels, such as adenosine triphosphate (ATP) or creatine phosphate (CP), or indirectly by the amount of oxygen consumed by the activity. In these situations, however, the mechanical work done is zero because there is no change in the mechanical energy of the body. Because the body moves along a level surface, there is no change in gravitational potential energy and because the body has a constant speed, there is no change in kinetic energy. (The cost of friction is considered to be negligible.)

Clearly, the person (or machine) can traverse a distance efficiently or inefficiently. For example, she could walk at a self-selected pace in a straight line from point A to B, or she could meander between the two points, hop from one foot to another, or walk with a severely disturbed pattern as a result of a poorly constructed prosthesis or a neurological disorder. In all cases, if the person starts and arrives at the same speed, no mechanical work can be said to have been done if only the external work is measured.

Of course, the physiological cost reflects any abnormal, inefficient movements during a locomotor task; thus, the most efficient gait patterns have the smallest costs, and it is likely that the self-selected walking gait in the example just discussed was the most efficient gait. But the physiological measurements cannot tell the researcher where the inefficiencies lie. To solve this problem, various researchers have included the internal work done in the numerator so that ME is defined as

$$ME = 100\% \times (\text{External Work} + \text{Internal Work})$$
$$\div (\text{Physiological Cost}) \qquad (6.35)$$

The numerator (external plus internal) is also called the total mechanical work and is best measured by summing the integrals of the absolute powers produced by the net moments of force, but it may also be estimated by summing the absolute values of the changes in total body mechanical energy. These measures, in addition to evaluating the efficiency of the locomotor pattern, aid in identifying where mechanical energy is produced and dissipated, and they potentially can identify where inefficiencies exist.

Although calculating the mechanical energy costs in this way is imperfect, it provides valuable insights into how people perform an activity. However, another source of difficulty with quantifying ME in biological systems is determining exactly how to calculate the physiological costs. Because it is very difficult to quantify the exact amount of biological fuel required to perform an activity, physiologists and biomechanists have used indirect calorimetry to estimate the energetic costs. It is assumed that the oxygen consumed is equivalent to the energetic costs. This works well as long as submaximal activities are investigated, but activities that exceed the *anaerobic threshold* result in the accumulation of lactic acid and an *oxygen debt*. The oxygen debt is the amount of oxygen that needs to be consumed to restore the body to its equilibrium state. This cost must be included as part of the physiological cost of the movement, but it is not included if the experimenter stops measuring oxygen cost at the end of the activity. To correctly assess the physiological cost, the researcher should quantify the oxygen cost until the debt is repaid. The difficulty lies in determining exactly when this occurs and the length of time it takes to restore equilibrium. Whereas the debt is created quickly, recovery is quite slow, and for vigorous activities it may take several hours. This makes the data collection expensive and limits the number of subjects who can be tested per day.

Another difficulty that must be considered is whether the whole oxygen cost should be used to quantify the physiological cost of an activity. Part of the oxygen cost is maintenance energy—the energy required to keep the nonmuscular tissues functioning (see figure 6.1). This energy can be measured from the basal metabolic rate (BMR), which is the metabolic cost of maintaining vital functions. These requirements are necessary to scientifically and reliably quantify BMR:

▶ The subject should have fasted for between 14 and 18 hours.

▶ The subject should be lying supine, quietly awake, following a restful night of sleep.

▶ The subject should not have exerted himself within the past 3 hours.

▶ The subject's body temperature should be within the normal range, and the ambient temperature should be thermoneutral.

Clearly, this measure is difficult and expensive to obtain. Some researchers have proposed that a more appropriate measure is the oxygen cost of the person's resting state immediately before the activity. For most activities, this is the cost of standing. Thus, to compute the physiological cost of walking, one would measure the oxygen cost of standing and then subtract this amount

from the oxygen cost during walking. Although this may be appropriate for some projects (because it increases sensitivity), it is not typical of evaluations of mechanical systems. It is up to the researcher to determine the best approach and to report precisely how efficiency was computed.

SUMMARY

This chapter detailed how to compute or measure mechanical energy, work, and power of planar human motions. Methods for computing whole body work, the work done on a single segment, and the work done by the moments of force at each joint were presented. One of the most useful tools in the biomechanist's toolkit was outlined—joint power analysis, which tells the researcher where mechanical energy is consumed, where it is transmitted within the body, and where it is produced. This analysis requires that an inverse dynamics analysis be done, but it adds important information about how the moments of force at each joint contribute to the energetics of the musculoskeletal system. Chapter 7 presents additional information that deals with 3-D motion. Fortunately, it is not difficult to extend the principles outlined here to 3-D motions.

FROM THE SCIENTIFIC LITERATURE

van Ingen Schenau, G.J., and P.R. Cavanagh. 1990. Power equations in endurance sports. *Journal of Biomechanics* 23:865-81.

This survey article "attempts to clarify the formulation of power equations applicable to a variety of endurance activities." The authors outline some of the approaches taken to investigate the energetic costs of locomotor activities. Their concern is not only with the biomechanical costs of motion but also with the considerations necessary for measuring the physiological costs. The authors point out that there still is no completely accepted way of "accurate[ly] accounting [for] the relationship between metabolic power input and the mechanical power output," but they do present the best currently available techniques. The paper reviews the research on the energetics of running, cycling, speed skating, swimming, and rowing.

This is an excellent overview of how equations based on Newtonian mechanics can be used to derive the mechanical power output of human locomotor activities. By lumping together all the external forces acting on the body, the authors derived an expression for the power production during locomotion. The expression equates

the summation of the moment powers for all the joints (on the left side of the equation) to the rate of change of segmental mechanical energy minus the power that is delivered to the environment through the external forces ($\vec{F}_{external}$). That is,

$$\sum M_j \omega_j = \frac{d\sum E_{tme}}{dt} - \sum \left(\vec{F}_{external} \times \vec{v}_{external} \right) \quad (6.36)$$

where M_j is the net moment of force at each joint, ω_j is the joint angular velocity, E_{tme} is the segment total mechanical energy, $\vec{F}_{external}$ is any external force acting on the body, and $\vec{v}_{external}$ is the velocity of the point of application of each external force. To the right-hand side of the equation could be added any powers resulting from external moments of force ($M_{external} \omega_{external}$), such as when a cyclist grips handlebars, but these types of forces are infrequently encountered in biomechanics.

SUGGESTED READINGS

Alexander, R.M., and G. Goldspink. 1977. *Mechanics and Energetics of Animal Locomotion.* London: Chapman & Hall.

Cappozzo, A., F. Figura, M. Marchetti, and A. Pedotti. 1976. The interplay of muscular and external forces in human ambulation. *Journal of Biomechanics* 9:35-43.

van Ingen Schenau, G.J., and Cavanagh, P.R. 1990. Power equations in endurance sports. *Journal of Biomechanics* 23:865-81.

Winter, D.A. 1987. *The Biomechanics and Motor Control of Human Gait.* Waterloo, ON: Waterloo Biomechanics.

Winter, D.A. 2009. *Biomechanics and Motor Control of Human Movement.* 4th ed. Toronto: Wiley.

Zajak, F.E., R.R. Neptune, and S.A. Kautz. 2002. Biomechanics and muscle coordination of human walking Part I: Introduction to concepts, power transfer, dynamics and simulations. *Gait and Posture* 16:215-32.

Chapter 7

Three-Dimensional Kinetics

W. Scott Selbie, Joseph Hamill, and Thomas M. Kepple

K*inetics* is the study of the forces and moments that cause motion of a body. As expressed by Vaughan and colleagues (1996), coordinated movement results from the activation of many muscles, and it is the tension in muscles acting across joints, in concert with the interaction of the body with the environment, that causes the kinematics we observe. The study of the kinetics, therefore, allows researchers to explore basic mechanisms of human movement.

In this chapter we present an introduction to methods of 3-D kinetics. Note that just as with 3-D kinematic analysis, multiple planar views of 2-D forces and moments should not be considered a 3-D analysis. In preparation for this chapter, the reader should become comfortable with the principles of vectors and matrices presented in appendixes D and E and the process of transforming vectors between coordinate systems (e.g., between the global coordinate system and a segment local coordinate system) as presented in chapter 2.

We illustrate 3-D kinetic analysis of human gait, specifically lower-extremity motion, based on the recording of a performance using an optical 3-D motion-capture system, and once again, we take a how-to approach that elaborates and extends the methods described in chapter 2.

In this chapter, we

- ► extend the representation of segments and link models from chapter 2,
- ► describe the process for a Newton-Euler inverse dynamics analysis of human locomotion,
- ► address important considerations in the presentation of 3-D joint moment data and potential source of error,
- ► calculate joint power, and
- ► discuss the interpretation of 3-D joint moment data through induced acceleration analysis and induced power analysis.

> ## VISUAL3D EDUCATIONAL EDITION
>
> The Visual3D Educational Edition software included with this text includes data sets that represent walking and running with full body sets of markers to help you further explore and understand the type of analysis presented in this chapter. Using the Visual3D Educational Edition software, you can experiment with all of the modeling capabilities of the professional Visual3D software by manipulating the model definitions, signal definitions, and basic signal processing in these sample data sets. To download the software, visit http://textbooks.c-motion .com/ResearchMethodsInBiomechanics2E.php.

SEGMENTS AND LINK MODELS

In this chapter, the study of kinetics is based on two fundamental assumptions: (1) anatomical segments are rigid bodies and (2) these rigid bodies are connected in a hierarchical chain to form a link model representation of the subject being studied.

Segments should be thought of as rigid bodies of indeterminate size composed of a fixed local coordinate system (LCS) and fixed anthropometric properties: mass, location of the center of mass, and principle moments of inertia. The LCS and these anthropometric properties are all that are required to define the segment. Although we often measure properties such as segment lengths, segment radii, and depth, these measures are only used in the anthropometric property estimates and are not intrinsic characteristics of the segments. Thus, because a segment consists of only an LCS and inertial properties, segments could be considered as bodies with no defined borders extending infinitely in all directions.

Having simplified the definition of a segment as a coordinate system with inertia, we next need to provide some realistic method of linking these segments to form a hierarchical model of the human body. In this chapter, linkage models connect the segments by joints in a manner that either allows complete rotational and translational motion between segments (termed *six degrees of freedom models* because three rotations and three translations are allowed at the joints) or specifies constraints at the joints to limit one or more of the rotational or translational degrees of freedom. These constraint-based linkage models were introduced in chapter 2 for the global optimization pose estimation algorithm. Regardless of which linkage method is used, the inverse dynamics calculations are identical, so the following methods apply to all pose estimations.

3-D INVERSE DYNAMICS ANALYSIS

In traditional dynamics, given a set of initial conditions and a set of forces as inputs, the resultant motion of the bodies can be computed over time. In biomechanics, this process is referred to as *forward dynamics*. Two simple examples of forward dynamics are computing the flight of a ball (projectile motion) and computing the motion of a pendulum. In both of these cases, gravity serves as the only external input, and the resultant motion is computed.

In biomechanics, we often compute in the other direction; that is, we use motion-capture systems to record the motion of rigid bodies (usually anatomical segments) and then compute the forces and moments that must have been responsible for the observed movement. The process is referred to as *inverse dynamics* because we are computing forces and moments from recorded motion, which is the opposite of traditional dynamics. Several methods can be used to calculate inverse dynamics. The method presented in this chapter is referred to as a Newton-Euler method. Figure 7.1 displays a representative 3-D motion-capture setup, including force platforms, that is typical of a laboratory that computes inverse dynamics.

Before delving into the details of 3-D inverse dynamics, we should highlight the indeterminacy of the solution to the inverse dynamics problem and the limitations that this imposes. Figure 7.2 is a schematic diagram of a foot segment. All structures (with the exception of the soft tissue), the moments that they exert, and the ground reaction force (GRF) are illustrated. The pose of the foot (as described by one rigid segment) and the GRF are recorded experimentally and used as inputs to the equations of motion. In a 3-D inverse dynamics analysis, a rigid body has six equations of motion relating to the six degrees of freedom (three translational and three rotational). If we wish to compute the contributions

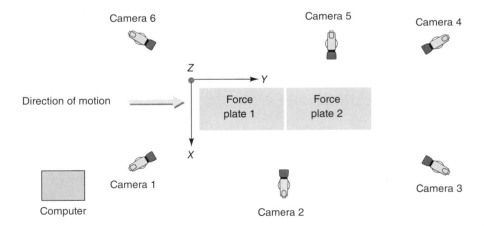

▲ **Figure 7.1** Typical multicamera setup for a 3-D kinematic analysis.

of all muscles acting on a segment, it should be obvious that there are many more unknowns than equations. The result is that we have an indeterminate solution. A straightforward way to reduce the number of unknowns to six is to replace all external forces acting on the foot with a single equivalent force that represents the sum of all the effective forces acting on the body and a single equivalent couple that represents the sum of all effective moments acting on the body. In this chapter, the inverse dynamics method represents all bone, muscle, and external forces by a single resultant joint reaction force vector with three components and a net joint moment vector (the more common biomechanical term for a couple) with three components.

The action of all muscles acting at an anatomical joint produces a net moment about the joint that we can estimate using the methods of this chapter and a compressive load on the joint surfaces, which we cannot estimate using the methods of this chapter. The joint reaction force computed at the joint is independent of this compressive load, and the net compressive load (joint reaction force plus muscle compressive load) cannot be computed from the methods of this chapter. In other words, the joint reaction force is not the force experienced by the anatomical joint.

To compute the joint reaction forces and net moments for any segments, including the lower extremity model presented in this chapter, we begin with the Newton-Euler equations of motion:

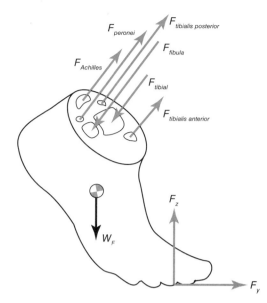

▲ **Figure 7.2** A free-body diagram of the foot with all external forces identified.

Reprinted from *Human Movement Science,* Vol. 15, C.L. Vaughan, "Are joint moments the holy grail of human gait analysis?" pgs. 423-443, copyright 1996, with permission of Elsevier.

$$\sum \vec{F} = \frac{d}{dt}\left(m\vec{v}\right) = m\vec{a} \tag{7.1}$$

$$\sum \vec{\tau} = \frac{d}{dt}\left(I\vec{\omega}\right) \tag{7.2}$$

Equation 7.1 (Newton's equation) states that the sum of all forces acting on a rigid body is equal to the rate of change of momentum of the body. Equation 7.2 (Euler's equation) states that the sum of all moments acting on a rigid body is equal to the rate of change of the angular momentum of the body.

Consider the one-link model shown in figure 7.3*a* representing a free body diagram of a foot in contact with a force platform (note that the force acting on the foot is recorded). The Newton-Euler equations used to define the force and moment at the proximal end of the foot segment can be expressed as

$$\vec{F}_a = m_f\left(\vec{a}_f - \vec{g}\right) - \vec{F}_{grf} \tag{7.3}$$

$$\vec{\tau}_a = \frac{d}{dt}\left(I_f\vec{\omega}_f\right) + \left(\vec{r}_{a.f} \times m_f(\vec{a}_f - \vec{g})\right) - \vec{\tau}_{grf} - \left(\vec{r}_{a.grf} \times \vec{F}_{grf}\right) \tag{7.4}$$

where $\vec{F}_a$ is the force at the proximal end of the foot (ankle reaction force), $\vec{\tau}_a$ is the net moment at the proximal end of the foot (net ankle moment), and $\vec{g} = (0,0,-9.81)$ is the gravity vector. The derivation of these equations will appear later in this chapter. For now we need to point out that the solution for the inverse dynamics requires several inputs: the segment's anthropometric properties, including mass (m_f), moment of inertia (I_f), and center of gravity location $(\vec{r}_{a.f})$; the kinematics, including the translational velocity $(\vec{v}_f)$, translational acceleration $(\vec{a}_f)$, angular velocity $(\vec{\omega}_f)$, and angular acceleration $(\vec{\alpha}_f)$; and the external ground reaction force data $(\vec{F}_{grf}, \vec{\tau}_{grf}, \vec{r}_{a.grf})$ (see figure 7.4).

We will solve equations 7.1 and 7.2 for a multisegment link model of the lower extremity starting with the foot segment from motion capture data using an inverse dynamics approach. In this example, we will use the following steps, some of which were covered in earlier chapters:

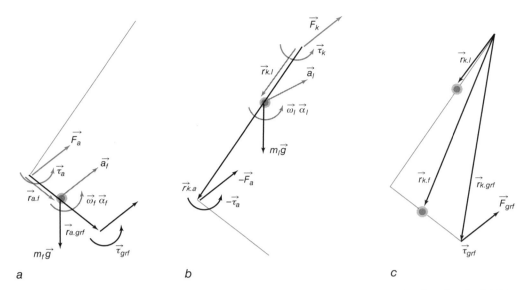

▲ **Figure 7.3** *(a)* Free body diagram for the foot. *(b)* Free body diagram for the shank. *(c)* Diagram displaying the three vectors relative to the proximal end of the proximal segment.

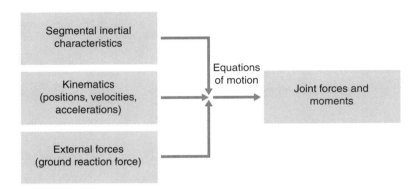

▲ **Figure 7.4** Flowchart of the inverse dynamics approach.

1. Define the segment local coordinate systems (LCS) (see chapter 2).

2. Estimate the pose of the model from recorded motion-capture data (see chapter 2).

3. Scale the segment anthropometry to the subject and identify the segment inertial characteristics (also chapter 4).

4. Compute kinematics (e.g., angular velocities and accelerations) from the pose estimates (see chapter 2).

5. Record and represent external forces acting on the body.

6. Compute the joint reaction forces, net joint moments, and joint powers.

In this chapter, the biomechanical model is a collection of rigid segments linked together hierarchically. One of the fundamental attributes of a segment is an LCS that is fixed within the segment and that, kinematically speaking, fully defines a segment. In chapter 2 we presented a method for defining the LCS of the lower-extremity segments from motion-capture data that has anatomical meaning. We will not review the method here, but our inverse dynamics calculations in this chapter use precisely the same approach to define the LCS for the foot, shank, and thigh segments.

In chapter 2 we also presented several methods for estimating the pose (position and orientation) of the LCS from motion capture data. We will not review these methods other than to point out that the inverse dynamics computations presented in this chapter are identical regardless of which pose estimation algorithm is used. The user should review chapter 2 before tackling this chapter because pose estimation and the transformation between coordinate systems are fundamental to the methods presented here.

Identifying Segment Inertial Characteristics

Anthropometry is described in chapter 3 of this book, but we will summarize the methods of estimation used in this chapter because they are intimately linked with our definitions of a segment. Segment anthropometry can be estimated by a number of different means, including; direct measurements (Brooks and Jacobs 1975; Zatsiorsky and Seluyanov 1985); regression equations derived from cadaver dissections (Chandler et al. 1975; Clauser et al. 1969; Dempster 1955); regression equations derived from skeletons (Vaughan et al. 1992); and geometrical representations of the segments (Hanavan 1964; Hatze 1980; Yeadon 1990a). All of these approaches are estimates, and all have merit in the appropriate context. The choice of method should be based on the population being studied and the level of accuracy that the researcher requires for a given experimental hypothesis. The use of any of these equations does not change the concepts or computation of inverse dynamics presented in this chapter. In the example of the lower extremity presented in this chapter, the mass of each segment is defined based on the study by Dempster (1955) and the moment of inertia is defined using geometric primitives based on the study by Hanavan (1964).

Segment Mass

In this chapter, the mass of a segment m_s is expressed as a percentage p_s of the total mass, M, of the subject using regression equations presented by Dempster (1955):

$$m_s = p_s M \tag{7.5}$$

For the lower-extremity segments used in this chapter, p_s of the foot = 0.0145, shank = 0.0465, and thigh = 0.1.

Segment Center of Mass

The center of mass of a segment is the point at which the segment's mass can be considered to be concentrated. When an object is supported at its center of mass, there is no net moment acting on the segment and it will remain in static equilibrium. Of the many methods for estimating the location of the center of mass of a segment, a straightforward physically based method is presented here that can be generalized for any segment. To customize the center of mass locations to an individual subject, we approximate the shape of a segment by a geometric primitive of uniform density (Hanavan 1964) scaled uniformly to the subject's segment length. The benefit of this geometric approach is that we can define any segment (any species even), not just the segments of the human body favored by the anatomists in the development of the published regression equations.

Many geometric shapes may be suitable for different segments, but for simplicity in this presentation, we represent all segments of our lower extremity model by a frustrum of a right circular cone (see figure 7.5). Three parameters define this shape: the length (L), the proximal radius $(R_{proximal})$, and the distal radius (R_{distal}). This shape has a symmetrical cross section, so the center of mass lies along the vector passing through the proximal and distal ends of the segment at a ratio distance, c, from the proximal end of the segment in the LCS. The vector from the origin to the center of mass in the default LCS used in this chapter is

$$\mathbf{r'} = \begin{bmatrix} 0 \\ 0 \\ -cL \end{bmatrix} \tag{7.6}$$

Note that the z-component is negative because the z-axis is defined from the distal end of the segment to the proximal end of the segment (see chapter 2). The value of c from equation 7.6 (relative distance from the proximal end of the segment to the center of mass of the segment) is computed as follows:

$$x = \frac{R_{distal}}{R_{proximal}} \tag{7.7}$$

$$c = \frac{1+2x+3x^2}{4(1+x+x^2)} \quad \text{for } R_{distal} < R_{proximal} \tag{7.8}$$

$$c = 1 - \frac{1+2x+3x^2}{4(1+x+x^2)} \quad \text{for } R_{proximal} < R_{distal} \tag{7.9}$$

Segment Moment of Inertia

Moment of inertia is the name given to rotational inertia, the rotational analog of mass for linear motion. The moment of inertia of a segment is not only related to its mass but also the distribution of the mass throughout the segment, so two segments of the same mass may possess different moments of inertia. The moment of inertia is specified with respect to a segment's principal axes of rotation (typically, and in the example in this chapter, these are the axes of the segment LCS).

In this chapter we assume that a segment is represented by a frustum of a right circular cone (see figure 7.5), with the long axis of the cone being collinear with the segment axis passing from the distal endpoint to the proximal endpoint of the segment (the z-axis) and the distribution of mass is symmetrical about this long axis. The inertia tensor I' in the LCS is represented by the following diagonal matrix:

$$I' = \begin{bmatrix} I'_{xx} & 0 & 0 \\ 0 & I'_{yy} & 0 \\ 0 & 0 & I'_{zz} \end{bmatrix} \tag{7.10}$$

The diagonal components of the moments of I' can be estimated as follows:

$$I'_{xx} = \frac{a_1 a_2 M^2}{\delta L} + b_1 b_2 ML^2 \tag{7.11}$$

$$I'_{yy} = \frac{a_1 a_2 M^2}{\delta L} + b_1 b_2 ML^2 \tag{7.12}$$

$$I'_{zz} = \frac{2 a_1 a_2 M^2}{\delta L} \tag{7.13}$$

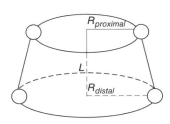

▲**Figure 7.5** A frustum of right cones is created by cutting the top off of a cone such that the cut is parallel to the base of the cone. The right pane shows the geometric representation overlaid on bones.

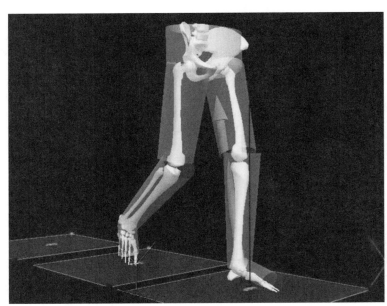

where

$$\delta = \frac{3M}{L(R^2_{proximal} + R_{distal}R_{proximal} + R^2_{distal})\pi}$$

$$a_1 = \frac{9}{20\pi}$$

$$a_2 = \frac{(1 + x + x^2 + x^3 + x^4)}{(1 + x + x^2)^2}$$

$$b_1 = \frac{3}{80}$$

$$b_2 = \frac{(1 + 4x + 10x^2 + 4x^3 + x^4)}{(1 + x + x^2)^2}$$

Computing Kinematics

The solution for the inverse dynamics requires as kinematic inputs the position of the center of gravity $(\vec{r}_s)$, translational velocity $(\vec{v}_s)$, translational acceleration $(\vec{a}_s)$, angular velocity $(\vec{\omega}_s)$, and angular acceleration $(\vec{\alpha}_s)$ of each segment. Translational kinematics are defined for the position, velocity, and acceleration of the center of mass of a segment in the global coordinate system (GCS). Expressing equation 7.6 for the fixed location of the center of mass of segment s in its LCS results in

$$\vec{r}_s' = \begin{bmatrix} 0 \\ 0 \\ -c_s L_s \end{bmatrix} \tag{7.14}$$

Note that the z-axis is directed from the distal end of the segment to the proximal end of the segment, so the location of the center of mass is in the negative z direction from the origin.

We can transform (see equation 2.53) the known location of the center of mass (r_s') of segment s (e.g., from joint j at the proximal end of segment s), defined in the LCS, into the location $r_{s,t}$ in the GCS at time t using the rotation matrix $R'_{s,t}$ and the origin of the segment computed in the pose estimation as follows:

$$\vec{r}_{s,t} = R'_{s,t}\vec{r}_s' + \vec{O}_{s,t} \tag{7.15}$$

The velocity $\vec{v}_{s,t}$ and acceleration $\vec{a}_{s,t}$ of the center of mass are computed from a finite difference method using times $t + 1$, t, and $t - 1$ as

$$\vec{v}_{s,t} = \frac{\vec{r}_{s,t+1} - \vec{r}_{s,t-1}}{2\Delta t} \tag{7.16}$$

$$\vec{a}_{s,t} = \frac{\vec{r}_{s,t+1} - 2\vec{r}_{s,t} + \vec{r}_{s,t-1}}{\Delta t^2} \tag{7.17}$$

As presented in chapter 2 (Equations 2.74-2.76), the angular velocity $\vec{\omega}_{s,t}$ of segment s at time t is computed from the transformation between the rotation matrix, $R_{s,t-1}$, at time $t - 1$ and $R_{s,t+1}$ at time $t + 1$. The angular acceleration of segment s at time t is computed using finite differences from the angular velocity $\omega_{s,t-1}$ at time $t - 1$ and $\omega_{s,t+1}$ at time $t + 1$.

Recording and Representing External Forces

At this point in our presentation of inverse dynamics, the biomechanical model is composed of segments defined by an LCS, decorated with inertial properties, and the kinematics have been computed from the pose of the segments as estimated from the 3-D motion capture data. The actions of muscles cause relative rotation of the segments of the body, but acceleration of the entire body through space requires interaction of the body with the environment.

Ground Reaction Forces

For inverse dynamics calculations of human locomotion, the most prominent external force, and often the only significant external force, is the ground reaction force (GRF). Although it is possible to model the interaction of the foot with the ground during gait, it is beyond the scope of this chapter and the methods herein to accomplish this. More important, it is usually more accurate and far easier to simply record these external forces. Most advanced motion-capture systems allow for simultaneous collection of kinematic and force platform data at compatible sampling rates.

The principles of computing the force applied to a rigid platform are described in a cursory fashion here, but it is important to emphasize that the signals recorded from the force platform allow the computation of a single force vector and that this signal represents the force in the force platform coordinate system (FCS) (see figure 7.6). In the methods presented here it is assumed that only one segment (often one foot) is in contact with the platform at any instant in time. If two feet are in contact with the platform at a given instance, the inverse dynamics presented here are invalid.

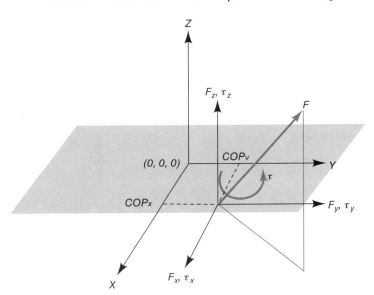

▲ **Figure 7.6** Schematic of a force platform showing the GRF ($\vec{F}$), the center of pressure ($\overline{COP}$), and free moment $\vec{\tau}$.

There are many types, and manufacturers, of force platforms used in biomechanics laboratories, and the number of signals used to compute the GRF differs between types. Regardless of the number of channels of data recorded from a force platform, however, the channels can be assimilated into the following six independent signals that are available for computing the GRF in the FCS: a net force vector

$$\vec{F}' = \left(F_x', F_y', F_z' \right) \tag{7.18}$$

and a net moment vector containing components about the principal axes of the platform

$$\vec{M}' = \left(M_x', M_y', M_z' \right) \tag{7.19}$$

From these six independent signals, nine parameters must be identified for a 3-D inverse dynamics analysis $(F_x, F_y, F_z, \tau_x, \tau_y, \tau_z, COP_x, COP_y, COP_z)$. The three components of the force vector are typically used in the form of equation 7.18. That leaves us with six unknown parameters and three equations.

The center of pressure $(\overline{COP})$ represents the intersection of the force vector with the surface of the platform such that the z-component of $\overline{COP}'$ is the surface on which the subject walks. For simplicity, we will assume it is the top surface of the platform with a height 0 in the motion capture volume:

$$COP_z' = 0 \tag{7.20}$$

The X and Y components of the COP can be calculated using the following formulas:

$$COP_x' = \frac{F_x d_z - M_y}{F_z} \tag{7.21}$$

$$COP_y' = \frac{F_y d_z + M_x}{F_z} \tag{7.22}$$

where d_z refers to the distance from the electrical origin of the force platform to the center of the top surface.

The free moment, $\vec{\tau}$, about the normal axis to the platform is defined as a measure of torque about the normal axis at the interface with the foot and the ground in the case of a gait analysis. It is calculated as

$$\tau'_z = M'_z - COP'_x F'_y - COP'_y F'_x \qquad (7.23)$$

We now have two components unaccounted for, and no equations left, so we must declare their values or estimate these values using on other information. If we assume that the subject cannot pull up on the corners of the platform (e.g., she cannot grasp the platform and does not have sticky feet), we can set the remaining components of the free moments to zero:

$$\tau'_x = \tau'_y = 0 \qquad (7.24)$$

If the subject can produce a moment on the platform, these assumptions are invalid, but this is rarely the case for gait analysis. The assumptions made for three of the necessary parameters should be evaluated by each gait laboratory to ensure that they are true. If, for example, a handle or a cuff were placed on the force platform, then these assumptions would not necessarily be true. For cases such as these, other measurement devices must be used to quantify one or more of the nine parameters.

Transforming the GRF From the FCS Into the GCS

In addition to computing the force vector, we must transform the GRF from the FCS into the GCS (via a rotation matrix) in which the rest of the data are resolved. A common method of locating the platform is to identify the location of the four corners $(\vec{c}_1, \vec{c}_2, \vec{c}_3,$ and $\vec{c}_4)$ of the top surface of the platform in the GCS. The order of the corners is defined by the manufacturer and type of platform. Note that using the following procedure does not require the platform to be flat on the floor or coplanar with any other platform in the laboratory. The top center of the platform (the origin) is assumed to lie at the average location of the corners as follows:

$$\vec{O}_{FP} = 0.25 * (\vec{c}_1 + \vec{c}_2 + \vec{c}_3 + \vec{c}_4) \qquad (7.25)$$

The rotation matrix can be computed by first defining a unit vector along the x-axis of the platform as follows:

$$\hat{i}' = \frac{\vec{c}_4 - \vec{c}_3}{|\vec{c}_4 - \vec{c}_3|} \qquad (7.26)$$

We then define another unit vector along the surface of the platform:

$$\hat{v} = \frac{\vec{c}_2 - \vec{c}_3}{|\vec{c}_2 - \vec{c}_3|} \qquad (7.27)$$

We can find the surface normal using a cross product:

$$\hat{k}' = \hat{i}' \times \hat{v} \qquad (7.28)$$

Finally, we can compute the last unit vector using the right-hand rule:

$$\hat{j}' = \hat{k}' \times \hat{i}' \qquad (7.29)$$

The rotation matrix from the laboratory coordinates to the force plate coordinates is

$$\mathbf{R}_{FP} = \begin{bmatrix} \hat{i}'_x & \hat{i}'_y & \hat{i}'_z \\ \hat{j}'_x & \hat{j}'_y & \hat{j}'_z \\ \hat{k}'_x & \hat{k}'_y & \hat{k}'_z \end{bmatrix} \qquad (7.30)$$

Thus, to transform the force platform from the FCS into the GCS, we compute as follows:

$$\vec{F} = R'\vec{F}' \tag{7.31}$$

$$\overrightarrow{COP} = R'\overrightarrow{COP}' + \vec{O} \tag{7.32}$$

$$\vec{\tau} = R'\vec{\tau}' \tag{7.33}$$

Note that because only center of pressure is a displacement, this is the only vector where it necessary to add the force platform origin to the transformation.

Verifying the GRF

Proper laboratory calibration includes the accurate determination of the positions of the force platforms in the GCS as well as the correct setting of force platform parameters related to the manufacturer and type. Any errors in the parameter settings that define the force platform signals in the data file can lead to incorrect values of kinetic variables calculated during a movement analysis. This spatial synchrony can, and should, be assessed regularly (Holden et al. 2003).

Testing the parameters is quite straightforward and takes very little time. Figure 7.7b shows an example of a machined rod with markers attached securely. The mechanical testing device has a pointed tip at each end and is used together with a handle and a test plate, each with machined conical depressions. This design allows the user to apply a force to the surface of the force platform without applying a moment to the rod. Rigid posts are used to attach five tracking targets to the testing rod. Data are sampled simultaneously from the force platform and the cameras, as forces are applied through the rod to the force platform. If the measurement systems are configured and functioning properly, the rod tip location (determined from the target locations measured by the camera system) should coincide with the center of pressure location (measured by the force platform and transformed into the laboratory coordinate system). In addition, the force platform–measured line of action of the operator-applied force (ground reaction force minus the testing device weight) should align with the kinematic-based estimate of the rod shaft orientation (figure 7.7a).

Problems with the GRF signal are rarely the result of the manufacturer's calibration of the platform but more likely errors in specifying the parameters by the user of the motion-capture software. Errors in specifying the force platform parameters can often be corrected after the data are collected, but this is not always the case.

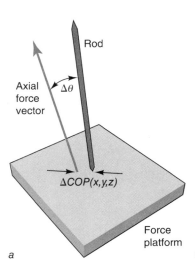

▲ **Figure 7.7** A mechanical testing device that allows a force to be applied to the surface of the force platform without any applied moment.

Computing Joint Reaction Forces and Net Moments

Once we have calculated segment inertial properties, kinematics, and external forces, we have all the inputs required to determine the joint reaction force and the net joint moment. The last major assumption that we will declare before we compute inverse dynamics is that the joint reaction forces are equal and opposite about the joint and that the net joint moments are equal and opposite about the joint. This means that we are assuming that there is no work done by the joint.

Computing the Joint Reaction Force

For segment s we can express Newton's equation (see equation 7.1) as the sum of all forces $\vec{F}_s$ acting on segment s:

$$\sum \vec{F}_s = m_s \vec{a}_s \tag{7.34}$$

If there is more than one unknown force $\vec{F}_s$, it is not possible to compute the individual forces without additional information, so we have to set up the equation such that there is only one unknown. If we go back to figure 7.3, we can compute the net reaction force at the ankle and the knee. For clarity we will add meaningful subscripts to the GRF (grf), to segments (f = foot; l = leg; t = thigh), and to joints (a = ankle; k = knee ; h = hip).

The ankle reaction force $(\vec{F}_a)$ (presented earlier as equation 7.3) is computed as

$$\sum \vec{F}_f = m_f \vec{a}_f \tag{7.35}$$

This equation may be expanded to give

$$\vec{F}_a + \vec{F}_{grf} + m_f \vec{g} = m_f \vec{a}_f \tag{7.36}$$

We can then solve for $\vec{F}_a$ as follows:

$$\vec{F}_a = m_f \left(\vec{a}_f - \vec{g} \right) - \vec{F}_{grf} \tag{7.37}$$

The knee reaction force $(\vec{F}_k)$ (assuming equal and opposite forces at both sides of ankle joint) is computed in a similar manner:

$$\sum \vec{F}_l = m_l \vec{a}_l \tag{7.38}$$

$$-\vec{F}_a + \vec{F}_k + m_l \vec{g} = m_l \vec{a}_l \tag{7.39}$$

$$\vec{F}_k = m_l \left(\vec{a}_l - \vec{g} \right) + \vec{F}_a \tag{7.40}$$

We can substitute the expression for $\vec{F}_a$ from equation 7.35 to get

$$\vec{F}_k = m_l \left(\vec{a}_l - \vec{g} \right) + m_f \left(\vec{a}_f - \vec{g} \right) - \vec{F}_{grf} \tag{7.41}$$

This formulation can be extended to a general expression for any linkage of m segments distal to a joint such that the reaction at the joint can be expressed as

$$\vec{F}_j = \left(\sum_{s=1}^{m} m_s (\vec{a}_s - \vec{g}) \right) - \vec{F}_{grf} \tag{7.42}$$

where we sum over all (m) segments distal to the proximal joint (including the segment for which the joint is associated). Note that the summation in this formulation of the inverse dynamics can be done in any order! That is, it does not need to be solved going from distal to proximal. This formulation is also ideal for handling complex linkages like those that have multiple branches. For example, when trying to compute the wrist reaction force from the motion data of the five fingers, we simply apply the above equation to all of the fingers distal to the joint in any order.

This method cannot solve for a closed chain, which occurs when there is no distal segment in the chain. For example, if a brace is attached securely to the shank and to the thigh and this arrangement provides a moment at the knee, it is not possible for the methods of this chapter to be used to compute the knee moment exerted by the subject unless a force sensor is added between the attachment point on each the thigh and the shank. The hip moment, however, can still be computed because the shank and brace are distal to the hip.

Computing the Joint Moment

Before describing the computation of the net moment acting on a segment, we review the computation of the moment of force $\vec{\tau}$ produced by a single force $\vec{F}$ acting on a segment as follows:

$$\vec{\tau} = \vec{r} \times \vec{F} \tag{7.43}$$

where $\vec{r}$ is the vector from the point at which the moment is computed (in this chapter this is typically the origin of the segment or the center of mass of the segment) and one point along the force vector $\vec{F}$. This equation holds in any segment LCS, but all three terms of the equation must be resolved into the same LCS, as will be demonstrated later in our model of the lower extremity.

Net Inertial Moment   The moments of inertia (I') described earlier in this chapter were constant in the segment LCS. The same moments of inertia (I) in the GCS are not constant because the segment is moving. We, therefore, make our life considerably simpler by computing the inertial contribution to the joint moment ($\vec{\tau}_s^{\,\prime I}$) in the segment LCS and then transforming back to the GCS ($\vec{\tau}_s^{\,I}$). To do this we must first transform the angular velocity and angular acceleration into the LCS:

$$\vec{\omega}_s^{\,\prime} = R_s(\vec{\omega}_s) \tag{7.44}$$

$$\vec{\alpha}_s^{\,\prime} = R_s(\vec{\alpha}_s) \tag{7.45}$$

We then compute the inertial contribution to the net moment $\vec{\tau}_s^{\,\prime I}$ as

$$\vec{\tau}_s^{\,\prime I} = \frac{d}{dt}\left(I_s^{\,\prime}\vec{\omega}_s^{\,\prime}\right) = I_s^{\,\prime}\vec{\alpha}_s^{\,\prime} + \vec{\omega}_s^{\,\prime} \times \left(I_s^{\,\prime}\vec{\omega}_s^{\,\prime}\right) \tag{7.46}$$

and then transform the inertial contribution back into the GCS:

$$\vec{\tau}_s^{\,I} = R_s^{\,\prime}\vec{\tau}_s^{\,\prime I} \tag{7.47}$$

Net Joint Moment   We can express the sum of the moments acting about the center of mass of the foot as equal to the inertial moment in the GCS:

$$\sum \tau_f = \tau_f^{\,I} \tag{7.48}$$

If we go back to figure 7.3 and derive the equations for the net moments at the ankle and the knee, we can generalize this equation to any hierarchical model.

Ankle Moment:

$$\vec{\tau}_a + \vec{\tau}_{grf} - \vec{r}_{a.f} \times \vec{F}_a + \left(\vec{r}_{a.grf} - \vec{r}_{a.f}\right) \times \vec{F}_{grf} = \vec{\tau}_f^{\,I} \tag{7.49}$$

where $\vec{r}_{a.f}$ is the vector from the ankle to the center of mass of the foot, and $\vec{r}_{a.grf}$ is the vector from the ankle to the center of pressure. Rearranging the terms we get

$$\vec{\tau}_a = \vec{\tau}_f^{\,I} - \vec{\tau}_{grf} - [(\vec{r}_{a.grf} - \vec{r}_{a.f}) \times \vec{F}_{grf}] + (\vec{r}_{a.f} \times \vec{F}_a) \tag{7.50}$$

Substituting for , we get

$$\vec{\tau}_a = \vec{\tau}_f^{\,I} - \vec{\tau}_{grf} - [(\vec{r}_{a.grf} - \vec{r}_{a.f}) \times \vec{F}_{grf}] + \vec{r}_{a.f} \times [m_f(\vec{a}_f - \vec{g}) - \vec{F}_{grf}] \tag{7.51}$$

Collecting the terms that cross $\vec{F}_{grf}$, we arrive at

$$\vec{\tau}_a = \vec{\tau}_f^{\,I} - \vec{\tau}_{grf} - \left[\left(\vec{r}_{a.grf} - \vec{r}_{a.f} + \vec{r}_{a.f}\right) \times \vec{F}_{grf}\right] + [\vec{r}_{a.f} \times m_f\left(\vec{a}_f - \vec{g}\right)] \tag{7.52}$$

$$\vec{\tau}_a = \vec{\tau}_f^l - \vec{\tau}_{grf} - \left[\vec{r}_{a.grf} \times \vec{F}_{grf}\right] + [\vec{r}_{a.f} \times m_f(\vec{a}_f - \vec{g})] \tag{7.53}$$

Knee Moment:

Using an equivalent force system (Zatsiorsky and Latash, 1993) to assume equal and opposite moments on both sides of the ankle, we can compute the knee moment as

$$\vec{\tau}_k - \vec{\tau}_a + (-\vec{r}_{k.l} \times \vec{F}_k) + [(\vec{r}_{k.a} - \vec{r}_{k.l}) \times -\vec{F}_a] = \vec{\tau}_l^l \tag{7.54}$$

where $\vec{r}_{k.l}$ is the vector from the knee to the center of mass of the leg, and $\vec{r}_{k.a}$ is the vector from the knee to the ankle. Rearranging the terms, we get

$$\vec{\tau}_k = \vec{\tau}_l^l + \vec{\tau}_a + (\vec{r}_{k.l} \times \vec{F}_k) + [(\vec{r}_{k.a} - \vec{r}_{k.l}) \times \vec{F}_a)] \tag{7.55}$$

Then, substituting for $\vec{F}_a$ and $\vec{F}_k$, we get

$$\vec{\tau}_k = \vec{\tau}_l^l + \vec{\tau}_a + \left\{\vec{r}_{k.l} \times \left[m_l(\vec{a}_l - \vec{g}) + m_f(\vec{a}_f - \vec{g}) - \vec{F}_{grf}\right]\right\}$$
$$+ \left\{(\vec{r}_{k.a} - \vec{r}_{k.l}) \times \left[m_f(\vec{a}_f - \vec{g}) - \vec{F}_{grf}\right]\right\} \tag{7.56}$$

First distributing the cross products for $\vec{r}_{k.l}$ and $\vec{r}_{k.a}$ and then collecting the terms associated with $m_f(\vec{a}_f - \vec{g})$ and $\vec{F}_{grf}$, we get

$$\vec{\tau}_k = \vec{\tau}_l^l + \vec{\tau}_a + \left[\vec{r}_{k.l} \times m_l(\vec{a}_l - \vec{g})\right] + \left[(\vec{r}_{k.a} - \vec{r}_{k.l} + \vec{r}_{k.l}) \times m_f(\vec{a}_f - \vec{g})\right] - [(\vec{r}_{k.a} - \vec{r}_{k.l} + \vec{r}_{k.l}) \times \vec{F}_{grf}] \tag{7.57}$$

Now we substitute for $\vec{\tau}_a$ from 7.52:

$$\vec{\tau}_k = \vec{\tau}_l^l + \vec{\tau}_f^l - \vec{\tau}_{grf} - \vec{r}_{a.grf} \times \vec{F}_{grf} + \vec{r}_{a.f} \times m_f(\vec{a}_f - \vec{g})$$
$$+ \vec{r}_{k.l} \times m_l(\vec{a}_l - \vec{g}) + \vec{r}_{k.a} \times m_f(\vec{a}_f - \vec{g}) - \vec{r}_{k.a} \times \vec{F}_{grf} \tag{7.58}$$

We rearrange and again collect terms associated with $m_f(\vec{a}_f - \vec{g})$ and $\vec{F}_{grf}$,

$$\vec{\tau}_k = \vec{\tau}_l^l + \vec{\tau}_f^l + \left[\vec{r}_{k.l} \times m_l(\vec{a}_l - \vec{g})\right] + \left[\vec{r}_{k.a} \times m_f(\vec{a}_f - \vec{g})\right]$$
$$+ \left[\vec{r}_{a.f} \times m_f(\vec{a}_f - \vec{g})\right] - \vec{\tau}_{grf} - \left[(\vec{r}_{a.grf} + \vec{r}_{k.a}) \times \vec{F}_{grf}\right] \tag{7.59}$$

If we let $\vec{r}_{k.grf}$ be the vector from the knee to the ground reaction force $(\vec{r}_{k.grf} = \vec{r}_{a.grf} + \vec{r}_{k.a})$ and $\vec{r}_{k.f}$ be the vector from the knee to the center of mass of the foot $(\vec{r}_{k.f} = \vec{r}_{a.f} + \vec{r}_{k.a})$, the previous equation becomes

$$\vec{\tau}_k = \vec{\tau}_l^l + \vec{\tau}_f^l + \left[\vec{r}_{k.l} \times m_l(\vec{a}_l - \vec{g})\right] + \left[\vec{r}_{k.f} \times m_f(\vec{a}_f - \vec{g})\right] - \vec{\tau}_{grf} - [\vec{r}_{k.grf} \times \vec{F}_{grf}] \tag{7.60}$$

Finally, this can be rearranged into a general form as follows:

$$\vec{\tau}_j = \left[\sum_{s=1}^{m} \vec{\tau}_s^l + \vec{r}_{j.s} \times m_s(\vec{a}_s - \vec{g})\right] - \vec{\tau}_{grf} - [\vec{r}_{j.grf} \times \vec{F}_{grf}] \tag{7.61}$$

Thus, we have developed a general equation to compute the net joint moments by summing the above expression over all *(m)* segments distal to the joint (in any order!) where $\vec{r}_{j.grf}$ is the vector from the joint to the ground reaction force and $\vec{r}_{j.s}$ is the vector from the joint to the center of mass of segment *s*. Like our above expression for the joint forces, this formulation for the joint moment can be solved in any order for the segments distal to the joint and is therefore also ideal for handling complex linkages, such as the hand, that have multiple branches.

The net joint moment just derived in the inverse dynamics calculations is referred to in the literature as the *internal moment*. In other words the moment created by the muscles and other tissues. The literature also references the *external moment,* which in gait analysis is the external moment produced by the ground reaction force that must be balanced by the internal moment produced by the muscles and ligaments. For example, the extensor muscles of the leg during the stance phase of gait counteract the GRF. The external moment is equal in magnitude but opposite in sign to the internal moment. The literature also references the *support moment,* which is the sum of all extensor moments at the hip, knee, and ankle. In this approach, all extensor moments must be assigned the same sign convention or the addition doesn't make sense. As a consequence, the hip and ankle moments are defined by the right hand rule and sagittal knee moment is defined by the right hand rule, but then multiplied by -1. Regardless of which sign convention has been used, it is important that biomechanists be aware of the sign convention when reading the literature and that they report clearly the representation they used in their own publications. It is highly recommended that the data are reported in anatomical notation (e.g., flexion/extension, abduction/adduction, and axial rotation) because these are not affected by different coordinate system definitions (e.g., y-up versus z-up conventions) and that the sign convention (e.g., flexion versus extension) is noted on the axes of the graphs.

PRESENTATION OF THE NET MOMENT DATA

The joint reaction forces and net joint moments that are calculated from an inverse dynamics procedure are 3-D vectors computed in the GCS but are typically resolved into an anatomically meaningful coordinate system when reported (figure 7.8). There is no standard resolution coordinate system, however, which means that readers of journal articles must pay close attention to the definition of the signals that are presented. The most straightforward representation would be to leave these signals in the GCS (Bresler and Frankel 1950; Winter et al. 1995). This was suggested on the basis that the individual's line of progression is generally aligned along one of the planes of the global reference system, and thus the sagittal plane of the lower extremity and the line of progression will be essentially the same. This interpretation is a 2-D centric view of gait analysis and may be useful for the sagittal plane (e.g., moments about the *x*-axis in our GCS definition), but it is difficult to interpret the other two components of the signals because the LCS is rarely aligned precisely with the GCS in every pose. It is our recommendation, therefore, that users refrain from using this 2-D representation and present the joint forces and moments in a coordinate system that can be readily understood regardless of the subject's (and individual segment's) position and orientation in the GCS.

There are four representations of the net joint moment and joint reaction force that have documented merit for 3-D analysis (see figure 7.9), and it is not clear what rules should be used for selecting from the four because they all have merit in some context but can sometimes differ dramatically in the results. Option 1 is to resolve the joint reaction force and net joint moment into the proximal segment LCS (Schache and Baker, 2007). Option 2 is to resolve the signals into the distal segment LCS (Kaufman et al. 2001). Option 3 is to resolve the signals into the joint coordinate system (Schache and Baker 2007; Astephen et al. 2008). Option 4 is ad hoc, in which the anterior-posterior and inferior-superior moments are projected onto the plane of progression (Mundermann et al. 2005). The confusion for the reader is that each of these representations is mathematically sensible and, depending on the parameters of interest, may be superior to the others in the appropriate context (Schache and Baker 2007; Schache et al. 2008). This confusion extends, also, to the international biomechanics societies in which no international standard has been proposed, let alone adopted. In this chapter, we refrain from advocating a particular option. The researcher, therefore, must take note of the coordinate system in which the moments are presented and be very careful in comparing results from different journal articles. It is unfortunately quite rare for authors to specify the resolution coordinate system for their joint moments, so much of the biomechanics literature contains questionable moment data.

This issue gets more confusing because options 1 and 2 result in the signals' being vectors, which is consistent with our representations of joint angular velocity, angular acceleration, and joint force, but inconsistent with a JCS representation of joint angles. Option 3 is a nonorthogonal representation, which means that the reaction force and net moments are not vectors, so this representation is consistent with joint angles (which makes it popular with clinical gait analysis) but not with angular velocity, angular acceleration, or joint force. The support moment (described in the previous section) cannot be computed using this approach because only vectors can be added sensibly. Option 4 is not consistent with any of our kinematic measures. Readers, therefore, must be careful that they are only comparing "like" signals: for example, comparing vectors with vectors, and comparing nonorthogonal representations with other consistent nonorthogonal representations.

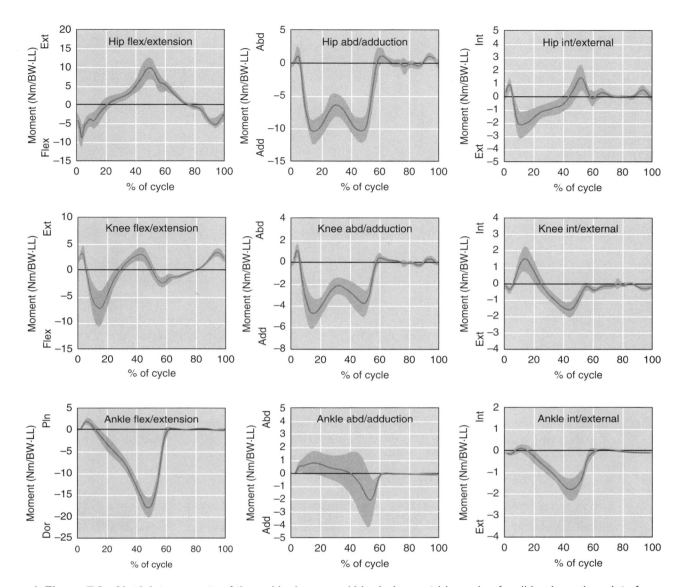

▲ **Figure 7.8** Net joint moments of the ankle, knee, and hip during a stride cycle of walking based on data from 13 subjects (the subject for the text book data is one of these subjects). The solid line represents the mean value, whereas the gray zone indicates ±1 standard deviation. All joint angles used an XYZ cardan sequence, and the joint moments (with segment mass) were resolved in the proximal segment's coordinate system. The sign conventions are as follows: hip joint moment (positive flexion, adduction, internal rotation); knee joint moment (positive flexion, adduction, internal rotation); ankle joint moment (positive flexion, adduction, internal rotation).

Just as there is no standard convention for choosing the coordinate system in which to represent joint reaction force and net moment, most published articles differ in choosing normalization (scaling) factors. Scaling the net joint moments is an attempt to remove between-subject differences. In a clinical setting in which an individual's data are being collected and a report is generated on that subject, there is perhaps no need to scale the data. In a research study, however, in which data are collected on multiple subjects in different conditions, it probably is appropriate to scale the data for purposes of comparison. The most-reported methods of scaling joint kinetic data are scaling to body mass, N·m/kg, (Vaughan 1996; Winter et al. 1995), or to body weight, BW, and leg length, N·m/(BW)(LL) (Meglan and Todd, 1994).

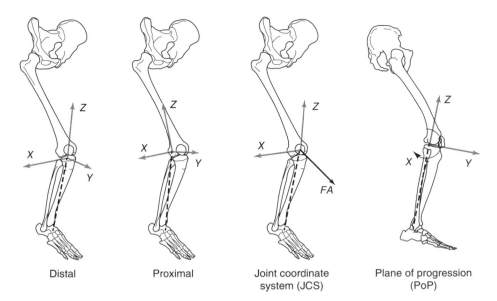

| Distal | Proximal | Joint coordinate system (JCS) | Plane of progression (PoP) |

▲ **Figure 7.9** Four anatomical reference frames for the net knee joint moment. 1) Resolve the net moment into the shank coordinate system. 2) Resolve the net moment into the thigh coordinate system. 3) Project the net moment vector onto the unit vectors of the Joint Coordinate System (see chapter 2; typically the flexion/extension axis of the thigh segment, the axial rotation axis of the shank segment, and an axis perpendicular to these two axes, which for gait is close to the abduction/adduction axis). 4) The plane of progression (PoP) frame fixes the flexion axis perpendicular to the plane of progression, whereas the adduction and internal rotation axes are taken from the distal frame and projected onto the plane of progression.

Reprinted from *Clinical Biomechanics,* Vol. 26(1), S.C. Brandon and K.J. Deluzio, "Robust features of knee osteoarthritis in joint moments are independent of reference frame selection," pgs. 65-70, copyright 2011, with permission of Elsevier.

FROM THE SCIENTIFIC LITERATURE

Brandon, S.C., and K.J. Deluzio. 2011. Robust features of knee osteoarthritis in joint moments are independent of reference frame selection. *Clinical Biomechanics* 26(1):65-70.

The physiological interpretation of 3-D joint moments can change when the moments are transformed to different body-fixed anatomical reference frames. The purpose of this paper was to identify features of the hip, knee, and ankle joint moment waveforms that, regardless of the choice of reference frame, were consistently different between control subjects and subjects with knee osteoarthritis. Subjects walked at self-selected speed, and external 3-D joint moments were calculated with a standard inverse dynamics approach. Moments were then expressed using the four alternative reference frames shown in figure 7.9: distal, proximal, joint coordinate system (JCS), and plane-of-progression (PoP). Finally, the primary features of variance across all four systems were extracted using principal component analysis.

The magnitude and shape of every joint moment were different between the four reference frames. However, regardless of the choice of reference frame, subjects with knee osteoarthritis exhibited decreased overall hip adduction moment magnitudes, increased overall knee adduction moment amplitudes, decreased knee internal rotation moment amplitudes, and increased early-stance ankle adduction magnitudes (figure 7.10). The authors concluded that these four robust waveform features are characteristics of the pathogenesis of knee osteoarthritis and are not merely artifacts of reference frame selection.

This study also demonstrated the importance of using an appropriate body-fixed anatomical reference frame for reporting joint moments. For example, when hip flexion moments were expressed in the proximal (pelvis) reference frame, subjects with knee osteoarthritis displayed significantly greater overall magnitudes than control subjects throughout stance. However, this feature was not significant (P = .07) using either the globally fixed flexion axis from the plane-of-progression reference frame or the distal-segment flexion axis (P = .83). Not only does the choice of reference frame affect the magnitude and shape of joint moments; it can affect joint moments differently for different subjects.

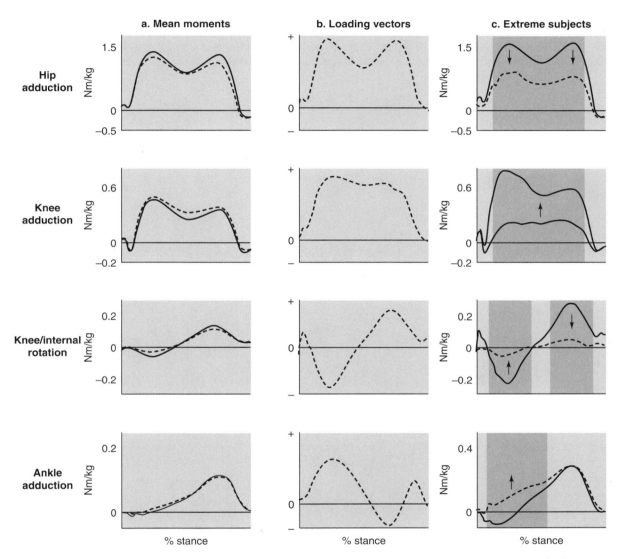

Figure 7.10 Four robust changes due to osteoarthritis that are independent of the choice of reference frame, identified using principal component analysis. *(a)* Mean waveforms, averaged across all anatomical reference frames for control (solid black line) and osteoarthritis (dashed gray line). *(b)* Loading vectors that indicate biomechanical changes in overall magnitude (hip and knee adduction), amplitude (knee internal rotation), and early-stance magnitude (ankle adduction). *(c)* Extreme subjects are shown for control (solid black line) and osteoarthritis (dashed gray line) groups to characterize robust group differences. Arrows give the direction of the osteoarthritis group relative to the control group throughout the shaded regions.

Reprinted from *Clinical Biomechanics,* Vol. 26(1), S.C. Brandon and K.J. Deluzio, "Robust features of knee osteoarthritis in joint moments are independent of reference frame selection," pgs. 65-70, copyright 2011, with permission of Elsevier.

JOINT POWER

Joint power represents the rate of work at which muscles add or remove energy from the system. A positive joint power indicates that muscles are adding energy as in a concentric contraction, whereas a negative power indicates that muscles are removing energy from the system as in an eccentric contraction. The joint power at a joint (P_j) is given by the following equation:

$$P_j = P_{proximal} + P_{distal} \tag{7.62}$$

where $P_{proximal}$ is the segmental power at the joint attributable to the segment on the proximal side of the joint and P_{distal} is the segmental power at the joint attributable to the segment on the segment on distal side of the joint (see figure 7.11). If we apply Newton's third law of equal and opposite forces and the simplification of equal and opposite moments on both sides of the joint system (Zatsiorsky and Latash 1993), this equation becomes

$$P_j = \left(\vec{F}_j \cdot \vec{v}_{proximal}\right) + \left(\vec{\tau}_j \cdot \vec{\omega}_{proximal}\right) + \left(-\vec{F}_j \cdot \vec{v}_{distal} - \vec{\tau}_j \cdot \vec{\omega}_{distal}\right) \tag{7.63}$$

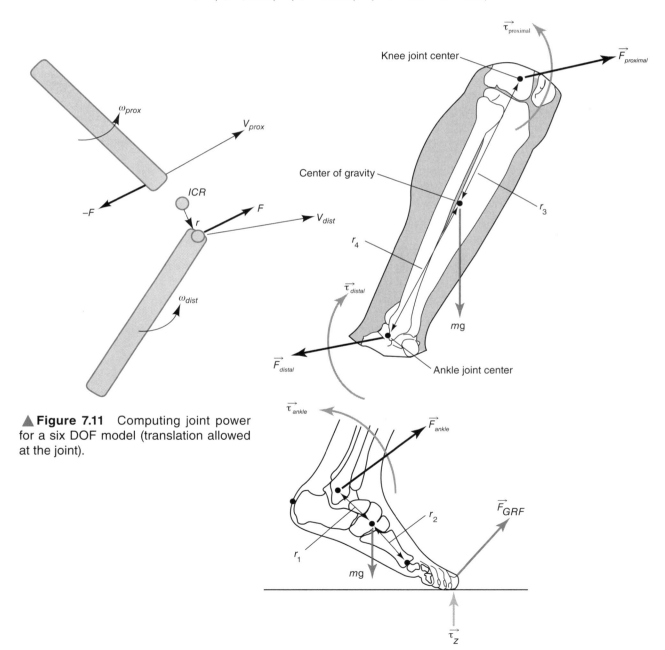

▲ **Figure 7.11** Computing joint power for a six DOF model (translation allowed at the joint).

where $\vec{v}$ is the translational velocity of the joint relative to the GCS. Now if we assume that the joint has no translation and thus $\vec{v}_{proximal} = \vec{v}_{distal}$ (which is an overassumption for six DOF joints), then this equation simplifies to

$$P = (\vec{\tau}_j \cdot \vec{\omega}_{proximal}) + (-\vec{\tau}_j \cdot \vec{\omega}_{distal}) \tag{7.64}$$

This equation can be further simplified to

$$P = \vec{\tau}_j \cdot \left(\vec{\omega}_{proximal} - \vec{\omega}_{distal}\right) = \vec{\tau}_j \cdot \vec{\omega}_j \tag{7.65}$$

where $\vec{\omega}_j$ is the joint angular velocity.

If we have a true six DOF joint and the velocity at the proximal end of the segment on the distal side of the joint does not equal the velocity at the distal end of the segment on the proximal side of the joint (see figure 7.11), then we restate equation 7.58 as

$$P_j = (\vec{F}_j \cdot \vec{v}_{proximal}) + (\vec{\tau}_j \cdot \vec{\omega}_{proximal}) + (-\vec{F}_j \cdot \vec{v}_{distal} - \vec{\tau}_j \cdot \vec{\omega}_{distal}) \tag{7.66}$$

where $\vec{v}_{proximal}$ is the translational velocity of this distal end of the segment on the proximal side of the joint and $\vec{v}_{distal}$ is the translational velocity of this proximal end of the segment on the distal side of the joint. The preceding expression can be reduced to

$$\vec{P}_j = \vec{\tau}_j \cdot \vec{\omega}_j + \vec{F}_j \cdot \Delta\vec{v} \tag{7.67}$$

where $\Delta\vec{v}$ is the difference between the translational velocities on both sides of the joint. If we introduce a vector $\vec{r}$ that goes from the joint's instant center of rotation (ICR) to the joint center that is located at the proximal of the distal segments (figure 7.12), the relationship between the moment at ICR and the moment at the joint will be given by

$$\vec{\tau}_{ICR} = \vec{\tau}_j + (\vec{r} \times F_j) \tag{7.68}$$

Rearranging this equation, we get

$$\vec{\tau}_j = \vec{\tau}_{ICR} - (\vec{r} \times \vec{F}_j) \tag{7.69}$$

In addition, the difference between the translational velocities ($\Delta\vec{v}$) can be expressed as

$$\Delta v = \vec{\omega}_j \times \vec{r} \tag{7.70}$$

Using the two prior equations, we can now express equation 7.64 as

$$\vec{P}_j = [\vec{\tau}_{ICR} - (\vec{r} \times \vec{F}_j)] \cdot \vec{\omega}_j + \vec{F}_j \cdot (\vec{\omega}_j \times \vec{r}) \tag{7.71}$$

We can simplify this via vector algebra to

$$\vec{P}_j = \vec{\tau}_{ICR} \cdot \vec{\omega}_j \tag{7.72}$$

This equation indicates that for a six DOF joint, if the difference between the translational velocities ($\Delta\vec{v}$) is accounted for, then the joint power will be equivalent to the joint power computed about the ICR (Buczek et al. 1994).

INTERPRETATION OF NET JOINT MOMENTS

Interpretation of the inverse dynamics results commonly centers on some form of pattern recognition based on deviations of signals from a normative equivalent. This strategy identifies differences from normal motion but rarely explains their causes. This is because it is extraordinarily difficult to infer the causal relationships between a force or moment and the resulting movement trajectory. For example, let us introduce a three-link planar model (figure 7.13) consisting of a shank, thigh, and combined head-arm-trunk segment (in this simplified model, the foot is fixed to the floor and thus ignored in our kinetic analysis). The model, a modified version of a model presented by Zajac and Gordon (1989), has two muscles, a soleus and a gastrocnemius. At the start of the movement in figure 7.13, the gastrocnemius and soleus contract and then the forces decrease to zero at the end of the movement. The top row of figure 7.14 shows the time history of the joint angles at the ankle, knee, and hip, whereas the bottom row shows the moment history generated by the combined action of the soleus and the gastrocnemius.

Allard, P., R. Lachance, R. Aissaoui, and M. Duhaime. 1996. Simultaneous bilateral able-bodied gait. *Human Movement Science* 15:327-46.

The purpose of this paper was to report on lower-extremity muscle power and mechanical energy during gait over two consecutive stride cycles. The walking speed of the subjects was 1.30 m/s. The relative duration of the stance phase for the right limb was not significantly different from that for the left limb. These authors suggested that the differences in the limb powers were mostly in the sagittal plane and reflected gait adjustments rather than asymmetry. In general, the differences occurred during the absorption portion of the stance phase. There was no difference in the total positive work between the limbs. However, the right limb developed significantly greater total negative work than the left limb did (see figure 7.12).

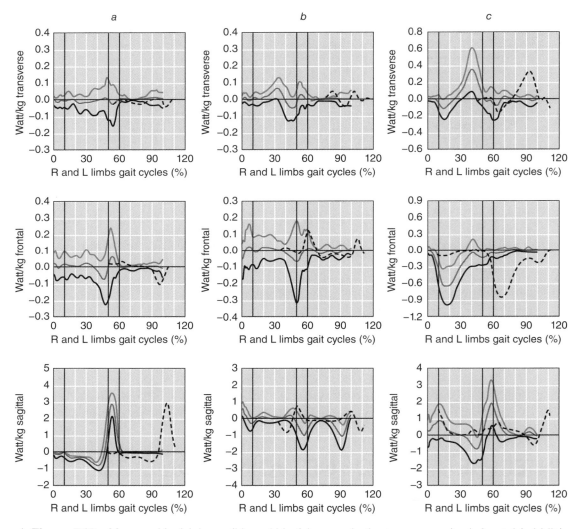

▲ **Figure 7.12**   Mean ankle *(a)*, knee *(b)*, and hip *(c)* power in the transverse (top), frontal (middle), and sagittal (bottom) plane with standard deviations for the right limb (solid line). Only the mean values are given for the left limb (dashed line). The right limb stance phase occurs from 0% to 61%. The swing phase of the left limb occurs from 11% to 61%, and the subsequent left limb toe-off occurs at 112%.

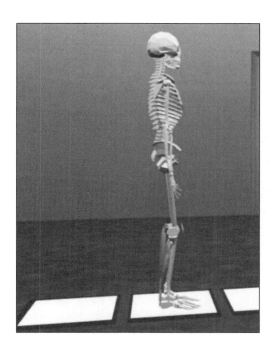
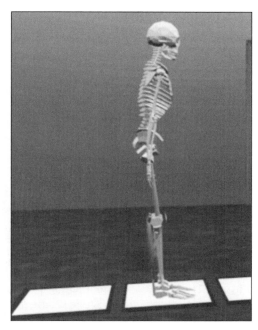

▲ **Figure 7.13** A three-link planar model. The subject starts in the above position (left) and contracts the soleus and gastrocnemius to reach the final position (right).

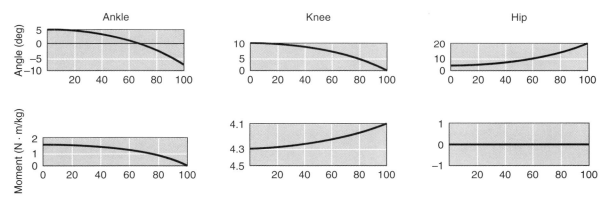

▲ **Figure 7.14** Kinematic and kinetic output from a computer simulation of the model's motion.

Note that at the knee we see movement from 10° of flexion at the start to full extension at the end despite the fact that the moment at the knee is generated by the gastrocnemius, which produces only a knee flexor moment. At the hip, the model moves into increased hip flexion despite the absence of any muscles crossing the hip. Thus, for our simple model, inverse dynamics do not offer a full explanation for the observed results. One method that can be used to supplement traditional inverse dynamics is *induced acceleration analysis*.

Induced Acceleration Analysis

Induced acceleration analysis (IAA) is devoted to the calculation of the instantaneous acceleration of each joint component, or generalized coordinate, $(\ddot{\vec{q}})$, due to one input torque $(\vec{\tau})$. To understand how IAA works, consider the two-segment inverted pendulum in figure 7.15, in which one end of the pendulum is constrained to the ground with a pin joint, and the two segments are constrained to each other by a pin joint. We will take advantage of the simplicity of a 2-D analysis to expand the equations of motion from the 3-D presentation earlier in the chapter, so that we can explicitly introduce the generalized coordinates.

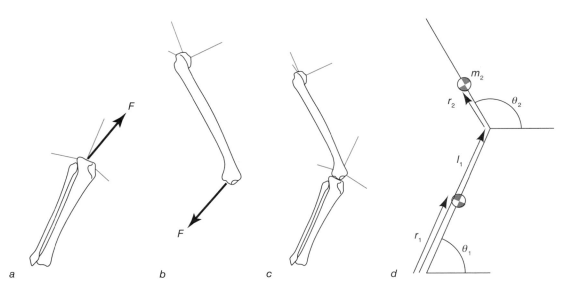

▲**Figure 7.15** A two-segment planar pendulum is used as a reference for the three analytical techniques presented. θ_1 and θ_2 refer to the orientation of each segment. Segment 1 is constrained to the ground with a pin joint; (b) segment 2 is constrained to segment 1 by a pin joint. Segment 2 is unconstrained at its distal end, so the net moment and force are calculated for this segment first.

Equations 7.42 and 7.53 can be expanded as follows:

$$F_{2x} = m_2 a_{2_{cgx}} = m_2\left(-l_1 sin\theta_1\left(\ddot{\theta}_1\right) + -l_1 cos\theta_1\left(\dot{\theta}\right)^2 - r_2 sin\theta_2\left(\ddot{\theta}_2\right) - r cos\theta_2\left(\dot{\theta}_2\right)^2\right) \tag{7.73}$$

$$F_{2y} = m_2 a_{2_{cgy}} = m_2\left[l_1 cos\theta_1\left(\ddot{\theta}_1\right) - l_1 sin\theta_1\left(\dot{\theta}_1\right)^2 + r_2 cos\theta_2\left(\ddot{\theta}_2\right) - r_2 sin\theta_2\left(\dot{\theta}_2\right)^2 + g\right] \tag{7.74}$$

$$\tau_2 = I_2\ddot{\theta}_2 + r_2 \times F_2 = I_2\ddot{\theta}_2 + r_2 cos\theta_2 F_{2y} - r_2 sin\theta_2 F_{2x} \tag{7.75}$$

The calculations for segment 1 use the results of equations 7.54 through 7.61, where the joint reaction force is equal and opposite, and the joint moment is equal and opposite.

$$F_{1x} = F_{2x} + m_1 a_{1_{cgx}} = F_{2x} + m_1 r_1\left(-sin\theta_1\left(\ddot{\theta}_1\right) - cos\theta_1\left(\dot{\theta}_1\right)^2\right) \tag{7.76}$$

$$F_{1y} = F_{2y} + m_1 a_{1_{cgx}} = F_{2y} + m_1\left[r_1\left(cos\theta_1\left(\ddot{\theta}_1\right) - sin\theta_1\left(\dot{\theta}_1\right)^2\right) + g\right] \tag{7.77}$$

$$\tau_1 = I_1\ddot{\theta}_1 + \tau_2 + m_1 r_1 cos\theta_1\left(r_1 cos\theta_1\left(\ddot{\theta}_1\right) - r_1 sin\theta_1\left(\dot{\theta}_1\right)^2\right)$$
$$+ m_1 r_1 cos\theta_1 g + m_1 r_1 sin\theta_1\left(r_1 sin\theta_1\left(\ddot{\theta}\right) + r_1 cos\theta_1\left(\dot{\theta}_1\right)^2\right) + l_1 cos\theta_1 F_{2y} - l_1 sin\theta_1 F_{2x} \tag{7.78}$$

Given that the position of the proximal end of segment 1 is fixed to the ground, the pose (position and orientation) of this model can be described completely by variables θ_1 and θ_2

$$\tau_1 = [I_1 + m_1 r_1^2 + m_2 l_1^2 + m_2 r_2 l_1 \cos(\theta_1 - \theta_2)]\ddot{\theta}_1$$
$$+ [I_2 + m_2 r_2^2 + m_2 r_2 l_1 \cos(\theta_1 - \theta_2)]\ddot{\theta}_2$$
$$+ m_2 r_2 l_1 \sin(\theta_1 - \theta_2)(\dot{\theta}_1^2 + \dot{\theta}_2^2)$$
$$+ (m_1 r_1 \cos\theta_1 + m_2 l_1 \cos\theta_1 + m_2 r_2 \cos\theta_2)g \tag{7.79}$$

$$\tau_2 = m_2 r_2 l_1 \cos(\theta_1 - \theta_2)\ddot{\theta}_1 + (I_2 + m_2 r_2^2)\ddot{\theta}_2$$
$$- m_2 r_2 l_1 \sin(\theta_1 - \theta_2)\dot{\theta}_1^2$$
$$+ m_2 r_2 g \cos\theta_2 \tag{7.80}$$

Equations 7.79 and 7.80 are more complicated than the six equations (7.71-7.76) of the Newton-Euler formalism, but the reduction to two equations containing only a minimum set of variables (θ_1 and θ_2) and their derivatives is required for induced acceleration analysis. This minimum set of variables required to establish the pose of the model are referred to as the *generalized coordinates*. By collecting all of the acceleration, velocity, and gravity terms together, we can use simple substitution to simplify the notation for equations 7.79 and 7.80:

$$\tau_1 = M_{11}\ddot{\theta}_1 + M_{12}\ddot{\theta}_2 + C_{11} + C_{12} + G_{11} \tag{7.81}$$

$$\tau_2 = M_{21}\ddot{\theta}_1 + M_{22}\ddot{\theta}_2 + C_{21} + C_{22} + G_{21} \tag{7.82}$$

We can further consolidate these equations into a convenient representation by adopting the following matrix notation

$$\begin{bmatrix} \tau_1 \\ \tau_2 \end{bmatrix} = \begin{bmatrix} M_{11} & M_{12} \\ M_{21} & M_{22} \end{bmatrix} \begin{bmatrix} \ddot{\theta}_1 \\ \ddot{\theta}_2 \end{bmatrix} + \begin{bmatrix} C_1 \\ C_2 \end{bmatrix} + \begin{bmatrix} G_1 \\ G_2 \end{bmatrix} \tag{7.83}$$

and introducing the generalized coordinates as a vector $\vec{q} = (\theta_1, \theta_2)$

$$\vec{\tau} = M(q)\ddot{q} + \vec{C}(q,\dot{q}) + \vec{G} \tag{7.84}$$

where M is the inertia matrix (we do not use the notation I because M may contain mass and moments of inertia), $\vec{C}$ is the vector of velocity-related terms, and $\vec{G}$ is the gravity vector and any external forces applied to the model. Details of this nomenclature can be found in most engineering mechanics textbooks.

Rewriting equation 7.84 gives us the following:

$$\ddot{\vec{q}} = M^{-1}(\vec{q})\vec{\tau} + M^{-1}(\vec{q})\vec{C}(\vec{q},\dot{\vec{q}}) - M^{-1}(\vec{q})\vec{G}(\vec{q}) \tag{7.85}$$

Equation 7.85 holds within it some very key biomechanical principles. Because the elements of the inverse mass matrix (M^{-1}) will always be fully populated, it is clear that the acceleration of a single moment at any joint will act to accelerate all of the joints of the body and not just the joint it crosses. Also, because the elements of M^{-1} contain the joint angular position (as well as the inertial properties), it should also be clear that the acceleration produced by a joint moment will be a function of both the magnitude of the moment and the pose of all of the segments of the body. Thus, if the body posture changes, the acceleration produced by a given moment will be either increased or decreased and in some cases even the direction of the acceleration produced can change. For example, during running, the ankle plantar flexors during midstance will generally act as knee extensors. However, as the runner approaches toe-off and the knee and hip are more flexed, the ankle plantar flexors may act as knee flexors.

It is clear that induced acceleration lies at the intersection (conceptually) of the field of forward dynamics (predictive analyses) and inverse dynamics (predominantly descriptive analyses used traditionally for clinical studies) because induced accelerations can be used to interpret experimental or simulated data.

The interpretation of movement analysis data is a critical step in the routine use of movement analysis. This interpretation allows the data in the clinical report to be presented in terms that are understandable to a clinician. These analyses for interpreting movement data have recently gained popularity in the biomechanics community. The basis of the analysis is to identify the contribution of a particular muscle group (e.g., plantar flexors) to an outcome measure (e.g., acceleration of the center of mass of the body). This analysis was named *induced acceleration analysis* by Zajac and Gordon (1989). They used a simple planar model to demonstrate that the gastrocnemius muscle, anatomically a knee flexor and ankle plantar flexor, can act as a knee extensor in certain circumstances.

Patients who have lost function in one or more muscle groups often use adaptive strategies that rely on the ability of other muscle groups to accelerate the joints they do not cross. Such patients can unknowingly use these principles to produce compensatory control strategies that may enable them to continue to walk. More research needs to be done as we attempt to understand and ultimately improve the gait of people with functional limitations. Kepple and colleagues (1998) used a sensitivity approach to examine the capacity of individual muscles for support and propulsion in crouch gait. Neptune and colleagues (2001) and Anderson and Pandy (2001) used a dynamic optimization approach that extended

FROM THE SCIENTIFIC LITERATURE

Siegel, K.L., T.M. Kepple, and S.J. Stanhope. 2007. A case study of gait compensations for hip muscle weakness in idiopathic inflammatory myopathy. *Clinical Biomechanics* 22:319-26.

The purpose of this case series was to quantify different strategies used to compensate in gait for hip muscle weakness. An instrumented gait analysis was performed of three women who had been diagnosed with idiopathic inflammatory myopathies, and this analysis was compared with that of a healthy unimpaired subject. Lower-extremity joint moments obtained from the gait analysis were used to drive an induced acceleration model that determined each moment's contribution to upright support, forward progression, and hip joint acceleration. Results showed that after midstance, the ankle plantar flexors normally provide upright support and forward progression while producing hip extension acceleration. In normal gait, the hip flexors eccentrically resist hip extension, but the hip flexor muscles of the impaired subjects (S1-S3) were too weak to control extension. Instead, S1-S3 altered joint positions and muscle function to produce forward progression while minimizing hip extension acceleration. S1 increased knee flexion angle to decrease the hip extension effect of the ankle plantar flexors. S2 and S3 used either a knee flexor moment or gravity to produce forward progression, which had the advantage of accelerating the hip into flexion rather than extension and decreased the demand on the hip flexors. This study showed how gait compensations for hip muscle weakness can produce independent (i.e., successful) ambulation, although at a reduced speed compared with normal gait. Knowledge of these successful strategies can assist in the rehabilitation of patients with hip muscle weakness who are unable to ambulate and potentially can be used to reduce these patients' disability.

these techniques in order to estimate the contributions of the individual muscles to propulsion and support in normal walking. These studies have increased our understanding of normal walking and advancing the capabilities of clinical movement analysis.

Induced Power Analysis

A net moment can contribute to the power of a segment (i.e., accelerate a segment) to which it is not applied through the intersegmental reaction forces (e.g., Fregly and Zajac 1996). Each net moment's contribution to the instantaneous segment powers can be determined at each instant in time from the current state of the system and the instantaneous acceleration induced by that moment. The segment power analysis provides a clear interpretation of a muscle's (or net moment's) influence on a segment because a linear transformation exists between segment power and acceleration (i.e., if the power is positive, the muscle or net moment acts to accelerate the segment, whereas if power is negative, the net moment acts to decelerate the segment).

Neptune and colleagues (2001) used an induced acceleration and induced power analysis to demonstrate the functional role of the ankle muscles during normal gait, a technique that has been controversial in the clinical movement analysis community. The induced acceleration and induced power analysis technique allowed the authors to demonstrate conclusively that at push-off, the ankle muscles contribute substantially to trunk support and forward progression and that the uniarticular and biarticular muscles can have distinctly different functional roles.

Figure 7.16 shows right-leg mechanical power produced by the soleus and gastrocnemius and its distribution to the leg and trunk during the gait cycle. The net power produced by the soleus and gastrocnemius is the sum of the power to or from the leg segments (right leg), trunk segments (trunk), and contralateral leg segments (small and not shown). Positive (negative) net power produced by the muscle indicates power generation (absorption). Positive (negative) power to the leg and trunk indicates that the muscle is accelerating (decelerating) the leg and trunk in the direction of movement. Note that how the uniarticular soleus and biarticular gastrocnemius generate power is in the opposite direction to and from the leg and trunk (Neptune et al. 2001).

Siegel and colleagues (2004) used an induced acceleration model to estimate the ability of net joint moments to transfer mechanical energy through the leg and trunk during gait. They reported that pairs of joint moments with opposite energetic effects (knee extensor vs. gravity, hip flexor vs. ankle plantar flexor) worked together to balance energy flow through the segments. This intralimb coordination suggests that moments with contradictory effects are generated simultaneously to control mechanical energy flow in the body during walking.

SOURCES OF ERROR IN THREE-DIMENSIONAL CALCULATIONS

Combining GRF, kinematic, and anthropometric data leaves many possible sources of error in the calculation of joint reaction forces, net moments, and induced accelerations and powers. Of these, the GRF data in the force platform coordinate system are the most accurate and reliable, but the transformation from the FCS to the GCS is susceptible to user error in the specification of the parameters that are used in computing this transformation. Although the segmental inertial parameters vary widely based on the technique used to derive them, they do not greatly contribute to the end result (at least for walking). The sources of error in estimating the kinematics were described in detail in chapter 2.

In the calculations in this chapter, the body segments are assumed to be rigid. This assumption, although useful, introduces potential sources of error. The foot consists of multiple segments rather than a single rigid body. Regardless of the number of rigid segments into which the foot is divided, the kinetics from the ankle remain the same, so unless the reader is exploring the kinetics of the foot specifically, there is nothing to be gained (from a kinetics perspective) from parsing the foot into several segments.

Although rigid segments are an obvious idealization, this can be justified to some degree for the leg segment but not for the foot or thigh segments. The thigh has a great deal of muscle tissue surrounding the femur, and some researchers use a wobbling mass model to represent the segment (Pain and Challis 2001). For walking or running, it is questionable whether the gains from treating the segment as a deformable structure are just gains in compensating for soft tissue artifact. The kinematic data needed to calculate joint moments are highly dependent on the accuracy of the pose estimation, which in turn is sensitive to soft tissue artifact.

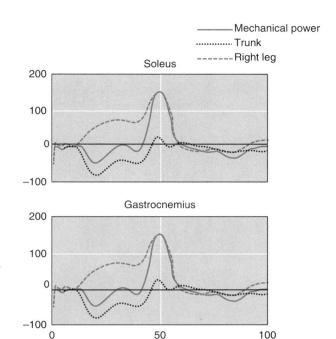

▲ **Figure 7.16** The right-leg mechanical power produced by the soleus (SOL) and gastrocnemius (GAS) and distribution to the leg and trunk is shown during the gait cycle.

Reprinted from *Journal of Biomechanics*, Vol. 34, R.R. Neptune, S.A. Kautz, and F.E. Zajac, "Contributions of the individual ankle plantar flexors to support, forward progression and swing initiation during walking," pgs. 1387-1398, copyright 2001, with permission of Elsevier.

SUMMARY

There are several different methods for computing 3-D kinetics, although only one, the Newton-Euler procedure, is discussed in this chapter, and there are several different approaches to presenting the results. A 3-D analysis involves merging three types of data: (1) external force data (usually the GRF), (2) the 3-D coordinates of markers describing the individual segments, and (3) the anthropometric data of the individual segments. Error associated with each of these types of data must be taken into consideration. From the calculation of net moments, we can further calculate joint power, which represents the rate of work at which muscles add or remove energy from the system. It is extraordinarily difficult to infer the causal relationships between a force or moment and the resulting movement trajectory. Joint moments can be presented in different reference frames, the most common of which is the reference frame of the proximal segment. Joint moments can also be expressed as internal or external moments. To further explain the effects of net moments on the kinematics of movement, induced acceleration and induced power analyses may be used.

SUGGESTED READINGS

Allard, P., A. Cappozzo, A. Lundberg, and C. Vaughan. 1998. *Three-Dimensional Analysis of Human Loco-motion*. Chichester, UK: Wiley.

Berme, N., and A. Cappozzo (Eds.). 1990. *Biomechanics of Human Movement: Applications in Rehabilitation, Sports and Ergonomics*. Worthington, OH: Bertec Corporation.

Nigg, B.M., and W. Herzog. 1994. *Biomechanics of the Musculo-Skeletal System*. Toronto: Wiley.

Vaughan, C.L., B.L. Davis, and J.C. O'Connor. 1992. *Dynamics of Human Gait*. Champaign, IL: Human Kinetics.

Zatsiorsky, V.M. 2002. *Kinetics of Human Motion*. Champaign, IL: Human Kinetics.

PART III

MUSCLES, MODELS, AND MOVEMENT

Whereas the previous chapters dealt with the body as a theoretical structure with single actuators (i.e., moments of force at each joint), part III is primarily concerned with the muscular side of the human musculoskeletal system. For instance, chapter 8 deals with the study of muscle activity as exhibited by the electrical output of the muscles that can be measured by the electromyograph. An electromyograph records the electrical signal that "leaks" out of a muscle after it is activated. Such recordings, called electromyograms, may be saved electronically for later mathematical analysis by tools introduced in this chapter. Chapter 9 presents methods for mathematically modeling human muscles using the Hill model and includes information on obtaining parameters to allow the Hill model to represent individual muscles in a subject-specific manner. Chapter 10 explains techniques for modeling the human body or selected parts and creating motion simulations based on initial physical conditions. Computer simulations permit the study of motions without requiring that a subject perform the motion, which allows researchers, physicians, therapists, or coaches to test novel motions without placing people at risk of injury. Finally, chapter 11 explores the use of musculoskeletal models in analyzing human movement, an area of growing interest that permits the study of muscle forces beyond that allowed by inverse dynamics. Note that appendixes F and G contain a method for integrating a double pendulum model of the lower extremity and the derivation of the double pendulum equations, respectively.

Chapter 8

Electromyographic Kinesiology

Gary Kamen

Electromyography, the study of muscle electrical activity, can be quite valuable in providing information about the control and execution of voluntary (and reflexive) movements. This chapter provides an overview of the use of **electromyography** in biomechanics. In this chapter, we

► determine the basis of the muscle fiber action potential and how it propagates along the muscle fiber,

► determine the characteristics of the electromyography (EMG) signal,

► explain the basic features of EMG electrodes,

► examine some of the technical issues that alter the characteristics of the EMG signal,

► determine what variables are usually implemented to describe the EMG signal, and

► review some examples to illustrate how EMG has been used to understand human movement.

PHYSIOLOGICAL ORIGIN OF THE ELECTROMYOGRAPHIC SIGNAL

Although researchers must understand several technical features to process the EMG signal, the signal itself has a physiological origin. The first section of this chapter briefly explains the physiological concepts that underlie the origin of the EMG signal and how the action potential propagates along muscle fibers.

Muscle Fiber Action Potential

To produce muscular force, muscle fibers must receive an impulse from a motoneuron. Once a motoneuron is activated by the central nervous system, an electrical impulse propagates down the motoneuron to each motor endplate. At this specialized synapse, ionic events occur that culminate in the generation of a muscle fiber action potential (AP).

Resting Membrane Potential

Even at rest, muscle is an excitable tissue from which electrical activity can be recorded. Normally, the inside of the muscle fiber has an electrical potential of about -90 mV. This voltage gradient results from the presence of different concentrations of sodium (Na^+), potassium (K^+), and chloride (Cl^-) ions across the sarcolemma. The resting membrane potential is about 9 to 15 mV more positive in slow-twitch fibers, apparently because they have greater Na^+ permeability and higher intracellular Na^+ activity than do fast-twitch fibers (Hammelsbeck and Rathmayer 1989; Wallinga-De Jonge et al. 1985). Moreover, the resting membrane potential is not necessarily fixed; it can be altered, for example, by exercise training (Moss et al. 1983).

Action Potentials

The AP is the neural messenger responsible for activating every segment of the muscle fiber so that each sarcomere contributes to the generation of muscular force. The process begins with a change in the muscle fiber membrane's permeability to Na^+. Because these Na^+ ions occur in relatively greater concentration outside the muscle fiber, any change in permeability results in an influx of Na^+ across the membrane. Eventually, sufficient Na^+ ions enter the cell to reverse the polarity of the membrane potential so that the inside of the muscle fiber becomes positive (by about 30 mV) with respect to the surrounding extracellular medium. As the membrane's potential polarity reverses, the permeability of the membrane to K^+ changes, causing K^+ to exit the cell.

It is largely this efflux of K⁺ that repolarizes the cell and restores the resting membrane potential.

To ensure complete electromechanical activation of the muscle fiber, an AP produced in one small segment adjacent to the neuromuscular junction must spread to adjoining sections. By a passive process, the AP propagates along each adjacent section of the muscle fiber and in both directions from the neuromuscular junction so that the entire muscle fiber is electrically activated. As the AP spreads, the membrane potential at each subsequent muscle fiber section changes from negative to positive and back to negative as each adjacent muscle fiber area is successively activated. The deeper portions of the muscle fiber also need to be electrically activated, and so, by means of the transverse tubule system, the AP propagates to the deeper sections as well. The terminal portion of some motoneuron axons is slightly longer for some muscle fibers than for others, so it may take slightly longer to begin activation in those fibers (figure 8.1).

Muscle Fiber Conduction Velocity

The varying rates of transmission of the propagated AP determine many of the characteristics of the EMG. APs moving at slow rates, for example, contribute low-frequency components to the surface EMG. Thus, understanding the factors that determine the rate at which this propagation occurs is important. Because AP generation is an ionic process, the velocity at which the AP is conducted along the muscle fiber is dependent on the rate at which these ions can be exchanged. The passive membrane permeability characteristics determine

part of this exchange rate, along with the active metabolic mechanism for pumping Na⁺ back out of the fiber.

Differences in conduction velocity among different muscle fibers can be attributed to both histochemical and architectural features of the muscle fiber. The amplitude of the muscle fiber AP tends to be larger in fast-twitch fibers. Moreover, the shapes of the APs of fast- and slow-twitch fibers differ, causing the fast-twitch fiber AP to occur more quickly (including depolarization and repolarization) than the corresponding slow-twitch AP. Consequently, fast-twitch fibers have faster conduction velocities than slow-twitch muscle fibers. Larger-diameter muscle fibers also produce larger APs than do smaller fibers (Andreassen and Arendt-Nielson 1987), partly because of the greater activity of Na⁺. Atrophied fibers have distinctly slower conduction velocities (Buchthal and Rosenfalck 1958). Increases in the length of muscle fibers tend to decrease conduction velocity, and this may result from other architectural changes that occur in the fiber (Dumitru and King 1999).

Motor Unit Action Potential

Each motoneuron typically innervates several hundred muscle fibers, although the number varies in different muscles from around 10 to as many as several thousand. This characteristic—termed the *innervation ratio*—is computed by determining the number of muscle fibers per motoneuron. The muscle fibers within a single motor unit tend to be distributed throughout the muscle, although some muscles may have a more focused distribution (Windhorst et al. 1989). The individual unit of

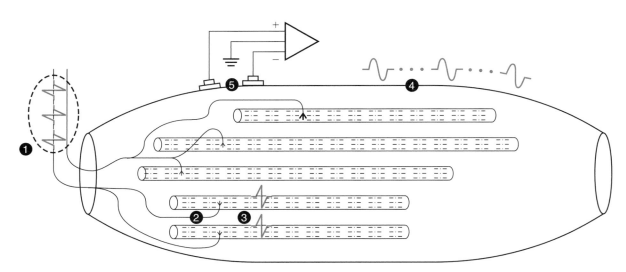

▲ **Figure 8.1**   The motoneuron AP initiates the process of muscle fiber excitation (1). The AP arrives at all of the motor endplates innervated by the motoneuron (2). By electrochemical processes, a muscle fiber AP is initiated and propagates along the length of the muscle fiber (3). The sum of all muscle fiber potentials activated by one motoneuron produces a motor unit AP (4), which can be recorded at the skin surface with amplifiers used specifically for biological signals (5).

motor action is the *motor unit*—one motoneuron and all of the muscle fibers innervated by that motoneuron.

The motor unit action potential (MUAP) represents the summated electrical activity of all muscle fibers activated within the motor unit. The amplitude of the MUAP is partly determined by the innervation ratio. In addition, motor units with more (or larger) muscle fibers have a larger MUAP. However, there are also some temporal dispersion issues that define the shape of the MUAP as measured by EMG—deeper muscle fibers contribute less to the surface EMG signal (figure 8.2).

Motor Unit Activation

The production of muscular force is controlled by the action of numerous muscles acting across a joint. In a single muscle, muscular force is initiated by activating an increasing number of motor units in a process termed *recruitment*. To produce almost any muscular action, smaller motor units are recruited first, and successively larger motor units are recruited as the force requirement increases. The nervous system also controls how frequently motor units are activated, the quantification of which is termed the motor unit *discharge* or *firing rate*. As the motor unit fires at faster rates, it produces an increasing amount of muscular force.

Muscular force can be altered in other ways as well. For example, at the onset of a muscular contraction, when considerable effort may be required to overcome

the inertia of the limb to be moved, motor units may fire in two short latency bursts before beginning a regular firing rate. The twin bursts are termed *doublets* and have the potential to produce larger forces than might be expected from the addition of two motor unit force twitches (Clamann and Schelhorn 1988). Two or more motor units frequently fire simultaneously in a process called *synchronization*. The exact role of motor unit synchronization in muscular force production is not clear, although it seems to occur more often than would be expected from chance alone.

In performing an action like wrist flexion, we tend to activate other muscles *(synergists)* that perform similar actions. Wrist flexion might be performed by activating the flexor digitorum profundus and superficialis, flexor carpi ulnaris and radialis, palmaris longus, and other wrist flexors. Later in this chapter, we show that electrode sensors placed over a skin surface serve as detectors of EMG activity over a potentially large volume. If neighboring muscles perform similar actions, then that activity is recorded by the surface electrodes.

Motor Unit Action Potential and the Electromyographic Signal

In the mildest muscle contraction, a single motor unit may be activated. This is recorded at the surface as the MUAP, followed by electrical silence until the unit's next firing. As the desired force increases, other motor units

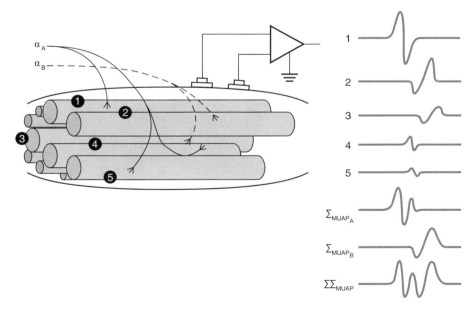

▲ **Figure 8.2** The contribution that each fiber's AP makes to the EMG signal depends in large part on the depth of the fiber; note that fiber 5 contributes a smaller AP than fiber 1. The temporal characteristics of the signal also depend on the electrode–motor endplate distance as well as the terminal lengths and diameters of the motoneurons. Two motor units are shown here, with the amplitude of each motor unit represented as the algebraic sum of the individual muscle fiber APs (Σ_{MUAP}). The overall signal is the algebraic sum of all motor units ($\Sigma\Sigma_{MUAP}$).

may be recruited and fire at ever-increasing frequency. At any point in time, the EMG signal is a composite electrical sum of all of the active motor units. A large peak in the EMG signal might be the result of the activation of two or more motor units separated by a short interval. Note that the signal has both positive and negative components. When the signal crosses the baseline, a positive phase of one MUAP is likely balanced by the negative phase of other MUAPs.

The amplitude of the surface-recorded EMG signal varies with the task, the specific muscle group under study, and many other features. Naturally, EMG amplitude increases as the intensity of the muscular contraction increases. However, the relationship between EMG amplitude and force frequently is nonlinear. Moreover, cocontraction activity from antagonists may require compensatory activity from the agonist muscle group from which EMG recordings are being made. Thus, one cannot assume that increases in EMG activity are indicative of parallel increases in force (Redfern 1992; Solomonow, Baratta et al. 1990). Issues regarding the relationship between EMG amplitude and force are discussed later in this chapter.

RECORDING AND ACQUIRING THE ELECTROMYOGRAPHIC SIGNAL

EMG activity can be recorded using either a monopolar or a bipolar recording arrangement (figure 8.3). In monopolar recordings, one electrode is placed directly over the muscle and a second electrode is placed at an electrically neutral site, such as a bony prominence. In general, monopolar signals yield lower-frequency responses and less selectivity than bipolar recordings. Although monopolar recordings are frequently used during static contractions (Ohashi 1995, 1997) and in a variety of clinical investigations involving needle electrodes (Dumitru et al.

1997), the monopolar recording is inherently less stable and would be a poor choice for measuring nonisometric contractions. Monopolar recordings are appropriate for the assessment of H- and T-reflexes and muscle M-waves, however (Mineva et al. 1993).

Bipolar (or *single differential*) recordings are considerably more common. In a bipolar recording arrangement, two electrodes are placed in the muscle or on the skin overlying the muscle, and a third neutral, or ground, electrode is placed at an electrically neutral site. This configuration uses a differential amplifier that records the electrical difference between the two recording electrodes. Thus, any signal that is common to the two inputs is greatly attenuated. The feature that allows the amplifier to attenuate these common signals is called *common-mode rejection*, and the extent to which signals common to both inputs are attenuated is described by the *common-mode rejection ratio* (CMRR). These CMRRs are expressed in either a linear or a logarithmic scale. A very good commercial amplifier might have a CMRR of 100 dB. One can convert CMRR in a decibel scale using the following formula:

$$CMRR_{(dB)} = 20 \ log_{10} \ CMRR_{(linear)} \quad (8.1)$$

where the CMRR is equivalent to 100,000:1.

In a typical laboratory or field environment, there may be considerable radio frequency (RF) and line activity from electrical outlets, lights, or other line signals. These signals are typically at power line frequency (50 or 60 Hz, depending on the recording location), and so they are directly in the frequency range represented by the EMG signal. There may also be ambient RF signals in the atmosphere at other frequencies. Because the differential amplifier reduces the signals appearing in-phase at both amplifier inputs, their influence is greatly reduced. There are instances in which both monopolar and bipolar recordings may be valuable, such as in investigating the geometry of muscle fibers during changes in muscle length (Gerilovsky et al. 1989).

Care must be used in setting the amplifier gain. If the amplitude of the EMG signal exceeds the amplifier's

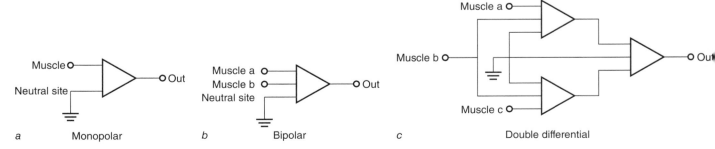

▲ **Figure 8.3** EMG signals can be acquired using either a monopolar (a) or bipolar (b) configuration. Some applications however, may require specialized amplifiers, such as the double differential recording technique (c).

gain, the amplifier is saturated and distortion occurs in the form of *clipping* (last waveform in figure 8.4). Clipping occurs when the amplitude of the output exceeds the bounds of the power supply. However, if the gain is set too low, the resolution of the signal after analog-to-digital (A/D) conversion will be small. Ideally, the gain should be set so that the amplitude of the signal is matched to the range of the A/D converter.

Electrode Types

A large variety of EMG electrodes are available (figure 8.5). The choice of electrode depends on the motor task to be explored, the nature of the research question, and the specific muscle from which recordings are to be made. This section discusses the use of EMG electrodes placed on the skin surface as well as the use of indwelling fine-wire and needle electrode designs. A few other electrode designs currently in research use are also noted.

Surface Electrodes

The first EMG electrodes were simple conducting surfaces made of various kinds of metal, including silver, gold, stainless steel, and even tin. These plate-type electrodes are commonly used for clinical applications such as assessing sensory and motor nerve conduction velocity, F-waves, M-waves, and H-reflexes (Oh 1993). However, some

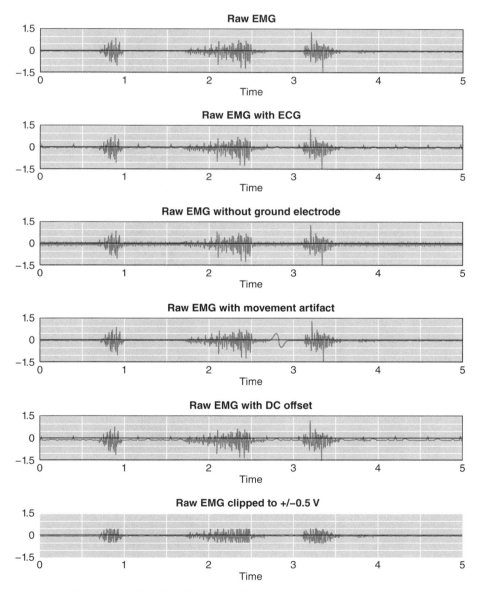

▲ **Figure 8.4** Various distortions of an EMG signal. The first panel is the true signal; the following panels illustrate various distortions.

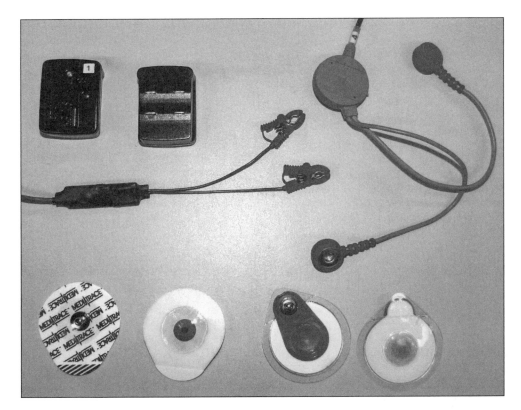

▲ **Figure 8.5** Some of the many surface electrodes commercially available.

technical problems may occur with these surface electrodes. There is normally a 30 mV potential between the inside and outside skin layers. When the skin is stretched, the potential decreases to about 25 mV, and the resulting 5 mV change is recorded as motion artifact. Obviously, motion artifact produced by skin transients can be a problem in recording situations in which appreciable movement is expected. However, these artifacts can frequently be minimized by using silver–silver chloride electrodes (Webster 1984) and by abrading the skin lightly to reduce the normal 50 kΩ impedance across the skin surface. These surface-applied electrodes serve as detectors of EMG activity in the underlying muscles, but they can also be excellent antennas for other RF activity in the ambient environment. Consequently, interelectrode impedance has to be minimized to minimize RF activity. After the electrodes are applied to the skin, an impedance meter can be used for this purpose. Before the electrodes are applied, impedance can be minimized by carefully preparing the skin, removing dead skin cells and skin oils, and increasing local skin blood flow. Minimizing cable distances, using shielded cables, and braiding individual electrode cables together also help to minimize RF interference.

One technique that successfully reduces these artifacts entails locating the first-stage amplifier as close to the electrode site as possible. For example, friction and movement in the cable insulation can produce artifacts. These artifacts can be reduced by using a unity gain operational amplifier at each electrode. Frequently called *active* electrodes, these devices position the amplifier very close to the surface sensor (see figure 8.5*d*). The signals from the first-stage amplifiers frequently have a higher signal-to-noise ratio (SNR) and thus are "cleaner" signals (Hagemann et al. 1985). Several authors have described the low-cost construction of on-site preamplified electrodes (Hagemann et al. 1985; Johnson et al. 1977; Nishimura et al. 1992). Disposable electrodes are also commercially available for applications that require them.

Surface electrodes have limited use in recording activity from deeper muscles or deeper portions of large muscles. The estimated effective recording area of surface electrodes ranges from 10 to 20 mm from the skin surface (Barkhaus and Nandedkar 1994; Fuglevand et al. 1992). It is also difficult to use surface electrodes to record from small muscles, because it can be difficult to discern whether the signal is arising from the underlying muscle or an adjacent muscle. This crosstalk is a serious problem that is discussed later in the chapter. There is some evidence that deeper motor units may be smaller than their superficial counterparts, so surface EMG recordings must be interpreted in the context of a possible bias toward recording from larger and more glycolytic motor units (Lexell et al. 1983).

Fine-Wire Electrodes

An alternative procedure that facilitates recording from deeper or smaller muscles is the use of fine-wire electrodes (Basmajian and Stecko 1962; Burgar et al. 1997). This configuration generally consists of two fine-diameter insulated wires (about 75 μm diameter) that are threaded through a hollow needle cannula. The tips of the wires constitute the recording surfaces, and either these are cut flush or the last millimeter or so of insulation is removed from the wire. The greater the amount of insulation that is removed, the greater is the recording volume. The distal centimeter of these wires is bent backward so that the wires and needle can be inserted into the muscle. After the needle is inserted, it is carefully removed, leaving just the wires as the recording electrodes. The cannula can be either removed or left off to the side during the experiment, and the ends of the wires are then connected to an amplifier.

Needle Electrodes

Many types of needle electrodes are commercially available, and they are frequently used to monitor the activity of one or more individual motor units rather than the EMG activity of the composite muscle. Concentric electrodes consist of a small wire placed in the middle of a hollow cannula. The wire is set in place with epoxy. The cannula is then cut at an acute angle (about 15°), leaving a shiny wire surface that is then referenced to the cannula for a bipolar recording.

Other Electrode Designs

Many other types of EMG electrodes have been designed for different purposes. Longitudinal array electrodes are used to record the features of propagating muscle fiber APs (Masuda et al. 1985; Merletti et al. 1999). Multiple 2-D array electrodes consisting of nine or more electrode surfaces arranged in a grid are used to investigate the architectural features of muscle (Thusneyapan and Zahalak 1989). One company (Delsys) has designed a wireless electrode (Trigno) that includes a 3-D accelerometer to measure the motion of the electrode attached to the skin surface.

Recording individual muscle fiber APs during high-force contractions requires specialized electrodes. There have been some excellent attempts to record individual APs using surface multielectrode arrays (Rau et al. 1997), but again, a surface detector is biased toward muscle fibers closer to the skin surface. APs from individual muscle fibers are usually recorded from wire electrodes (as just described), multiple wire electrodes (Hannerz 1974; Shiavi 1974), or needle electrodes (Kamen et al. 1995; Sanders et al. 1996).

Electrode Geometry and Placement

The importance of correctly placing surface electrodes over the muscle cannot be overemphasized. Certainly, the electrodes must be placed in a position from which APs from the underlying muscle fibers can be recorded. In general, this means placing the electrodes away from highly tendinous areas. The motor point (the area where the nerve enters the muscle) is not a good location for surface electrodes. For studies requiring several measurement sessions, the endplate zone has the potential to produce the most variable EMG signals. Because MUAPs propagate in both directions from the neuromuscular junction, signals recorded over the motor point are frequently subjected to algebraic subtraction from a differential amplifier, resulting in the cancellation of EMG signals common to both electrodes. Standard locations for electrodes can be found in various sources (LeVeau and Andersson 1992; Zipp 1982). However, the orientation of the electrodes with respect to the muscle fibers is also important. If the electrode is not placed parallel to the muscle fibers, the amplitude of the signal may be reduced by as much as 50% (Vigreux et al. 1979). The frequency content of the EMG signal is also affected by off-parallel electrode placement. Consequently, although the muscle fiber pennation angle can be difficult to determine, every effort should be made to orient the electrodes parallel to the muscle fibers. Anatomical atlases are available for assistance (Cram et al. 1998).

Interelectrode geometry is also an important consideration. Although the evidence is equivocal, the amplitude of the EMG signal may be affected by the interelectrode distance—the distance between electrode pairs. One researcher found that an interelectrode distance of about 60 mm produces the greatest EMG amplitude when surface electrodes with a 7 mm diameter recording area are used (Vigreux et al. 1979). However, more recently, Jonas and colleagues (1999) failed to find any difference in compound muscle AP amplitude using different surface electrode types, different recording areas, and different interelectrode distances. EMG frequency characteristics are also amenable to change with electrode spacing, with higher spectral frequencies obtained from electrodes placed closer together (Bilodeau et al. 1990; Moritani and Muro 1987). When you are placing

FROM THE SCIENTIFIC LITERATURE

Masuda, T., H. Miyano, and T. Sadoyama. 1992. The position of innervation zones in the biceps brachii investigated by surface electromyography. *IEEE Transactions on Biomedical Engineering* 32:36-42.

EMG techniques are sometimes useful in revealing morphological features in muscle. In this paper, Masuda and colleagues describe the construction of a linear surface electrode array that is placed along the surface of the biceps brachii muscle. Muscle fiber APs are detected by the multiple electrodes and followed and tracked as they propagate along the muscle. When the array is placed in the vicinity of the innervation zone (the so-called motor point), the AP polarity reverses, reliably and accurately marking the location of the motor point. The authors describe a computer program that automatically calculates this position using regression techniques. In this way, surface EMG techniques are quite useful to muscle morphologists.

EMG electrodes, it is best to keep constant as many conditions as possible between recording sessions and among subjects to minimize variability from any of these factors.

Electromyographic Signal Processing

Advances in both analog and digital electronics and in signal-processing techniques, as well as continued advances in our understanding of the EMG signal, have changed the way we process and analyze the signal. This section focuses on the most common techniques available for acquiring and filtering the signal. We also consider some of the technical problems inherent in processing the EMG signal.

Application of Analog and Digital Filtering

Knowledge of the inherent frequency characteristics and prevailing sources of RF noise and interference of any biological signal allows some decisions to be made regarding the kinds of filtering techniques to be used. Analysis of the EMG signal frequency spectrum indicates that little (if any) of the signal is contained at frequencies below about 10 Hz or above 1 kHz. Indeed, for the surface EMG signal, the upper frequency limits of the signal are even more bandwidth-limited, with the highest frequency components found at about 400 Hz. Indwelling signals (recorded within muscles) contain higher-frequency content (Gerleman and Cook 1992). Consequently, it is frequently recommended that the surface EMG signal be acquired using a high-pass filter set at about 10 Hz, with an antialiasing low-pass filter set at about 1 kHz. However, opinions vary on the exact band-pass characteristics to be used. Software filters can be used after recording to apply further digital signal processing.

Analog-Digital Data Acquisition

Knowledge of the inherent frequency characteristics of the EMG signal is also important in choosing a sampling rate. Certainly, the Nyquist limit must be considered, requiring as it does a sampling rate that is at least twice that of the highest-frequency component in the signal. For the surface EMG signal, this generally requires a sampling rate of at least 1000 samples per second, and sometimes higher. A low-pass, antialiasing filter set at the Nyquist limit (the highest frequency in the signal) should then be implemented to prevent signals above this frequency from distorting the true signal.

Many manufacturers offer 50 to 60 Hz notch filters to attenuate RF activity from lights or other equipment. However, there is considerable EMG activity at these line frequencies. Consequently, although the signal may look "cleaner," a significant portion of the EMG signal is eliminated with these notch filters. If necessary, techniques other than the use of notch filters should be used to eliminate line-frequency interference. This may require changing the surrounding equipment environment, using alternative electrodes, improving the grounding configuration, or improving the electrode-skin interface.

Influences on the Electromyographic Signal

The large number of influences on the EMG signal are both physiological and technical in origin. Knowledge of the myriad sources that can contribute to the surface-recorded EMG signal is necessary to ensure correct interpretation.

Electrodes

As described earlier, electrode characteristics that can affect the frequency and amplitude content of the EMG signal include the type of electrode (e.g., surface, metal plate, silver–silver chloride, indwelling), electrode size, and the interelectrode distance. The electrode configuration (monopolar vs. bipolar) is also a determinant.

The characteristics of the underlying tissue also affect the EMG signal. A poor electrode-skin interface increases electrode impedance, contributing to a poorer SNR. A good skin-electrode interface is particularly important for unamplified electrodes. EMG amplitude decreases as a power function with increasing distance between surface electrodes and muscle (Roeleveld et al. 1997). Consequently, the volume of subcutaneous fat can increase the electrode-muscle distance and also affect the EMG signal. Not surprisingly, lower levels of subcutaneous fat are associated with higher SNR. In obese individuals, the subcutaneous fat layer can be 400% to 500% greater than in lean individuals (Petrofsky 2008). The thickness of the subcutaneous fat layer can explain more than 50% of the variance in the EMG signal; one recommendation is to attempt to normalize the EMG signal if the subcutaneous fat layer is very thick (Nordander 2003).

Blood Flow and Tissue Influences

The tissue between the muscle and surface electrodes has a dramatic low-pass filtering effect. Consequently, the high-frequency characteristics of individual muscle fiber APs, particularly those from deeper fibers, are attenuated appreciably at the surface. This decay occurs exponentially with distance, so at even small distances from the muscle fiber, the higher-frequency characteristics of the AP are sharply attenuated. Even 100 μm away from the muscle fiber, about 80% of the signal strength is lost (Andreassen and Rosenfalck 1978). Low-pass filtering may also be induced by increased muscle blood flow, which increases dramatically during contraction. Thus, it is possible that EMG signal characteristics can be altered by factors unrelated to muscle electrical activity. Because the EMG signal can be affected by changes in Na^+, situations that involve appreciable fatigue, dehydration, or interruption in muscle blood flow may affect the EMG.

Muscle Length

Our knowledge of changes in muscle length during dynamic contraction has been extended considerably by the ultrasound studies conducted by Kawakami and colleagues (Ito et al. 1998). The rate of propagation of the muscle fiber AP changes as the muscle changes in length. This is demonstrated as a decrease in muscle fiber conduction velocity with muscle lengthening (Morimoto 1986). Also, the amplitude of the AP declines with increasing length (Gerilovsky et al. 1986; Hashimoto et al. 1994), likely as a result of morphological changes in the neuromuscular junction or the sarcolemmal membrane (Kim et al. 1985). Moreover, the frequency characteristics are affected, with increasing muscle length shifting the spectrum toward the lower frequencies (Okada 1987). These changes in the characteristics of the EMG signal are one reason that it is difficult to interpret the signal during dynamic contractions.

Muscle Depth

Because the AP decays so rapidly as the distance from the recording electrodes increases, deeper muscle fibers are considerably more difficult to record from the surface than are more superficial muscles. As noted earlier, surface electrodes record signals only from these more superficial muscle fibers. Moreover, there is some evidence that the fiber type may vary with muscle depth. The deeper muscle fibers seem to have more slow-twitch characteristics, and so they are recruited at lower forces, whereas the more superficial fibers may be more fast-twitch and recruited at somewhat higher muscle forces. Consequently, the surface EMG may be biased in favor of recording from the more superficial, fast-twitch muscle fibers, which also generate the larger APs.

Crosstalk

Because muscle is a volume conductor, electrical signals are transmitted indiscriminately through it, regardless of the origin of the signal. Thus, for example, EMG activity produced in one plantar flexor muscle (e.g., gastrocnemius) can readily be detected by electrodes placed over an adjacent plantar flexor (e.g., soleus). This EMG crosstalk can be considerable (Morrenhof and Abbink 1985) and is likely frequently underestimated, although the issue is not without controversy. Some investigators view the extent of crosstalk from adjacent muscles to be quite small (Winter et al. 1994). Nevertheless, whenever crosstalk may be a problem, the investigator must determine the extent to which it exists.

Crosstalk is affected by a number of factors. For example, it increases with the size of the subcutaneous fat layer (Solomonow et al. 1994). Thus, crosstalk may be more frequently observed in muscles surrounded by large layers of subcutaneous fat (e.g., gluteus maximus) and in women and infants.

Crosstalk can be greatly minimized by using more selective electrodes. Wire electrodes that have a smaller recording area than surface electrodes considerably minimize crosstalk. A rehabilitation technique involving proprioceptive neuromuscular facilitation has been used to demonstrate the risks of EMG crosstalk (Etnyre and Abraham 1988). Earlier studies had shown that muscle EMG activity increases when a muscle is stretched during antagonist muscle contraction. When subjects performed plantar flexion followed by dorsiflexion, surface electrodes recorded the soleus muscle EMG activity during the dorsiflexion phase. However, no EMG activity was observed when wire electrodes were used (Etnyre and Abraham 1988). What had been thought to be cocontraction of agonist and antagonist muscles turned out to be crosstalk produced by the antagonist dorsiflexor muscles.

Crosstalk can be identified using several techniques. The motor nerve innervating the antagonist muscle can be stimulated and the EMG inspected visually for evidence of crosstalk activity (Koh and Grabiner 1992). Cross-correlation analysis of two EMG signals (in two different muscles) can be conducted to determine whether the two signals are strongly related at some time lag (Etnyre and Abraham 1988).

EMG activity from three surface electrodes placed in a series can be recorded using two conventional EMG amplifiers. The signal from these two amplifiers can then serve as input to a third differential amplifier, resulting in a *double-differential signal* (see figure 8.3). This double-differential recording technique may attenuate signals from distant sources and thus reduce the potential for crosstalk (Koh and Grabiner 1992). Wire electrodes that have a small area can be used to obtain slightly more selective recordings from each pair of wires.

ANALYZING AND INTERPRETING THE ELECTROMYOGRAPHIC SIGNAL

The two important characteristics of the EMG signal are amplitude and frequency. Amplitude is an indicator of the magnitude of muscle activity, produced predominantly by increases in the number of active motor units and the frequency of activation, or the firing rate. The frequency of the signal is also affected by these factors. When more motor units are activated, the number of spikes and turns in the surface EMG signal increases. Changes in firing rate also change EMG frequency characteristics. However, as discussed earlier, a number of other factors, both technical and physiological, affect both the amplitude and frequency of the EMG signal. In this section, we discuss the major variables used to analyze the EMG and provide some information regarding their interpretation.

Electromyographic Amplitude

The major variables used to define EMG amplitude include peak-to-peak (p-p) amplitude, average rectified amplitude, root mean square (RMS) amplitude, linear

envelope, and integrated EMG. These variables are discussed next.

Peak-to-Peak Amplitude

The peak-to-peak (p-p) amplitude is one of the simplest ways to describe the magnitude of the EMG signal. This variable is particularly useful when the signal is highly synchronous—composed of multiple simultaneously firing motor units. For example, when a peripheral motor nerve is stimulated, most or all of the motoneurons are activated simultaneously to produce a synchronous signal called the M-wave (figure 8.6). When the intensity of the stimulation is increased sufficiently, all motoneurons are activated, resulting in the maximal EMG activity that the muscle is capable of producing. This maximal-amplitude M-wave can be described by calculating the negative-peak-to-positive-peak amplitude (p-p amplitude).

▲ **Figure 8.6** The M-wave (M) reflects the synchronous electrical activity of all muscle fibers following an electrical stimulus (S). The H-reflex (H) is produced by the activation of Ia afferents.

Another example is the H-reflex, which is evoked by delivering a low-intensity electrical stimulus to the peripheral motor nerve in an effort to activate only the primary muscle spindle afferents. The orthodromic (normal axon direction) AP from these Ia afferents synapses on α-motoneurons, resulting in motoneuron discharge. The resultant signal is similar in appearance to the M-wave, although smaller in magnitude, because it is rare for all of the motoneurons to be activated by this technique. The amplitude of the H-reflex can also be described using p-p amplitude.

Average Rectified Amplitude

The normal EMG (also called the *interference pattern*) is an alternating current (AC) signal, varying in both the positive and negative voltage directions. Unless there is some voltage offset in the signal acquisition and amplification system, the mean of the signal is zero. Consequently, the mean value is not a valid indicator of EMG amplitude.

To compute a representative averaged amplitude measure over a period of time, the signal must first be rectified. Rectification involves converting the negative voltages to positive values (i.e., absolute values). After this is done, the average of the values is a nonzero amplitude measure termed the *average rectified amplitude* (figure 8.7*b*).

Root Mean Square Amplitude

One alternative that does not require rectification is to compute the RMS amplitude as follows:

$$RMS\left[EMG(t)\right] = \left[\frac{1}{T}\int_{t}^{t+T} EMG^2(t)\,dt\right]^{1/2} \quad (8.2)$$

where *EMG* is the value of the EMG signal at each moment of time *(t)*, and *T* represents the duration of the analyzed signal. Because RMS amplitude incorporates the squared values of the original EMG signal, it does not require full-wave rectification.

Linear Envelope

Because the EMG is a time-varying signal with a zero mean, the value of the signal at any instant is not an indicator of the overall magnitude of EMG activity. However, an estimate of the "volume" of activity can be obtained using a variable called the *linear envelope*, which is computed by passing a low-pass filter through the full-wave rectified signal (figure 8.7*c*). The linear envelope, then, is a type of moving average indicator of EMG magnitude. The exact selection of a frequency to be used for the cutoff is somewhat arbitrary, and the appropriate cutoff frequency depends on the application. Cutoff frequencies of 3 to 50 Hz have been suggested. Shorter-duration activities benefit by a higher cutoff frequency, but generally a frequency of 10 Hz gives satisfactory results; however, the resolution of the high-frequency characteristics of the signal is attenuated. Consequently, the resolution available in computing onsets and offsets is reduced when the linear envelope is used for this purpose.

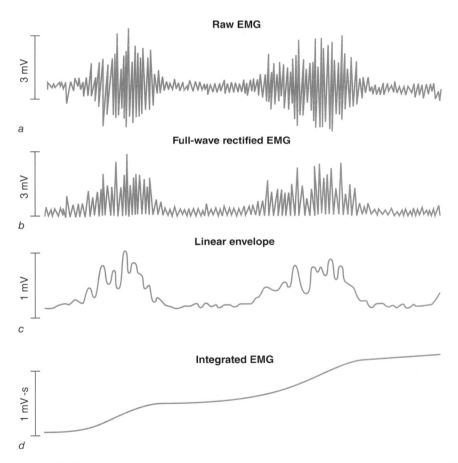

▲ Figure 8.7 *(a)* Raw EMG. *(b)* Average rectified signal. *(c)* Linear envelope. *(d)* Integrated EMG.

Reprinted from *Archives of Physical Medicine and Rehabilitation,* Vol. 75, G.F. Harris and J.J. Wertsch, "Procedures for gait analysis," pgs. 216-225, copyright 1994, with permission from The American Congress of Rehabilitation Medicine and the American Academy of Physical Medicine and Rehabilitation.

Integrated Electromyography

An integrator is an electronic device (or computer algorithm) that sums and totals activity over a period of time so that the total accumulated activity can be computed for a chosen time period (illustrated in figure 8.7*d*). If the device is not reset, the totals continue to accumulate. Consequently, at a preset time, the output of the integrator is reset to zero and integration begins again. The term *integrated EMG* has a strict definition and is frequently misused and mistaken for average rectified EMG amplitude or RMS amplitude.

Electromyographic Frequency Characteristics

After amplitude analysis, the next most common analytical method involves characterizing the frequency characteristics of the EMG signal. This can be accomplished by defining so-called turning points and zero crossings or identifying the median or mean frequency, as well as by other techniques that are discussed later.

Turning Points and Zero Crossings

One of the simplest ways to describe the frequency characteristics of the EMG signal is by counting spike peaks. Each time the signal changes direction, a new turning point is created. The number of turning points in peaks per unit time in the EMG is one estimate of the frequency content of the signal. Similarly, the number of times the signal crosses the zero baseline can be counted. The number of zero crossings is also a valid estimate of frequency content. Turning points and zero crossings are frequently used clinically to describe potential neuromuscular pathologies (Hayward 1983; Ronager et al. 1989). Moreover, the number of zero crossings is well correlated with other frequency variables, like those obtained from spectral analysis (Inbar et al. 1986).

Mean and Median Frequency

Spectral analysis techniques are often used to describe the EMG frequency characteristics. In general, the surface-recorded EMG frequency spectrum is positively skewed, with a mean value of approximately 120 Hz and a median value of about 100 Hz (figure 8.8). These frequency variables are often indicative of changes occurring in muscle fiber conduction characteristics, and thus they may be better interpreted as markers of peripheral muscular changes than as markers of neural or central drive.

A word of caution: EMG frequency characteristics are misinterpreted and overinterpreted all too often. For example:

■ An increase in frequency does not necessarily indicate that more fast-twitch motor units are active. It may indicate a higher firing rate for slow-twitch motor units, activation of muscle fibers with higher conduction velocities, decreased motor unit synchronization, additional activation of synergist muscles, or other possibilities.

■ Similarly, a decrease in frequency does not necessarily indicate an increase in motor unit synchronization. It could indicate a decrease in the total number of active motor units, a decrease in motor unit firing rate, a slowing of conduction velocity, or a change in the intramuscular milieu.

■ The analysis of EMG spectral frequency characteristics during dynamic contractions is particularly difficult. To compute spectral frequency content, we assume that the signal is stationary—that is, that the frequency content does not change over the analysis interval. During isometric contractions, the stationarity assumption is reasonably well met, particularly for short time intervals. However, the EMG signal obtained during dynamic contractions generally violates the stationarity assumption. The extent to which the stationarity of the signal is violated depends on the task. In rapid cycling,

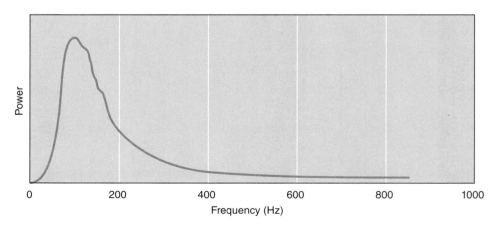

▲ **Figure 8.8** Typical frequency spectrum obtained from the surface EMG.

for example, there might be considerable violation of the stationarity assumption. One solution is to consider the analysis of short epochs, during which the signal would be *quasi-stationary* or *wide-sense stationary* (Hannaford and Lehman 1986). Other solutions involve alternative algorithms, such as the Choi-Williams distribution (Knaflitz and Bonato 1999) and wavelet analysis (Karlsson et al. 2000). Consequently, analyzing and interpreting the frequency characteristics of the EMG signal during dynamic activity requires particular caution.

Other Electromyographic Analysis Techniques

Amplitude and frequency analyses are the most common ways of interpreting the EMG signal, but many other techniques are also used. If only the start and end of muscle activity are needed, onset-offset analysis is suitable. This method plus several others, including the use of polar and phase plots, is presented next.

Onset-Offset Analysis

Frequently, electromyographers are interested in determining when muscle electrical activity begins and terminates. One criterion for determining onsets and offsets is to ensure that the high-frequency components of the signal have not been filtered or otherwise attenuated to any appreciable extent. Filtering can delay the identification of the onset and offset times, with the delay varying depending on the high-frequency content at the time of analysis (figure 8.9). Many of the algorithms that have been suggested for determining when the signal begins and ends are subjective and produce figures that require the reader's interpretation. Other methods are more objective, using a threshold EMG activity or a change in the rate of EMG activation (for a review, see Hodges and Bui 1996). Li and Caldwell (1999) introduced a novel procedure to identify onsets and offsets with respect to kinematic events by using cross-correlation analysis.

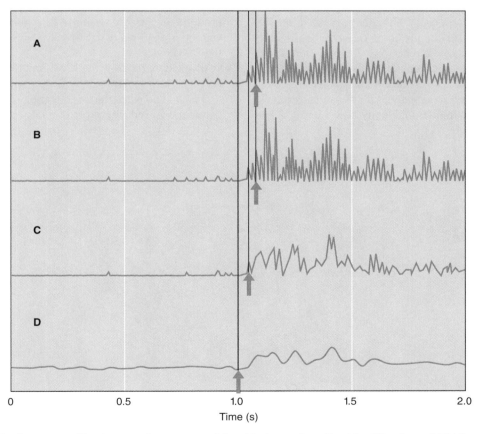

▲ **Figure 8.9** Low-pass filtering can have a considerable impact on the identification of EMG onset time. *(a)* Full-wave rectified raw EMG. Note that the onset identification becomes increasingly inaccurate as the signal is increasingly smoothed using low-pass filters from 500 Hz *(b)* to 50 Hz *(c)* and finally to 10 Hz *(d)*.

Reprinted from *Electroencephalography and Clinical Neurophysiology* Vol. 101, P.W. Hodges and B.H. Bui, "A comparison of computer-based methods for the determination of onset of muscle contraction using electromyography," 511-519, copyright 1996, with permission of Elsevier.

Polar Plots and Phase-Plane Diagrams

At times, it is desirable to illustrate EMG changes with respect to a performance measure, such as changes in muscular force or joint displacement. Phase-plane diagrams (figure 8.10) attempt to link the kinematic characteristics of a movement with the resultant EMG activity of both agonist and antagonist muscles (Carrière and Beuter 1990). Polar plots are another alternative. For example, Dewald and colleagues (1995) asked healthy and hemiparetic subjects to perform an isometric task that required combinations of elbow flexion-extension, shoulder abduction-adduction, and forearm pronation-supination. Their polar plots illustrated that more EMG activity occurred in the contralateral than in the impaired side in the patients. Many other examples illustrate the use of polar plots for EMG analysis (Buchanan et al. 1986, 1989; Chen et al. 1997).

Other Analysis Techniques

Many other analysis techniques have been used to describe and interpret the EMG signal, including wavelet analysis (Karlsson et al. 1999), autoregressive models (Sherif et al. 1981), analysis of cepstral coefficients (Kang et al. 1995), neural network classification (Liu et al. 1999), period-amplitude analysis (Betts and Smith 1979), recurrence quantification analysis (Filligoi and Felici 1999), and fractal analysis (Gitter and Czerniecki 1995). These techniques and many others being devel-oped have merit and should be considered with regard to the analytical requirements of the specific application.

Normalization of the Electromyographic Signal

EMG data from different subjects, muscles, and days can be compared. In light of the characteristics that determine EMG magnitude, discussed earlier, it is likely that both technical and physiological factors will contribute to variations in EMG magnitude among these conditions. Investigators interested in changes in the EMG signal over the course of some treatment condition should **normalize** the signal and report changes in the normalized EMG signal; otherwise, inappropriate interpretations can result (Lehman and McGill 1999).

Many normalization techniques are available. Most frequently, subjects are asked to perform a maximal isometric contraction, and the magnitude of the EMG signal of interest is normalized to the value obtained during this maximal contraction (e.g., Mathiassen et al. 1995). Another approach is to use the maximal M-wave amplitude. The muscle is electrically stimulated to produce the largest response (the maximal M-wave), and the p-p or area of the M-wave is used to normalize the EMG data.

The technique of using EMG amplitude from maximal isometric contractions for normalization may be less than ideal when the investigation requires dynamic contractions. Indeed, errors in interpretation can result

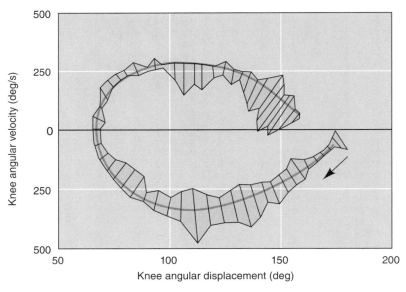

▲ **Figure 8.10** Phase plots provide an opportunity to link the kinematic and electromyographic characteristics of a movement.

Reprinted from *Human Movement Science,* Vol. 9, L. Carriere and A. Beuter, "Phase plane analysis of biarticular muscles in stepping," pgs. 23-25, copyright 1990, with permission of Elsevier.

from normalizing to maximal isometric contractions (Mirka 1991). One alternative for dynamic contractions is to use the maximal EMG activity during some reference part of the contraction. In gait, for example, the cycle can be partitioned into a number of logical epochs that correspond to phases of the gait cycle. The maximal EMG amplitude, then, is represented as 100%, and the EMG amplitudes in other portions of the gait cycle are normalized to this maximal value.

Normalization of the EMG signal is not always required. In a simple investigation designed to assess EMG activity in one muscle group over several days, reporting the absolute value of the EMG magnitude is acceptable. Many studies have demonstrated that absolute value reporting is valid and reliable (Finucane et al. 1998; Gollhofer et al. 1990) and can be more meaningful than a relative score derived using normalization methods. EMG amplitude measurements obtained from submaximal isometric contractions appear to be more reliable than those from maximal contractions (Yang and Winter 1983), suggesting that researchers should also consider the EMG amplitude from submaximal contractions when performing normalization.

APPLICATIONS FOR ELECTROMYOGRAPHIC TECHNIQUES

As discussed in the previous section, the nature of the research question determines the selection of electrodes, amplifiers, and filters; the A/D data-acquisition require-

ments; and the subsequent analysis procedures. In this section, several research areas requiring the use of EMG analysis are cited as examples of the kinds of procedures that might be conducted using EMG.

Muscle Force versus EMG Amplitude Relationships

Electromyography can be used to determine the extent of muscular force based on the amplitude of the EMG signal. When a person throws a ball, for example, what external forces are produced, and what forces are placed on the muscles involved? In controlling prosthetic limbs, how much electrical activation should be used to achieve the desired level of force? Such research questions require knowledge of the relationship between EMG amplitude and muscular force.

Under isometric conditions, the EMG–muscular force relationship is frequently linear: Incremental changes in muscular force are produced by linearly related incremental changes in EMG amplitude (Bouisset and Maton 1972; Jacobs and van Ingen Schenau 1992; Milner-Brown et al. 1975). These increases in EMG amplitude are presumably produced by a combination of motor unit recruitment and increases in motor unit firing rate. However, there are many exceptions to this linear EMG-force relationship. For example, a curvilinear relationship is frequently observed (figure 8.11). Small increments in force at the low- and high-force end of the scale may be accompanied by large increases in EMG amplitude (Clamann and Broecker 1979). Because the EMG-force relationship can be affected by the techniques used to

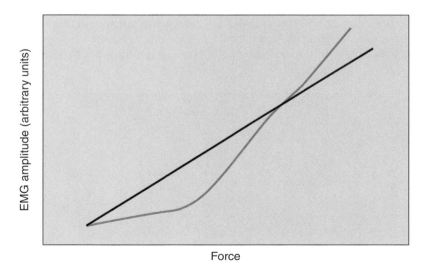

Force

▲ **Figure 8.11** A linear relationship between EMG amplitude and external muscular force is often observed (black line). However, there are many exceptions in which there is a curvilinear relationship (green line) between the two variables.

process the EMG signal (Siegler et al. 1985), it is important to report the details of the technique used to acquire and process the EMG activity.

Our knowledge of the EMG-force relationship is built predominantly on experiments using isometric contractions. When one produces a rapid elbow extension movement, for example, EMG amplitude during the first 100 ms of the movement is linearly related to kinematic features such as peak velocity and acceleration (Aoki et al. 1986). However, in cycling, the EMG-force relationship may be linear in one plantar flexor—the soleus—and nonlinear in another, such as the gastrocnemius (Duchateau et al. 1986).

During repeated contractions involving changes in muscle temperature or fatigue, the EMG-force relationship can be altered (Dowling 1997). These observations make the assessment of the EMG-force relationship a viable research tool for assessing both central and peripheral contributions to muscular force. For example, the EMG-force relationship can be affected by conditions such as hemiparesis (Tang and Rymer 1981).

The relationship between muscular force and EMG frequency characteristics is usually nonlinear (figure 8.12). In general, the mean and median power frequency increase rapidly with increases in muscular force to about 20% to 30% of maximal voluntary contraction (Hagberg and Ericson 1982).

Thus, analysis of the EMG-force relationship requires an assessment of the neuromuscular and movement features. Issues to be considered include the type of muscle contraction (isometric vs. dynamic), the size of the muscles involved, the potential role of various agonists and antagonists, and the extent to which the EMG recording is representative of muscle electrical activation. Note that the vast majority of the extant literature has considered only the relationship between EMG activity and external force production. Although there have been many investigations of in situ forces in both

human and animal models (Gregor et al. 1987; Gregor and Abelew 1994; Landjerit et al. 1988), we lack a full understanding of the relationship between these internal forces and muscle electrical activation.

Gait

Human gait is a frequent subject for EMG investigation. The order of activation and the relative magnitude of EMG activity are frequently of interest. However, accurate determination of activation order may require an appropriate algorithm and analysis technique for onset analysis. In such a case, the raw EMG signal may be useful in determining EMG onset times. EMG amplitude analysis using the linear envelope may require one of the normalization techniques. One advantage of studying gait is that it is repetitive and cyclical, lending itself to analysis over a number of different gait cycles.

Sprinting

The assessment of global EMG activity is more challenging during sprinting than during walking. One problem is that a high-velocity cyclical activity like running can produce considerable artifact by skin and electrode movement. This is particularly evident at heel-strike or during gait cycle phases when the skin is considerably stretched. However, reliable EMG results can be obtained at the highest running velocities when some care is used in the recording procedure. Wireless **telemetry** instrumentation is considerably more useful for overground (as opposed to treadmill) sprinting situations (Mero and Komi 1987). Cyclical EMG activity with sharp onset and offset times representing distinct periods of activation and inhibition or inactivation is quite apparent in the signals obtained from trained sprinters (Jönhagen et al. 1996; Mero and Komi 1987). As in other situations, however, EMG signals obtained during sprinting can be prone to crosstalk, and so the

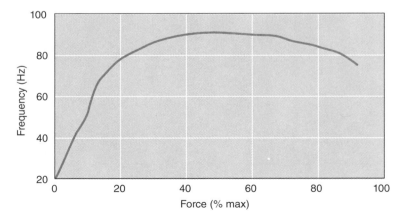

▲ **Figure 8.12** The relationship between EMG mean power frequency and muscular force.

recommended procedures need to be used to ensure that crosstalk is not a problem.

Both surface and indwelling electrodes can be used to record activity from lower-limb muscles during sprinting (Chapman et al. 2008; O'Connor & Hamill 2004). The choice of indwelling (typically fine-wire) versus surface electrodes is particularly important when we are recording from some of the small or deep muscles, like iliacus, psoas, and tibialis posterior (Andersson et al. 1997; Reber et al. 1993), since surface EMG recorded signals here would likely be contaminated by the activity of other neighboring muscles. Also, it is recommended that indwelling electrodes be considered when accurate estimates of muscle onset-offset times are important, since activation durations can be as much as 80% longer with surface electrodes than with indwelling wire electrodes (Bogey et al. 2000).

Many gait researchers have chosen to position the electrodes according to readily available anatomical landmarks. This electrode location procedure is useful in clinical situations and is probably acceptable to answer the vast majority of research questions involving sprinting. Specific recommendations can be found in several sources and have recently been revised by the European-sponsored Surface Electromyography for the Non-Invasive Assessment of Muscles (SENIAM) project. This group's recommended electrode positioning locations can be viewed online at www.seniam.org.

However, EMG signals obtained from bipolar electrodes placed at different portions of the same muscle can differ markedly (Sacco et al. 2009). If the electrodes are placed along the innervation zone or tendon zones, considerable variability can be observed because of decreased reliability (Merletti et al. 2001). This is most serious if EMG electrodes are applied on multiple days and the electrode site varies from day to day. Accurate comparisons between subject groups could also depend on the quality of electrode placement. Although it may be a bit time-consuming, the most accurate and reliable EMG signals can be obtained by attending to electrode positioning, avoiding innervation zones and tendon zones. This can be achieved by using a stimulation procedure to assess the innervation zone (Sacco et al. 2009; Saitou et al. 2000; Walthard and Tchicaloff 1971).

A research question of interest concerning injuries in running involves the magnitude of hamstrings EMG activity during the eccentric phase of sprinting. One suggestion is that this EMG activity may be responsible for the high frequency of hamstrings strain injuries. The data do seem to support the idea that the EMG activation observed during muscle lengthening may contribute to these injuries, but one technical factor that can confound this issue concerns the analysis technique used for the EMG data. The timing of hamstrings activation and the duration of activity during the gait cycle are quite important in determining injury risk, and these characteristics are best observed in raw EMG recordings rather than averaged or filtered EMG signals (see Jönhagen et al. 1996; Mero and Komi et al. 1987).

Developmental Gait Issues

The study of gait development using EMG affords the opportunity to examine some interesting conceptual issues involving motor control, biomechanics, and human development. However, when we record EMG signals from infants, children, or older adults, there are important technical issues to consider.

Figure 8.13 shows the typical pattern of muscle activation observed across development. In adults, it is quite evident that muscle activation occurs during distinct intervals, with little or no EMG activity during periods of inactivation. The biceps femoris, for example, begins a brief period of activation just before heel-strike and again around heel-off. Similarly, the lateral gastrocnemius exhibits a brief period of low-amplitude muscle activity just after heel-strike with somewhat greater EMG activity toward the end of stance phase.

This adult pattern differs from that seen in infants and children. Perhaps the most interesting observation can be made about the time infants begin to walk (figure 8.13). In this figure, we can see extended periods of activation in most muscles with several examples of agonist-antagonist cocontraction.

Interpretation of these results requires attention to the issues that affect the **electromyogram**. From a technical viewpoint, there is relatively more subcutaneous fat in infants than in adults (Frantzell et al. 1951) and, as discussed earlier, this has the potential to render the EMG signals more difficult to record because of increased impedance between electrodes. However, Okamoto and Okamoto (2003) were able to reduce the impedance between surface electrodes to less than 5 kΩ in infants, demonstrating that technical issues can be overcome in these subjects. Consequently, the observation of more frequent EMG activity in infants may indeed be a reliable observation. Subcutaneous fat also increases in older adults (Ishida et al. 1997), posing additional concerns for gait studies involving older individuals.

It is not unexpected to observe a coactivation type of EMG pattern in adults in other situations. The challenge of learning new motor tasks is frequently marked by the activation of many muscles, including some that may not be necessary to complete the task (Kamon and Gormley 1968; Vorro and Hobart 1981). This "turn everything on" pattern later changes to a more distinct pattern of muscle activation as the performer learns which muscles to activate, how long they should be activated, and

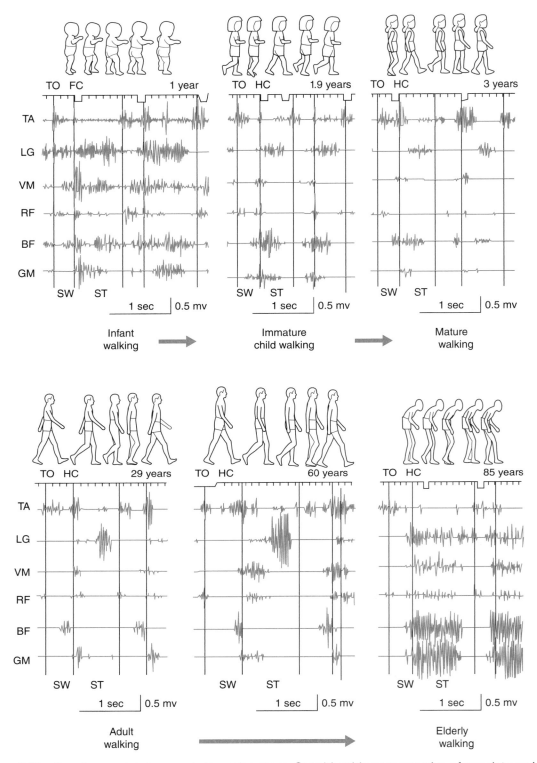

▲ Figure 8.13 Developmental changes in the gait pattern. Considerable cocontraction of agonists and antagonists can be observed in the infant, whereas brief periods of well-defined EMG activity can be seen in the adult. Again, muscle cocontraction returns in some older adults as a stabilizing feature. TO = toe-off; FC = foot contact; HC = heel contact; SW = swing phase; ST = stance phase; TA = tibialis anterior; LG = lateral gastrocnemius; VM = vastus medialis; RF = rectus femoris; BF = biceps femoris; GM = gluteus maximus.

Reprinted, by permission, from T. Okamoto and K. Okamoto, 2007, *Development of gait by electromyography: Application to gait analysis and evaluation* (Osaka, Japan Walking Development Group).

when they should be turned off. Not unexpectedly then, when infants begin to walk, they activate many muscles including some not required for the task of walking. As they become more skilled at what adults accept as an automatic activity, the nature of muscle activation, as verified by the EMG activity, changes to a more refined pattern (Okamoto and Okamoto 2007).

Gait Disorders

In the assessment of pathological gait, we may want to determine whether there are distinctive normal or abnormal gait patterns. EMG analysis can be used to discern a subtle gait disorder, such as in the pattern analysis that Shiavi and colleagues (1992) used to describe some of the features accompanying anterior cruciate ligament injury. Other decision-making tools used by electromyographers to detect abnormal EMG patterns include neural network models, applied cluster analysis, and other EMG "expert systems" (Pattichis et al. 1999).

Foot drop is a common gait problem that lends itself to clinical EMG analysis. Failure to sufficiently activate the tibialis anterior prior to heel-strike leads to this foot "slapping." In such cases, surface EMG records reveal that there is insufficient or no EMG activity in the tibialis anterior (Kameyama et al. 1990). Thus, EMG analysis can be used to identify or confirm the origin of abnormal gait behavior.

Electromyography in Ergonomics

Electromyography is one of the important tools in the **ergonomics** armamentarium. Ergonomists have used EMG to explore overuse injuries such as carpal tunnel syndrome and epicondylitis, as well as sudden acute injuries, most notably, injuries to the lower back. In this section, we'll explore some of the techniques and problems inherent in assessing workplace ergonomic issues using EMG.

EMG and Carpal Tunnel Injuries

Repetitive stress injuries have become a topic of increasing concern for biomechanists, ergonomists, and other professionals in rehabilitation science. Injuries to the carpal tunnel have been traced to many types of light-manufacturing tasks (soldering, hammering, assembly tasks, and others). The use of computer keyboards and positioning devices (such as computer mice) can also place excessive pressure on the carpal tunnel in the wrist, and recognition of this problem has led to new designs and the need to identify individuals at risk for repetitive stress injuries. Carpal tunnel problems are often seen in wheelchair users and require modification

of the wheelchair or changes in propulsive techniques (Veeger et al. 1998). Manual laborers exposed to excessive upper-limb vibration or hammering frequently present with symptoms of carpal tunnel syndrome (Brismar and Ekenvall 1992). Biomechanists can use many EMG tools to identify tasks that threaten the integrity of the carpal tunnel.

Sensory and Motor Conduction Velocity

When excessive pressure is placed on any mixed nerve, the ability of both sensory and motor nerve fibers to conduct APs may be threatened. Sensory and motor conduction velocities in the median nerve can be readily assessed via noninvasive EMG techniques. To measure sensory conduction velocity, low-intensity stimuli are applied to the median nerve at the wrist, about 3 cm proximal to the distal wrist crease. These square-wave stimuli are used to antidromically elicit APs in the sensory nerves innervated by the median nerve. Ring electrodes placed around the second or third digit record these APs. Because the signals can be quite small (typically 20-50 μV), the average response of multiple trials must be computed using a technique called *spike-triggered averaging* (figure 8.14). In this technique, multiple stimuli are delivered, and each response is averaged with all of the previous responses to obtain an averaged signal across many trials. Spike-triggered averaging assumes that the noise present in each trial is random. Therefore, over a number of stimuli, the noise averages to zero, enabling the low-amplitude signal to be studied with improved resolution. When the stimuli are applied, the time between stimulus onset and nerve AP (sensory nerve latency) is determined. As well, the distance between the stimulation site and the recording site is measured so that the nerve conduction velocity (in units of meters per second) can be computed. In addition to sensory nerve diseases, compression of the median nerve at the wrist can deleteriously affect motor nerve conduction velocity. Similar stimulation techniques can be used to activate the motor neurons in the mixed median nerve to identify the presence of a conduction delay.

If these stimuli fail to produce an AP, then a complete conduction block exists between the stimulation and recording sites. Many researchers have published "normal" latency and conduction velocity values to help biomechanists discern the presence of peripheral nerve disorders or disease (see DeLisa and Mackenzie 1982). These conduction measures are important indicators of injuries or disorders in electronic-assembly personnel (Feldman et al. 1987).

Other Applications Surface EMG can be useful for optimizing the use of the wrist in tasks requiring manual manipulation. For example, in determining the appropriate stiffness for a computer keyboard, Gerard and colleagues (1999) observed that subjective comfort

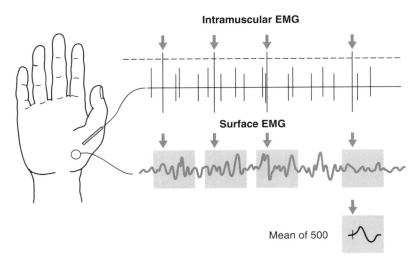

Intramuscular EMG

Surface EMG

Mean of 500

▲ **Figure 8.14** The spike-triggered averaging technique is used to obtain an electrophysiological representation of motor unit size.

Reprinted from *Electroencephalography and Clinical Neurophysiology,* Vol. 39, R.G. Lee et al., "Analysis of motor conduction velocity in the human median nerve by computer simulation of compound muscle action potentials," pgs. 225-237, copyright 1975, with permission of International Federation of Clinical Neurophysiology.

was greatest for keyboards that had a key activation force of 0.83 N. EMG techniques were used to verify the subjective reports. Other studies comparing subjective reports to surface EMG activity have suggested that fewer upper-extremity disorders might be produced by using a centrally located trackball rather than a mouse positioned on either side of the keyboard (Harvey and Peper 1997). Procedures for counting the number of viable motor units in muscle have been used to determine the severity of carpal tunnel types of disorders (Cuturic and Palliyath 2000). EMG biofeedback techniques can be useful in rehabilitating individuals recovering from carpal tunnel syndrome and other occupational disorders (Basmajian 1989; Reynolds 1994). Similar EMG techniques have been applied in the diagnosis and treatment of back pain disorders (Ambroz et al. 2000; Ikegawa et al. 2000; Lariviere et al. 2000).

EMG and Low Back Pain

The diagnosis and treatment of low back pain are important issues. In the United States, for example, back pain accounts for two-thirds of all workers compensation costs. Peripheral muscle problems may be one issue; atrophy of the multifidus is frequently observed in patients with low back pain (Hides et al. 1994). EMG is becoming an increasingly useful tool for the diagnosis and treatment of low back pain.

The trunk musculature plays an important role in tasks such as lifting and throwing. During these kinds of tasks as well as in other activities that might threaten the postural control of the individual and perturb balance,

the nervous system implements strategies to activate trunk muscles while performing voluntary movements involving remote muscle groups. However, the prevalence of back pain in the general population, and the need to identify ergonomically efficient and safe ways to use back muscles, require that we understand the nature of activation in muscles controlling the trunk. Here, too, EMG is an important tool.

A number of issues render EMG recordings from the trunk musculature difficult to obtain and interpret. From an anatomical viewpoint, the architecture of the back muscles is complex. For example, it is generally accepted that injury or disorder of the multifidus frequently results in back pain. Both multifidus and erector spinae have distinct superficial and deep portions (Bustami 1986; Macintosh et al. 1986) with different histochemical and biomechanical characteristics (Bogduk et al. 1992; Dickx et al. 2010). In some respects, it is fortunate that most of the muscle mass of the multifidus lies in the more superficial portion, so that surface EMG has a greater possibility of characterizing the fibers that produce large amounts of force. However, the multifidus superficial portion has a different role than the deeper fibers. Fibers in the superficial portion cross numerous spinal segments and are thus in a better position to produce back extension. However, the deeper fibers are relatively short, crossing perhaps one or two segments, and thus protect segments of the lumbar spine from inappropriate shear or torsion torques (Macintosh and Valencia 1986).

In addition to issues concerning the activation of the back extensor muscles, a number of important questions require resolution by EMG analysis in which considerable **electrocardiographic** (ECG) artifact may be present. ECG artifact is particularly problematic during relatively low force contractions, exaggerating the ratio between the signal of interest and the interfering ECG signal. Recording EMG signals from abdominal, knee extensor, and other trunk muscles during normal human movements, for example, can result in the placement of electrodes well within recording range of the electrocardiogram. The relatively large ECG signal can exaggerate the amplitude of EMG activity, and the low-frequency characteristics of the ECG wave can also alter the frequency characteristics of the recorded EMG activity.

For example, the absence of sufficient trunk activity can produce instability resulting in low back pain (van Dieën et al. 2003). However, the amplitude of EMG activity required to ensure stability is small and thus can be affected by the ECG signal (Cholewicki et al. 1997). The best solution is to place the electrodes in a location from which no or minimal ECG artifact can be recorded. However, this is frequently difficult if not impossible. Consequently, numerous algorithms have been suggested to remove the ECG artifact.

One frequently implemented technique to remove this artifact is to compute a template of the ECG signal and subtract it from the electromyogram. This has been proved to be a reasonably successful procedure and has been implemented in studies recording from the diaphragm (Bartolo et al. 1996) as well as rectus abdominis (Hof 2009). Hof (2009) described a technique in which

the ECG signal is recorded simultaneously with the EMG signal of interest, and template subtraction is then implemented (figure 8.15).

Digital filtering is another useful alternative for ECG artifact removal. Drake and Callaghan (2006) used a FIR (finite impulse response) filter with a hamming window, although they concluded that the most efficient filtering result could be obtained using a somewhat simpler fourth-order, 30 Hz high-pass cutoff Butterworth filter. Template subtraction improved extraction of the ECG signal but required a large amount of time.

Alternative ECG removal techniques include the use of adaptive filters (Lu et al. 2009; Marque et al. 2005), wavelet-independent component analysis (Taelman et al. 2007), and wavelet-based adaptive filters (Zhan et al. 2010). Irrespective of the technique used for ECG signal artifact removal, it is apparent that the resultant improvement in signal interpretation can be important. Hu and colleagues (2009), for example, found that the use of independent component analysis to remove artifact resulted in an improved ability to discriminate between patients with low back pain and normal subjects during both sitting and standing tasks.

Analysis of the EMG activity in back muscles has produced some interesting results, in part reflecting the unique anatomical issues discussed here. As long ago as 1962, Morris and colleagues noted that "the three muscles of the erector spinae group considered here . . . do not always show parallel activity, and one may be active while the other two are inactive" (p. 519). This may

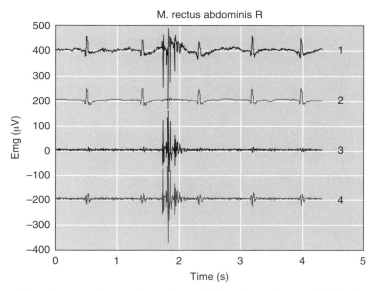

▲**Figure 8.15** ECG artifact frequently can be substantially reduced in the EMG signal by using a template subtraction technique. Line 1: 1 Hz high-pass filtering; Line 2: ECG signal; Line 3: EMG signal with ECG contamination removed and 20 Hz high-pass filtering; Line 4: 20 Hz high-pass filtering only.

Reprinted from *Journal of Electromyography and Kinesiology,* Vol. 19, A.L. Hof, "A simple method to remove ECG artifacts from trunk muscle EMG signals," pgs. e554-e555, copyright 2009, with permission of Elsevier.

suggest that the mere demonstration of greater or lesser EMG amplitude may not be indicative of abnormality in a particular muscle. Potentially more important than EMG amplitude is the timing of EMG activity. Deep and superficial portions of the multifidus are differentially active during arm movements (Moseley et al. 2002). One suggestion is that the manner in which the multifidus is differentially controlled in people with back pain compared with those without may be the source of chronic back pain (MacDonald et al. 2009).

Similarly, the analysis of EMG frequency characteristics in patients with back pain compared with control subjects has suggested that there may be differences between muscles and even differences within muscles. Patients with low back pain frequently exhibit alterations in the electromyographic response to fatigue in the back extensor muscles (Biedermann et al. 1991). This is one area in which spectral analysis of the EMG signals has been helpful. As in other muscles, median EMG frequency decreases in the trunk extensors with fatigue (Demoulin et al. 2007). It is interesting to note that Kramer and colleagues (2005) found that the magnitude of median frequency decrease was greater in healthy subjects than in individuals with back pain. Support for importance of electrode placement is obtained from the observations of Sung and colleagues (2009). Normal subjects and patients with back pain underwent an isometric contraction protocol that induced fatigue of the back extensors. EMG frequency measurements made in both the thoracic and the lumbar portions of the erector spinae indicated that the thoracic portion had a significantly lower median frequency than the lumbar portion in patients with low back pain. However, median frequency was lower in the lumbar portion than in the thoracic portion in control subjects. Thus, in the analysis of trunk musculature that might be involved in the development of back pain, EMG studies using wire electrodes frequently may be required to record from different portions of the muscle.

Electrophysiological Assessment of Muscle Fatigue

Fatigue frequently accompanies short-term, high-intensity motor activity. Even low-intensity activity can be fatiguing if conducted for prolonged intervals. Consequently, the identification of fatigue is important in the workplace. Because fatigue can result from either peripheral (muscular) or central (neural) mechanisms, the exact site of the fatigue can be determined with EMG techniques. EMG analysis may be a useful tool in redesigning workplace techniques to minimize fatigue.

Electromyographic Activity During Prolonged Isometric Contractions

At the onset of maximal isometric contractions, all motor units are active to the maximal extent. During a sustained maximal contraction, EMG amplitude decreases in parallel with the decline in muscular force. The question arises as to what extent this decline is a result of either peripheral or central mechanisms. One way of identifying the site of fatigue is to measure the response of muscle to electrical stimulation, thereby eliminating the influence of the central nervous system. When a high-intensity electrical stimulus is applied to a mixed nerve, the motoneurons are orthodromically activated, resulting in muscle fiber contraction. This EMG response of muscle to electrical stimulation (the M-wave) is frequently used as a measure of maximal muscle electrical activity. The size of the M-wave varies with the muscle size, the number of muscle fibers, and a number of other characteristics.

With prolonged muscle activation, the amplitude of the M-wave usually declines, particularly during long-duration, low-intensity contractions. In some muscle fibers, the AP may cease to propagate along the full length of the fiber, which would explain part of the M-wave decline during fatigue (Bellemare and Garzaniti 1988).

During submaximal isometric contractions, EMG amplitude initially increases (see Krogh-Lund 1993), likely because of the increase in the number of motor units needed to sustain the contraction as contractile failure ensues. Also, the firing rate of newly recruited motor units increases (Maton and Gamet 1989). Thus, it appears that the central nervous system can adapt to changing conditions, as evidenced by the recruitment of new motor units and increased motor unit firing rate.

Changes in the frequency characteristics of the EMG signal during fatigue are considerably more complicated. The median frequency of about 100 Hz declines during a fatiguing exercise task, as does the number of zero crossings (Hägg 1981), indicating a greater predominance of lower-frequency activity. There are many sources of this frequency decline, which demonstrates why surface EMG data alone may not be sufficient to identify the mechanism underlying the fatigue phenomenon.

First, muscle fiber conduction velocity declines during fatiguing muscular contractions (Eberstein and Beattie 1985). The decline in muscle fiber conduction velocity results from a number of metabolic factors, including intramuscular pH and changes in Na^+ (Juel 1988). However, many changes in neural control also occur with fatigue, and these cannot be identified on the basis of the surface EMG alone. Although the motor unit firing rate decreases with prolonged muscle contractions (Bigland-Ritchie et al. 1983), both simulation and experimental studies have concluded that changes in firing rate exert little influence on the EMG power spectrum (Hermens et al. 1992; Pan et al. 1989; Solomonow, Baten et al. 1990). Some researchers have suggested that motor units may be activated in bursts in a pattern called *synchronization*, which could have some advantages. For example, the relationship between motor unit firing rate and muscular force is nonlinear, and a few brief bursts could result in large increases in muscular force (Clamann and Schelhorn 1988). It is possible that synchronization would decrease the EMG power spectrum toward lower frequencies. However, there is insufficient evidence that increased motor unit synchronization occurs with fatigue.

In long-duration tasks, subjects may be able to rotate the required activity among different synergists. In animal experiments, activating muscle compartments sequentially rather than synchronously decreases fatigue (Thomsen and Veltink 1997). During a 1-hour experiment requiring 5% of muscular effort, Sjøgaard and colleagues (1986) found that different magnitudes of activity occurred in the human quadriceps muscles over the 1-hour period, although this observation requires additional examination. Clearly, recording among numerous muscle synergists requires the use of rather selective recording techniques, such as intramuscular wire electrodes.

The idea of using EMG analysis to discern the relative contributions of slow-twitch and fast-twitch muscle fibers during a fatiguing contraction is intriguing. However, because of the myriad factors discussed, it is not possible to state with certainty that the contribution of slow-twitch or fast-twitch fibers is lesser or greater with fatigue.

Changes in EMG activity also follow dynamic contractions. For example, individuals who operate industrial sewing machines exhibit changes in EMG frequency characteristics during the course of a workday, suggesting the onset of fatigue (Jensen et al. 1993). EMG analysis was also used to determine that carrying excessive loads in one hand can result in fatigue and possible occupational injury (Kilbom et al. 1992).

In similar studies, EMG analysis performed during fatiguing contractions has been useful in understanding postpolio syndrome (Cywinska-Wasilewska et al. 1998; Sandberg et al. 1999), metabolic diseases (Mills and Edwards 1984), chronic fatigue syndrome (Connolly et al. 1993), and other disorders. Using EMG analysis, Milner-Brown and Miller (1990) documented important therapeutic benefits from a pharmacologic treatment in patients with myotonic dystrophy.

SUMMARY

The EMG signal is physiologically rooted in the ionic charges at the muscle fiber membrane. Neural activation produces transient changes in these ionic concentrations, allowing the muscle fiber AP to propagate along the sarcolemma. Many electrodes, amplifiers, and analysis techniques are available for detecting, processing, and describing the signal. Users must be aware of a number of technical issues that can interfere with the characteristics of the EMG signal. Examples of EMG use in areas such as physical rehabilitation, clinical medicine, dentistry, gait analysis, biofeedback control, motor control, ergonomics, and fatigue analysis demonstrate the wide variety of research questions amenable to study using these techniques.

SUGGESTED READINGS

Basmajian, J.V., and C.J. De Luca. 1985. *Muscles Alive: Their Functions Revealed by Electromyography.* 5th ed. Baltimore, MD: Williams & Wilkins.

Loeb, J.E., and C. Gans. 1986. *Electromyography for Experimentalists.* Chicago: University of Chicago Press.

Luttmann, A. 1996. Physiological basis and concepts of electromyography. In *Electromyography in Ergonom-* ics, ed. S. Kumar and A. Mital, 51-95. London: Taylor & Francis.

U.S. Department of Health and Human Services. 1992. *Selected Topics in Surface Electromyography for Use in the Occupational Setting: Expert Perspectives.* Washington, DC: National Institute for Occupational Safety and Health.

Chapter 9

Muscle Modeling

Graham E. Caldwell

In chapter 8, we discussed how the control signals used by the nervous system to elicit muscle activity could be monitored using electromyography (EMG). These EMG signals allow us to examine how the nervous system effects purposeful movement, yet they tell only a portion of the story of how human motion is produced. In this chapter, we

▶ examine how muscles respond to the nervous system's signals to produce the forces that result in skeletal motion,

▶ introduce the Hill muscle model, named after the illustrious British Nobel laureate A.V. Hill,

▶ consider the basic mechanical properties of muscle tissue that dictate how much force a muscle will produce for a given level of nervous stimulation,

▶ examine the dynamic interaction between Hill muscle model components,

▶ discuss how such models can be used to represent specific individual muscles by finding muscle-specific parameters, and

▶ introduce other muscle models that are useful in biomechanics research.

THE HILL MUSCLE MODEL

Muscle is unique because it can convert signals from the nervous system into force, which in turn can cause movement of the skeletal system. Most readers of this text will be familiar with the contractile function of skeletal muscle, which is based on anatomical structure that can be described at the level of whole muscle, muscle fascicles, muscle fibers, or individual sarcomeres. At the heart of muscle force production within the sarcomeres are the contractile proteins actin and myosin that are found in the thin and thick filaments, respectively. Force production arises from neural signals that ultimately result in mechanical coupling of the thick and thin filaments through the attachment of myosin heads

to actin binding sites and the subsequent rotation of these crossbridges. There are several excellent texts and reviews that explain basic muscle physiology, and we will assume that the reader has some familiarity with this topic. In this chapter we discuss muscle force production from a mechanical rather than biological perspective.

It is unfortunate that one of the first terms we learn is "muscle contraction," from which we infer that the muscle shortens in response to nervous input. In fact, the muscle may shorten, lengthen, or remain at a constant length, depending on other internal and external forces acting on the skeleton. One fact is incontrovertible under any loading situation: In response to a nervous signal, the muscle produces force. It would be expedient if the quantitative relation between the nervous signal input and the muscle force output were linear (i.e., *x* units of nervous stimulation yield *y* units of muscular force under any conditions). However, in the first half of the 20th century it became obvious to scientists studying muscle that this was not the case. Muscular tissue could produce different amounts of force for a given level of nervous input, depending on specific experimental conditions. Perhaps the best known of the early scientists is A.V. Hill, who developed a simple but powerful conceptual model of muscle function. Although many advances have been made in our knowledge of muscular structure and function since this early work, the Hill model is still appropriate for describing basic muscle mechanics and is the model of choice for modeling specific muscle behavior when we are attempting to understand voluntary human movements. Therefore, in this chapter we explore the Hill model in some depth before turning our attention to other muscle models that are useful in biomechanics research.

The basic Hill model (figure 9.1) consists of the *contractile component* (CC), the *series elastic component* (SEC), and the *parallel elastic component* (PEC). Each component has mechanical characteristics that explain specific phenomena seen in experimental studies. It is

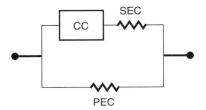

▲ **Figure 9.1** The three-component Hill muscle model consists of a contractile component (CC), series elastic component (SEC), and parallel elastic component (PEC).

important to realize that the model represents muscle behavior rather than structure. This means that anatomical correlates to the individual model components do *not* exist, although in some cases we can see how certain muscle structures are related to particular model components. First, we describe the model components separately and then examine how dynamic interactions between the components aid our understanding of muscular force production.

Contractile Component

In the Hill model, the CC is the "active" element that turns nervous signals into force. The magnitude of the CC force produced depends on its mechanical characteristics, which can be expressed in four separate relationships: stimulation-activation (SA), force-activation (FA), force-velocity (FV), and force-length (FL).

Stimulation-Activation

The first of the CC's mechanical properties concerns how the nervous system signal (stimulation or excitation) is related to the muscle's intrinsic force capability or potential (activation). Physiologically, this property reflects the excitation-contraction coupling process, in which alpha motor neuron action potentials trigger motor unit action potentials (MUAPs) that travel along muscle fibers. These MUAPs are carried inward through the transverse tubule system to the sarcoplasmic reticulum, where they cause the release of calcium ions into individual sarcomeres. This portion of the excitation-contraction coupling sequence can be considered the stimulation, because it is independent of the actual force production mechanism in the sarcomere at the level of the crossbridges, which link the thick and thin filaments containing the contractile proteins myosin and actin, respectively. The actin-myosin complex responds to the calcium ion influx by changing from its resting state (no crossbridge attachment and no force potential) to an activated state in which force production can occur. This latter sequence is the activation part of the stimulation-

activation process. Note that the stimulation represents the input to the process and the activation represents the response, or output.

How much activation is produced for a given input of stimulation? This simple question cannot be easily answered, because it is difficult to quantify either stimulation or activation. As described in chapter 8, the level of stimulation is altered through the processes of motor unit recruitment and rate coding. How does one quantify the "amount" of stimulation that a muscle receives based on these two nonlinear mechanisms? Likewise, in response to neural stimulation, the sarcoplasmic reticulum releases calcium ions that result in activation (force potential) at the level of the crossbridges. Activation should therefore be measured as the number of attached crossbridges that can produce force. But how does one measure this quantity? To get around these quantification difficulties, for modeling purposes both stimulation and activation are placed on relative scales that range from 0% to 100%. The exact shape of the relationship between stimulation and activation is, of course, difficult to ascertain, but for the present we can consider that it is a direct linear relation, although there is evidence from the neural literature that it is in fact nonlinear.

When a motor unit is initially activated, there is a time delay between the onset of the neural action potential and the activation at the crossbridge level. This time delay has two components, the first of which is the transit time for the MUAP to travel from the myoneural junction to the sarcoplasmic reticulum. The second component is the length of time for the calcium ions to be released from the sarcoplasmic reticulum and become attached to the thin filaments, a process that, when completed, removes the inhibition for crossbridge attachment imposed by the troponin-tropomyosin complex. When the force response from the motor unit is no longer necessary, the alpha motor neuron stops sending impulses. However, for a brief period, there is still a supply of calcium ions within the sarcomeres, allowing the crossbridges to remain activated even in the absence of stimulation. The duration of this *deactivation* process is longer than the activation process, and it is dictated by the time it takes for the sarcoplasmic reticulum to reabsorb the free calcium ions within the sarcomeres. The time periods for both activation and deactivation are shown schematically in figure 9.2.

Force-Activation

In the previous discussion, the term *force potential* was used to emphasize the fact that activation is a state in which force can be produced, rather than an actual force level. The importance of this distinction will become clear later, when we consider that the actual force level

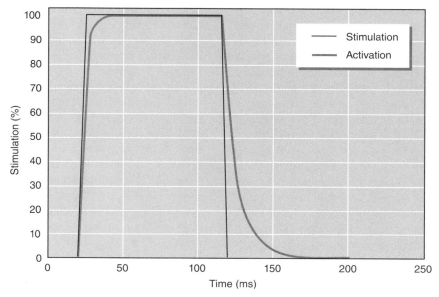

▲ **Figure 9.2** Temporal relationship between stimulation and activation in the CC. Stimulation shown as black line, activation as green line.

in newtons depends not only on activation but also on the kinematic state of the CC. However, we need to convert the state of activation (in percentage of force potential) to an actual force level, expressed either in newtons or as a percentage of a particular muscle's maximal force. To keep this force independent from any specific kinematic state, the force-activation relation must be conceptual only; it is impossible to measure the force produced in the absence of a specific CC kinematic state. Therefore, the force-activation relation is direct and linear (e.g., 10%, 20%, or 50% activation represents 10%, 20%, or 50% force, respectively).

Force-Velocity

Perhaps the most important CC mechanical property is the influence of CC velocity on force production, a fact well established in the first half of the 20th century. This relationship, shown in figure 9.3, is expressed mathematically by the famous Hill (1938) equation for a rectangular hyperbola:

$$(P + a)(v + b) = (P_o + a)b \tag{9.1}$$

In this equation, P and v represent the CC force and velocity at a given instant in time, respectively. P_o represents

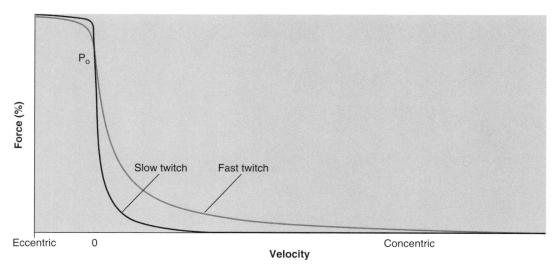

▲ **Figure 9.3** The CC relationship between force and velocity at maximal activation for both slow- and fast-twitch muscles.

the force level the CC would attain at that instant if it were isometric. The muscular dynamic constants a and b were originally conceived by Hill to represent constants of energy liberation. These dynamic constants are muscle- and species-specific and dictate the exact shape of the rectangular hyperbola and its intercepts on the force (P_o) and velocity (V_{max}) axes. For example, predominantly slow-twitch muscles have different dynamic constants than those with a high proportion of fast-twitch fibers, with the maximal shortening velocity V_{max} being roughly 2.5 times greater in fast-twitch fibers than in slow-twitch fibers from the same species. Edman (1988) showed that a double hyperbolic shape better represents the concentric force-velocity relation, with a clear deviation from the single Hill hyperbola in low-velocity–high-force concentric situations. This presents no major conceptual problem for the Hill model, but it does require the use of a different equation to represent the CC's force-velocity characteristics.

Note that the Hill equation refers only to isometric or concentric CC velocities. The relationship can be extended to include CC eccentric (lengthening) conditions (figure 9.3), but the Hill equation must be modified. Finally, the relationship shown in figure 9.3 represents full activation of the muscle and therefore the maximal force output possible across the range of CC velocities.

Force-Length

Another important CC property is the dependence of isometric force production on CC length, described in 1940 by Ramsey and Street. The best-known descrip-tion of the force-length relationship is in the paper by Gordon and colleagues (1966), which proved to be the cornerstone of the well-known sliding filament theory of muscular contraction. The basic shape of the FL relation (figure 9.4) illustrates that isometric force production is greatest at intermediate CC lengths and declines as the CC is either lengthened or shortened. The highest isometric force level is referred to as P_o, which can be confusing because the same acronym found in the Hill force-velocity equation has a slightly different meaning. For the FL relationship, P_o is said to occur at the muscle length L_o, the optimal length for force production. In the FV relation, P_o is the maximal force at zero velocity (i.e., isometric) for a given CC length at which all FV data points occur, regardless of whether that happens to be L_o. Thus, P_o in the FV relation is actually a function of CC length, as will become obvious in the next section.

Contractile Component Properties Viewed Together

The four CC properties (SA, FA, FV, and FL) must be viewed together to understand CC function. When force output is required, the central nervous system (CNS) initiates this process by sending a stimulation signal to the muscle. This stimulation signal causes the activation of the CC, according to the stimulation-activation relationship and the temporal characteristics illustrated in figure 9.2. The CNS modifies the stimulation signal as needed to regulate the CC's activation level and therefore its force capability. Thus, the CNS exerts control over

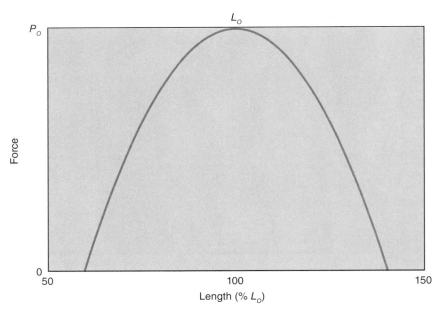

▲ **Figure 9.4**   The CC relationship between force and length at maximal activation.

the CC FL through its regulation of stimulation, but this control is somewhat indirect because of the intermediate influence of the SA and FA relations. This control of force is more complex than it may seem, given that the actual force is dictated by the kinematic state of the CC that results from the FV (figure 9.3) and FL (figure 9.4) relationships. The combined effect of these two mechanical characteristics can be seen in figure 9.5, which depicts a force-length-velocity surface. Individual points on this surface represent the force output by the CC for any combination of CC length and velocity if the CC activation level is maximal (100%). In many cases, the CNS operates at submaximal levels, with the actual force for a given CC length and velocity being less than the force depicted on this surface. One way to visualize these submaximal conditions is to consider an activation line segment joining the surface (100%) and the floor (0%) of this three-dimensional (3-D) plot. All submaximal activation levels lie somewhere along this line segment (figure 9.5). The exact CC force at a given time can be determined by first finding the CC length and velocity on the surface and then following the line segment to the current level of activation. Evidence suggests that figure 9.5 is too simplistic in that submaximal activation changes the shape of FL characteristics, with L_o shifting toward greater muscle lengths (Rack and Westbury 1969; Huijing 1998).

Series Elastic Component

Muscle includes materials that display a degree of elasticity and are related to passive connective tissue rather than the active contractile proteins that produce the CC force in response to CNS stimulation. The existence of the series elastic component—elastic elements in series with the active force-producing structures in the muscle—has been known since the days of Hill's experiments on frog muscle more than 50 years ago. Any force that the CC produces is expressed across the SEC. One obvious contributor to the SEC elasticity is the tendon that joins the muscle fibers to the skeleton. However, other structures also contribute to the elasticity, including the aponeurosis, or "inner tendon," which connects the external tendon to the muscle fibers, and connective elements within the muscle fibers (e.g., Z-lines). For example, Kawakami and Lieber (2000) demonstrated that the aponeurosis contributes substantially to the series elasticity. Thus, although some authors characterize the CC as the equivalent of muscle fibers and the SEC as the equivalent of the tendon, those assertions are not quite true. Remember also that the SEC is a behavioral model component, and an exact correspondence with anatomical structures is unnecessary; SEC elasticity results from all elastic elements in series with the active force-generating elements in the muscle.

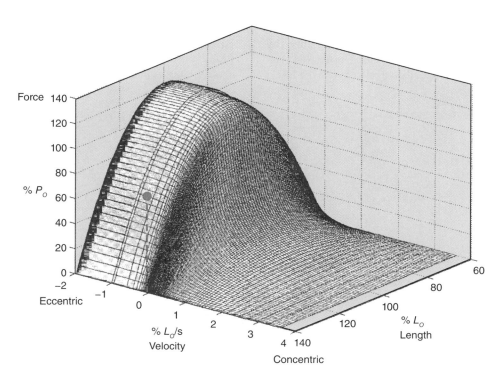

▲ **Figure 9.5** The CC force-length-velocity 3-D surface, representing 100% activation. The green dot (•) indicates force for a particular CC length and velocity. Submaximal activation results in forces beneath the surface; the green broken vertical line indicates forces with a continuously decreasing activation level.

The SEC elastic behavior is described by its force-extension (F∆L) relationship (Bahler 1967), shown in figure 9.6. In physics, the elasticity of a material is usually quantified by its stiffness k, which is calculated as the change in applied force divided by the resulting change in the length of the material ($k = \Delta F/\Delta L$). If the material displays linear F∆L characteristics, the stiffness is the slope of the F∆L line. However, figure 9.6 illustrates that the SEC F∆L relation is highly nonlinear, which is common for biological materials. The slope of the F∆L association increases as the SEC extends, meaning that at low force levels the SEC is quite compliant (low stiffness), whereas at higher force levels the SEC becomes stiffer and unit increases in force produce less extension. This nonlinear SEC elasticity is an extremely important feature of the Hill model, one that imparts a powerful influence on the muscular force response. Some experiments indicate that the SEC is lightly damped, meaning that the rate at which the force is applied changes its elastic nature. The damping characteristic of the SEC has a relatively small influence on the overall model behavior and will be ignored in the present discussion.

Parallel Elastic Component

Muscles display elastic behavior even if the CC is inactive and producing no force. If an external force is applied across an inactive, passive muscle, it resists but stretches to a longer length. This resistance is not a response of the SEC, because no force is being produced by the inactive CC. Instead, this inactive elastic response is produced by structures that are in parallel to the CC. The parallel elastic component (PEC) is usually correlated with the fascia

that surrounds the outside of the muscle and separates muscle fibers into distinct compartments. Like the SEC, the F∆L relationship of the PEC is highly nonlinear in nature, with increasing stiffness as the muscle lengthens.

The PEC elasticity is considered a passive response, yet it can play a role during active force production. In an active isometric force situation, the measured force response is the summation of the active CC force and the passive PEC force associated with the isometric length of the muscle (figure 9.7). Figure 9.7 depicts the FL characteristics of the active CC, the passive PEC, and the summed CC and PEC responses. At shorter lengths, the PEC is not stretched and thus the muscle force response will be entirely caused by the active CC. As the muscle is placed at longer lengths, the PEC is stretched and its force response is added to the active CC response. The exact shape of the summated force-length response depends on the stiffness of the PEC and the CC length at which the PEC first generates force in relation to the optimal length of the active FL relation, L_o.

Component Interactions During Active Force Production

We can gain a real understanding of the Hill model only by considering the dynamic interactions among the components during an active force-producing situation. For this purpose, we examine a simplified version of the Hill model that includes the CC and SEC but omits the PEC. This is akin to performing muscle experiments at shorter lengths in which the passive PEC plays no role.

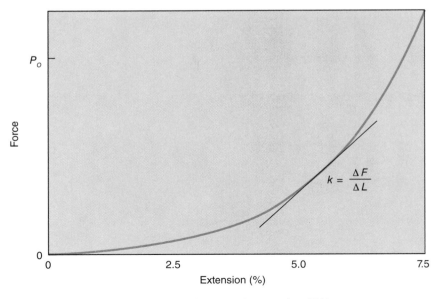

▲ **Figure 9.6** The SEC relationship between force and extension (∆L).

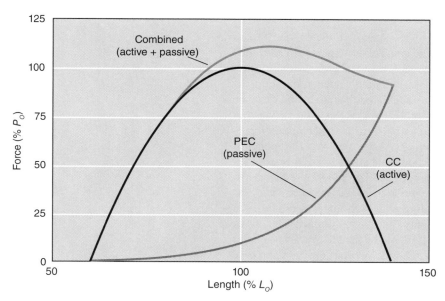

▲**Figure 9.7** The contractile component (CC) and parallel elastic component (PEC) force-length (FL) relationships, separate (black [CC] and gray [PEC] lines) and combined (green line).

FROM THE SCIENTIFIC LITERATURE

Alexander, R.M. 1990. Optimum take-off techniques for high and long jumps. *Philosophical Transactions of the Royal Society of London B* 329:3-10.
Alexander, R.M. 1992. Simple models of walking and jumping. *Human Movement Science* 11:3-9.
Selbie, W.S., and G.E. Caldwell. 1996. A simulation study of vertical jumping from different starting postures. *Journal of Biomechanics* 29:1137-46.

Originally, the Hill model was developed to help elucidate the basic mechanical characteristics of striated muscle function. More recently, it has been used to replicate the mechanical properties of individual muscles or muscle groups during studies of specific human movements. Muscle models can play different roles in these investigations, but in all cases they are used to provide physiological information concerning the muscles involved. The exact manner in which the models are used depends on the specific aims and questions addressed by the study. In some cases, the muscle models are rotational in nature, representing the total effect of all muscles contributing to a resultant joint moment. In other instances, the models represent individual muscles or a group of synergists.

A good example of a relatively simple muscle model is found in the work of Alexander (1990, 1992). In that work, a rigid trunk mass supported by a massless, two-segment (thigh and leg) lower extremity represented the human body (figure 9.8). The trunk center of mass was located at the hip, and a single knee extensor "muscle" controlled the motion of the model. This rotational muscle model generated torque based on the angular velocity of the knee joint and was fully activated at all times. Therefore, the muscle model lacked SA dynamics and FL characteristics. Alexander used this model to simulate the push-off phase of both long jumping and high jumping. Two initial conditions concerning the start of the contact

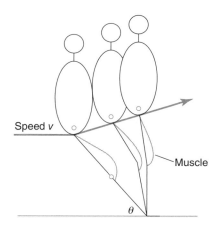

▲**Figure 9.8** Alexander's jumping model (1990, 1992).

(continued)

(continued)

phase before takeoff were specified: the center of mass horizontal velocity and angle of the almost fully extended lower extremity. The simulation calculated the torques and motion of the knee joint and center of mass during the contact phase and calculated the height and distance of the subsequent flight phase. After confirming that the model would predict realistic ground reaction forces and jump performances, Alexander used the model to study the effects of systematic alteration of the approach speed and touchdown angle. The model predicted different optimal characteristics for the two jumps and was useful in understanding why high jumpers might use a lower approach velocity than a long jumper. The physiological constraints imposed by the FV characteristics of the knee extensors limited the approach speed to provide the appropriate combination of vertical and horizontal impulses in the two jumping tasks.

Selbie and Caldwell (1996) also used rotational muscle models as torque generators in simulations of vertical jumping. They modeled the body as four linked rigid segments with torque generators acting at the hip, knee, and ankle joints (figure 9.9). At each joint, the muscle model obeyed torque-velocity, torque-angle, and SA relationships but did not include elastic characteristics. The jumping motion was initiated from a given static posture, and the simulation produced segmental and center-of-mass kinematics throughout the jumping motion until the moment of takeoff. The model kinematics at takeoff were used to calculate vertical jump height using projectile motion equations. This forward dynamics model was optimized to find the combination of hip, knee, and ankle stimulation-onset times that produced the highest

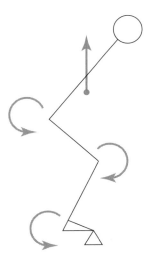

▲ **Figure 9.9** Selbie and Caldwell's vertical jumping model with three joint torque actuators (1996).

jump from a given initial posture. The model was able to produce realistic jump heights and segmental kinematics. The researchers' question was whether the model could generate jumps to similar heights from widely varying initial postures, and the results illustrated that it could from most of the 125 initial positions simulated. However, some discrepancies from actual jumping performances were noted, perhaps because of the simplicity of the muscular representation.

For more information on forward dynamics simulations and the use of muscle models within the framework of musculoskeletal models, the reader is referred to chapters 10 and 11, respectively.

We begin by considering two often-studied muscular responses, the *isometric tetanus* and *isometric twitch*. These two muscular responses are easily produced in isolated muscle preparations and have been used to characterize muscles as either slow-twitch or fast-twitch, depending on the time course of the force response. The experimental setup consists of an isolated muscle kept viable in a bath of Ringer's solution, with the muscle fixed at one end and attached to a force transducer at the other (figure 9.10), and thus kept at a constant fixed length. Stimulation in the form of pulses of electric current is applied to the muscle through an electrode, with the pulse width, magnitude, and frequency determined by the muscle and protocol under examination. A single brief pulse is applied for the isometric twitch, while a continuous train of pulses is applied for the isometric tetanus. The characteristic force responses elicited by these stimulation patterns are shown in figure 9.11. Note that for the tetanus condition, the frequency of stimulation pulses determines whether the force response is an incomplete summation of individual twitches (low stimulation frequency) or a smooth, complete tetanus (high stimulation frequency).

The shapes of the tetanic and twitch force responses raise a number of questions. Why do the force responses have these characteristic shapes? Why are the force responses for slow-twitch and fast-twitch muscles different? Why does the peak force in the isometric twitch reach only a fraction of the force level produced in the isometric tetanus? Why does the isometric twitch force peak occur so long after the end of the single stimulation pulse? The Hill muscle model can greatly help our understanding of these (and other) force responses.

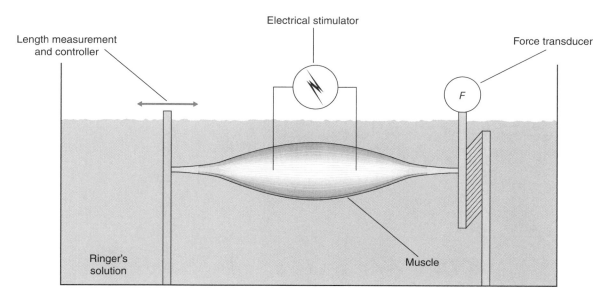

▲ **Figure 9.10** Schematic of isolated muscle preparation.

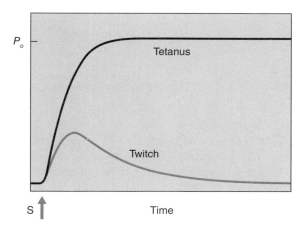

▲ **Figure 9.11** Isometric twitch (green line) and tetanus (black line) force responses.

CC-SEC Interactions During an Isometric Twitch

In response to the brief maximal impulse of stimulation, the muscle force exhibits a relatively slow rise to a submaximal peak, followed by an even slower decline back to zero force (figure 9.11). Whereas the stimulation pulse may last only a few milliseconds, the time to peak twitch force is 25 to 50 ms. This discrepancy in time course results from CC and SEC dynamic interaction and the interplay of their mechanical properties. At the beginning, before the arrival of the stimulation pulse, the CC is inactive and producing no force and the SEC is at its unloaded length (figure 9.12a). The stimulation pulse causes the CC to become active and begin to produce force, according to the CC SA, FA, FV, and FL relations. This force is expressed across the SEC, which responds by extending according to its FΔL relationship. This basic CC-SEC interaction of CC force changes causing SEC length changes continues throughout the isometric twitch condition, following a sequence of events associated with the magnitude and time course of the stimulation pulse. Note that the relative input stimulation and output force values are always different, except for the beginning (figure 9.12a) and end (figure 9.12d), when both equal zero.

When the stimulation increases from zero to maximal, the CC activation rises according to the time course associated with the transport of calcium ions from the sarcoplasmic reticulum to the crossbridge binding sites on the thin filaments (figure 9.2). As activation rises, the CC produces force according to its instantaneous values of activation, velocity, and length (figures 9.3, 9.4, and 9.5). The force rises throughout the early part of the twitch, meaning that the SEC is continually lengthening until the peak twitch force is achieved (figure 9.12b). Consequently, the CC must be shortening by a concomitant amount, because the total muscle length (CC length plus SEC length) is being held constant (isometric). After the peak twitch force, the force continually falls (figure 9.12c), meaning that the SEC is recoiling (shortening) and the CC is therefore lengthening. Only at the peak, where $dF/dt = 0$, are the SEC and CC instantaneously in an isometric state, neither lengthening nor shortening.

It is possible that the discrepancy between the rise time in stimulation (almost instantaneous) and in force

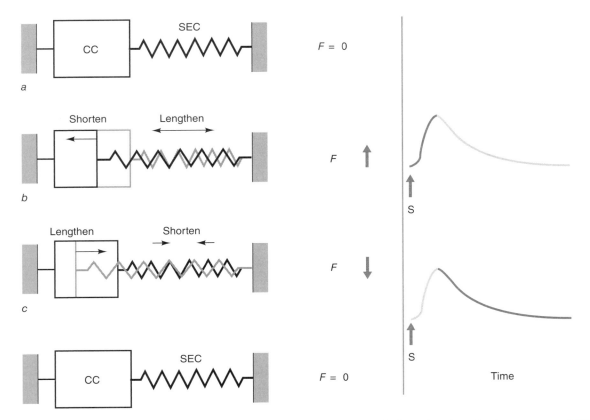

▲ **Figure 9.12** Dynamic interaction between contractile component (CC) and series elastic component (SEC) during isometric twitch.

(25-50 ms) could be related either to the time course of calcium dynamics that dictates the SA temporal association or to the ability of the CC to produce force according to its kinematic conditions (length and velocity). During the rising phase of force the CC shortens (concentric), and during the falling force phase the CC lengthens (eccentric). Only at the instant of peak twitch force is the CC isometric. The SEC does not directly influence the amount of force produced but merely responds to the CC force by changing its length according to its FΔL relationship.

Active State

Hill found that it was possible to delineate the effects of calcium dynamics versus those from CC kinematics by describing the time course of the active state throughout the isometric twitch. Hill defined *active state* as the instantaneous force capability of the CC if it is neither shortening nor lengthening (i.e., it is isometric). While the CC is isometric, only the level of activation and the instantaneous CC length influence the force level attained. Keeping the CC length on the FL plateau eliminates FL effects, meaning that force is dictated by the activation alone. By using experimental protocols known

as *quick release* and *quick stretch*, researchers were able to follow the time course of the active state (figure 9.13*a*) and thus account for the effects of SA dynamics throughout the isometric twitch. Comparing the time course of the active state with that of the isometric twitch force clearly shows that if the CC is isometric, it is indeed capable of producing large forces shortly after the onset of the stimulation pulse. The rapid rise in the active state with the onset of stimulation means that the CC is fully capable of producing maximal force almost immediately. Until the time of peak twitch force, the actual force produced by the CC is less than its capability if it were isometric. By deduction, we are left to conclude that the shortening of the CC, associated with the rising force and lengthening of the SEC, imposes a strong depressive effect on CC force production. To fully understand the shape of the twitch force, we must examine more closely the dynamic interaction of the CC and SEC.

Importance of CC-SEC Dynamics

The keys to understanding the shape of the twitch force profile are the sequence of CC-SEC dynamics shown in figure 9.12, the CC FV relationship (figure 9.3), and the SEC FΔL relationship (figure 9.6). As the stimulation

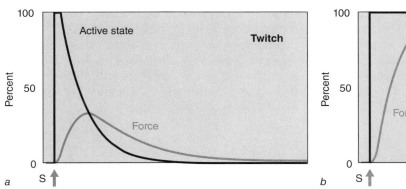

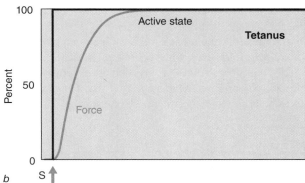

▲ **Figure 9.13** Contractile component (CC) active state (black line) and force (green line) response during *(a)* an isometric twitch and *(b)* an isometric tetanus.

pulse causes the CC activation to rise, the CC begins to produce force across the SEC. In response to this force, the SEC elongates according to its FΔL characteristics. Because of its highly compliant nature at low force levels, the SEC stretches very rapidly. The CC in turn shortens at the same high velocity, as the CC and SEC velocities must offset each other to keep the total muscle in an isometric state. As a result of the nature of the concentric FV property, this high shortening velocity keeps the actual force production far below the CC isometric capability. As the force rises slowly, the SEC undergoes more extension, and because of its nonlinear stiffness characteristics, it becomes less compliant. This slows its stretching velocity, which in turn slows the CC shortening velocity and allows the force to rise.

In the isometric tetanus, the activation of the CC rises rapidly to its maximal level and remains there (figure 9.13*b*) in response to the continual train of stimulation pulses. Therefore, the force continues to rise, bringing the SEC to the stiffest portion of its FΔL curve. This dramatically slows the SEC lengthening and brings the CC shortening velocity toward zero. Eventually, the CC and SEC length changes cease, bringing the CC to its isometric value and the SEC to the constant length associated with that force level. As long as the stimulation pulses keep the CC fully activated, the force will remain at this isometric plateau level.

In the isometric twitch, the story differs in that the activation of the CC begins to decay after the end of the single stimulation pulse, as seen in the falling phase of the active state profile (figure 9.13*a*). Despite the falling activation, the twitch force continues to rise in response to the slowing of CC shortening. However, the falling activation results in a slower force rise, and at the peak of the twitch force the active state and twitch force intersect. At this peak, the force is neither rising nor falling, the SEC ceases to lengthen, and both the CC and the SEC become instantaneously isometric. After the peak in the

force profile, the SEC starts to shorten as the force begins to fall. Because of the reciprocal CC-SEC interaction, the CC begins to lengthen and produce force according to its eccentric FV relationship. This keeps the force higher than the active state that reflects the CC isometric capability. However, because the activation continues to fall, the force level also drops. This submaximal, eccentric CC state continues with falling activation and force, until they both return to zero at the end of the twitch contraction.

General CC-SEC Interactions

When a muscle is stimulated, either externally in an experiment or in vivo by the CNS, the CC-SEC interaction has a direct bearing on how much force is generated. In the description of the isometric twitch and tetanus, we noted that the CC-SEC interaction can be understood using the equation

$$\text{Muscle Length} = \text{CC Length} + \text{SEC Length} \quad (9.2)$$

In these isometric conditions, the muscle length is constant, emphasizing the inverse relation between the CC and SEC lengths during isometric muscle contractions. Another kinematic equation that expresses the CC-SEC interaction is

$$\text{Muscle Velocity} = \text{CC Velocity} + \text{SEC Velocity} \quad (9.3)$$

In the isometric case, muscle velocity equals zero, meaning that the CC and SEC have equal but opposite velocities during an isometric situation and emphasizing the importance of the nonlinear SEC FΔL and CC FV relationships in dictating muscle force-time profiles.

These two equations hold true during all muscle contractions, both isometric and dynamic. In the dynamic case, the constraints of constant muscle length and zero muscle velocity are removed and the CC and SEC kinematics lose their inverse, equal-but-opposite relationship. The SEC always mediates the CC velocity so that, in general, the muscle velocity is not equal to the CC velocity. However, when the overall muscle changes length, it is more difficult to envision the exact nature of the CC velocity than in the isometric case. For example, it has been shown experimentally that during eccentric muscle contractions, the muscle fibers (an anatomical representative for the CC) may either shorten or lengthen depending on the exact state of the elastic elements in series (Biewener et al. 1998; Griffiths 1991). One of the strengths of the two-component Hill muscle model is that it allows estimation of CC length and velocity during active force production.

EXAMPLE 9.1

MUSCLE MODEL ALGORITHM

A Hill-type muscle model can be implemented in several ways. The most flexible arrangement is to write a software subroutine that contains the code for the muscle model in a general form. The model for any specific muscle is then implemented by writing a calling subroutine that contains the model parameters (e.g., P_o, FL information, a and b dynamic constants, SEC elasticity) for that particular muscle. The calling subroutine passes the specific parameters to the general model subroutine so that the model output represents that individual muscle. This permits the general model subroutine to be used for more than one specific muscle, which is advantageous when we are using a musculoskeletal model in which several different muscles are represented and are required to produce force simultaneously. This scheme can use the muscle model algorithm and software language of the researcher's choice.

The Hill model iterative algorithm presented here (Baildon and Chapman 1983; Caldwell and Chapman 1989, 1991) is similar in concept (but not necessarily implementation) to others found in the literature (Bobbert et al. 1986a; Pandy et al. 1990; van den Bogert et al. 1998; Winters and Stark 1985; Zajac 1989). The algorithm has explicit equations representing the CC and SEC prop-

- For each time step Δt

  Need predicted muscle forces for N different muscles ($i = 1, 2, \ldots, i + 1, \ldots N$)

  - For muscle i

    Get muscle parameters for CC and SEC
    Get muscle length and stimulation inputs
    Get CC activation and length from previous time step
    Get SEC length and force from previous time step

    Call muscle model subroutine

    1. SEC force = f (SEC length)
       CC force = SEC force
    2. CC activation = f (stimulation)
    3. CC velocity = f (CC force, activation, length)
    4. CC length = $\int$ CC velocity dt
    5. SEC length = muscle length — CC length

    Repeat for next muscle $i + 1$

  Repeat for each time step Δt

▲ **Figure 9.14** Flowchart of muscle model algorithm.

erties and should give the user a sense of how the CC and SEC interact dynamically (figure 9.14). The description of the algorithm indicates exactly where functions representing the CC and SEC properties are needed, and table 9.1

Table 9.1 Model Equations and Parameters

Model component	Relation	Equation	Parameters
CC	FV concentric	$(P + a)(v + b) = (P_o + a)\,b$	P_o = maximal isometric force a = Hill dynamic constant b = Hill dynamic constant
	FV eccentric	$S = b(P_s - P_d)/(P_o + a)$ $P = P_s - S(P_s - P_d)/(S - v)$	P_s = eccentric force saturation level (% of P_o)
	FL (parabola)	$RF = [c0(RL - 100)^2] + 100$	c0 = parabola width coefficient
SEC	FΔL	$RF = 0.0258\,[\exp(\text{stiff})\,(RLS)] - 0.0258$	stiff = nonlinear stiffness coefficient

CC = contractile component; P = force; RF = relative force; RL = relative CC length; RLS = relative SEC length; SEC = series elastic component; v = CC velocity.

Reprinted from *Human Movement Science*, Vol. 10, G.E. Caldwell and A.E. Chapman, "The general distribution problem: A physiological solution which includes antagonism," pgs. 355-392, copyright 1991, with permission of Elsevier.

gives examples of specific equations for these functions (Caldwell and Chapman 1991). Researchers building their own models may want to use different equations for these component properties.

The model subroutine needs two types of inputs: (1) model parameters to specify the particular muscle of interest and (2) time histories of the muscle kinematics and stimulation for a given movement situation. For example, we can attempt to simulate an isometric twitch experiment for a rat gastrocnemius or human tibialis anterior. In either case, we need specific muscle parameters (more easily obtained directly for the rat gastrocnemius), such as its FV dynamic constants (a and b), maximal isometric force output (P_o), and so on. For the isometric twitch conditions, the overall muscle length is held constant at a specified length for the entire time of simulation (perhaps 200 ms). The stimulation history is represented by a series of zeros interrupted for only a few milliseconds during the supramaximal pulse (see figure 9.11), when the stimulation value equals 100%. The model subroutine uses the muscle parameters and the kinematic and stimulation inputs to calculate histories of the force output and the CC and SEC dynamics that produced this force profile.

CALLING SUBROUTINE

1. Muscle parameters:
2. CC: P_o, L_o, width of FL parabola, FV dynamic constants a and b, activation constants for temporal SA relationship
3. SEC: FΔL relation
4. Experimental conditions: muscle kinematics and stimulation inputs

For each time point of the simulated contraction, specify the muscle length and input stimulation value. For this example, we will assume that we have 3000 time steps so that time t will progress from $t = 0$ to $t = 3000$ in steps of 1, each time step indicating a change of 0.1 ms for a total simulation time of 300 ms.

MUSCLE MODEL SUBROUTINE

For each time point of the simulated contraction, starting from $t = 0$:

1. Use the SEC FΔL relationship to predict the instantaneous CC force. Because the CC and SEC are in series, the CC and SEC forces are equal. The SEC FΔL equation must therefore be written such that SEC force is a function of its length, or

$$\text{SEC Force} = f(\text{SEC Length}) \qquad (9.4)$$

Note that for $t = 0$, the stimulation and activation equal zero, the CC is producing no force, and the SEC is at its unloaded length. Therefore, the SEC force will be predicted to be zero, consistent with CC "reality."

2. Use the CC FV relation to predict the CC velocity from the force predicted in step 1. Recognize that the CC force depends on both the CC FL and FA relationships (see figure 9.4). Therefore, first calculate and account for the (possibly) submaximal activation level of the CC, using the input stimulation value. Remember that there is both a magnitude and a temporal association (figure 9.2) between stimulation and activation. Also account for the CC length and its effect on the P_o value used in the CC FV relation. Thus, the CC FV equation must be written in the form

$$\text{CC Velocity} = f(\text{CC Force, Activation,}$$
$$\text{and Length}) \qquad (9.5)$$

Again, for $t = 0$, the stimulation and activation equal zero, so the CC is "passive" and producing no force. The equation will therefore work on the "zero activation" CC FV curve and predict a CC velocity of zero. Because the SEC has zero force exerted across it and the experimental conditions call for a constant muscle length, the SEC and CC are both isometric.

3. Use the predicted CC velocity from step 2 to predict the CC length at the next time increment. This numerical integration allows the simulation to progress from one time point to the next (in this case, from $t = 0$ to $t = 1$). For the simulation to be accurate, the time increments must be small (0.1 ms in this example). This is the same integration process described in chapter 10 for movement simulation.

$$\text{CC Length} = \int \text{CC Velocity } dt \qquad (9.6)$$

With the CC velocity at zero at the beginning of the contraction, the CC length change is also equal to zero.

4. Use the predicted CC length from step 3 to estimate a new SEC length for the next time point ($t = 1$). Here, use the muscle length input from the calling subroutine along with the predicted CC length. Because the CC and SEC are in series and make up the total muscle length,

$$\text{SEC Length} = \text{Muscle Length} - \text{CC Length} \qquad (9.7)$$

(continued)

EXAMPLE 9.1 (CONTINUED)

At $t = 1$, given the CC and SEC dynamics described in steps 1 through 3, this step will predict the same SEC length as before (SEC unloaded, force = zero).

5. With the new SEC length for the next time point from step 4, return to step 1 with the time increment shifted to $t = 1$. The steps in the model algorithm are repeated continuously until $t = 3000$ to give values for CC force, CC length, CC velocity, and SEC length at each time step throughout the entire contraction sequence. These values are consistent with the equations that dictate the behavior of each component of the Hill model.

How does this algorithm produce force profiles consistent with experimental results such as those in an isometric twitch or tetanus? As the time moves forward from $t = 0$, soon the time increment representing the stimulation pulse is reached (stimulation becomes 1 instead of zero). The stimulation pulse produces a nonzero activation level that puts the CC on a submaximal FV relationship, and the equation in step 2 now predicts a nonzero CC shortening velocity. When the CC velocity is integrated (step 3), the CC length becomes shorter. In step 4, this shorter CC length results in SEC lengthening. The longer SEC length is fed back into step 1 and therefore predicts a nonzero SEC force. The CC force (equal to the SEC force) goes forward into the CC FV equation in step 2, along with the new activation level. This predicts a new CC velocity that in turn produces more shortening of the CC (step 3) and a longer SEC length (step 4). The cycle repeats as stimulation and activation change, and CC force climbs as the activation level increases. In the tetanus, the activation becomes maximal and stays there, resulting in a continuous rise in force along with a continually lengthening SEC and shortening CC (as force rises). When the force plateaus at the P_o level for the current CC length, the CC and SEC both become static, with the SEC extended and the CC at a shortened length. For the twitch condition, the stimulation becomes zero after the pulse, which causes the activation to fall back to zero. This activation drop causes another set of changes through the submaximal FV relationships seen in step 2, resulting in CC and SEC dynamics that produce the characteristic twitch force response.

Implement this technique by writing the computer code for the general muscle model and then testing the model using a calling subroutine set up to mimic a specific muscle under isometric twitch and tetanus conditions. The model subroutine should replicate these well-known force responses. Once the model has passed these validation tests, other well-documented conditions should be implemented (e.g., trains of stimulation pulses at different frequencies to produce unfused and fused tetanus, isovelocity contractions at different shortening and lengthening speeds, inertial contractions with different inertial loads). The model should also react correctly when the model parameters are altered (i.e., slow-twitch vs. fast-twitch dynamic constants, alteration of P_o, location of L_o, and so on). After demonstrating the validity of the model, move on to the more challenging task of implementing the model subroutine within a musculoskeletal model in which each muscle of interest is represented by a specific version of the general Hill muscle model (see chapter 11 for more information on musculoskeletal modeling).

MUSCLE-SPECIFIC HILL MODELS

In chapter 5, the concept of resultant joint moments was presented in the description of inverse dynamics movement analysis. During a movement sequence, the joint moments represent the calculated rotational kinetics that must have been present to generate the observed motion of body segments. This calculation implies that a separate entity is responsible for the moment at each joint involved in the motion. In reality, the human body consists of a multitude of individual muscles that span one or more joints of the rigid, bony skeleton. For example, the human lower extremity can be represented as a series of articulating bones (pelvis, femur, tibia, fibula, calcaneus, talus, and other bones of the foot). The bones form the basis for modeling the body as a set of linked rigid segments. Overlying these bones are individual muscles (such as vastus lateralis, tibialis anterior, and soleus) that pro-duce the forces that cause the resultant joint moments and, therefore, the movement of the skeleton and body segments. As discussed in chapter 8, muscle forces are produced in response to control signals arising from and delivered by the nervous system. Inverse dynamics modeling allows one to estimate the resultant moment at a joint that arises in response to the sum total of all muscle forces, but it cannot resolve the joint moment into individual muscular forces.

If the body is viewed as a mechanical system, it is clear that the system is redundant in that it has more muscles than are necessary to produce any given lower-extremity joint moment. For example, only one muscle is needed to produce a flexor moment at the knee, but the human body has multiple muscles (biceps femoris, semitendinosus, semimembranosus, gastrocnemius, popliteus) that directly produce this moment. Exactly why the system is overdetermined is a matter of some debate among movement scientists in the fields of biomechanics and motor control. Other questions of interest include why some muscles are biarticular (span two adjacent joints) whereas others are monoarticular (van Ingen Schenau 1989); why some muscles have long, thin tendons whereas others have short, thick ones (Biewener and Roberts 2000; Caldwell 1995); and what contributions individual muscles make during specific movements (Anderson and Pandy 2003; Zajac et al. 2002, 2003).

To fully understand skeletal movement during a movement sequence, it would be useful to measure the muscle control signals, the resulting muscle forces and the moments they provide at each joint, and the movement of each body segment. Although EMG data can be used to estimate the muscle control signals in some cases and body segment motion is routinely measured, with today's technology one cannot easily measure muscle forces directly in humans. In this context, muscle models can enhance our understanding of musculoskeletal function in several ways. Rotational muscle models representing the properties of all muscles crossing a joint can be useful either in understanding calculated joint moment patterns or as simple torque generators in forward dynamics simulation models (chapter 10). Alternatively, muscle models can be used to represent the function of specific individual muscles, including biarticular muscles that may be important for coupling the actions of adjacent joints. Direct muscle force measurement is possible in animals, and this approach has been used to investigate the force predictions from muscle models (e.g., Herzog and Leonard 1991; Perreault et al. 2003). There are isolated human examples of direct muscle force measurement (e.g., Gregor et al. 1991; Komi et al. 1996), but such invasive techniques are unlikely to find wide use in human subjects.

Therefore, researchers have learned to construct musculoskeletal models of the entire movement system (e.g., Anderson and Pandy 2003; Gerritsen et al. 1998), with Hill models representing specific individual muscles of importance. Each individual Hill model must be tailored to the anatomical and functional characteristics of the specific muscle it represents. Some of these characteristics are geometrical in nature, associated with the location of the muscle with respect to the skeleton and joints that it crosses (e.g., origin, insertion, movement path, moment arms). Other characteristics determine the force generation potential of specific muscles, dictated by muscle morphological characteristics and exact mechanical properties. Here we focus on the latter, with the geometrical aspects of muscle modeling being covered in more depth in chapter 11, Musculoskeletal Modeling.

Muscle Architecture

Each muscle is unique in its morphological characteristics and its ability to produce force under different situations. A variety of studies have linked the morphological and physiological characteristics of muscles with their mechanical properties, so estimates of appropriate values for muscle models can be made from measurable physiological characteristics. Together, the characteristics that influence the mechanical force-producing properties of muscle are known as *muscle architecture*. Certain architectural parameters can be used to characterize individual muscles for modeling purposes, including pennation angle, physiological cross-sectional area, fiber length, fiber-type composition, tendon length, and tendon elasticity. These architectural elements are important because they affect the mechanical properties of the Hill model CC and SEC.

Pennation Angle

Muscles are sometimes characterized according to the direction in which their fibers are oriented. The fibers of fusiform muscles run in the same direction as their tendon, parallel to a line joining the muscle origin and insertion. In contrast, pennate muscles have fibers oriented at an angle to the tendon, and therefore they produce force with vector components both parallel and perpendicular to the tendon. The obvious geometric consequence of this pennation angle, θ_p, is that when the fibers produce X newtons of force, only $X \cos \theta_p$ newtons are transmitted via the tendon to the connected bones. If the angle θ_p is small, this effect is not dramatic (e.g., cos 10° = 0.985). However, in some situations (short fiber lengths, very compliant aponeurosis), the fibers rotate and the pennation angle increases dramatically as fiber force increases, so the ability of the fibers to provide force in the intended direction may be severely compromised. Also important in pennate muscles is the phenomenon of fiber packing. A pennate muscle has shorter fibers than a fusiform muscle of equal volume, but more of the fibers are arranged in parallel. Thus, the pennate muscle can produce more force than a similar fusiform muscle, which suggests that pennate muscles are designed for high force production.

Cross-Sectional Area

The concept of fiber packing can be understood by considering the arrangement of sarcomeres within a muscle. Sarcomeres arranged in series, such as in a single myofibril, transmit force only to the extent that each individual sarcomere can support. If each sarcomere produces 1 N of force, 100 sarcomeres in series would produce a total of 1 N. In contrast, sarcomeres arranged in parallel each transmit their own force so that the more sarcomeres there are in parallel, the greater the force transmitted. The same 100 sarcomeres arranged in parallel could produce a total force of 100 N. A muscle's cross-sectional area (CSA) indicates the number of fibers in parallel and thus can be used as an indicator of the muscle's maximal force capacity. Several factors must be considered when one is predicting a muscle's force capability in this way.

One of these factors is muscle shape, because the muscle belly's girth is often larger than the girth at either end. At what point along the length of the muscle do you measure the CSA? One possibility is to take several girth measures in an attempt to get an average CSA. However, this method assumes that all of the fibers run in parallel and are oriented with respect to the muscle origin and insertion, which is not true for pennate muscles. To account for pennation angle and the fact that not all fibers run the entire length of a muscle belly, the term *physiological cross-sectional area* (PCSA) was introduced (Haxton 1944). PCSA is calculated by dividing muscle volume by fiber length. Originally, PCSA was applicable only to animal or cadaver studies in which these measurements were possible. However, the advent of new medical imaging techniques makes it possible to obtain this information from living human subjects as well; later we briefly address these methods.

Another factor to consider when using the PCSA is the muscle FL relationship. When a muscle fiber changes length, its isometric force capabilities are modified by the alteration in the myofilament overlap. At what muscle length should measurements be taken for the calculation of PCSA? The obvious answer is the length at which the fibers are at L_o, the optimal force-producing length. However, it is difficult to know the optimal length for any given muscle, and PCSA data taken from cadavers are obtained from muscles in rigor in the anatomical position, making it highly unlikely that fibers from a given muscle would be at optimal length. Considering all of the factors outlined in the preceding discussion, it is obvious that the maximal isometric force (P_o) capabilities of individual human muscles are not well established, despite the straightforward observation that force production is dependent on muscle size.

Fiber-Type Composition

It is well known that muscle fibers and motor units vary in histochemical and mechanical characteristics. For many musculoskeletal models, it is sufficient to recognize two types: slow-twitch (ST) and fast-twitch (FT) fibers. A further distinction between fatigable and fatigue-resistant FT fibers may be warranted in models that consider the consequences of movements of longer duration. The importance of the ST-FT dichotomy arises from their separate activation time constants and FV characteristics and the modifications in force-time profiles that result from these differing characteristics. The time for force to rise during an isometric tetanus is longer for an ST muscle than for a comparable FT muscle. In isometric twitch conditions, an ST muscle also exhibits a reduced peak twitch force as a consequence of the delayed time at which peak force occurs. Differences between ST and FT fibers are also seen in the rates of

mechanical and metabolic energy expenditure that may be computed with metabolic cost models (e.g., Umberger et al. 2003), as described in a later section.

Fiber-Tendon Morphological Characteristics

Each muscle-tendon unit in the human body is unique in the relative length of the muscle fibers compared with its tendon. This architectural detail dictates the relative excursion of each portion during a movement sequence. The number of sarcomeres in series determines the length of the fibers. Because each sarcomere can shorten across its FL range by about 2.2 μm (from 3.65 to 1.45 μm), the total number of sarcomeres in series dictates the maximal excursion the fiber can undergo. If a fiber contains more sarcomeres in series, it can undergo a greater total excursion. The fiber length also determines the absolute fiber velocity during a movement and therefore has a significant impact because of the FV characteristics.

The length of the tendon is important for similar reasons, because of the amount it will extend as force is applied across its length. Some tendons are short and thick, whereas others are long and slender. These morphological features affect overall tendon elasticity and seem to be correlated with muscle-tendon function. For example, more distal extensor muscles of the lower extremity often have long, compliant tendons in series with relatively short fibers. This muscle-tendon morphological characteristic results in relatively larger changes in tendon length and therefore smaller changes in fiber length. Thus, for a given joint motion and muscle-tendon excursion, the fibers would undergo less overall excursion than if the tendon were shorter and stiffer. This design means that the fibers can operate at lower contractile velocities and therefore can produce a given required force at a lower activation level than at a higher velocity (figure 9.15). In a repeated cyclic activity such as locomotion, the lower activation level could result in reduced metabolic cost. In contrast, more proximal muscles tend to have shorter, thicker tendons attached to relatively long fibers that have large excursion capabilities. It has been suggested that these muscles are ideally designed for producing large amounts of mechanical work, but because of higher fiber velocities and a tendency for increased active muscle mass, the mechanical work may come at a higher metabolic cost.

Estimating Model Parameters

These differences in morphological characteristics and fiber-type composition dictate that individual muscles have unique mechanical characteristics. The values used within muscle models to represent specific properties are known as *muscle parameters*. Hill model CC, *FL* and *FV*,

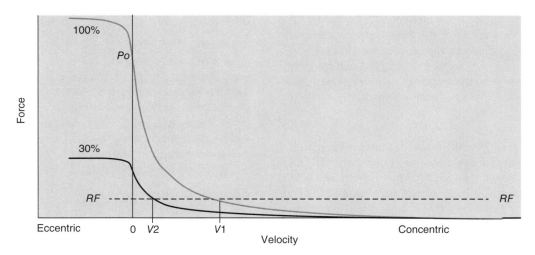

▲ **Figure 9.15** Contractile component (CC) force-velocity diagrams illustrate the advantage of slower contractile velocities. For a required force RF, the higher contractile velocity V1 would require the muscle to be activated at 100% (green line) to attain the needed force. However, for the slower contractile velocity V2, the needed force could be generated at a submaximal activation level of 30% (black line). Series elasticity can modify the CC velocity to permit lower contractile velocities during human movements.

and SEC $F\Delta L$ properties for specific human muscles are generally inaccessible due to the difficulty in measuring individual muscle forces and the lack of correspondence between muscle anatomical structures and the Hill model components. Nevertheless, modeling studies in the literature report these properties for individual muscle actuators, usually estimated from muscle morphological and architectural considerations. In this section, we examine how the muscle model parameters described previously can be estimated for human muscles.

Direct Joint Measurements

For muscle groups crossing specific joints, the CC properties are sometimes expressed as the torque-angle ($T\theta$) and torque-angular velocity ($T\omega$) relations. Measuring the torque capabilities under different kinematic conditions can establish these isometric and dynamic relationships for specific isolated joints. Such measurements are routinely done with an isovelocity dynamometer that can control the isometric position or dynamic joint velocity while a subject applies maximal effort torque. Proper methods for collecting torque-angle and torque–angular velocity data are explained in chapter 4. The subsequent $T\theta$ and $T\omega$ data can be used to compute model CC parameters for a rotational Hill model representing that specific joint. The general strategy is to fit the measured data to equations that will be used within the rotational muscle model. For example, the $T\omega$ data can be fit with a rotational Hill equation:

$$(T + a)(\omega + b) = (T_o + a)b \qquad (9.8)$$

where T and ω are the measured pairs of torque and angular velocity data points, T_o is the peak isometric torque found in the isometric $T\theta$ data set, and a and b are the unknown muscle dynamic Hill parameters. Application of this equation allows one to find the a and b coefficients that give the best fit to the measured $T\omega$ data, and these coefficients can be used as the CC model parameters representing the $T\omega$ relation in the rotational Hill model. A similar data-fitting procedure can be used with an isometric $T\theta$ data set using a parabolic equation:

$$T = c\ \theta^2 + d\theta \qquad (9.9)$$

where T and θ are the measured pairs of torque and angle data points, and c and d are the unknown muscle $T\theta$ parameters that can be found for a specific joint through a best fit procedure. Note that any appropriate equation can be chosen to represent these $T\theta$ and $T\omega$ characteristics, with the equation coefficients serving as the muscular parameters. The unknown coefficients for the $T\theta$ and $T\omega$ equations can be found easily with common software packages as long as there are sufficient known $T\theta$ and $T\omega$ data points; the minimum number of data points required will depend on the exact equation being used.

The parameters representing the SEC characteristics can be found in the form of individual joint torque-extension ($T\Delta\theta$) relationships, using an isovelocity dynamometer under high-speed controlled or quick release conditions (de Zee and Voigt 2001; Hof 1998). The concept here is that the muscles are loaded in tension under isometric conditions and then released to shorten rapidly. In the initial loaded phase, the SEC elements are stretched by the force produced by the active CC. When suddenly released, the force decreases abruptly and the SEC returns to the unloaded or slack length. After proper corrections for PEC, inertial, and apparatus compliance effects, the resulting $T\Delta\theta$ data can be fit with an appropriate equation, such as a quadratic:

$$T = k\Delta\theta^2 \qquad (9.10)$$

where T and $\Delta\theta$ are the measured torque and angular extension data, and k is the unknown SEC stiffness parameter. Recall that the SEC force-extension ($F\Delta L$) relationship is nonlinear in nature, with a compliant region at low force giving way to a much stiffer higher force region, so it is no surprise that a similar nonlinear shape is observed in the joint $T\Delta\theta$ data. In some cases a two-parameter model with quadratic and linear sections gives a better fit to the $T\Delta\theta$ data. PEC $T\Delta\theta$ properties can be found by measuring the passive, nonactivated torque response as the dynamometer arm and joint are slowly moved through the complete joint range of motion. Again an appropriate equation (e.g., quadratic or exponential) is then fit to the data to give the PEC model parameters.

Although isolated joint measurement techniques have found widespread use since the 1950s (e.g., Wilkie 1950), many of the same limitations that apply to inverse dynamics calculations (chapter 5) are found here as well. Multiple muscles and ligaments cross each joint, so it is difficult to know which internal tissues contribute to the measured joint torque at any given time, position, or velocity. The measured torque is the sum total from all contributions, and important concepts like synergistic and antagonist cocontraction cannot be addressed from these measurements alone. Interpretation of the $T\theta$ relationship is therefore complicated because it is influenced by a combination of the muscle FL characteristics, moment arm–angle patterns, and relative activation level for each contributing muscle. Thus, although the optimal angle for isometric torque production at a joint can be measured, the optimal length for force production (L_o) for specific human muscles is generally unknown. Similarly, the $T\omega$ relations for many joints are routinely measured, but these are influenced by a combination of the FV, moment arm, and activation characteristics for each contributing muscle. The FV relationships for individual muscles therefore cannot be attained directly from such joint measurements. Considering that contributing muscles to a given joint torque usually have unique fiber type composition and skeletal attachment points, they are likely to be at different contractile velocities and at different places on their FV relations at any measured $T\omega$ data point. Finally, this isolated joint approach assumes that each joint has independent torque generation properties and thus ignores the action of biarticular muscles that couple the torque capabilities at adjacent joints.

Individual Muscle Parameters

With these difficulties, how can specific individual human muscle parameters be found? Because of the limitations in the rotational joint measurement approach, methods for determining the mechanical properties and parameters of single human muscles have been investigated. Because these efforts cannot rely on direct force and length measurements, they all can be classified as modeling procedures. Here we group them into analytical, imaging, and computational techniques.

Analytical Techniques

One method used to estimate specific muscle parameters entails extrapolating information from animal studies and measurable morphological characteristics. For example, we can estimate FL characteristics by considering the number of sarcomeres in series in a fiber and the known lengths of thick and thin filaments and then extrapolating the sarcomere FL relation to the fiber as a whole (Bobbert et al. 1986a). This method has the advantage of predicting the optimal CC force length, L_o. However, there are relatively few estimates of sarcomere number for any given human muscle, and this value may widely vary between individuals. Another method for estimating FL parameters is to extend animal results (Woittiez et al. 1983) to humans based on the *index of architecture* (Ia) of a muscle, which is defined by the ratio of fiber length to muscle belly length. In general, the pennation angle dictates the length of muscle fibers relative to the muscle belly length. Woittiez and colleagues found a strong association between the width of a muscle's FL characteristics and its Ia, with higher Ia (i.e., less pennate) values indicating a wider FL relationship. However, with this method the exact L_o length and the physiological range of the FL relationship in which the muscle operates are both unknown. In similar fashion, FV properties can be estimated from morphological and fiber-type considerations. Animal studies indicate that the dynamic constants a and b that shape the Hill rectangular hyperbola are related to the fiber-type profile of a muscle. This can be observed by contrasting the force rise-time profile in muscles composed predominantly of slow-twitch or fast-twitch fibers. Researchers have therefore extrapolated from animal studies and used fiber-typing to estimate the values of a and b for specific human muscles (Bobbert et al. 1986a).

A different approach is to capitalize on the fact that biarticular muscles contribute to torque at adjacent joints. For example, by measuring isometric joint torque in hip flexion and knee extension across a range of different joint positions, one can estimate the FL characteristics of the rectus femoris muscle that contributes to both joint torques (Herzog and ter Keurs 1998). Theoretically, this approach can be applied for any muscle that crosses two joints. However, sets of multiple biarticular muscles that link not two but three adjacent joints can complicate the procedure, such as the gastrocnemius and hamstrings muscles, which both contribute to knee flexion but also participate in ankle plantar flexion and hip extension, respectively. Although this approach may have limited experimental application, it is the conceptual foundation for computational techniques (described subsequently) that use data from multiple joints to find the parameters from an ensemble set of muscles that cross these joints (e.g., Garner and Pandy 2003).

Imaging Techniques

A relatively recent development in our ability to predict subject- and muscle-specific model parameters is the use of medical imaging technology such as magnetic resonance imaging (MRI) and ultrasound imaging. The research applications of these methods have increased as these instruments have become more commonplace and accessible, and their uses are likely to expand in the future. These imaging tools are critical for advancing the use of muscle models into clinical practice where subject-specific model characteristics are a necessity.

MRI can be used to obtain subject-specific images of internal musculoskeletal anatomy such as bones, ligaments, and muscles (e.g., Hasson, Kent-Braun, and Caldwell, 2011). Unlike older methods such as X-ray technology, MRI gives excellent resolution of soft tissues

and does not put the subject at risk as a result of radiation exposure. Sagittal, frontal, and transverse plane images of segmental anatomy allow a host of measurements useful for characterizing muscle morphological features and musculoskeletal geometry. These images typically use a field of view ~200 to 300 mm in length and width that is captured in a 512 × 512 or 1024 × 1024 pixel grid, yielding high-resolution digital images for analysis (figure 9.16). A series of transverse or axial images can be used to measure specific muscle length, volume, and average CSA. Typically these axial images are separated by ~5 to 10 mm and are taken throughout the length of a body segment or limb. Specific muscles of interest are identified and outlined with digital computer techniques to allow calculation of the muscle CSA on that particular image. This procedure is repeated for the entire set of axial images that span the complete length of the muscle. By computing the centroid of the muscle area in each axial image, one can easily compute the muscle length from the entire set of images. Likewise, the muscle volume can be calculated from the set of individual CSA values and the interslice spacing. The volume and length data can then be used to compute the average cross-sectional area of the muscle and can be combined with fiber length information to compute PCSA. An added feature of MRI is the ability to differentiate between muscle and fat tissue based on pixel intensity, which allows the calculation of fat-free mass, area, and volume and gives a better indication of the amount of contractile

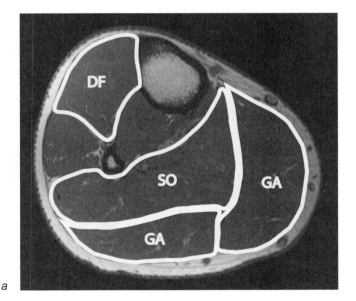

a

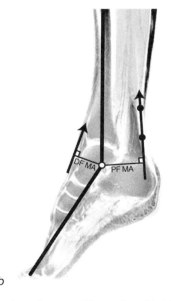

b

▲ **Figure 9.16** Examples of magnetic resonance images. *(a)* Transverse plane image of lower leg, highlighting the cross-sectional area of the dorsiflexor muscles (DF), soleus (SO), and both heads of the gastrocnemius (GA). *(b)* Sagittal plane image of ankle joint, with dorisiflexor and plantarflexor moments arms (DF MA, PF MA) indicated.

Reprinted, by permission, from C.J. Hasson and G.E. Caldwell, 2012, "Effects of age on mechanical properties of dorsiflexor and plantarflexor muscles," *Annals of Biomedical Engineering* 40: 1088-1101.

tissue available for force production. Other MRI-based techniques like diffusion tensor imaging (DTI) permit one to envision 3-D paths of muscle fascicles. Such techniques may allow the development of more precise 3-D anatomical models and could be used to give more accurate pennation angle data for existing Hill models. Finally, sagittal and frontal MRI images are useful for measuring subject-specific musculoskeletal geometry such as muscle moment arms or for scaling model data from computer software such as SIMM to individual subjects.

Another technology that is effective for identifying subject-specific muscle characteristics is ultrasound imaging (e.g., Hasson, Miller, and Caldwell, 2011). Internal longitudinal images of skeletal muscles can be recorded with an ultrasound surface probe and then recorded as digital video files for subsequent measure-

ment and analysis. Ultrasound images have been used to measure muscle thickness, fascicle length and velocity, and fiber pennation angle, all of which are useful for assessing CC characteristics. For example, muscle thickness can be used to indicate the CC isometric strength, because it is highly correlated with muscle volume (Miyatani et al. 2004). Currently the most prevalent use of ultrasound in biomechanics is to assess the compliance of the series elastic structures, such as tendon and aponeurosis (figure 9.17). In older literature, simple measurements of tendon CSA and length were used to estimate SEC compliance, even though it is incorrect to consider the tendon as the sole contributor to the SEC. Ultrasound imaging allows in vivo dynamic measurement of internal muscle fiber, aponeurosis, and tendon length changes with alterations in external force. Series elasticity can therefore be studied in detail by observing

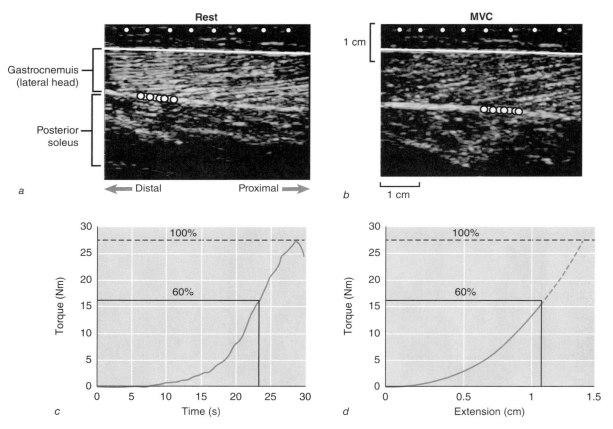

▲ **Figure 9.17** *(a)* Ultrasound image of ankle plantar flexors at rest. Digital reference markers at the top do not move when the underlying muscles produce force, but similar markers on the aponeurosis in the middle are displaced to the right as contractile force deforms series elastic elements. *(b)* At maximum voluntary force (MVC), the aponeurosis markers have been displaced by more than 1 cm. *(c)* During a slow voluntary isometric contraction, torque is measured simultaneously with the ultrasound measurements, which permits construction of a torque-extension relation that characterizes series elasticity, shown in *d*.

how selected points on the deep aponeuroses or tendon move as the muscular state changes from inactive to maximum voluntary force in an isometric contraction on a dynamometer. Ultrasound imaging has also been useful in assessing muscle fiber and tendon length changes during movements such as postural sway (Loram et al. 2004) and has indicated that tendon and aponeurosis may undergo different amounts of elongation but similar strain when isometric torque is produced (Arampatzis et al. 2005).

Computational Techniques

The analytical and imaging techniques described so far are useful for determining some subject-specific muscle parameters, but not all muscle model parameters can be assessed with these methods. Thus, some model parameters can be estimated with a degree of confidence, whereas others are in general unknown for specific human muscles. One way to overcome this limitation is to use an optimization approach to find a set of model parameters that can replicate measured experimental data (Garner and Pandy 2003; Gerritsen et al. 1998; Hasson and Caldwell, 2012). The concept is straightforward: use experimental techniques to define joint $T\theta$, $T\omega$, and $T\Delta\theta$ relationships for a subject, and then use a musculoskeletal model to replicate these relations, using computational optimization algorithms to find the set of muscle model parameters that give the best fit between the experimental and musculoskeletal model data.

For example, assume that five muscles are responsible for the moment produced at a given joint. Using a musculoskeletal model that includes Hill models to represent each muscle, we can find force and moment estimates for each muscle model for a given isometric joint position. Summing the five muscle model moments yields a model joint torque estimate for that position. By repeating this process in other isometric positions across a full range of joint motion, we can form a model torque-angle ($T\theta_M$) relationship. Then, an optimization algorithm and the actual measured joint torque-angle data can be used to find the best values for the FL parameters (e.g., P_o, L_o, FL parabola width) for each muscle model. The optimization procedure would be formulated to find the model FL parameter values for each muscle that gives the best fit between measured and model torque-angle relationships (i.e., one that minimizes the difference between $T\theta$ and $T\theta_M$). This same method can be used to find other muscle parameters by fitting model $T\omega_M$ and $T\Delta\theta_M$ relations to experimental joint $T\omega$ and $T\Delta\theta$ data, respectively. Here the parameters to be optimized would be the dynamic constants a and b that influence the model $T\omega_M$ relationship and the SEC stiffness characteristics that determine the $T\Delta\theta_M$ relation.

This computational approach can be augmented by the use of data drawn from imaging or analytical techniques. Such a "hybrid" method can take several forms. For example, some parameters can be based directly on MRI and included as known values in the optimization algorithm that seeks to find the remaining unknown parameters. Alternatively, data from imaging or analytical techniques can be used to form boundaries for the possible solutions from the optimization algorithm. Examples of selected muscle parameters drawn from the literature are shown in table 9.2. These data originate from a number of different sources, which is undoubtedly one of the reasons for the wide range of values found for some parameters. This large variation establishes the importance of developing methods to attain subject-specific parameter values. Furthermore, simulation models and muscle force estimates have been shown to be sensitive to some muscle model parameters (Scovil and Ronsky 2006). Sensitivity analysis is discussed in more detail in chapter 11.

BEYOND THE HILL MODEL

Although the Hill model is arguably the most prevalent model used today in biomechanics research, many important aspects of muscular function have been uncovered that cannot be explained by, or that even contradict, the predictions of the Hill model. These shortcomings

Table 9.2 Human Muscle Parameters

Muscle	P_o (N)	L_o (mm)	Pennation θ (°)	Tendon, unloaded length (mm)
Gluteus maximus	1050-4490	144-250	0	90-150
Hamstrings	2000-3900	104-107	9	350-390
Vasti	4500-6375	84-93	3-10	160-225
Rectus femoris	925-1700	75-81	5-14	323-410
Soleus	3550-4235	24-55	20-25	238-360
Gastrocnemius	1375-3000	45-62	12-17	48-425

of the Hill model are partly due to its scope, because it is meant to describe behavior for whole muscle or single fibers under "physiological" contractile lengths and speeds. Therefore, it is not surprising that the Hill model is unable to give insight into transient phenomena where the time course and magnitude of length changes are so small that crossbridges could not attach or detach; such experimental data should be addressed with more specific models of crossbridge mechanics (e.g., Huxley and Simmons 1971). And although the Hill model is consistent with FL and FV characteristics that can be linked to the sliding filament theory and the underlying generation of force by independent crossbridges, it is phenomenological in nature and does not rely on the verity of either sliding filaments or crossbridge mechanics.

Beyond the Hill model, several topics of muscle structure and function are of great interest to students of muscle mechanics. One such topic is contractile history, as it has been shown that prior stretch can enhance force production (e.g., Edman et al. 1978), whereas prior shortening can depress the ability of muscle fibers to produce force (e.g., Edman 1975; Herzog et al. 2000). These empirical phenomena fall outside the predictions of the normal Hill model, although there have been attempts to include these characteristics in Hill-type models. Possible reasons for these force variations with prior contractile state have been discussed extensively in the muscle physiology literature. Potential explanations for these and other force phenomena lie at the level of the fibers, sarcomeres, myofilaments, or crossbridges. For example, one explanation of the stretch-induced force enhancement phenomenon is based on the heterogeneity of sarcomeres in series and has led to the development of an intersarcomere dynamics model (Morgan 1990).

Another topic of interest is the transmission of force from the crossbridges to the skeleton, which is often considered a serial transmission along the active sarcomeres of a given fiber to the muscle origin and insertion via the tendon. This serial transmission model is based on independent fibers acting in parallel with others. However, there is experimental evidence that lateral force transmission (perpendicular to the fibers) also occurs, meaning that active force can be transmitted by adjacent fibers, including ones that are passive (e.g., Huijing 1999; Monti et al. 1999). Improved resolution and advances in microscopic techniques have highlighted the role of the structural proteins desmin and titin in force transmission (e.g., Granzier and Labeit 2006), and there is considerable interest in understanding their role in both healthy and pathological muscle force production.

One of the tenets of the sliding filament theory is that active force production takes place at the independent attachment sites where the myosin heads form crossbridges to binding sites on actin. Therefore, many researchers have studied the dynamics of crossbridge formation and subsequent force production (e.g., Huxley and Simmons 1971). Much of this early work inferred the behavior of a single crossbridge using molecular models. However, advances in the field of single molecule biophysics now allow researchers to directly observe the behavior of a single crossbridge. These techniques can directly quantify the step size and force-generating capacity of a single myosin molecule (Finer et al. 1994). The results confirm some aspects of earlier molecular models but also reveal novel molecular behaviors that might underlie many fundamental properties. One particularly exciting area is the ability to directly determine the effect of load on the mechanics and kinetics of a single crossbridge (Veigel et al. 2003). These and similar techniques are beginning to uncover the molecular basis of muscle's force-velocity relationship (Debold et al. 2005) as well as the aforementioned force enhancement observed following a stretch (Mehta and Herzog 2008).

Although these and other topics concerning muscle behavior are of great interest to muscle physiologists and biomechanists alike, they form a substantial body of knowledge that is outside the scope of this chapter and book. Here we limit our discussion to four types of muscle models of specific interest to researchers in biomechanics.

3-D Structural Models

Lumped parameter models like the Hill model use simplified geometry in determining muscle line of action, length, and moment arm changes when used to represent specific muscles in a musculoskeletal model (see chapter 11). The models represent the muscle line of action by a series of straight-line segments that link the origin and insertion while traveling over via points and wrapping surfaces. These models assume that all fibers undergo the same length changes and have the same moment arm, which may not be true in real skeletal muscles that exist in a 3-D world. More detailed information on muscle geometry will enhance our understanding and lead to the development of more accurate muscle force and torque predictions. Such a detailed understanding of muscle function is the goal behind 3-D models of skeletal structure based on high-resolution imaging and engineering modeling techniques.

As mentioned earlier in the section on MRI, diffusion tensor imaging can trace the 3-D paths of individual muscle fascicles (Lansdown et al. 2007). Such detailed 3-D information is valuable for visualizing and understanding the paths that muscle fibers traverse within a muscle and can form the basis of volumetric models of muscle that allow quantitative analysis of muscle fascicle and fiber length. One such modeling method is finite element analysis (FEA), a mathematical technique based on numerical solution of partial differential equations that is useful for examining the structural distribution of 3-D solids. In the FEA approach, the muscle is represented as a very large number of small finite subregions or elements. The elements are given mechanical characteristics (e.g., shear and compressive stiffness, or viscoelastic properties) that are representative of the muscle and tendon being modeled. The quasi-static model can be used to indicate the distribution of stresses and displacements within the muscle as the joint or joints that it crosses are placed in different positions or as the muscle force conditions change.

Construction of an FEA muscle model starts with a representation of the muscle 3-D surface and shape determined from a high-resolution imaging technique such as MRI. Because muscle shape in vivo is partially determined by the surrounding anatomical structures (typically FEA models of bones), ligaments and other muscles are constructed simultaneously to form the basis for changes in muscle shape in different joint positions. The 3-D muscle shape then undergoes a meshing procedure that sections the muscle into discrete solid elements of known geometrical shape (e.g., hexahedrons) with internal nodes that link the elements together. Elements are given structural elastic or viscoelastic properties, with different stress-strain or strain rate characteristics for elements that represent muscle and tendon tissue. The model can then be solved under different force conditions to examine how the individual elements react in terms of stress and strain distribution. Further insight can be gained by linking a set of individual elements that form the direction and path of muscle fibers from the real muscle specimen (Blemker and Delp 2005, 2006). By quantifying the position and length of these virtual

model fibers at different joint positions, we can study the heterogeneity of within-muscle fiber distribution. This approach also allows the determination of moment arms at a joint for different fibers within the muscle and the study of variation in these moment arms with changes in joint position (figure 9.18).

Linear Engineering Muscle Models

A common engineering modeling approach uses rheological elements with relatively simple properties to model materials with more complex structure and mechanical response characteristics. One common model element is the Hookean spring, a purely elastic body whose properties are completely described by its stiffness k, which is the slope of its linear force-extension relationship. Another element is the Newtonian body, commonly called a dashpot or damper, characterized by its viscosity coefficient β, which is the slope of its linear force-strain rate relation. These model elements and their characteristic properties (figure 9.19) can be used in a simple muscle model that can produce realistic force-time histories under certain kinematic and activation conditions. Mimicking the Hill model, this engineering model has a contractile component (CC) in series with an elastic component (SEC), but these components have linear properties rather than the nonlinear relationships found in the Hill model.

In this model, the CC consists of an active force generator F_m in parallel with a viscous dashpot (figure 9.20). The force generator F_m represents the CC active state capability, or maximal isometric CC force. If unopposed, when it produces force the CC tends to shorten, causing reduction of the CC length x. The dashpot will resist this

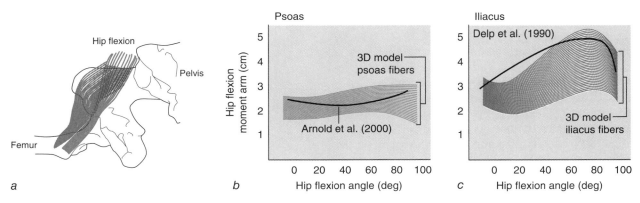

▲ **Figure 9.18** (a) 3-D model of hip flexor muscles psoas (green fibers) and iliacus (gray fibers). (b) Hip flexion moment arms for psoas and (c) iliacus as predicted by the model for different fiber bundles. Solid black lines represent moment arm prediction from "single line of action" models (Arnold et al., 2000; Delp et al., 1990).

Courtesy Silvia Blemker, based on Blemker and Delp 2005.

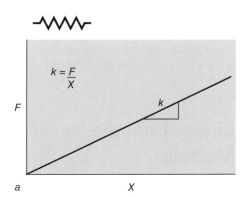

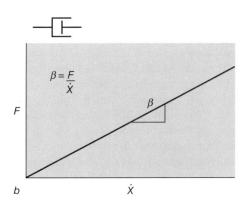

▲ Figure 9.19 (a) Hookean spring and its characteristic linear force-extension relation with stiffness coefficient k. (b) Newtonian body and its characteristic linear force-strain rate relation with viscous coefficient β.

shortening, with the amount of resistive force dependent on the dashpot damping coefficient β and the CC velocity v_{CC}. The result of the F_m and dashpot characteristics in parallel produces a linear CC force-velocity relation, and the CC force at any time can be computed from

$$F(t) = F_m - \beta v_{CC} \qquad (9.11)$$

Note that the value of damping coefficient β will dictate the exact slope of the FV relation and can be altered to represent the fiber type of a specific muscle, with different β values for slow-twitch and fast-twitch muscle. The passive SEC is represented by a Hookean spring with stiffness coefficient k, the slope of its linear force-extension (FΔL) relationship. At any time the force across the SEC is given by

$$F(t) = kx_{SEC} \qquad (9.12)$$

where x_{SEC} is the SEC length relative to its unloaded length (i.e., extension). Because of the series configuration of the CC and SEC, the force at any time as expressed in equations 9.11 and 9.12 must be the same, so

$$F(t) = F_m - \beta v_{CC} = kx_{SEC} \qquad (9.13)$$

In the isometric case where the total muscle length is constant, changes in CC and SEC length and velocity are equal in magnitude and opposite in direction. Therefore, equation 9.13 can be expressed completely with respect to CC kinematics by substituting $-x_{CC}$ for x_{SEC}. If we assume that F_m undergoes a step input "activation" from rest (zero) to its maximal value, this second-order differential equation can be solved using Laplace transformation to give

$$F(t) = F_m (1 - e^{-t\tau}) \qquad (9.14)$$

where the time constant $\tau = k/\beta$ This equation has the form of a saturation exponential and can be used to compute the time course of force under isometric tetanus

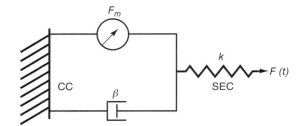

▲ Figure 9.20 Linear muscle model with contractile component (CC) and series elastic component (SEC). The CC is composed of an active force generator F_m in parallel with a damper (Newtonian body) with viscous coefficient β. The CC force is expressed across a series Hookean spring with stiffness k.

conditions (i.e., isometric with a step input of stimulation). Changes in muscle specific fiber type and SEC compliance can be incorporated easily by changing the values for β and k, respectively, thus altering the value of τ and the shape of the force-time profiles (figure 9.21).

The advantage of this engineering model approach is a significantly reduced computational cost compared with Hill model numerical algorithms. The disadvantages are that the model is limited to isometric, fully activated situations and that there are slight inaccuracies in the early force profile just after F_m is switched from passive to full activation, as the linear SEC FΔL association does not account for the high compliance of the SEC at low force levels.

Huxley Model

Along with the Hill model, the most influential muscle model of the 20th century is undoubtedly that published in 1957 by A.F. Huxley. The Huxley model was formulated to incorporate known facts regarding skeletal muscle at that time, and it became one of the cornerstones

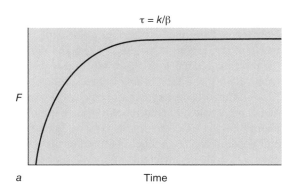

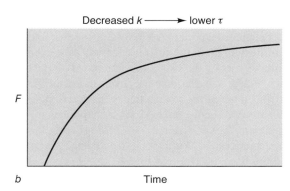

▲ **Figure 9.21** Force-time histories from the linear model with different combinations of damper viscosity β and spring stiffness k. The ratio of stiffness to viscosity forms the constant τ, which dictates how quickly the force rises during an isometric tetanus. In (a) the β and k values represent a moderately fast contractile component (CC) with a relatively stiff SEC. In (b), the constant τ has been reduced by decreasing the stiffness k, which results in a slower rise in force. A similar effect could have been achieved by increasing β to represent a slower CC.

of the sliding filament theory of muscle contraction. It was based on the known structure and function of the muscle sarcomere, with assumptions regarding myofilament sliding and thick filament crossbridge attachment and detachment. One of the strengths of the Huxley model is its ability to predict both mechanical and metabolic characteristics of muscle that are consistent with measured empirical data, which add credence to the model assumptions.

At the heart of the model is the notion that the crossbridges act as independent force generators, each delivering force when they are attached to the thin filament. This idea is based on two pieces of evidence from the well-known FL and FV relationships. The first is that the shape of the FL relation is dependent on the amount of thick and thin filament overlap, meaning that the amount of force depends on the number of crossbridges that could be attached at a given length. The second is that the FV maximal concentric velocity V_{max}, above which the muscle cannot sustain force, is independent of the thick and thin filament overlap, suggesting that this velocity is related to individual crossbridge dynamics.

Each crossbridge works in a cyclic pattern of (1) attachment to a thin filament binding site, (2) production of force while the crossbridge rotates through its working stroke, (3) detachment at the end of the stroke, and (4) repositioning to its original position in anticipation of the next such cycle. The energy for the crossbridge cycle is supplied by adenosine triphosphate (ATP), which is hydrolyzed into adenosine diphosphate (ADP) and an inorganic phosphate molecule (P_i) during the cycle.

The model that Huxley proposed, which is based on independent crossbridges, is shown schematically in figure 9.22. The model assumes that the muscle is activated maximally, that there is no elasticity, that the population of crossbridges is constant and unchanging (i.e., on the plateau of FL curve), and that each crossbridge goes through a complete cycle. Computations can take place under conditions of prescribed constant velocity. The model has a fixed thick filament situated next to a freely moving thin filament. Projecting from the thick filament is a sidepiece M (representing a crossbridge) secured by two opposing springs and moving back and forth under Brownian motion. In the middle of

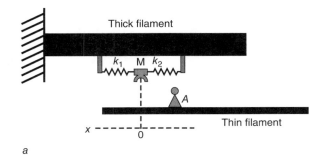

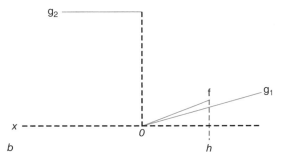

▲ **Figure 9.22** (a) Huxley model structure. (b) Attachment and detachment rate constant maps. See text for details.

the motion range is the equilibrium point where spring forces balance and equal zero in total. On either side of the equilibrium point the sidepiece M will be subjected to a force $F = kx$ by the longer (i.e., stretched) of the two springs, where x is the distance from the equilibrium point. This force can be transmitted to the thin filament if the sidepiece M is attached to attachment point A on the thin filament. The functional part of the Huxley model lies in the process of M attaching to and detaching from A and the force transmitted to the thin filament when A and M are attached at different distances x from the equilibrium point.

Two critical parameters of the Huxley model are the attachment and detachment rate constants f and g, respectively. These rate constants will determine the likelihood of a sidepiece M (crossbridge) being attached and unattached, in the same way that the marriage and divorce rates will dictate the likelihood of a person being married or single. Increasing the attachment rate constant f will result in greater likelihood of a crossbridge being attached, much as an increase in the marriage rates would lead to more married men. Conversely, a greater detachment rate constant g will result in greater likelihood of a crossbridge being unattached, analogous to higher divorce rates leading to more single men. To promote the likelihood of crossbridge attachment in positions that would lead to tensile force production, Huxley created a rate constant map with different values on either side of the equilibrium point (figure 9.22b). On the right side, the attachment constant f is larger than the detachment constant g_1, leading to a higher likelihood of a sidepiece M being attached in this region, up to a hypothetical maximal length h. A sidepiece M that is attached to a binding site A on the right side of the equilibrium point will be acted on by a force from the stretched spring on the left (k_1) and thus will tend to cause movement of the thin filament in that direction. On the left side, there is no chance of attachment $(f = 0)$ and a high rate of detachment (g_2). Thus, an attached sidepiece M that is swept across the equilibrium point from the right will have a high likelihood of detachment as it travels on the left of the equilibrium point. A sidepiece M that remained attached on the left would be acted on by a force from the stretched spring on the right (k_2), opposite to the direction of concentric shortening.

From these considerations, Huxley proposed a differential equation involving $n(x)$, defined as the fraction of crossbridges attached at displacement x:

$$dn(x)/dt = [1 - n(x)] f(x) - [n(x) g(x)] \quad (9.15)$$

where $n(x)$ is equivalent to the probability that a given crossbridge is attached at displacement x. This equation can be used to compute the probability of crossbridges

being attached across the spectrum of different x displacements from the equilibrium point under different constant contractile velocities. The model will predict the force associated with the crossbridge distribution at each velocity and thus is able to predict the concentric FV relationship at any given set of attachment and detachment rates. Furthermore, because the metabolic energy cost is associated with crossbridge attachment and detachment, the model can predict the amount of energy liberated at different contractile speeds. For both mechanical (FV) and metabolic (energy liberation) phenomena, the Huxley model can give realistic predictions of experimental skeletal muscle data.

Metabolic Cost Models

Although the Huxley model is impressive in its ability to predict both mechanical and metabolic aspects of muscle function, it is not suitable for general use in dynamic situations or when series elasticity should be considered. There are several examples of models that predict the metabolic energy cost of muscular contraction, and some of these are coupled with Hill-type models that can estimate mechanical events (e.g., Umberger et al. 2003). Such models are useful for estimating mechanical and metabolic energy expenditure during whole body movements that can be studied with musculoskeletal models (see chapter 11), particularly when the movement's objective may be linked to minimizing energy expenditure (e.g., walking at preferred speed). The linking of mechanical and metabolic models is appropriate because contractile mechanics must be taken into account when calculating total muscle energy expenditure.

Since the first part of the 20th century, muscle physiologists have tried to account for all energy liberated during muscle force production. In general, the energy liberation takes the form of heat and mechanical work, and indeed the dynamic constants a and b in the Hill FV equation were originally conceived as heat constants. The total energy expenditure has three components, related to (1) mechanical work, (2) heat associated with the sarcoplasmic reticulum during activation, and (3) heat produced from actin and myosin coupling. The actin and myosin interaction can be further subdivided into heat of maintenance and heat of dynamic shortening or lengthening. These components can be combined to form an energy rate equation such as that expressed by Umberger and colleagues (2003):

$$dE/dt = dW_{CC}/dt + dH_A/dt + dH_M/dt + dH_D/dt \quad (9.16)$$

where E is total metabolic energy, W_{CC} is mechanical work of the CC, H_A is activation heat, H_M is maintenance heat, and H_D is dynamic heat. Consideration of these terms separately demonstrates the efficacy of coupling the energy cost model with a Hill-type model, as knowledge of CC dynamics is necessary to compute the energy rate terms. For example, the rate of W_{CC} is computed directly from CC force and velocity. The dynamic heat rate associated with myofilament shortening and lengthening can also be computed from CC force and velocity and depends on muscle fiber type and maximal CC shortening velocity V_{max}. Heat rates due to activation and maintenance both depend on muscle fiber type composition, with the fast-twitch combined activation and maintenance heat rate roughly six times greater than for slow-twitch muscle. The heat rates display some dependence on CC length, and energy expenditure associated with heat liberation depends on whether the force production is fueled by aerobic or anaerobic processes. Finally, heat rates change under submaximal conditions, which can be implemented by using the same stimulation and activation parameters that are used in a Hill-type model.

Recent models have shown promise in predicting energy expenditure in human movements ranging from single joint to whole body movement (Anderson and Pandy 2001; Umberger et al. 2003). A challenge with modeling the energetic costs of skeletal muscle is in obtaining accurate parameter values for the many heat- and work-related terms for specific human muscles. Current models perform well in reproducing the qualitative nature of energy expenditure under various conditions, but the lack of accurate parameter values can cause quantitative discrepancies when compared with empirical measurements. Finally, as with muscle model force predictions, there are limited opportunities to make comparative measurements in human subjects to validate the output of these metabolic cost models.

SUMMARY

This chapter introduced the basic concepts of muscle mechanical properties and modeling. The first section of the chapter described the major mechanical characteristics of muscle within the framework of the Hill muscle model and provided an algorithm for implementation of the Hill model. The second section discussed how the general Hill model could be parameterized to represent specific muscles within the body and described methods for estimating these parameters for individual muscles. The third part of the chapter introduced important aspects of muscle mechanics that are not within the purview of the Hill model. This final section described four other muscle models useful to biomechanists. Interested readers should note that many issues and topics concerning muscle mechanics and modeling were touched on only briefly in this chapter. Following is a list of Suggested Readings to help broaden your knowledge in various areas of muscle mechanics and modeling.

SUGGESTED READINGS

Muscle Mechanics

Chapman, A.E. 1985. The mechanical properties of human muscle. *Exercise and Sport Sciences Reviews* 13:443-501.

Chapman, A.E., G.E. Caldwell, and W.S. Selbie. 1985. Mechanical output following muscle stretch in forearm supination against inertial loads. *Journal of Applied Physiology* 59:78-86.

Ettema, G.J.C. 1996. Contractile behaviour in skeletal muscle-tendon unit during small amplitude sine wave perturbations. *Journal of Biomechanics* 29:1147-55.

Ettema, G.J.C., and P.A. Huijing. 1990. Architecture and elastic properties of the series element of muscle tendon complex. In *Multiple Muscle Systems*, ed. J.M. Winters and S.L.-Y. Woo, 57-68. New York: Springer-Verlag.

Hasson, C.J., and G.E. Caldwell. 2012. Effects of age on mechanical properties of dorsiflexor and plantarflexor muscles. *Annals of Biomedical Engineering* 40:1088-101.

Hasson, C.J., Miller, R.M., and G.E. Caldwell. 2011. Contractile and elastic ankle joint muscular properties in young and older adults. *PLoS One* 6(1):e15953.

Heckman, C.J., and T.G. Sandercock. 1996. From motor unit to whole muscle properties during locomotor movements. *Exercise and Sport Sciences Reviews* 24:109-33.

Herzog, W., and H.E.D.J. ter Keurs. 1988. Force-length relation of in-vivo human rectus femoris muscles. *Pflugers Archive: European Journal of Physiology* 411:642-7.

Hill, A.V. 1970. *First and Last Experiments in Muscle Mechanics*. Cambridge, UK: Cambridge University Press.

Hof, A.L. 1998. In vivo measurement of the series elasticity release curve of human triceps surae. *Journal of Biomechanics* 31:793-800.

Hof, A.L., J.P. van Zandwijk, and M.F. Bobbert. 2002. Mechanics of human triceps surae muscle in walking, running and jumping. *Acta Physiologica Scandinavica* 174:17-30.

Joyce, G.C., P.M.H. Rack, and D.R. Westbury. 1969. The mechanical properties of cat soleus muscle during controlled lengthening and shortening movements. *Journal of Physiology* 204:461-74.

Lieber, R.L., G.J. Loren, and J. Friden. 1994. In vivo measurement of human wrist extensor muscle sarcomere length changes. *Journal of Neurophysiology* 71:874-81.

Pollack, G.H. 1990. *Muscles and Molecules: Uncovering the Principles of Biological Motion.* Seattle: Ebner & Sons.

Rack, P.M.H., and D.R. Westbury. 1969. The effects of length and stimulus rate on tension in the isometric cat soleus muscle. *Journal of Physiology* 204:443-60.

Rack, P.M.H., and D.R. Westbury. 1984. Elastic properties of the cat soleus tendon and their functional importance. *Journal of Physiology* 347:495.

Wilkie, D.R. 1950. The relation between force and velocity in human muscle. *Journal of Physiology* 110:249-80.

Winters, J.M., and L. Stark. 1988. Estimated mechanical properties of synergistic muscles involved in movements of a variety of human joints. *Journal of Biomechanics* 21:1027-42.

Zajac, F.E. 1989. Muscle and tendon: Properties, models, scaling, and application to biomechanics and motor control. *CRC Critical Reviews in Biomedical Engineering* 17(4):359-411.

Hill Muscle Model

Audu, M.L., and D.T. Davy. 1985. The influence of muscle model complexity in musculoskeletal motion modeling. *Journal of Biomedical Engineering* 107:147-57.

Bobbert, M.F., and G.J. van Ingen Schenau. 1990. Isokinetic plantar flexion: Experimental results and model calculations. *Journal of Biomechanics* 23:105-19.

Garner, B.A., and M.G. Pandy. 2003. Estimation of musculotendon properties in the human upper limb. *Annals of Biomedical Engineering* 31:207-20.

Hof, A.L., and J. van den Berg. 1981a. EMG to force processing I: An electrical analogue of the Hill muscle model. *Journal of Biomechanics* 14:747-58.

Hof, A.L., and J. van den Berg. 1981b. EMG to force processing II: Estimation of parameters of the Hill muscle model for the human triceps surae by means of a calf ergometer. *Journal of Biomechanics* 14:759-70.

Hof, A.L., and J. van den Berg. 1981c. EMG to force processing III: Estimation of parameters of the Hill muscle model for the human triceps surae muscle and assessment of the accuracy by means of a torque plate. *Journal of Biomechanics* 14:771-85.

Hof, A.L., and J. van den Berg. 1981d. EMG to force processing IV: Eccentric-concentric contractions on a spring flywheel set up. *Journal of Biomechanics* 14:787-92.

Lieber, R.L., C.G. Brown, and C.L. Trestik. 1992. Model of muscle-tendon interaction during frog semitendinosus fixed-end contractions. *Journal of Biomechanics* 25:421-8.

Winters, J.M., and L. Stark. 1987. Muscle models: What is gained and what is lost by varying model complexity. *Biological Cybernetics* 55:403-20.

Muscle Architecture

Alexander, R.M., and A. Vernon. 1975. The dimensions of knee and ankle muscles and the forces they exert. *Journal of Human Movement Studies* 1:115-23.

Biewener, A.A. 1991. Musculoskeletal design in relation to body size. *Journal of Biomechanics* 24:19-29.

Gans, C., and A.S. Gaunt. 1991. Muscle architecture in relation to function. *Journal of Biomechanics* 24:53-65.

Hasson, C.J., Kent-Braun, J.A., and G.E. Caldwell. 2011. Contractile and non-contractile tissue volume and distribution in ankle muscles of young and older adults. *Journal of Biomechanics* 44:2299-306.

Huijing, P.A. 1981. Bundle length, fibre length and sarcomere number in human gastrocnemius (Abstract). *Journal of Anatomy* 133:132.

Huijing, P.A. 1985. Architecture of human gastrocnemius muscle and some functional consequences. *Acta Anatomica* 123:101-7.

Huijing, P.A., and R.D. Woittiez. 1984. The effect of architecture on skeletal muscle performance: A simple planimetric model. *Netherlands Journal of Zoology* 34:21-32.

Kaufman, K.R., K.N. An, and E.Y.S. Chao. 1989. Incorporation of muscle architecture into the muscle length-tension relationship. *Journal of Biomechanics* 22:943-8.

Otten, E. 1988. Concepts and models of functional architecture in skeletal muscle. *Exercise and Sport Sciences Reviews* 16:89-139.

Spector, S.A., P.F. Gardiner, R.F. Zernicke, R.R. Roy, and V.R. Edgerton. 1980. Muscle architecture and force velocity characteristics of cat soleus and medial gastrocnemius: Implications for motor control. *Journal of Neurophysiology* 44:951-60.

Musculoskeletal Geometry

Fukashiro, S., M. Itoh, Y. Ichinose, Y. Kawakami, and T. Fukunaga. 1995. Ultrasonography gives directly but noninvasively elastic characteristic of human tendon in vivo. *European Journal of Applied Physiology* 71:555-7.

Fukunaga, T., Y. Ichinose, M. Ito, Y. Kawakami, and S. Fukashiro. 1997. Determination of fascicle length and pennation in a contracting human muscle in vivo. *Journal of Applied Physiology* 82:354-8.

Fukunaga, T., M. Ito, Y. Ichinose, S. Kuno, Y. Kawakami, and S. Fukashiro. 1996. Tendinous movement of a human muscle during voluntary contractions determined by real-time ultrasonography. *Journal of Applied Physiology* 81:1430-3.

Herzog, W., and L. Read. 1993. Lines of action and moment arms of the major force-carrying structures crossing the human knee joint. *Journal of Anatomy* 182:213-30.

Kawakami, Y., T. Muraoka, S. Ito, H. Kanehisa, and T. Fukunaga. 2002. In vivo muscle fibre behaviour during counter-movement exercise in humans reveals a significant role for tendon elasticity. *Journal of Physiology* 540(Pt. 2):635-46.

Kellis, E., and V. Baltzopoulos. 1999. In vivo determination of the patella tendon and hamstrings moment arms in adult males using videofluoroscopy during submaximal knee extension and flexion. *Clinical Biomechanics* 14:118-24.

Maganaris, C.N., V. Baltzopoulos, and A.J. Sargeant. 1999. Changes in the tibialis anterior tendon moment arm from rest to maximum isometric dorsiflexion: In vivo observations in man. *Clinical Biomechanics* 14:661-6.

Murray, W.M., S.L. Delp, and T.S. Buchanan. 1995. Variation of muscle moment arms with elbow and forearm position. *Journal of Biomechanics* 28:513-25.

Rugg, S.G., R.J. Gregor, B.R. Mandelbaum, and L. Chiu. 1990. In vivo moment arm calculations at the ankle using magnetic resonance imaging (MRI). *Journal of Biomechanics* 23:495-501.

Spoor, C.W., and J.L. van Leeuwen. 1992. Knee muscle moment arms from MRI and from tendon travel. *Journal of Biomechanics* 25:201-6.

Visser, J.J., J.E. Hoogkamer, M.F. Bobbert, and P.A. Huijing. 1990. Length and moment arm of human leg muscles as a function of knee and hip-joint angles. *European Journal of Applied Physiology* 61:453-60.

Yamaguchi, G.T., A.G.U. Sawa, D.W. Moran, M.J. Fessler, and J.M. Winters. 1990. A survey of human musculotendon actuator parameters. In *Multiple Muscle Systems*, ed. J.M. Winters and S.L.-Y. Woo, 717-73. New York: Springer-Verlag.

Muscular Force Production

Gregoire, L., H.E. Veeger, P.A. Huijing, and G.J. van Ingen Schenau. 1984. Role of mono- and biarticular muscles in explosive movements. *International Journal of Sports Medicine* 5:301-5.

Gregor, R.J., R.R. Roy, W.C. Whiting, R.G. Lovely, J.A. Hodgson, and V.R. Edgerton. 1988. Mechanical output of cat soleus during treadmill locomotion: *In vivo* vs *in situ* characteristics. *Journal of Biomechanics* 21:721-32.

Herzog, W. 1996. Force-sharing among synergistic muscles: Theoretical considerations and experimental approaches. *Exercise and Sport Sciences Reviews* 24:173-202.

Ingen Schenau, G.J. van. 1994. Proposed actions of biarticular muscles and the design of hindlimbs of bi- and quadrupeds. *Human Movement Science* 13:665-81.

Jacobs, R., M.F. Bobbert, and G.J. van Ingen Schenau. 1993. Function of mono- and biarticular muscles in running. *Medicine and Science in Sports and Exercise* 25:1163-73.

Jacobs, R., and G.J. van Ingen Schenau. 1992. Control of an external force in leg extensions in humans. *Journal of Physiology* 457:611-26.

Prilutsky, B.I. 2000. Coordination of two- and one-joint muscles: Functional consequences and implications for motor control. *Motor Control* 4:1-44.

Prilutsky, B.I., and R.J. Gregor. 1997. Strategy of coordination of two- and one-joint leg muscles in controlling an external force. *Motor Control* 1:92-116.

Chapter 10

Computer Simulation of Human Movement

Saunders N. Whittlesey and Joseph Hamill

Modeling of physical systems can be divided into two categories. *Physical modeling* is a process in which we construct tangible scale models that look very much like the real system. The Greeks and Romans long ago recognized the advantages of building small models of ships, buildings, and bridges before constructing the large, real-life structures. In the past century, consider the model wings flown by the Wright brothers, the model hydrofoil boat by Alexander Graham Bell, the laboratory crash-testing of automobiles, and the animal models that have significantly influenced the development of disease treatments and artificial joints. However, scale models and crash-test dummies require a great deal of time and resources to develop, and there are limits to what can be learned from them. *Behavioral* or *mathematical modeling* is a more abstract system used for studying a research question that does not necessarily lend itself to physical modeling. In the modern era, this means that researchers construct a set of equations and run them on a computer. Behavioral models are used for researching weather systems, animal populations and the spread of disease, and economic conditions to help governments set interest rates. In these models, the system is simplified by limiting the number of components so that they represent the net effect of many parts. For example, a model of a country's economy does not have millions of individuals; rather, it has a relatively small number of components representing the behavior of different socioeconomic and geographical groups. Biomechanists take the same approach in their research: Instead of modeling every bone and motor unit in the body, they simplify the body to have only as many parts as needed to answer their research question.

The pioneers in human movement studies, such as E. Muybridge, E.-J. Marey, and A.V. Hill, were largely experimentalists. They formulated theories by collecting large amounts of subject data because they were not able to construct physical or computer models of the human body. Behavioral models such as Hill's equation came into common use with the advent of digital computing. Computers, of course, changed human movement study forever, because models of human movement could be constructed, tested, and studied within a reasonable amount of time. In this chapter, we explore a variety of topics related to movement simulation:

- ▶ how to use simulation for research,
- ▶ why simulation is a powerful tool,
- ▶ the general procedure for creating and using a simulation,
- ▶ free-body diagrams,
- ▶ differential equations of motion,
- ▶ numerical integration,
- ▶ control theory,
- ▶ examples of human movement simulation, and
- ▶ limitations of computer simulations.

OVERVIEW: MODELING AS A PROCESS

Readers of this chapter will learn about the tools they will need to construct their own computer models of a human movement. However, an equally important—and perhaps more fundamental—purpose of this chapter is to convey the fact that modeling is a process used to enhance our understanding of human movement. We demonstrate that modeling is both enormously powerful and greatly limited as a tool. As with any tool, the most important thing is to use it correctly.

How do we use mechanical models in human movement science? The answer, of course, is that researchers have their own styles, but nonetheless, a basic process should occur. This process, diagrammed in figure 10.1, is a variant of the scientific method. It starts with a research question, proceeds to constructing the model, and then progresses through repeatedly improving the model. Eventually, the model is capable of making predictions that in turn lead to new understanding and new questions. We now discuss these steps in order.

All studies in human movement should start with a research question. Modeling in particular does not stand up well to testing if there is not a specific question to test for. Not only does the basic purpose of the model become unclear, the proper design for the model will also be unclear because it depends on the question asked. For example, there is no all-encompassing model of a bicyclist. If we want to determine the effect of body mass when a subject is climbing hills on a bicycle, the body can be modeled as a point mass. In contrast, if we

want to know how the activation of two-joint muscles changes between cycling on level ground and up a steep hill, we need a complex model that includes multiple lower-extremity segments and muscles. Every model must start with a particular research question so that the relevant parameters are included in the model. This point cannot be overemphasized.

Once the research question has been formulated, the next steps are to create a mechanical model of the movement and program it into a computer. These steps are detailed later in this chapter. Then, the model is evaluated to see whether it seems to work. This is a three-step process that involves using the model to generate data, collecting corresponding experimental data, and then comparing the two data sets to establish whether the simulation data are reasonable. There are no set criteria for this, and the evaluative process is more of an art than a science. In addition, the data needed to verify a model

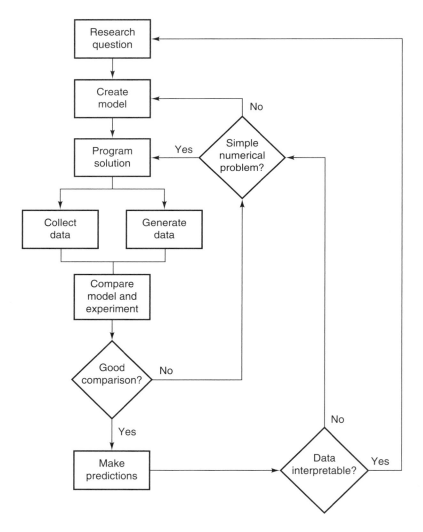

▲ **Figure 10.1** Flowchart of the mechanical modeling process.

vary widely: Sometimes a full, three-dimensional (3-D) kinematic data set is necessary, but in other cases, the results from a recent athletic competition will suffice. If the model data are not realistic, it must be determined whether there is a problem with the model itself or with the way the model program runs. In either case, development must begin again.

Once a model can replicate experimental data, it is ready to make predictions about the human movement to answer the research question. Researchers again have to ask whether the data generated by the model are reasonable; it might be necessary to compare these data with experimental data. An entire series of studies may have to be performed to establish the model's validity, and in many cases models have been debated for years in the literature. For instance, the model of the running human as a bouncing ball (Cavagna et al. 1977) has been revisited for more than two decades. When model data are not reasonable, development must be undertaken again until the predictions appear to be reasonable and lead to new questions.

There are different categories of human movement simulations. *Rigid-body* models represent the human system in whole or part as a set of rigid segments controlled in their movements by joint moments. Rigid-body models are the most common in human movement studies. *Mass-spring* models comprise one or more masses linked to one or more springs; they are commonly used to model running, hopping, and other repetitious movements. *Wobbling-mass* and *mass-spring-damper* models consist of linked masses, springs, and dampers (also called *dashpots*). Mass-spring-damper models are most often used to model human movements in which impacts occur. *Musculoskeletal models* also have rigid segments, but instead of having simple moments at each joint, they have submodels of individual muscles. The details of muscle models are discussed in chapter 9 of this text.

Simulating human movement requires drawing from a variety of technical areas, especially from such mathematical techniques as differential equations and numerical analysis but also from control theory, advanced dynamics, and computer programming. In this chapter, we explore each of these areas separately and then integrate them. Readers should be able to construct their own simulations by using the information in this chapter. However, effective modeling does not depend exclusively on the mastery of these disciplines. As with any tool, proper implementation is essential. Modeling is riddled with approximations and interpretations, and thus the proper use of models depends on recognizing their limitations. Before we delve into such details, however, we first discuss the applications of computer simulation.

WHY SIMULATE HUMAN MOVEMENT?

The best way to obtain human movement data, it seems, would be to collect it from human subjects. In other words, no data can be more accurate, more real than those collected from human subjects, so why would we want to generate data artificially? The most general answer is that human subjects are extremely complex and have intrinsic limitations. They are randomly variable, are subject to fatigue, have limits of strength and coordination, and must be treated with due consideration for safety and ethics. A computer model eliminates these constraints, and so there are many studies for which computer simulation is a good tool. Consider these examples:

■ Suppose we wanted to determine the best position in which to perform an athletic task such as jumping or cycling. We would have many different positions to test, and we would have to collect several trials of each position because of subject variability. In this study, the best subject performance probably would occur early in the data collection, before the subjects became fatigued. Subjects also tend to perform best under the conditions in which they normally train (Selbie and Caldwell 1996; Yoshihuku and Herzog 1990).

■ One exciting use for simulations is in virtual surgery, in which multiple surgical outcomes can be reviewed. Suppose a surgeon has a patient with certain spastic muscles (like people with cerebral palsy, for example). A common procedure is to lengthen or relocate the insertion points of these muscles. A sophisticated computer simulation could predict the proper amount to lengthen specific muscles, eliminating the trial and error from the surgical process (Delp et al. 1990, 1998; Lieber and Friden 1997).

■ Risks to workers in many occupations could be alleviated by using simulations. Military paratroopers, for example, have trained for landings by jumping from platforms of up to 4.3 m high. Suppose that a military unit wanted to establish safe limits for jumping from various heights with backpacks of various weights onto surfaces such as concrete, soil, and sand, and into shallow water. If human subjects were used, the study could establish limits only by injuring the subjects. Clearly, that is unacceptable, so this study is feasible and ethical only as a computer simulation.

Computer models also allow researchers to experiment with conditions that cannot be tested on human subjects because the conditions exceed practical limits. For example, locomotion patterns under an increased and decreased gravitational constant (9.81 m/s^2) could be tested to study the effects of gravity on human movement. The mass of model limbs could be changed to study the effect of limb mass on a movement (Bach 1995; Tsai and Mansour 1986). A single parameter in a model also can be changed and immediately tested. For example, a researcher who wants to study how increasing elbow extensor strength might improve throwing performance can change the model thrower and test it immediately, whereas a human subject would have to undergo a training regimen in which strength might improve in other muscles as well. Another example is to test the effects of torso mass on ground reaction forces. This, too, is a matter of changing a single model parameter. However, if we attempted this study with human subjects, the results might be confounded by concomitant changes in walking velocity, stride length, and perhaps even footwear function.

Another important and commonly used feature of computer simulations is that they can be adapted to search for optimal solutions. This is very useful for athletic performance, clinical applications, and other large issues in human movement. For example, we could determine the optimal movement pattern for a football kick, the most economical walking cadence for individuals with different body types, and the optimum flexibility of a prosthetic foot given the body weight, age, and stride length of a person with an amputation.

Models are often used simply to test our understanding of movement. This is a key reason for the popularity of rigid-body models in biomechanics despite the fact that none of them are perfect. If we understood a movement perfectly, we could replicate it on a computer. However, models eventually fall short in their predictive capabilities, and we are left to assess why. The search for answers often develops into a new movement theory. This topic is discussed in more detail later in this chapter. Now, we will discuss the how-to of computer simulation.

GENERAL PROCEDURE FOR SIMULATIONS

In general, a simulation is a computer program that generates a movement pattern (kinematics). In some cases, a simulation might also generate the kinetics of the movement. The procedure for developing a simulation is to

▶ create and diagram the mechanical model,

▶ derive the equations of motion for the mechanical model,

▶ program a numerical solution of the equations of motion,

▶ determine the starting and ending conditions for the simulations,

▶ run the program, generating model kinematics, and

▶ interpret the model data and compare them with experimental data.

This oversimplifies the modeling process by reflecting only the most important steps. The first step is to diagram the model so that it approximates the human system. These free-body diagrams (FBDs) are then used for the second step, the derivation of model equations, which are used in the third step, and so on.

Free-Body Diagrams

Free-body diagrams are discussed at length in chapters 4 and 5 of this book. For the purposes of this chapter, FBDs are used as visual aids to derive the equations of motion and to establish the type of movement that the model can perform. Specifically, researchers must answer these questions:

▶ How many components does the system have?

▶ Does each system component move linearly, rotationally, or both?

▶ Can the system's movements be described by a one-dimensional or two-dimensional (2-D) model, or must a 3-D representation be used?

▶ How many coordinates (degrees of freedom, or DOF) are needed to locate each component's position?

▶ Are there any means to reduce the number of constraints?

The complexity of the computer model depends on the answer to each question. If a system has n components and each component requires k DOF, then there will be n times k equations of motion unless there are factors (called *constraints*) that reduce their number. In this regard, the answers to the last two questions affect the size of the computer simulation code.

One-dimensional (1-D) models have been used to study locomotion cadence (e.g., Holt et al. 1990), impacts between a runner and the ground (Derrick et al. 2000; McMahon et al. 1987), deformation of the leg (Mizrahi and Susak 1982), and even the influence of the viscera on breathing patterns (Minetti and Belli 1994). However, most models are either 2-D or 3-D, because most human movements occur in multiple directions with multiple body segments. Three-dimensional models remain less common because of the complexity of controlling the model. Models of aerial movements such as diving and gymnastics (e.g., Yeadon 1993) have been constructed in 3-D because of the multiaxial nature of the movements. Certain walking models (such as Pandy and Berme 1989) have also been constructed in 3-D to include the effects of 3-D pelvic movements on gait.

EXAMPLE 10.1

Consider the following double pendulum representing the lower extremity.

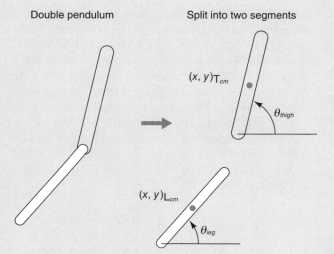

Double pendulum   Split into two segments

In response to the five questions:

1. *Components:* The system has two components, a leg and thigh.

2. *Motion:* Each segment moves both linearly and rotationally.

3. *Dimensions:* This can be a planar model. It clearly is not a 1-D model, and the planar movement is analogous to our use of 2-D lower-extremity kinematics.

4. *Coordinates:* Like any 2-D segment, each component requires an *(x, y)* coordinate and an angular position to completely describe its position in 2-D space. In

this case the *(x, y)* coordinates are the coordinates of the centers of mass of the thigh (T_{CM}) and the leg (L_{CM}), respectively.

5. *Constraints:* This is tricky, but the answer is that we can simplify the coordinates of the system. The constraint is that the leg must join the thigh at the knee; wherever the thigh is located, we know where the knee is in relation to it and thus where the leg is located. In other words, the coordinates of the leg mass center are functions of the thigh mass center position, thigh angle, and leg angle. Therefore, we can completely describe the position of the system with four coordinates—the *(x, y)* position of the thigh, the angle of the thigh, and the angle of the leg. There is no set procedure for identifying constraints; instead, it is a process learned by example. Our double pendulum can now be drawn like this:

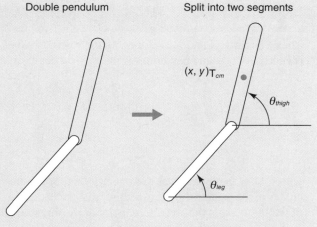

Double pendulum   Split into two segments

Differential Equations

Most physics textbooks deal with the motions of very simple systems, such as projectiles, pendulums, and springs. The positions of these simple systems may be described as polynomial or trigonometric functions. For example, the trajectory of a stone thrown from a cliff is a parabola. The position of a simple pendulum as a function of time is a sinusoid. In human movement simulations, simple systems are rare. We cannot describe most movements with polynomials, trigonometric functions, or any of the common functions. Instead, we have to describe how the positions in the system change. Specifically, the motions of body segments are described with *differential equations* that quantify the segments' changes in position over time. In other words, every kinetic factor that alters the movement of a body

segment must be expressed in the differential equation of motion for that segment. For example, a typical body segment is acted upon by muscle forces, joint forces, gravity, and frictional forces. Each of these factors must be expressed mathematically.

Students of calculus may wish for a more precise reason that we must express human movements as differential equations rather than as the more understandable functions of time. The answer is that the equations of motion for body segments are too complex to be solved into positions as functions of time. Although the movement of one segment alone is often solvable, even that requires simplification (i.e., $\sin \theta \approx \theta$). A system of two segments (a double pendulum) becomes hopelessly complex, because the segments interact. When we simulate an entire human body, the equations become very complex, indeed. Consider the following examples.

The height of a projectile launched from the ground can be written as

$$y = y_o + v_o t + 0.5gt^2$$

where y is the height above ground, y_o is the starting height over the ground, v_o is the velocity at which the projectile was launched, g is the gravitational constant, and t is the time from launching.

Alternatively, the differential equation for the height of this projectile is

$$\frac{d^2 y}{dt^2} = g$$

subject to initial conditions y_o and v_o. Both of these equations are short, and the latter is readily integrated twice to yield the former.

The differential equation of motion for the lower segment (i.e., leg) of a double pendulum is

$$\frac{d^2 \theta^2}{dt^2} = \frac{1}{I_L}\left\{ M_K + m_L d_L \begin{bmatrix} L_T a_T \cos(\theta_L - \theta_T) \\ + L_T \omega_T^2 \sin(\theta_L - \theta_T) \\ + a_{Hx} \cos\theta_L \\ + (a_{Hy} + g)\sin\theta_L \end{bmatrix} \right\}$$

where M is the moment at the knee; θ_L and θ_T are the angles of the leg and thigh, respectively; L_T is the length of the thigh; ω_T is the angular velocity of the thigh; and a_{Hx} and a_{Hy} are the horizontal and vertical linear velocities of the hip.

Because this equation cannot be solved by any known analytical technique, we cannot write an equation for the segment's position. Interested readers can refer to Putnam (1991) and Cappozzo and colleagues (1975).

Model Derivation: Lagrange's Equation of Motion

The equations of motion of simple systems often can be derived via visual examination. However, multisegmented human systems typically defy such analysis because of the complex interactions between segments. There are several advanced techniques used to perform such derivations, of which *Lagrange's equation of motion* is the most popular. Unlike Newton's second law, Lagrange's equation uses the mechanical energies of the system; its formulation is

$$\frac{d}{dt}\frac{\partial KE}{\partial V_i} - \frac{\partial KE}{\partial Y_i} + \frac{\partial PE}{\partial Y_i} = F_i \qquad (10.1)$$

where KE and PE are, respectively, the translational and rotational kinetic energies of the entire system being simulated, V_i and Y_i are, respectively, the velocity and position of the ith component of the system in the y direction, and F_i is the external force acting on the ith component. This equation can be written for the x and z directions as well. This equation may at first appear to be a messy set of partial derivatives, but it is, in fact, a more general form of Newton's second law, $F = ma$. Consider a simple projectile moving in the vertical direction that is acted on by the force of gravity and wind resistance, as shown in figure 10.2.

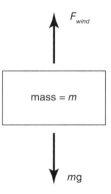

▲ **Figure 10.2** Free-body diagram of an object in free fall.

Applying Newton's second law to this FBD yields

$$F_{wind} - mg = ma \qquad (10.2)$$

Let us now apply Lagrange's equation to this same system. The energies of the complete system are given by

$$KE = 0.5mv^2 \text{ and } PE = mgy \qquad (10.3)$$

These are simple, of course, because the system has only one component. Applying the terms of Lagrange's equation,

$$\frac{\partial PE}{\partial Y_i} = mg \qquad (10.4)$$

$$\frac{\partial KE}{\partial Y_i} = 0 \qquad (10.5)$$

$$\frac{\partial KE}{\partial V_i} = mv \Rightarrow \frac{d}{dt}\frac{\partial KE}{\partial V_i} = ma \qquad (10.6)$$

Collecting these terms into Lagrange's equation yields $ma - 0 + mg = F_i$, or

$$\vec{F}_i - mg = ma \qquad (10.7)$$

The only F_i in our FBD is F_{wind}. Substituting this for F_i in these last equations yields the same equation of motion that we obtained by applying Newton's second law to the FBD.

The middle term of Lagrange's equation is usually zero for single-component systems. However, it is usually nonzero in multicomponent systems in which components interact with each other, so it is referred to as the *interaction term*.

The angular form of Lagrange's equation is

$$\frac{d}{dt}\frac{\partial KE}{\partial \omega_i} - \frac{\partial KE}{\partial \theta_i} + \frac{\partial PE}{\partial \theta_i} = M_i \qquad (10.8)$$

where ω is an angular velocity, θ is an angular position, and M_i is a moment. Because most simulations of human movement involve body segments that rotate, it is this angular form of Lagrange's equation that is most commonly used. The moment M is generally the sum of the moments acting on a segment. In the case of the thigh, for example, M would equal the hip moment minus the knee moment.

To implement Lagrange's equation, one must first establish the DOF needed to completely describe the position of the system. This is something of an art, but in general, angular coordinates should be used to describe each segment's position, and one linear coordinate should be identified to locate either the most proximal or most distal joint of the entire system.

Lagrange's equation of motion is an extensive area of inquiry by itself. Interested readers should refer to books such as *Schaum's Outline* (Hill 1967) and texts by Kilmister (1967) or Marion and Thornton (1995).

Numerical Solution Techniques

Generating model kinematics is a complex process generally referred to as *solving* the model equations. It is simply a glorified extrapolation process in which a current position, velocity, and acceleration are used to calculate a new position and velocity. For example, if you were in your car and driving at 100 km/h, you could estimate that you would reach a city 200 km away in 2 h. Note that your estimate would not be exact, but that the closer you got to the destination, in general, the better you would be able to estimate your arrival time. The same principles operate in human movement simulation. We use the current kinematics of the system to estimate future kinematics, and in general, we are more accurate when we estimate into the very near future. As shown in figure 10.3, the velocity at P_1 (given by the slope of the line V_1 tangent to the position-time profile) is a good estimator of the position at P_2. It is slightly less accurate in determining the position at P_3, so we use the velocity V_3 to find P_4. Neither velocity V_1 nor V_3 will generate a reasonable estimate of position P_5.

An analytical solution to a differential equation can be determined in the form of equations, as demonstrated earlier in the simple projectile motion example. A numerical solution, as the name suggests, is developed using

---

EXAMPLE 10.4

A suspended bar (physical pendulum).

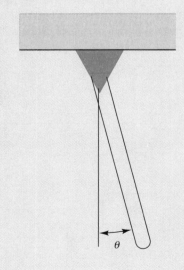

The position of a pendulum can be completely described by one angle, which for convenience can be referenced to the vertical. The pendulum has mass m, moment of inertia I about its pivot, and mass center location d. The energies of the system are $KE = 0.5I\,\omega^2$ and $PE = mgd(1 - \sin\theta)$.

Applying Lagrange's equation yields

$$\frac{\partial PE}{\partial \theta_i} = -mgd\cos\theta \qquad \frac{\partial KE}{\partial \theta_i} = 0$$

$$\frac{\partial KE}{\partial \omega_i} = I\omega \qquad \frac{d}{dt}\frac{\partial KE}{\partial \omega_i} = I\alpha$$

Compare this short derivation to the two-segment double pendulum derivation in appendix G. When referring to it, note the dramatic increase in complexity. Also note that the term $\frac{\partial KE}{\partial \theta}$ is now nonzero.

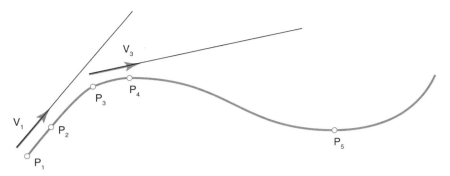

▲ **Figure 10.3** An arbitrary path of an object, with velocity vectors noted along tangent lines.

numerical values. It is an iterative process in which the differential equations are solved over and over again to estimate how much the position of the body will change over small periods of time. A flowchart of this process is presented in figure 10.4. It starts with the initial conditions, the starting values for the position and velocity of each segment. These specify the total starting energy in the system. The values are inserted into the differential equations, and the result is a new set of positions and velocities. These new positions and velocities are inserted into the differential equations, and the process is repeated for the next instant in time. This continues until one or more of the values being calculated reaches the desired ending value. This process is often referred to as a *forward simulation, forward solution,* or *numerical integration.*

An important part of the numerical solution is determining the starting and ending conditions. Typically, choosing initial conditions is fairly simple. For example, when simulating walking, we could use the conditions of the body segments at the start of a gait cycle. When simulating a jump, we would use a logical starting posture. However, end conditions are not as simple, because the exact position of the simulated body becomes less and less predictable as the simulation proceeds. For

example, heel contact may seem like a logical ending condition for a walking simulation. However, there can be times when a simulated heel drags the ground early in the swing phase or a toe contacts the ground before the heel. Thus, emphasis must be placed on determining a robust algorithm for properly ending the simulations.

In this example, the initial position of the object is its height above the ground. The ending condition could be any number or combination of variables. For example, we might stop the simulation when the object hits the ground, when it has fallen a certain distance, when it reaches a given velocity, or some other condition. The ending condition is chosen according to the goal of the simulation.

Complete expositions on a variety of numerical solution techniques appear in most textbooks on differential equations or numerical analysis. The simplest technique, called the *Euler method*, often is not precise enough for human movement simulation. A more precise and popular technique is the *Runge-Kutta algorithm*. Because it is a four-step process, it runs slower than the Euler method in computer code, but it returns more precise data.

Numerical solutions require that the *time step* be defined. Often denoted as Δt or h, the time step is the time between data points in the simulation. The size of

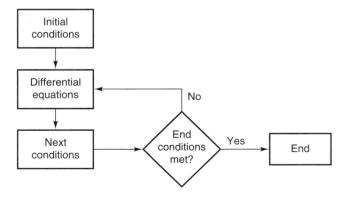

▲ **Figure 10.4** Flowchart of an iterative solution process.

EXAMPLE 10.5

The differential equation of motion that includes wind resistance is

$$F(v) - mg = ma$$

Note that the external force F—the wind resistance—has been written as a function of the object's velocity. This equation must be solved for the acceleration, a:

$$a = \frac{F(v)}{m} - g$$

The forward solution code then takes the following form:

► Set the initial value for the position of the object.

► Set the initial value for the velocity of the object.

► Set the time step of the iteration.

► Set the ending condition of the simulation.

► Repeat the following:

► Calculate the acceleration (a).

► Estimate a new velocity (v) for some small time step later.

► Estimate a new position at this small time step later.

► Record the new position and velocity.

► If the ending condition is not met, continue iterating.

EXAMPLE 10.6

Again, the differential equation is

$$a = \frac{F(v)}{m} - g$$

Returning to the example of the object in free fall (figure 10.2), let $F(v) = -0.2v$. Note that the right-hand side of this equation is negative because the force of the air resistance (F_{wind}), like all frictional forces, opposes the direction of movement. In mathematical or computer code, the Euler solution would be written as

$$g = 9.81 \text{ (acceleration due to gravity)}$$

$$m = 10 \text{ (mass of object in kg)}$$

$$h_o = 100 \text{ (initial height, 100 m over ground)}$$

$$v_o = 0 \text{ (initial velocity, zero)}$$

$$\Delta t = 0.01 \text{ (time step)}$$

while $h > 0$, do (stop when object hits ground)

$$a = F(v)/m - 9.81,$$

(calculate acceleration; note value of g)

$$v = v_o + a \, \delta t \text{ (calculate new velocity)}$$

$$h = h_o + v \, \delta t + 0.5 \, a \, \delta t^2$$

(calculate new position)

$h_o = h$ (set starting position for next iteration)

$v_o = v$ (set starting velocity for next iteration)

end while loop.

Carrying this example out numerically, the first iteration is

$$F(v) = -0.2(0 \text{ m/s}) = 0 \text{ N}$$

$$a = 0 \text{ N/10 kg} - 9.81 \text{ m/s}^2 = -9.81 \text{ m/s}^2$$

$$v = 0 \text{ m/s}^2 + (-9.81 \text{ m/s}^2) \, 0.01 \text{ s} = -0.0981 \text{ m/s}$$

$$h = 100 \text{ m} + (-0.0981 \text{ m/s}) \, 0.01 \text{ s}$$
$$+ 0.5(-9.81 \text{ m/s}^2)(0.01 \text{ s})^2 = 99.99853 \text{ m}.$$

In the first hundredth of a second, the object has fallen only 1.5 mm. The second iteration uses the velocity and position values calculated in the first iteration:

$$F(v) = -0.2(-0.0981 \text{ m/s}) = 0.01962 \text{ N}$$

$$a = 0.01962 \text{ N/10 kg} - 9.81 \text{ m/s}^2 = -9.80804 \text{ m/s}^2$$

$$v = -0.0981 \text{ m/s}^2 + (-9.80804 \text{ m/s}^2) \, 0.01 \text{ s}$$
$$= -0.19618 \text{ m/s}$$

$$h = 99.99853 \text{ m} + (-0.19618 \text{ m/s}) \, 0.01 \text{ s}$$
$$+ 0.5(-9.80804 \text{ m/s}^2)(0.01 \text{ s})^2 = 99.99608 \text{ m}.$$

Now the object has fallen 3.9 mm. Because this process uses small time steps, the distance traversed between each step is very small. Iterations can be carried out in a spreadsheet or with computer code. The first 10 iterations return the following values for wind resistance, acceleration, velocity, and height.

$F(v)$	a	v	h
0	−9.81	0	100
0	−9.81	−0.0981	99.99853
−0.01962	−9.80804	−0.19618	99.99608
0.039236	−9.80608	−0.29424	99.99264
0.058848	−9.80412	−0.39228	99.98823
0.078456	−9.80215	−0.4903	99.98284
0.098061	−9.80019	−0.58831	99.97646
0.117661	−9.79823	−0.68629	99.96911
0.137258	−9.79627	−0.78425	99.96078
0.15685	−9.79431	−0.88219	99.95147
0.176439	−9.79236	−0.98012	99.94118

the time step depends on the complexity of the model equations, but normally it is between 0.01 and 0.001 s. Too large a time step can lead to an inaccurate result after a number of iterations. Increasingly smaller time steps generally do not cause problems because most computers use extended-precision numbers, but the time step should remain large enough so that the simulation does not take an inordinate amount of time to run.

A feature common to human movement simulations is that they start with a differential equation for the acceleration. If we know the acceleration of the object at some point in time, then we can estimate how much the velocity will change over some small time step. Similarly, we can then estimate how much the object's position will change over that period of time. The remainder of the solution is then just a matter of repetition. This is demonstrated in example 10.6.

The iteration process is more complicated for multisegmented systems, because each iteration must be performed for each segment with the more sophisticated Runge-Kutta algorithm. However, it is still analogous to the example presented previously. Readers may refer to appendix F for a complete set of equations and steps for a two-segment pendulum.

CONTROL THEORY

There are many different ways to model the forces and moments acting on the human body. This becomes a concern as models become more sophisticated, because

in effect we begin to model the manner in which the human body controls itself. *Control theory* is a branch of engineering that studies the control of systems, both natural and human-made. Its applications are broad: autopilot controls for airplanes, temperature regulation in a nuclear power plant, antilock brakes for cars, and robotic controls on a factory production line. Human movement researchers explore the means by which a particular task is controlled and thus draw from control theory. We mention it in this chapter only to alert readers to its complexities and the elements that it shares with human movement simulation.

One aspect of control theory that is of particular interest is *feedback*. Feedback is information collected by the organism about either itself or the environment. When we walk, for example, we receive many different forms of feedback, including sensory information from our vision, proprioception, and vestibular system, to name but a few of the channels. We use this information to determine how to control our gait, altering our control as it becomes necessary. This type of control is called *closed-loop control*. In contrast, a control strategy in which a preprogrammed routine is enacted without using feedback is referred to as *open-loop* or *feedforward control*. In computer simulations, it is difficult to implement closed-loop control. Typically, researchers program the model controls in advance, run the simulation, and then change the controls afterward. The simplicity of this control may seem unrealistic, but every tool is limited, and we just have to use them while respecting their

FROM THE SCIENTIFIC LITERATURE

Mochon, S., and T.A. McMahon. 1980. Ballistic walking. *Journal of Biomechanics* 13:49-57.

A classic computer simulation was the ballistic walking model of Mochon and McMahon (1980) (see also McMahon 1984). This model had three segments representing the human body: a straight, rigid stance limb that supported a swinging thigh and leg (figure 10.5). The mass of the head, trunk, and arms was lumped together as a point mass at the hip. The model had no muscles. It started "walking" at the instant of right toe-off, and the simulation ended at right heel contact. This model was, of course, too simple to predict many features of human walking, but the authors noted that their goal was only to make simple statements. In addition to noting that their model could locomote without falling down, the authors found that their model reflected the change in walking speed that occurs with increasing body height. Also, as is the case with humans, the model's speed was limited by the need for the swing leg to clear the ground.

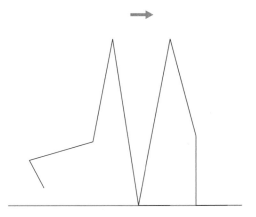

▲ **Figure 10.5** Stick figure of the walking model of Mochon and McMahon (1980).

FROM THE SCIENTIFIC LITERATURE

Onyshko, S., and D.A. Winter. 1980. A mathematical model for the dynamics of human locomotion. *Journal of Biomechanics* 13:361-8.

Onyshko and Winter developed a more advanced model than that of Mochon and McMahon in this 1980 paper. As shown in figure 10.6, it had seven segments corresponding to a torso with two thighs, legs, and feet. Six joint moments controlled this system, and it could perform a full gait cycle. Although this model appears to be similar to that of Mochon and McMahon, it was in fact much more complicated. The equations of motion, derived via Lagrange's method, required 22 anthropometric constants and had literally dozens of terms. The equation of motion for each segment also required joint moments determined from experimental data. Moreover, the model had to operate differently depending on the phase of the gait cycle. In other words, single support on the right foot required the equations of motion opposite to those for single support on the left foot, and these were different from the double-support equations. In effect, the simulation included four separate programs run by a master program that constantly determined what phase of the gait cycle the model was in.

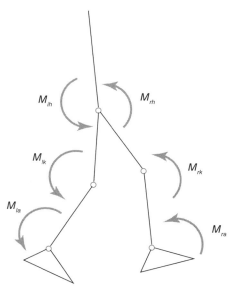

▲ **Figure 10.6** Stick figure of the walking model of Onyshko and Winter (1980), showing its joint moment controls.

limitations. The following examples of human movement simulations show that there is still much to learn even without perfect controls.

LIMITATIONS OF COMPUTER MODELS

No matter how sophisticated a computer model becomes, many limitations remain, and they arise from a number of factors:

▶ the numerical imperfections that exist in the solution process,

▶ the difficulty of accurately modeling impacts,

▶ the approximation of complex human structures,

▶ the enormous adaptability of the human system, and

▶ the inevitability of having to make assumptions, because we can never account for all of the human and environmental factors.

Numerical errors are often slight and can often be resolved (see, e.g., Risher et al. 1997). Nonetheless, models become increasingly sensitive as the number of components increases. In addition, small errors that arise early in a simulated movement can propagate and grow to measurable levels by the end of the simulation. Also, models can have parameters that differ by many orders of magnitude, a situation referred to as a *stiff system* of equations. This can lead to a sensitive model that generates inconsistent results.

Impacts are very difficult to model. One reason for this is that they incur sharp changes in the accelerations of the body segments, resulting in discontinuities that lead to numerical errors. Also, impacts generally violate the assumption that the human body consists of rigid segments. During walking, for example, the feet deform, storing and dissipating mechanical energy in a manner that is extremely difficult to model. When the human body runs or performs a drop landing, the impacts make computer simulation very difficult, indeed. Often, these issues can be resolved by reducing the time step of the solution, but sometimes it is necessary to increase the complexity of the model. For example, Derrick and colleagues' (2000) model used a time step of 0.0001 s. However, modeling heel contact in running and walking is still a very difficult problem, especially for different locomotion speeds and subject anthropometries. Refer to Gilchrist and Winter (1997) for more details.

Human structures themselves are much more complicated than our models of them. For example, most simulations use simple representations of joints and assume that segments are rigid; for most purposes, these work well. However, joints are not simple pivots, muscles are not simple torque actuators, and segments are not rigid. Thus, researchers must be alert to the effects of such simplifications. For example, if a discrepancy were observed in the ankle moment data of a running model, a researcher might first question whether the problem stemmed from the assumption that the foot segment was rigid. A second question might be whether the model ankle reasonably approximated the orientation of the ankle (subtalar) joint.

When modeling, we must remember that we are simulating a human system in which the sensory and control systems and their organization are extraordinarily complex. A complete model of the moving human—even if we understood the entire human system to begin with—would be too much for the most sophisticated computers to handle. Therefore, every model of the human system is an oversimplification. This is an accepted part of the science. For example, as was mentioned earlier in this chapter, virtually all walking simulations have been performed on level ground, yet researchers have difficulty completely modeling the movement in even this simple situation. In addition, humans can adapt to myriad surfaces, turns, steps, bumps, and other conditions while walking, and each of these would require a different—and likely complex—adjustment to the model. Our models also neglect the vast array of senses that influence our movements with their messages. Even if we put aside these concerns, the issue of adaptability under simple test conditions is also pertinent. In short, replicating a human movement is generally straightforward, but predicting how that movement will change under different test conditions is difficult. Suppose, for example, that we have a highly accurate simulation of normal walking. What will happen when the torso mass increases, as it would if we had to account for a backpack? Will the stride lengthen or shorten? Will the torso angle change? Will the load be too heavy altogether?

In light of these limitations, it is clear that models should be interpreted with caution. It is more instructive to look at the larger results of the study than at small differences in joint moments or segment kinematics. Interpretations must also be limited to the specific circumstances of a model, even when the risks of extrapolating results may seem inconsequential. For example, a model of normal walking should not be used to infer the gait of a person with an amputation, nor should a simulation performed at one speed of movement be assumed to apply to all speeds. Environmental conditions and constraints strongly affect subject behavior, as do such subject characteristics as age, gender, physical training, and body type; all of these can introduce limitations. Obviously, one cannot be too careful in deciding how much a model says about the real world. The beauty of simulation lies in its great flexibility and controllability, but it is because of those factors that we cannot draw direct inferences to the real world. All conclusions are implied and subject to further experimentation. In short, although model interpretation may be more art than science, its importance cannot be overemphasized. When interpreting model data, use the following guidelines:

▶ The exact magnitudes of quantitative data should be treated conservatively, because of their sensitivity to model assumptions.

▶ Analogies between model and human structures should remain loose.

▶ Never state that the way in which the model works is the way in which the human system works; discuss the model and humans separately.

Regarding the first point, it is shown throughout this text that many assumptions go into our calculations of such data as joint moments and segment kinematics. These same problems exist in computer models. In fact, models have additional assumptions, and so we need to be even more reserved in making our quantitative assessments. In chapters 2, 5, and 12, we learned that anthropometric estimates and various treatments of data such as filtering and differentiation limit us to about a 10% threshold of interpretation of joint moment data. So, for example, for a jumping model that generates knee-joint moment data, we must consider that 10% threshold of interpretation. However, the simulation model also makes other assumptions about how the subject generated that joint moment. Perhaps our model would produce slightly different joint moments if we imposed different constraints on the subject or if we used different optimization techniques or numerical integration methods.

In reference to the second guideline, we generally should not state that a particular model component is a human structure. Almost any physical system in the world can be modeled as a second-order differential equation, and a system of masses, springs, and dampers can be constructed to model almost any human movement. This should not lead us to state that some part of the human body *is* a mass, spring, or damper; instead, state that the body part *acts* like a mass, spring, or damper in our model. For example, do not state that the spring of a mass-spring running model is the lower extremity but rather that the spring behaves like the lower extremity. This may seem to be only a matter of semantics, but it is not. Biomechanists construct behavioral models of human movement, and therefore it is best to discuss a model separately from the real world.

FROM THE SCIENTIFIC LITERATURE

Derrick, T.R., G.E. Caldwell, and J. Hamill. 2000. Modeling the stiffness characteristics of the human body while running with various stride lengths. *Journal of Applied Biomechanics* 16:36-51.

Sometimes the simplest models to implement are the hardest to interpret. Earlier, we mentioned mass-spring models of human running, in which a mass corresponding to body mass bounces on a large spring. Interested readers should refer to McMahon and colleagues (1987) for further reading. Alexander and colleagues (1986) noted that mass-spring models for animal gait generate ground reaction force (GRF) curves shaped like inverted parabolas, whereas human GRFs are more complex, typically showing an initial sharp peak followed by a larger, more rounded peak. Therefore, Alexander and colleagues proposed a more sophisticated mass-spring-damper model of running. In this model, a large mass rested atop a spring, which in turn rested on a second, smaller mass. The smaller mass then rested on a spring and a damper (or dashpot), which is an energy-dissipation element much like a shock absorber in a car. Alexander and colleagues demonstrated that this model produced the more lifelike two-peak GRF. Diagrams of the two model types and their GRFs are shown in figure 10.7.

Derrick and colleagues (2000) used a mass-spring-damper model to study the influence of different stride characteristics on GRF characteristics. In the model, the spring between the two masses tended to influence the time course of the GRF (i.e., a stiffer spring had less contact time with the ground), and the lower spring tended to influence only the first impact peak. In the study, 10 experimental subjects ran at their preferred stride length and at ±10% and ±20% of their preferred stride length. Then, for each experimental condition, the investigators adjusted the stiffness of the two model springs using an optimization algorithm to match the model GRF to the experimental. In these experimental curves, as stride length decreased, the GRF impact peak decreased by 33% and stance time decreased by 7%. In the model parameters, the upper spring was twice as stiff with the short strides as it was with the long strides. The lower spring had the opposite effect, being about 80% as stiff with the short strides.

Interpretation of mass-spring-damper data is not straightforward, because the model is very

abstract. The lower mass was 20% of body mass, which might mean that it corresponded to the mass of the stance limb. Therefore, we can think of the upper spring as a measure of whole-body compliance. As much as we might like to, we cannot say that the upper spring represents the stiffness of a specific joint, such as the knee. The fact is that multiple joints are involved in absorbing the impacts of running, so we cannot draw such a direct analogy. The correct interpretation is that when human subjects run with longer strides, their bodies become more compliant. Both Derrick and colleagues (2000) and McMahon and colleagues (1987) noted that human subjects achieved greater compliance with greater degrees of knee flexion. However, neither is this something we can extract from the mass-spring-damper model, simply because it does not have knees.

The mass-spring-damper model demonstrates clearly that models must be interpreted carefully. The lower extremity is not a spring, but we can glean some useful information by modeling it as a simple spring.

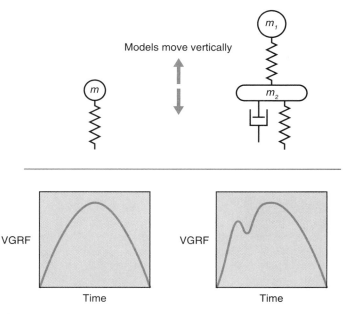

▲ **Figure 10.7** Mass-spring and mass-spring-damper models with their respective vertical GRF curves.

SUMMARY

This chapter began with a discussion of why one would want to simulate human movement on a computer. It concluded by discussing how limited our current models

are, which might seem to suggest that simulation is too grossly limited to be of much use. This is not the case, however. The fact is that we do not understand how we perform even a "simple" task like walking. Simulation is a fertile ground for testing and developing what we understand, a powerful tool that has great promise for developing many practical applications in the medical and industrial fields. The key, as with all tools, is to use it properly. Constructing a model is relatively straightforward, but we must heed the limits of interpretation.

SUGGESTED READINGS

Bobbert, M.F., K.G. Gerritsen, M.C. Litjens, and A.J. van Soest. 1996. Why is countermovement jump height greater than squat jump height? *Medicine and Science in Sports and Exercise* 28:1402-12.

Gerritsen, K.G.M., A.J. van den Bogert, and B.M. Nigg. 1995. Direct dynamics simulation of the impact phase in heel-toe running. *Journal of Biomechanics* 28:661-8.

Hill, D.A. 1967. *Schaum's Outline of Theory and Problems of Lagrangian Dynamics*. New York: McGraw-Hill.

Hull, M.L., H.K. Gonzalez, and R. Redfield. 1988. Optimization of pedaling rate in cycling using a muscle stress-based objective function. *International Journal of Sports Biomechanics* 4:1-20.

Kilmister, C.W. 1967. *Lagrangian Dynamics: An Introduction for Students*. New York: Plenum Press.

Marion, J.B., and S.T. Thornton. 1995. *Classical Dynamics of Particles and Systems*. 4th ed. New York: Harcourt Brace College.

van Soest, A.J., M.F. Bobbert, and G.J. van Ingen Schenau. 1994. A control strategy for the execution of explosive movements from varying starting positions. *Journal of Neurophysiology* 71:1390-402.

Chapter 11

Musculoskeletal Modeling

Brian R. Umberger and Graham E. Caldwell

In chapter 9 we discussed how muscle models can be used to represent the force-generating capabilities of muscles, and in chapter 10 we discussed the process for creating simulation models of the human body. In this chapter, we combine these two areas and discuss musculoskeletal models that represent the anatomic, geometric, and dynamic characteristics of human body segments and joints. In this chapter, we

- ▶ introduce the processes of using musculoskeletal models to perform inverse dynamics and forward dynamics analyses;
- ▶ identify the basic components of musculoskeletal models, including skeletal geometry, passive joint properties, muscle kinematics and moment arms, and interactions with the environment;
- ▶ discuss neural control of musculoskeletal models, including static and dynamic optimization techniques;
- ▶ work through an example of induced acceleration analysis used with a musculoskeletal model; and
- ▶ introduce other analysis techniques used to interpret and validate musculoskeletal models.

MUSCULOSKELETAL MODELS

From a biomechanical perspective, the human body can be thought of as a system of anatomical segments that are connected by joints and acted upon by muscles, ligaments, and other internal structures. This system interacts with the environment around it and is controlled in such a way as to produce purposeful movement. Musculoskeletal modeling is the process that takes all of these factors into account in a systematic effort to better understand human movement. Among their many uses, musculoskeletal models provide the best means for estimating internal loading of anatomical structures, because opportunities for direct measurement of muscle, bone, and joint forces in humans are severely limited. Given that muscles are the motors that drive human movement and act as the intermediaries between the nervous system and the skeletal system, musculoskeletal modeling also holds great promise for advancing our understanding of the energetics and control of movement. However, developing, evaluating, and using musculoskeletal models is a labor-intensive process, fraught with approximations, simplifications, and uncertainties. Given the steady increase in the number of musculoskeletal modeling studies in the scientific literature, it is critical for students of human movement to understand the opportunities and limitations associated with the use of musculoskeletal models. This can best be accomplished by understanding the procedures involved in creating and using musculoskeletal models to address questions in human movement.

The first decision to be made in any modeling study is the level of detail that will be necessary in the model. This decision should be driven by the nature of the research question to be addressed. Consideration should be given to issues discussed in chapter 10, such as whether the model will be two-dimensional (2-D) or three-dimensional (3-D), the number of body segments to include, and the way in which the joints will be modeled. Consideration must also be given to issues discussed in chapter 9, such as which muscle model to use, how many muscles to include, and how the muscles will be controlled. Examples of musculoskeletal models from the literature range in complexity from one degree of freedom (DOF), with a single segment acted upon by a single muscle, to multiple DOF models containing 10 or more body segments and dozens of muscles. In all cases, the model should represent the anatomy of the skeleton, the inertial characteristics of the body segments, and the properties of the muscular and neural systems to a level of detail and accuracy necessary to address the specific research question being investigated.

As with the Hill muscle model (chapter 9), the constituent parts of a musculoskeletal model are represented in software through mathematical expressions. For example, movement of the skeleton is described by the dynamic equations of motion (chapter 10), where segmental motion is produced by moments acting at the joints. If the model is used in an inverse dynamics approach (chapters 5 and 7), segmental motion is specified and used to calculate net joint moments. Additional modeling is then necessary to distribute the joint moments into individual forces from muscles, ligaments, and joint surfaces (figure 11.1). Forward dynamics models (chapter 10), which simulate body segment motion from given kinetic inputs, are driven either by joint moment actuators or by individual muscle actuators (figure 11.1). In either case, muscle models of some description can represent the physiological attributes of these kinetic actuators.

It is clear that net joint moments are mathematically associated with the (unknown) forces produced by the individual muscles. These unknown muscle forces can be estimated with a musculoskeletal model that uses Hill-type models to represent specific muscles. The mathematical relation between a given muscle force and a given joint moment is dictated by the geometrical relation between the muscle line of action and the center of rotation of the joint in question. Force output from an individual muscle model is determined in part by the stimulation (also referred to as *excitation*) it receives from the nervous system, the so-called *control signal*. This control signal can be generated from a measured electromyography (EMG) signal, derived from a theoretical control algorithm, or obtained via optimization techniques. Once a simulation of the desired movement has been generated, a musculoskeletal model can be used to gain insights into the mechanics, energetics, and control of movement that are not provided by experimental techniques alone.

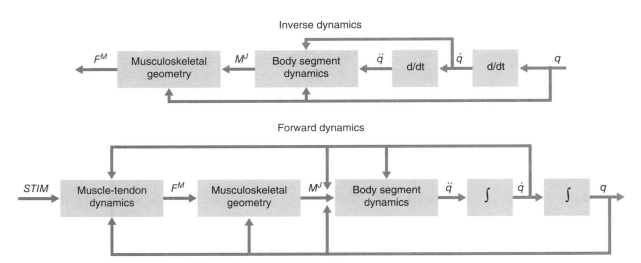

▲ **Figure 11.1** Overview of musculoskeletal model-based inverse dynamics and forward dynamics analyses. In inverse dynamics, the inputs are body segment positions *(q)* and the outputs are muscle forces *(F^M)*. In forward dynamics, the inputs are muscle stimulation patterns *(STIM)* and the outputs are body segment positions *(q)*. In both cases, musculoskeletal geometry defines the transformations between muscle forces *(F^M)* and joint moments *(M^J)*.

Modified from Pandy 2001.

In the following sections, we outline the components of musculoskeletal models. In later sections, we discuss the approaches used to control the movements of musculoskeletal models and some of the ways in which they can be used to study human movement.

Skeletal Geometry and Joint Models

From an anatomical viewpoint, the skeleton is the framework that physically supports the muscles. An accurate skeletal model permits the correct assignment of muscle origin and insertion locations as well as the paths that muscles follow between their origin and insertion sites. The surface geometry of individual bones can be obtained with a handheld digitizer, a laser scanner, or imaging modalities such as computed tomography or magnetic resonance imaging (MRI). Figure 11.2 shows a scanned femur from a musculoskeletal model of the human lower limb (Arnold et al. 2010). The wireframe image on the left (figure 11.2*a*) reveals the actual points and polygons that define the surface of the bone, whereas the image on the right (figure 11.2*b*) shows a smoothed version of the bone. Because of the effort involved in creating a complete skeletal model, usually a single generic model is used, which may be scaled in size to represent a specific individual. This scaling process can lead to inaccuracies in the model due to (1) unknown variations in skeletal shape between individuals, (2) unknown variations in muscle origin and insertion locations between individuals, and (3) inaccurate scaling. We can overcome these problems by using imaging techniques to scan individual subjects and thereby create true subject-specific skeletal models (Blemker et al. 2007). Currently, creating subject-specific models requires a great investment of time, effort, and scanning costs, but with technological advances this approach could become routine in the future.

From a mechanical viewpoint, the skeletal geometry is important because the forces transmitted by the muscles to the skeleton have direct bearing on the linear and angular motion of the skeleton and individual body segments. As the muscles run their course from origin to insertion, they pass over the joints formed by adjacent bones. The mechanical nature of each joint is determined by the local skeletal geometry of the contact surfaces between the bones. For example, at the hip, the head of the proximal femur and the pelvic acetabulum form a stable union that allows rotation of the femur with very little translation of the femoral head. Mechanically, this can be modeled in 2-D as a hinge joint with one rotational DOF or in 3-D as a ball-and-socket joint with three rotational DOF.

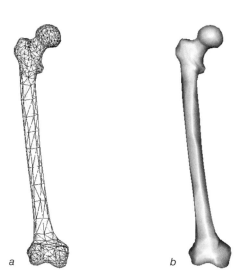

▲ **Figure 11.2** A scanned femur shown using wireframe *(a)* and smoothed *(b)* rendering. Bony geometry is from the musculoskeletal model described by Arnold et al. 2010 and was rendered using the open-source modeling package OpenSim (Delp et al. 2007).

From a modeling perspective, the way in which joints are represented is important because it has been shown to affect the prediction of muscle forces (Glitsch and Baumann 1997). The structure of the articulation between bones dictates the mechanical nature of surrounding soft tissues such as ligaments and muscles because these tissues must create forces that provide for both skeletal motion and joint integrity. For example, the relatively flat surface of the proximal tibia and the curved surface of the distal femur provide little bony stability at the knee, and it is no surprise that the ligaments and muscles there must provide a much higher degree of joint stability than do their counterparts at the hip. In 3-D, the knee should be viewed as having six DOF (three rotations and three translations), and models of the knee that seek to provide insights on tissue loading typically include this level of detail. However, because knee motion is restricted in some directions, models used to study gross body movement frequently model the knee as a planar hinge with specified amounts of translation in the anterior-posterior and proximal-distal directions or even as a simple hinge with no translation of the axis of rotation. In both of these cases the joint would have only one DOF, because the translations in the former case are prescribed as functions of the knee joint angle. Using a one DOF knee joint greatly simplifies the equations of motion in a model of the whole body. However, if a simple hinge joint is used, care must be taken in defining the muscle lengths and moment arms about the knee joint, because the true relative displacements of the muscle origins and insertions are not only functions of joint rotation but also depend on translations of the body segments that accompany joint rotation.

Passive Joint Properties

During most human movements, active muscles contribute the primary forces that drive joint motion. However, passive structures also exert moments across the joints. Ligaments, capsular tissue, fascia, cartilage, and parallel muscle elasticity may all contribute to the net moment at a joint. These effects have often been ignored in musculoskeletal models but generally should be taken into account to ensure accurate model results. Thus, there are really two major components to the net joint moment: the active joint moment, arising from active muscle forces, and the passive joint moment, arising from forces in passive tissues. The active and passive joint moments sum to equal the net joint moment, which is the result calculated from an inverse dynamics analysis (chapter 5). Given the design of most major diarthrodial joints in the human body, the passive moments tend to be small for much of the normal range of joint motion, leading to their exclusion from some musculoskeletal models. However, the passive moments can become quite large in magnitude as the end-range of joint motion is approached, and sometimes they make meaningful contributions to the net joint moment during normal movements. For example, the passive component represents more than one-third of the magnitude of the net hip flexor moment during the second half of the stance phase in walking (Whittington et al. 2008). In addition to ensuring more accurate simulation results, the inclusion of passive joint moments in a musculoskeletal model helps prevent the joints from achieving highly unrealistic postures during forward dynamics simulations.

The way in which passive joint properties are included in a musculoskeletal model depends on its intended use. For a detailed 3-D model of the knee joint that will be used to predict joint contact stresses, the actual lines of action and material properties for each major passive structure may need to be explicitly represented. A simpler approach, commonly used in models for studying gross movement, is to incorporate a single passive joint moment relation for each rotational DOF. Several such equations have been published in the literature and frequently take the form of a so-called double exponential curve (Audu and Davy 1985). An example of a double-exponential equation for passive joint moment (M_{pas}) as a function of joint angle (θ) is

$$M_{pas} = k_1 e^{-k_2(\theta - \theta_1)} - k_3 e^{-k_4(\theta_2 - \theta)} \qquad (11.1)$$

where k_1 through k_4, θ_1, and θ_2 are constants that determine the shape of the passive moment curve. Specifically, θ_1 and θ_2 set the range of joint angles over which the passive moment has a low magnitude, whereas k_1 through k_4 determine the degree to which the passive moment increases outside the joint angle range set by θ_1 and θ_2. This approach to modeling passive joint restraints provides a restoring moment that opposes the direction of joint angular displacement, as shown in figure 11.3a. For instance, as a joint is moved toward the limits of flexion, there is a passive extensor moment that grows exponentially in magnitude with the degree of joint flexion. For most joints, the passive moment relation will be asymmetrical, as different anatomical structures contribute to the passive moment at each extreme of the joint range of motion.

An important distinction among the passive joint moment models found in the literature is whether they assume that the passive moment at a joint depends only on the angle of that joint, or also on the angles of the adjacent joints. This issue is directly related to how passive forces generated by biarticular muscles are included in the overall musculoskeletal model. For example, the contribution of the parallel elastic component (PEC) of the biarticular hamstrings to the hip joint passive moment depends not only on the hip joint angle but also on the knee joint angle. As shown in figure 11.3b, when the knee joint is extended and the hamstrings are stretched, the passive hip extension moment is larger in magnitude than when the knee is flexed. Passive joint moment equations that depend on a single joint angle (e.g., Audu and Davy 1985) will only be able to adequately represent nonmuscular passive contributions and are best paired with muscle models that include a PEC, to properly account for the contributions of biarticular muscles. However, passive joint moment expressions that include the effects of adjacent

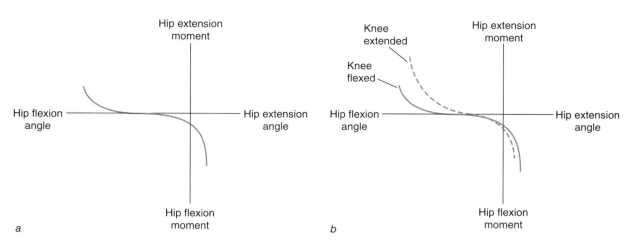

▲ **Figure 11.3** Passive joint moment curves. *(a)* The passive hip joint moment, assuming dependence only on the hip joint angle. *(b)* Way in which the passive hip joint moment is affected by changing the knee joint angle.

joint angles (e.g., Silder et al. 2007) will implicitly account for the PEC of any muscles that cross the joint. In these cases, it is appropriate to use a muscle model without a PEC, so that passive muscular contributions are not counted twice. The latter approach has the advantage that the total passive moment at a joint can be readily measured on a subject-specific basis. However, this approach requires more terms in the passive moment equations than are shown in equation 11.1. In contrast, the former approach involves partitioning out the muscular and nonmuscular contributions, which is difficult in practice.

Formulating passive joint moment expressions for a 2-D sagittal plane musculoskeletal model is relatively straightforward. However, the situation for 3-D joint models (e.g., hip or shoulder) is more complicated. One approach is to include a passive joint moment equation for each rotational DOF at the joint, corresponding to the three Cardan angles (chapter 2) that describe the orientation of the joint. However, the axes of these Cardan angles are not orthogonal, and they change their relative orientations during the course of a movement. Thus, this approach would miss any interactions among the three contributions to the total passive moment at the joint. For example, the passive moment associated with flexing and extending the hip joint may differ depending on the angle of abduction-adduction and internal-external rotation. These interdependencies could be captured using an extensive set of passive moment measurements and equations that include interaction terms, but examples of this are rare in the literature (e.g., Hatze 1997).

Muscle Kinematics and Moment Arms

Muscles change length as a function of angular displacements of the joint or joints they cross. It is easy to see that isolated extension of the knee joint causes the gastrocnemius to lengthen (figure 11.4*b*) or that isolated ankle plantar flexion causes both the gastrocnemius and soleus to shorten (figure 11.4*c*). However, it is harder to predict what will happen to gastrocnemius length with simultaneous knee extension and ankle plantar flexion (figure 11.4*d*). The amount of length change in a muscle-tendon unit depends on the geometry of the bones that form the joint and the amplitude of joint angular displacement. There are perfectly reasonable scenarios for knee extension and ankle plantar flexion that would cause the gastrocnemius to lengthen, shorten, or not change length at all. The ability to accurately model these muscle length changes is important for determining how much force a given muscle can produce, as discussed in chapter 9.

Muscle Length Prediction Equations

One common technique is to express the length for each muscle as a function of the angular displacement for all joints spanned by that muscle. For example, Grieve and colleagues (1978) used measurements from cadavers to estimate the muscle length changes for the gastrocnemius at a range of given angular

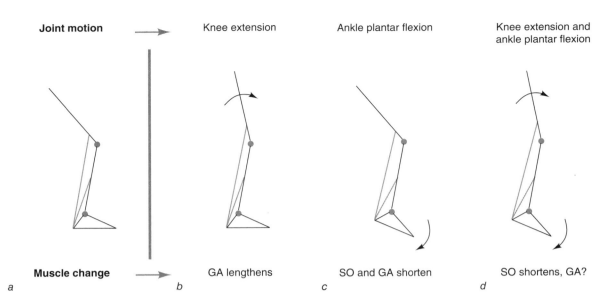

▲ **Figure 11.4** Muscle lengths change as a function of joint angles. GA = gastrocnemius; SO = soleus.

positions for both the ankle and knee joints. Their results were used to form equations that predicted muscle length, normalized to segment length, as a function of joint angles. Their equations are easily scaled to other subjects based on measured segment lengths. Another method is to measure the muscle length at a series of joint angles and form subject-specific equations directly with an imaging technique such as MRI. This method has the advantage of directly accounting for variability in individual subject anatomy rather than assuming similar muscle length-joint angle relations for all subjects.

The prediction equation approach for representing muscle lengths has a number of distinct advantages. It is a compact and computationally efficient way to represent muscle lengths in a musculoskeletal model. In addition, the predicted muscle lengths can easily be scaled to individuals based on relative segment lengths or by changing the coefficients in the equations. The primary limitation of this approach is that the lines of action of the muscles are not directly represented. So, even if muscle forces are known, it is not possible to use them to solve for joint contact forces.

Muscle Attachment Points

A different approach to define muscle lengths during movement is to use the instantaneous positions of the muscle origin and insertion points on the modeled skeletal segments. Identification of these points can be based on markings on the bones, reference to published muscle maps, or digitization of muscle attachment sites during a dissection. Once the origin and insertion points are defined, we can compute the positions of these points during a movement sequence by mathematically embedding the modeled bones into the reference frames that define the body segments in the musculoskeletal model. Because the segment positions in space are defined throughout a movement, we can calculate the bone and muscle attachment positions using the transformation techniques described in chapters 1 and 2. With this approach, major consideration has to be given to deciding how to handle muscles that have a broad area of attachment. Many investigators partition such muscles into two or more individual sections, such as an anterior, middle, and posterior portion for gluteus medius. However, the basis for defining these individual compartments is often arbitrary. An additional limitation when there is a broad area of attachment is that the origin or insertion point is usually assigned to the centroid of the area of attachment, which is not necessarily the center of force application.

With muscle attachment points defined in their respective skeletal segments, the simplest way to model muscle kinematics is with a straight line from the origin to the insertion. This leads to a good representation of the length of several muscles that follow a simple path, such as soleus (figure 11.5a). This approach fails, however, when the path of a muscle overlies other muscles, wraps around a bony prominence, or is constrained by a retinaculum. Deviations from a straight line between origin and insertion can be achieved by introducing intermediate *via points* (Delp et al. 1990) that constrain the path of the muscle in areas where anatomical structures cause them to change direction, such as where the tibialis anterior passes under the extensor retinaculum (figure 11.5b). If a muscle is modeled with a single via point between origin and insertion, then the path will be represented as two straight line segments and the muscle length will be the sum of the lengths of the two line segments. The extension to more than one via point is straightforward. If a more sophisticated representation of musculoskeletal geometry is required, it is also possible to define *wrapping surfaces* (van der Helm et al. 1992), which smoothly deflect the path of the muscle only where it comes in contact with the surface (e.g., proximal end of gastrocnemius, shown in figure 11.5c). In some cases, the muscle path may interact with a particular wrapping surface for only a portion of the joint range of motion. In this case, total muscle length will be the sum of the straight line and curved portions of the muscle path. Compared with using the muscle length prediction equations described earlier, modeling the actual path of each muscle is more involved but allows the point of force application and line of action to be represented for each muscle. Therefore, it provides the basis for computing bone-on-bone joint contact forces, as the mechanical effects of muscle forces can be fully accounted for in free-body diagrams of the segments that form a joint.

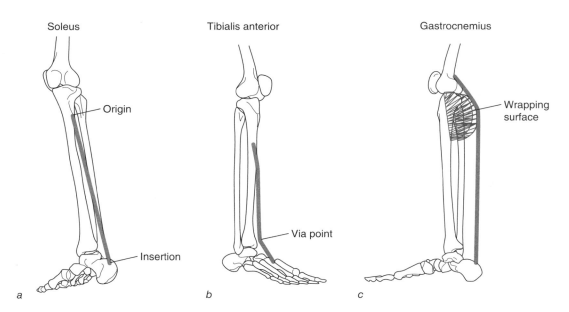

▲ **Figure 11.5** Muscle paths may be defined using points indicating the origin and insertion *(a)*. Deflection of the muscle path can be modeled using via points *(b)* and wrapping surfaces *(c)*. From the musculoskeletal model described by Arnold et al. 2010.

Muscle Moment Arms

Musculoskeletal geometry in the local vicinity of a joint will also determine the mechanical consequence of muscular force production. If the joint structure permits some translation, the direction of the muscle force vector determines how the adjacent bones linearly accelerate with respect to each other. If the joint allows rotation, the location of the muscle force line of action dictates how much torque (τ) is produced about the joint axis of rotation. In the 3-D case, this relation is given by the vector expression

$$\vec{\tau} = \vec{r} \times \vec{F} \qquad (11.2)$$

where $\vec{r}$ is a position vector from the axis of rotation of the joint to the line of action of the muscle force, and $\vec{F}$ is the muscle force vector. In the simpler 2-D case, the relation reduces to the scalar equation

$$\tau = Fd \qquad (11.3)$$

where F is the magnitude of the muscle force, and d is the perpendicular distance between the joint center of rotation and the line of action of the muscle force. The quantity d in equation 11.3 is referred to as the *muscle moment arm*. Muscle origin and insertion points are fixed on the adjacent bones that make up a joint; therefore, as the joint angle changes, the magnitude of the moment arm will usually also change. This partially accounts for the variation in joint moment potential as a function of joint angle that is commonly seen in human muscle strength studies (chapter 4).

There are various ways to determine moment arms for use in musculoskeletal models. Perhaps the simplest approach would be to take a single moment arm measurement on a cadaver specimen, or from an MRI image of a subject, and then assume that the moment arm of that muscle is constant at all joint angles. Although this may be reasonable when the joint range of motion is limited, in general the moment arm changes as joint angle is modified. As with muscle length, the changes in moment arm could be quantified from a series of measurements from a subject or cadaver while the joint is manipulated through a range of angles. There is, in fact, a unique relation between total muscle-tendon length (L_{MT}), muscle moment arm (d), and joint angle (θ), which is the basis for so-called *tendon excursion* experiments (An et al. 1984). This relation can be expressed as

$$d = \partial L_{MT} / \partial \theta \qquad (11.4)$$

which states that muscle moment arm is equal to the partial derivative of muscle length with respect to joint angle. The equation involves a partial derivative because muscle length may depend on more than one joint angle (e.g., gastrocnemius length depends on knee and ankle joint angles). This relation works quite well in conjunction with the muscle length equations described earlier. If muscle length is stored in a polynomial or other simple mathematical function, then this function need only be differentiated with respect to joint angle to determine an expression for moment arm that is also a function of joint angle. Alternatively, if a moment arm function is known, say from experimental data, a muscle length expression can be obtained by integrating moment arm with respect to joint angle. The unknown constant of integration will be equal to the muscle length when the joint angle is at the posture corresponding to zero radians. The approach used to determine muscle moment arms is different if the muscle paths in the model are represented by their origins, insertions, and possibly via points and wrapping surfaces. The most direct way to determine moment arm magnitudes in this case is from the cross product of a vector from the joint center to the line of action of the muscle, with a unit vector directed along the muscle line of action.

Model-Environment Interactions

Often, the equations of motion for a musculoskeletal model will include, either directly or indirectly, forces exerted on the model by the environment. For example, a model used to simulate walking, running, or jumping must account for ground reaction forces. Likewise, to simulate pedaling, the forces at the pelvis-seat and foot-pedal interfaces must be included in the model. A simple and common approach to achieve this is to rigidly constrain the motion of one or more points on the body. Many 2-D models used to simulate jumping have the foot segment fixed to the ground using a one DOF pin joint (figure 11.6a). This provides free rotation about a point approximating the metatarsophalangeal joint (MTPJ) but limits translation of the foot in any direction and thus precludes using the model to simulate the airborne phase of jumping. This is not a major limitation, as the jump displacement can be determined from the model kinematics at the end of the propulsion phase (i.e., when the ground reaction force at the MTPJ drops to zero). The ground reaction forces generated during simulations of vertical jumping using this simple approach are usually in reasonable agreement with experimental data (e.g., Pandy et al. 1990). This *rigid constraint* approach has also seen frequent use in models for simulating pedaling, with the hip joint free to rotate but fixed to the seat (no translation), consistent with the minimal translation of the hip joint center in actual pedaling. The foot is modeled as being rigidly fixed to the pedal, so that it rotates about the pedal spindle and translates following a path prescribed by the bicycle crank arm and pedal kinematics. This reasonably assumes no slipping or twisting between the foot and pedal. Similar to the case in jumping, realistic pedal reaction forces can be generated using this approach (e.g., Neptune et al. 2000).

In contrast, the use of a rigid constraint approach to simulate the full stance phase in walking would be more complicated. Immediately following heel-strike, the foot segment could be pinned to the ground to allow only for rotation about the heel. As the rest of the foot makes contact with the ground, the entire foot would need to be rigidly fixed to the ground (no translation or rotation). Toward the end of the stance phase, when the heel rises, the foot could be pinned to the ground at the MTPJ, which would again allow only for rotation. Although this approach is relatively straightforward, the number of independent DOF in the model would change as the foot transitions from rotating about the heel to the being rigidly fixed to the ground and then again with the transition to rotating about the MTPJ. This would require either switching to a different model for each of these three cases (i.e., the equations of motion would change) or introducing and removing additional constraint equations across the stance phase. Double limb support would be even more complicated, because both feet would need to be constrained to the ground. Some logic would be required to determine when exactly to switch from one state to the next, such as when the foot changes from rotating about the heel to being fixed to the ground. So, although using rigid constraints can be simple and effective for representing model-environment interactions in some applications, they are not universally applicable.

Another relatively simple approach to account for model-environment interactions is to use measured forces as direct inputs to the model, assuming that such experimental data are available. Although this

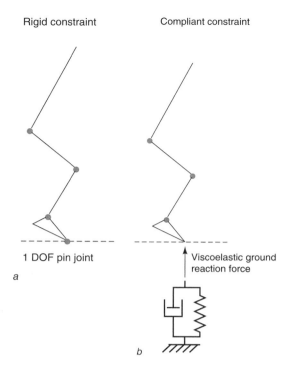

▲ **Figure 11.6** Models of the foot-ground interaction in jumping. One approach is to constrain the foot to the ground using a pin joint *(a)*. An alternative approach is to model the ground reaction force using a viscoelastic element *(b)*.

approach is straightforward, there are important limitations. Models never perfectly replicate the subjects they are meant to represent, and experimental data are never error-free. Thus, the experimental forces are likely to be dynamically inconsistent with the musculoskeletal model. Consider the simple case of having a model stand completely still, using ground reaction force data measured from a subject. If the mass of the subject was even slightly greater or less than the body mass in the model, then the model would not stand still but rather would experience a vertical acceleration. The same would be true if the masses of the subject and model were exactly the same but there were errors in the measured ground reaction forces. The nature and extent of these dynamic inconsistencies are hard to predict and will vary from one case to the next. Steps can be taken to address these issues (Thelen and Anderson 2006), but a detailed discussion is beyond the scope of this chapter.

Another popular approach for simulating ground reaction forces has been to model contact between the foot and ground as a viscoelastic interaction. Here, one or more contact points are defined on the plantar aspect of the foot segment in the model. Whenever any of these points attains a negative vertical position (i.e., below the zero ground level), a force is applied to the foot segment at the contact point (figure 11.6*b*). The force at each contact point is proportional to the amount by which the contact point has penetrated the ground and to the velocity of penetration. Although allowing the foot to penetrate the ground may seem inappropriate, this actually represents the deformation of the soft tissues and materials in the foot and shoe and typically is of a magnitude less than 1 cm. The viscoelastic approach to modeling ground contact represents a *soft,* or *compliant constraint* on the motion of the musculoskeletal system. An advantage of modeling ground contact using viscoelastic elements is that the foot is free to make and break contact with the ground during the course of a simulation such as in gait, thus elegantly solving the multiple rigid constraint problem described earlier. Although many different ground contact models exist in the literature, a representative example for computing the vertical force *(F$_v$)* is given by

$$F_v = ap^3(1 - b\dot{p}) \tag{11.5}$$

where p is the penetration distance into the ground of a contact point on the foot, $\dot{p}$ is the velocity of penetration, a is a stiffness parameter, and b is a damping parameter (Gerritsen et al. 1995). The forces computed in this manner can be summed over all of the contact points on the foot segment and compared with measured vertical ground reaction forces.

Horizontal forces (anterior-posterior, medial-lateral) can also be determined using a viscoelastic model. However, because horizontal forces represented frictional effects, they can just as effectively be determined using a model of Coulomb friction (i.e., dry friction). Because of the discontinuous nature of Coulomb's law between the static and dynamic states, a numerical approximation is usually required for use in computer simulations. One such approach (Song et al. 2001) for simulating horizontal ground reaction forces (F_h) is given by

$$F_h = -cF_v \ \text{htan}\frac{\dot{h}}{\gamma} \tag{11.6}$$

where c is the coefficient of friction, $\dot{h}$ is the horizontal sliding velocity of the ground contact point, and γ is a parameter that determines how closely the model approximates true Coulomb friction.

CONTROL MODELS

The human body is mechanically redundant in that the number of muscles at each joint exceeds the number of DOF. This raises questions: What rules concerning selection of muscle activations does the central nervous system (CNS) use? How should researchers formulate muscle model control signals in their musculoskeletal models? The answer to the first question is a matter of great interest to motor control scientists but is beyond the scope of this text. The answer to the second question is central to the use of musculoskeletal models, and in this section we discuss possible approaches.

The type of control invoked depends on the nature of the research question under investigation and the structure of the specific musculoskeletal model being used. In many cases, the goal is to supply the modeled muscles with patterns of stimulation that mirror the control signals that the CNS sends to individual muscles during an actual movement. Alternatively, the goal might be to supply the modeled muscles with control signals that produce an optimal movement (e.g., jump as high as possible) or fulfill some hypothesized movement goal (e.g., pedal with minimum energy). Thus, *control* refers to the process by which the muscle forces that produce the desired movement are determined. Depending on the specific application and model used, this process could involve solving for muscle stimulation or activation patterns. In other cases, the muscle forces are solved for directly, such as when a musculoskeletal model is used to distribute net joint moments obtained from an inverse dynamics analysis into individual muscle forces. The development and evaluation of algorithms for generating control signals for musculoskeletal models constitute an active area of research; most approaches can be grouped into three general categories: (1) models that use measured EMG signals, (2) theoretical neural models, and (3) optimization models. Hybrid models, combining facets of techniques from these three categories, are also plentiful in the literature.

Electromyographic Models

The most straightforward approach, it would seem, is to measure the "real" CNS control signals with EMG during a movement performance and then use these EMG signals as the model control signals. This approach has a number of advantages, not the least of which is the possibility to automatically account for the actual control strategies (i.e., actual patterns of muscle excitations) used by individual subjects. Using an EMG-based control model is also more computationally efficient than some other techniques, such as dynamic optimization (described later). However, using measured EMG signals as the basis for controlling a musculoskeletal model is not without difficulties. One limitation is that many deep muscles are impossible to monitor with surface electrodes. Indwelling electrodes can be used, but they sample from a limited volume and may not provide an accurate assessment of the activity level of the muscle as a whole. Furthermore, indwelling electrodes are invasive and may interfere with a subject's ability to perform the movement. Even with surface electrodes, technical considerations make it questionable whether the measured signal truly represents the CNS control signal (see chapter 8).

Another issue with the use of EMG as the basis for control signals is that musculoskeletal models often simplify reality and only explicitly represent some muscles. For example, 2-D models of human

Bobbert, M.F., P.A. Huijing, and G.J. van Ingen Schenau. 1986a. A model of the human triceps surae muscle-tendon complex applied to jumping. *Journal of Biomechanics* 19:887-98.

Studies of vertical jumping have used muscle models to examine the roles played by individual muscles during the jumping movement. Bobbert and colleagues (1986a) used Hill-type models to represent the soleus (SO) and gastrocnemius (GA) muscles during vertical jumping (figure 11.7). Their study examined the reasons why the maximal ankle plantar flexion velocity from isovelocity dynamometer studies is much lower than the velocities seen during natural movements such as jumping. The SO and GA muscle models included force-length, force-velocity, and series elastic characteristics. Kinematics from actual jumping performances were imposed on the SO and GA models, with muscle lengths and moment arms at the ankle calculated using equations based on cadaver studies. The SO and GA muscle force predictions were then used to estimate the plantar flexor moment at the ankle joint throughout the push-off phase prior to takeoff. The model ankle moment demonstrated good agreement with the moment calculated using inverse dynamics from the actual jumping performances. Analysis of the SO and GA muscle forces and contractile kinematics uncovered two reasons the ankle muscles could produce larger plantar flexor torques at velocities much higher than in dynamometer studies. The first was that the series elasticity of the muscles kept the contractile component shortening velocities much lower than the measured ankle velocity. Second, the GA is a biarticular muscle that acts as both a knee flexor and an ankle plantar flexor. In jumping, simultaneous ankle plantar flexion and knee extension result in a much lower GA shortening velocity than had occurred in dynamometer studies of isolated plantar flexion. This study demonstrates clearly that musculoskeletal modeling can lead to a greater appreciation of muscle function than that provided by inverse dynamics analysis alone.

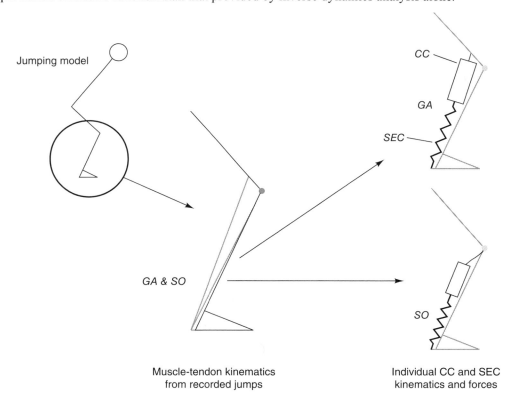

Figure 11.7 Schematic of the model developed by Bobbert and colleagues (1986a) to study how the gastrocnemius (GA) and soleus (SO) muscles contribute to the plantar flexion moment in vertical jumping. Each muscle is represented by a two-component Hill model consisting of a contractile component (CC) and a series elastic component (SEC).

jumping commonly use fewer than 10 individual muscle models (figure 11.8) to represent the muscular capabilities of the entire lower limb, when in fact there are more than 40 such muscles. Consequently, a single model control signal might need to represent the activity of several actual muscles, and it is unclear how measured EMG signals should be combined to form a control signal for one muscle model. For example, if all three heads of the vasti group are represented with a single Hill-type muscle model, should the EMG-based control signal be drawn from vastus lateralis, vastus medialis, vastus intermedius (which would require indwelling electrodes), or some weighted combination of all three? Finally, raw EMG signals have high-frequency content and must be preprocessed to form control signals for Hill-type muscle models (figure 11.8). This process must transform the EMG signal—a manifestation of CNS motor unit recruitment and rate coding—into a simplified, low-frequency control signal that varies from 0.0 (inactivity) to 1.0 (full activity). This transformation is relatively complex and has been the topic of much discussion in the literature (e.g., Buchanan et al. 2004). Even with appropriate processing, there is still the issue that a musculoskeletal model is not a perfect representation of reality; therefore, processed EMG signals may not represent ideal model control signals.

Despite these many issues, the potential advantages of the EMG-based approach have encouraged numerous researchers to tackle the challenges associated with using EMG as the basis for musculoskeletal model control signals. Hof and van den Berg (1981) pioneered the use of EMG signals as inputs to Hill-type muscle models for the prediction of muscle forces in a simple 2-D model of ankle plantar flexion. More recent examples of EMG-driven models incorporating greater numbers of muscles include a 3-D model of the lumbar spine (Cholewicki and McGill 1996) for predicting low back stability and

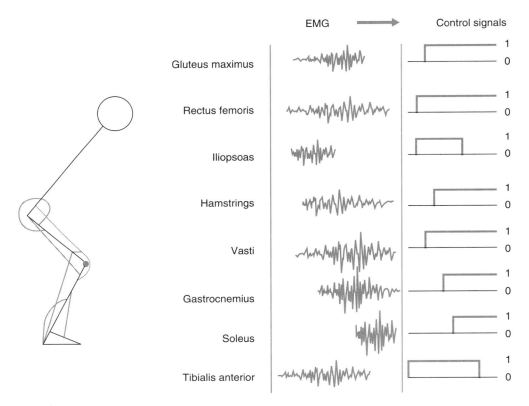

▲ **Figure 11.8** Vertical jumping model with eight musculoskeletal actuators. EMG can be recorded from each muscle as the basis for control signals to activate the eight muscle model actuators. The EMG signals are used to determine the onset time when each muscle actuator control signal changes from 0 (rest) to 1 (full excitation). In some cases, the cessation of EMG indicates when the control signal should be "turned off" and return to 0. In this example of maximal vertical jumping, the control signals can be only at rest or fully active, but in submaximal movements such as walking the EMG signals can be used to generate control signals that vary incrementally between 0 and 1.

a 3-D model of the knee joint (Winby et al. 2009) for predicting joint contact loads. It is expected that this approach will continue to see common use.

Theoretical Control Models

Another approach to finding appropriate control signals for musculoskeletal models is to use a neural control model that is based on a theoretical understanding of how the CNS controls and coordinates movement. These models might be based on first principles (i.e., they might attempt to predict the signals that the CNS would send to each muscle by modeling brain and spinal neuronal action) or on some motor control principle (e.g., a general principle of proximal-to-distal sequencing in lower-extremity extensions). Another possibility is to predict stimulation to muscle models based on some hypothetical control basis. For example, Pierrynowski and Morrison (1985) generated lower-extremity muscle stimulation patterns during walking by considering which muscles were in the best anatomical position to create the required joint torques in 3-D at each joint. Similarly, Caldwell and Chapman (1991) used a neural model with control "rules" based on EMG measurements to predict forces in the individual muscles controlling the elbow joint. This differed from the approaches described in the last section, in that EMG signals were not transformed directly into control signals for the muscle models but rather served as the basis for developing control rules. An alternative approach was taken by Taga (1995), who used a model of a spinal central pattern generator with proprioceptive feedback to control a bipedal walking model. Some distinct advantages to this more realistic representation of CNS control were that the simulated gaits were resistant to perturbations applied to the body, and locomotor speed could be varied by changing a single, nonspecific input to the central pattern generator.

Although there are advantages to directly including aspects of the neural system in the control of musculoskeletal models, one challenge is that in general, the manner in which the brain formulates control signals is still unknown. How can we check that the output signals from a neural model are reasonable? One method is to compare the timing (onset and offset) and patterns of predicted control signals with measured EMG signals, but as we have mentioned, the use of EMG can be problematic and requires caution in interpreting these signals. Despite the problem of confirming whether the control model predicts the "correct" signals, this theoretical approach is potentially powerful because it allows testing for the consequences of different putative control strategies. For example, a model of arm motion that is driven by an open-loop control mechanism could be built and compared with a model that relies on proprioceptive feedback to guide the muscle stimulation patterns. Comparisons between the two models might provide insight into the role of peripheral feedback in a specific motor task.

Optimization Models

A popular method for generating control signals in musculoskeletal models is to use some form of optimization. In some cases, when the optimization criterion is derived from motor control principles, this is tantamount to developing a theoretical control model. In other cases, the optimization criterion may result in a coordinated movement pattern yet have little physiological basis. In general, the optimization approach is used in an effort to determine which set of model control signals will produce a result that optimizes (minimizes or maximizes) a given criterion measure. The criterion measure is known as a *cost function, objective function,* or *performance criterion.* The cost function can be relatively simple (e.g., find the solution that yields minimal muscle force) or complex (e.g., determine a nonlinear combination of maximal muscle force at minimal metabolic cost). It can be directly related to muscle function (e.g., minimize muscular work) or to some aspect of the motion under study (e.g., maximize vertical jump height). Alternatively, the cost function may be formulated to minimize the differences between model outputs (e.g., joint angles, ground reaction forces) and corresponding experimental measures. This latter case is often referred to as a *tracking problem* in that the goal is to find a solution that causes the model to follow, or track, the experimental data.

In all cases, the cost function serves as a guiding constraint that determines the selection of one particular set of optimal muscular controls from among many different possible solutions. Two major optimization approaches that have been used to control musculoskeletal models are referred to in the literature as *static optimization* and *dynamic optimization.* The meaning of *static* in this context is that the cost function is evaluated at each time step during a movement, independent of any prior or subsequent time steps. In contrast, dynamic optimization is *dynamic* in the sense that the entire movement sequence must be simulated to determine the value of the cost function.

Static Optimization

Initial attempts to predict individual muscle forces used static optimization models in conjunction with inverse dynamics analysis of an actual motor performance. The inverse dynamics analysis permits the calculation of net joint moments at incremental times throughout the movement (see chapter 5). In most applications, numerical optimization is used to find the set of muscle forces that balances these joint moments while also satisfying a selected cost function (figure 11.9). Thus, with this approach the muscle forces themselves, rather than neural signals, are the controls. The time-independent nature of static optimization allows solutions to be obtained with relatively little computational cost, but there are some drawbacks. One issue in early applications was sudden, nonphysiological switching on and off of muscle forces caused by the independence of solutions for sequential time increments (i.e., the optimization model balanced the joint moments separately for each time interval using very different sets of muscles). This problem can be avoided by careful selection of the initial guess in the optimization problem, such as by using the solution from one time step as the initial guess for the next time step. A stronger approach to address this issue is to use muscle models that invoke physiological realism through force-length and force-velocity relations and time-dependent stimulation-activation dynamics (chapter 9). When more detailed muscle models are included in a static optimization model, muscle activations become the control variables, rather than muscle forces.

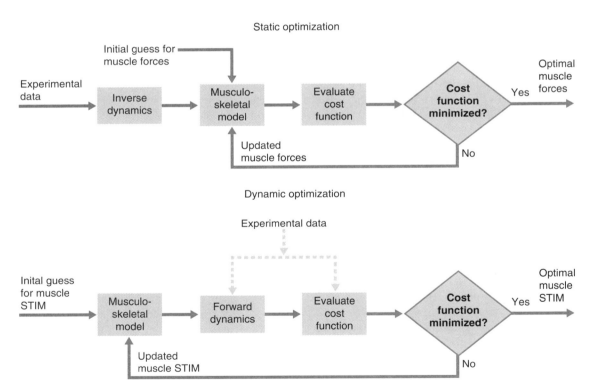

▲ **Figure 11.9**   Overview of static and dynamic optimization approaches. Static optimization usually results in predictions of optimal muscle forces. Dynamic optimization usually results in predictions of optimal muscle control signals (STIM). Note that experimental data are the primary input with static optimization but are not required with dynamic optimization. However, experimental data are frequently used with dynamic optimization (dashed lines) to define initial conditions for forward dynamic simulations and in the case of a tracking problem are used to evaluate the cost function (see text for more detail).

Early studies using static optimization were plagued also by two other kinds of nonphysiological results. The first was the prediction of model forces that were too high for actual muscles to produce. This difficulty was easily addressed by defining physiologically valid maximal force constraints for each muscle in the musculoskeletal model. The second problem was that solutions would often select only one muscle to balance the net joint moment rather than choose a more realistic muscular synergy. Depending on the exact formulation, the muscle with the largest moment arm (if minimizing muscle force) or a favorable combination of moment arm and muscle strength (if minimizing muscle stress) would be selected to fully balance the joint moment, with zero force predicted in other synergist muscles. Mathematically, synergism can be produced by using nonlinear cost functions (e.g., minimizing the sum of squared or cubed muscle forces), although the physiological rationale for specific nonlinear cost functions is not always clear. A widely used nonlinear cost function proposed by Crowninshield and Brand (1981a) involves minimizing the sum of cubed muscle stresses. It was originally argued that this particular cost function would lead to solutions that maximize muscle endurance, making it appropriate for predicting muscle forces in submaximal tasks such as walking. However, the Crowninshield and Brand criterion has since been used to solve for muscle forces in a range of activities, some of which are unlikely candidates for maximizing muscle endurance (e.g., jumping, landing).

An additional issue with static optimization models is that they are essentially a decomposition of experimental joint moments into individual muscle forces. Therefore, the predicted muscle forces will be subject to any errors in the experimental joint moments in addition to any shortcomings associated with the approximation made in creating the musculoskeletal model. To complete this section, we present a static optimization example motivated by a classic review article by Crowninshield and Brand (1981b), which demonstrates the process of distributing an empirically determined elbow joint moment across a set of muscles spanning the elbow joint.

The net joint moment to be balanced in example 11.1 was a flexor moment, and only muscles with flexor moment arms were included in the model (figure 11.10). If an elbow extensor muscle had been included, there is no cost function of the type presented here that would predict force in an antagonist muscle. Any force in an elbow extensor would itself contribute to a higher cost function value and would also require greater elbow flexor forces to balance the target 10 N·m joint moment, further increasing the cost function value. The total lack of coactivation in this example is contrary to the common observation that during heavy activation of the elbow flexors, there is some activity in the triceps muscle

EXAMPLE 11.1

The human upper limb is depicted in figure 11.10, with the arm and forearm segments articulating at a one DOF elbow joint, which is crossed by three muscles representing the brachialis, brachioradialis, and biceps brachii. Numerical values for individual muscle parameters used in this example are provided in table 11.1. We will assume that an inverse dynamics analysis revealed a net elbow flexor moment $(\vec{M}^j)$ of 10 N·m. Static optimization will be used to solve for the forces in the three muscles that gave rise to this measured joint moment. To solve this otherwise indeterminate problem, we will seek the combination of muscle forces that minimize the sum of the cubed muscle stresses.

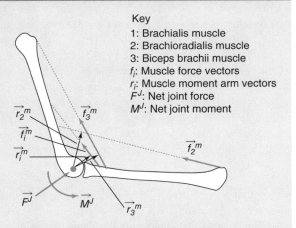

Key
1: Brachialis muscle
2: Brachioradialis muscle
3: Biceps brachii muscle
f_i: Muscle force vectors
r_i: Muscle moment arm vectors
F^J: Net joint force
M^J: Net joint moment

▲ **Figure 11.10** A simplified one DOF model of the human elbow joint with three flexor muscles. See text and table 11.1 for additional details. (Adapted from Crowninshield and Brand 1981b.)

(continued)

EXAMPLE 11.1 *(CONTINUED)*

Table 11.1

	Muscle 1 Brachialis	Muscle 2 Brachioradialis	Muscle 3 Biceps brachii
f^0 (N)	1000	250	700
A^m (m²)	0.0033	0.0008	0.0023
r^m (m)	0.02	0.05	0.04

f^0 = maximum muscle force; A^m = physiological cross-sectional area; r^m = muscle moment arm.

In this musculoskeletal system, the net joint *(J)* forces and moments determined using inverse dynamics are related to the internal muscle *(m)*, ligament *(l)*, and articular contact *(c)* forces *(f)* by the following equations:

$$\vec{F}^J = \sum_{i=1}^{m} \vec{f}_i^m + \sum_{i=1}^{l} \vec{f}_i^l + \sum_{i=1}^{c} \vec{f}_i^c \tag{11.7}$$

$$\vec{M}^J = \sum_{i=1}^{m} \left(\vec{r}_i^m \times \vec{f}_i^m \right) + \sum_{i=1}^{l} \left(\vec{r}_i^l \times \vec{f}_i^l \right) + \sum_{i=1}^{c} \left(\vec{r}_i^c \times \vec{f}_i^c \right) \tag{11.8}$$

which indicate that the net joint force $(\vec{F}^J)$ from inverse dynamics is equal to the vector sum of the muscle, ligament, and articular contact forces (equation 11.7), whereas the net joint moment $(\vec{M}^J)$ is equal to the vector sum of the moments generated by muscle, ligament, and articular contact forces (equation 11.8). If we make the common simplifying assumptions that (1) ligament forces are small enough to be ignored when the joint is in the middle of its range of motion and (2) the articular contact forces pass through the joint center of rotation, then equations 11.7 and 11.8 reduce to

$$\vec{F}^J = \sum_{i=1}^{m} \vec{f}_i^m + \sum_{i=1}^{c} \vec{f}_i^c \tag{11.9}$$

$$\vec{M}^J = \sum_{i=1}^{m} \left(\vec{r}_i^m \times \vec{f}_i^m \right) \tag{11.10}$$

Equation 11.9 provides a basis for determining joint contact loads, if they are of interest in a particular application, but we will not discuss this further here. In solving the static optimization problem, we will use equation 11.10 as a constraint to ensure that the muscle forces we determine via static optimization reproduce the measured joint moment. In this context, equation 11.10 is known as an *equality constraint* and must be satisfied during the solution process. It is also common to place *boundary constraints* on the muscle forces, such that muscles only generate tensile forces and do not exceed a designated maximum value (f_i^0). The upper bound for each muscle is usually determined by multiplying muscle physiological cross-sectional area (A^m) by an assumed value for muscle-specific tension. The boundary constraints can be expressed as

$$0 \le f_i^m \le f_i^0 \tag{11.11}$$

Note that there are an infinite number of combinations of muscle forces $\vec{f}_i^m$ that will satisfy equation 11.10 for a particular net joint moment. For example, given the moment arm (r^m) values listed in table 11.1, the following three potential candidate solutions $(f_1^m = 500 \text{ N}, f_2^m = 0 \text{ N}, f_3^m = 0 \text{ N})$, $(f_1^m = 0 \text{ N}, f_2^m = 200 \text{ N}, f_3^m = 0 \text{ N})$, and $(f_1^m = 0 \text{ N}, f_2^m = 0 \text{ N}, f_3^m = 250 \text{ N})$ all balance the net moment of 10 N·m. However, none of these three candidate solutions would appear to be physiologically reasonable, as only one of three synergistic muscles is selected. To achieve a distribution of forces among the three muscles, we will seek the solution that reproduces the

10 N·m joint moment while simultaneously minimizing the nonlinear function U, which can be expressed as

$$\text{Minimize } U = \sum_{i=1}^{3} \left(\frac{f_i^m}{A_i^m} \right)^3 \tag{11.12}$$

Equation 11.12 will serve as the cost function in the optimization problem. The quotient inside of the parentheses represents the individual muscle stresses, which are raised to the third power. If we assume that muscle forces are nonnegative, then the solution to equation 11.12 alone is ($f_1^m = 0$ N, $f_2^m = 0$ N, $f_3^m = 0$ N). However, if we require that the solution simultaneously satisfy equations 11.10, 11.11, and 11.12, then muscle forces greater than zero will be predicted in all muscles when the joint moment is non-zero.

Despite the relative simplicity of this example, obtaining an analytical solution by hand is quite challenging. Fortunately, any number of general purpose optimization algorithms can be used to obtain a numerical solution to equation 11.12, subject to the constraints expressed in equations 11.10 and 11.11. For this example, a commonly used technique known as sequential quadratic programming was used. The solution that was obtained was

$$f_1^m = 160.38 \text{ N}$$

$$f_2^m = 30.27 \text{ N}$$

$$f_3^m = 131.97 \text{ N}$$

The reader can easily confirm that this solution balances the measured joint moment and none of the individual muscle forces fall outside of the specified bounds. The reader can also verify, if not exhaustively, that other solutions that satisfy the constraints, such as the three potential candidate solutions mentioned previously, result in greater values of the cost function U.

A common variant of the approach presented here has been to minimize the sum of squared, rather than cubed, muscle stresses. Solving this quadratic, rather than cubic, problem results in a qualitatively similar, but numerically different solution, which for the present example is ($f_1^m = 151.04$ N, $f_2^m = 22.19$ N, $f_3^m = 146.74$ N). If, however, the exponent in equation 11.12 is set to 1 (i.e., a linear cost function), then the solution is ($f_1^m = 0$ N, $f_2^m = 0$ N, $f_3^m = 250$ N), which is one of the unrealistic candidate solutions mentioned previously. In the quadratic case, the muscle forces are distributed across all three muscles but in a different manner than with the cubic cost function. In the linear case, the moment is borne entirely by the biceps brachii, which has the most favorable combination of moment arm and A^m. Recall that the rationale for using a cubic cost function was that it was supposed to result in a distribution of muscle forces that maximize endurance (Crowninshield and Brand 1981a). Although the quadratic and cubic solutions may appear to be reasonable, it is important to recall that the opportunities to validate muscle forces predicted via static optimization, or any other technique, have thus far been quite limited (e.g., Prilutsky et al. 1997).

group. The extensor coactivation likely helps stabilize the joint during heavy exertion but will not be predicted using traditional static optimization techniques in simple one DOF models. Although the example presented here focuses on obtaining numerical results for muscle forces, the interested reader is referred to the review by Crowninshield and Brand (1981b), which presents an interesting graphical interpretation of the solution to this static optimization problem.

Dynamic Optimization

Ongoing work with static optimization models has led to the development of dynamic optimization or optimal control models, which are applied in conjunction with forward dynamics models of human motion. In contrast to the inverse dynamics approach, which uses experimental data from an actual performance to calculate net joint moments, forward dynamics analyses simulate the motion of the body from a given set of joint moments or muscle forces (see chapter 10). Dynamic optimization models are therefore often designed to find the muscle stimulation patterns that result in an optimal motion (figure 11.9). As mentioned earlier, the optimal motion may be one that maximizes (or minimizes) a performance criterion, such as maximizing jump height; or, the optimal motion may be defined as one that best reproduces a set of experimental data. Regardless, the variables that are optimized are usually the muscle stimulation patterns that control the motion of the musculoskeletal model. These stimulation patterns are used as the inputs to muscle models that predict individual muscle forces, which are then multiplied by the appropriate moment arms to compute the active muscle moments. The active muscle moments are combined with passive moments to compute the net joint moments, which actuate the skeletal model and produce movement. Thus, dynamic optimization leads to the synthesis of body segment kinematics associated with optimal performance, which is fundamentally different from the static optimization approach.

The optimal kinematics predicted from a dynamic optimization can be compared with experimental motion data; however, the experimental data are not required to obtain the solution, as they are with static optimization. This feature of dynamic optimization permits one to address questions for which no experimental data exist, such as testing the feasibility of possible forms of locomotion in extinct species (e.g., Nagano et al. 2005). Moreover, because dynamic optimization uses forward dynamics models that simulate the whole movement performance and provide complete muscle force time histories (rather than solutions at independent time increments), many of the problems associated with static optimization models are overcome. However, these advantages come at a computational cost, because dynamic optimization often requires a more detailed musculoskeletal model and the entire movement must be simulated for each possible solution. Thus, solving a dynamic optimization problem can easily require an order of magnitude more time than is required to solve a comparable static optimization problem. Although computational costs are difficult to compare in an objective manner, it is not uncommon for static optimization solutions to take seconds or minutes to obtain, whereas dynamic optimization solutions can take hours, days, or even weeks, when run on standard computer workstations.

In many cases, users of dynamic optimization must define the goal of the movement mathematically. This definition is easiest for movements in which the performance criterion to be optimized is clear in a mechanical sense. In vertical jumping, for example, the cost function can be stated as maximization of the vertical displacement of the body center of mass during the flight phase. If the model is constructed with appropriate constraints (realistic muscle properties and joint range of motion), jump heights and body segment motions approximating those of human jumpers can be attained (Pandy and Zajac 1991; van Soest et al. 1993). However, for many human movements the performance criterion is not as clear. In walking, for instance, the goal may seem to be getting from point A to point B in a certain amount of time. Unfortunately, this does not provide enough information to solve for a set of muscle stimulation patterns that will result in realistic walking. Successful simulations of walking have been generated using a cost function that minimizes the metabolic cost of transport (Anderson and Pandy 2001; Umberger 2010); however, the cost function formulation is considerably more complicated than for vertical jumping. Minimum-energy solutions also require the inclusion of an additional model for predicting the metabolic cost of muscle actions (e.g., Umberger et al. 2003). In cases where the performance criterion is not clear, dynamic optimization can serve as an effective approach for testing various theoretical criteria to see how well each can produce the desired movement patterns.

Movements for which the underlying criterion is difficult to define, such as walking or pedaling, have often been simulated by formulating and solving a tracking problem. This approach has a similar appeal to the EMG-based techniques described earlier, in that the resulting motion should closely approximate the way in which humans actually move. However, it is unclear which of several possible experimental measures (kinematics, kinetics, EMG) are the most important for the model to track. Also, because of errors in the experimental data and differences between the musculoskeletal model and the experimental subjects, it may be impossible for the model to track the data perfectly. This approach also limits much of the predictive ability of the dynamic optimization approach, as it is only possible to consider conditions for which experimental data are available.

The tracking approach has seen frequent use in hybrid solution algorithms, which seek to retain the advantages of forward dynamics and dynamic optimization while achieving the computational efficiency of static optimization. Two examples are computed muscle control (Thelen et al. 2003) and direct collocation (Kaplan and Heegaard 2001). Computed muscle control works by solving a static optimization problem at each time step within a single forward dynamics movement simulation. By using feedback on the kinematics and muscle activations at each time step of the numerical integration, computed muscle control can generate a forward dynamics simulation that optimally tracks a set of experimental data without solving a dynamic optimization problem that would require thousands of forward dynamics simulations. In contrast, direct collocation works by converting the differential equations of motion into a set of algebraic constraint equations and treating both the control variables (muscle stimulations) and state variables (positions and velocities) as unknowns in the optimization problem. Although they differ considerably in implementation, both techniques appear to be able to produce results that are similar to dynamic optimization tracking solutions, with a computational cost closer to that of static optimization.

FROM THE SCIENTIFIC LITERATURE

Pandy, M.G., and F.E. Zajac. 1991. Optimal muscular coordination strategies for jumping. *Journal of Biomechanics* 24:1-10.

van Soest, A.J., A.L. Schwab, M.F. Bobbert, and G.J. van Ingen Schenau. 1993. The influence of the biarticularity of the gastrocnemius muscle on vertical-jumping achievement. *Journal of Biomechanics* 26:1-8.

Earlier, we described the use of muscle models to understand the isolated function of the plantar flexor muscles in vertical jumping (Bobbert et al. 1986a). Subsequent modeling studies of vertical jumping investigated the overall segmental motion and coordination of muscles, which required a more complete representation of the jumper. Both Pandy and Zajac (1991) and van Soest and colleagues (1993) presented dynamic optimization studies of simulated jumping in which the segmental motion was driven by Hill-type models representing specific individual muscles (figure 11.11). In contrast to models that use independent joint moment actuators, individual muscle models can be used to uncover the effects of biarticular muscles that contribute to moments at adjacent joints. The optimization problem in both studies was to find the muscle stimulation patterns that produced the maximal jump height.

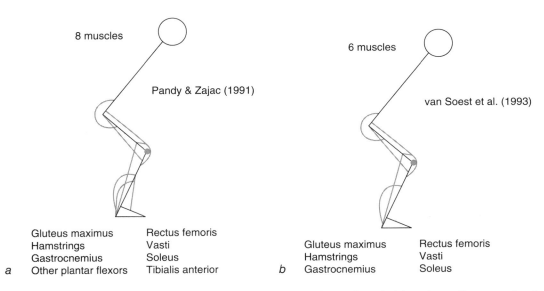

▲ **Figure 11.11** Vertical jumping models of Pandy and Zajac (1991) *(a)* and van Soest and colleagues (1993) *(b)*. The Pandy and Zajac model included eight muscle models, whereas the model by van Soest and colleagues included only six muscle models, with the tibialis anterior and the "other plantar flexor" muscles excluded.

(continued)

(continued)

Although both models produced realistic jump heights and segmental kinematics, their interpretations of results were quite different. Van Soest and colleagues (1993) used six muscle models to provide motion, representing both mono- and biarticular muscles of the lower extremities. They emphasized the contributions to jumping performance made by individual muscles, giving particular emphasis to the ability of biarticular muscles to transport energy between the adjacent joints that they span. The investigators considered the energy transport mechanism to be an important feature that helps determine the overall muscular coordination during the jump and contributes substantially to a proximal-to-distal energy flow.

In contrast, Pandy and Zajac (1991) characterized the energy flow as distal-to-proximal and identified the contributions of eight individual muscle models to the increased energy of the body's center of mass. These investigators discounted the importance of biarticular energy transport, and to emphasize this point they performed "virtual surgery" by moving the origin of the normally biarticular GA from the distal femur to the proximal tibia, limiting the GA model action to ankle plantar flexion alone. Simulations with the now monoarticular GA demonstrated only slight reductions in the model's jump-height ability, casting doubt on the importance of its biarticular nature. It was subsequently suggested that the discrepancies between the van Soest and Pandy models might be the result of differences in the GA knee flexion moment arm relations used by each group, with the Pandy model underestimating the moment arm, particularly at highly extended knee angles. This underestimation would limit the ability of their GA model to participate in the energy transport mechanism, and therefore the switch to a monoarticular GA would have relatively little effect.

Examination of the similarities and differences in results obtained using these two jumping models illustrates several important modeling issues. First, both models demonstrated results for individual muscle use during jumping that could not be obtained with experimental techniques—one of the main strengths of this musculoskeletal modeling approach. Second, both groups used their models to investigate questions about movement coordination rather than to merely describe muscle-specific results. Therefore, they both increased our understanding of the biomechanics of human jumping. However, the discrepancies between the two models highlight the importance of musculoskeletal model development. Should the lower extremity be represented with six muscles or eight? What are the consequences of the model parameters chosen, such as muscle moment arm relations at a specific joint? There is a degree of uncertainty in model parameters used to represent any given subject, and it behooves researchers to investigate the sensitivity of their model results to errors in these parameters.

Although static and dynamic optimization have both become popular means for controlling musculoskeletal models, several issues concerning their use must be addressed. Do humans actually produce movements based on one given performance criterion, or does the performance objective change during the movement? If the optimized model results differ from those of a human performer, is it because the model is too simple or not constrained properly, or is the human not performing optimally? Finally, EMG data have illustrated various degrees of simultaneous antagonistic cocontraction during some movement sequences. Optimization models tend not to yield solutions predicting muscle antagonism around a joint, although some antagonism is predicted when models contain muscles that contribute to more than one joint moment. However, optimization models that seek to minimize or maximize specific cost functions will not predict antagonistic cocontraction associated with joint stability or stiffness. In the case of walking, for example, there is no single cost function that could simultaneously predict the relatively minimal muscle coactivation in healthy subjects and the substantial antagonism exhibited by patients with cerebral palsy. Despite these drawbacks, optimization models have increased the understanding of human biomechanics and will continue to do so in the future.

ANALYSIS TECHNIQUES

After creating a musculoskeletal model and simulating a movement of interest, we can use the model to investigate many interesting biomechanical questions beyond what can be inferred using traditional experimental procedures alone. Sometimes, the process of generating the simulated movement may directly yield the desired insight by providing estimates of muscle or joint contact forces. In other cases, it may be necessary to vary some aspect of the model or perform additional analysis in order to answer the research question under consideration. The literature contains many examples of analysis techniques that can be performed using musculoskeletal models; here we provide three examples: induced acceleration, segmental power, and sensitivity analyses. We then conclude this section with a discussion of model validation.

Induced Acceleration

Induced acceleration techniques, which were introduced in chapter 7, have seen frequent use with musculoskeletal models. This is a potentially powerful approach, as it provides insight into how each specific muscle force (or any other force in the model) contributes to the acceleration of any joint or segment. Such techniques have helped illuminate some fundamental aspects of movement, such as the concept that muscles may generate accelerations at joints they do not cross. For example, in a standing posture, force generated by the soleus muscle will accelerate not only the ankle joint that it crosses but also the knee and hip joints, which it does not cross. The accelerations produced in distant joints can potentially be greater than in the joint crossed by the muscle.

There have been other, counterintuitive findings as well, such as the potential for the biarticular gastrocnemius to accelerate the knee into extension or the ankle into dorsiflexion, even though this muscle generates knee flexor and ankle plantar flexor moments (Zajac and Gordon 1989). The basis for these phenomena lies in the interactions or dynamic coupling of the segments that comprise the model of the body. For example, activating the soleus results in joint reaction forces at the ankle, knee, and hip, causing all three joints to experience angular accelerations. Investigating dynamic coupling does not necessarily require one to perform a forward simulation, but it does require the derivation of the equations of motion for the entire system under study, as discussed in chapter 10 (see also appendix G). Therefore, it is not possible to perform an induced acceleration analysis using the segment-by-segment inverse dynamics approach that was covered in chapter 5.

Although firmly grounded in the dynamic equations of motion, induced acceleration analyses can generate results that are difficult to verify. It has also been shown that the results of induced acceleration analyses are sensitive to model characteristics such as the number of segments and DOF (Chen 2006). However, there have been attempts to use functional electrical stimulation in able-bodied subjects as a way to empirically test the results of induced acceleration modeling studies. Although this line of validation is still in its infancy, the accelerations generated by isolated electrically stimulated muscles tend to support the muscle function predictions made with the induced acceleration approach (e.g., Stewart et al. 2007).

The basic components of induced acceleration analysis are best illustrated by an example; here we demonstrate how muscle-induced accelerations are computed and interpreted in a simple model of leg swing in walking.

EXAMPLE 11.2

The human lower limb has frequently been modeled as a double pendulum. Figure 11.12 depicts the lower limb with thigh and shank segments, free to rotate about one DOF joints at the hip and knee, respectively. The pelvis is fixed in space, and the mass of the foot is included in the shank segment. This is very similar to the double pendulum model derived in appendix G but differs in that the hip position is fixed and the generalized coordinates are the joint angles. The model is actuated by a uniarticular hip flexor muscle representing iliopsoas, and a biarticular hip flexor–knee extensor muscle representing rectus femoris. The notation and numerical values used for this example are provided in table 11.2. The subscripts T, S, H, and K stand for thigh, shank, hip, and knee, respectively. Hip flexion (θ_H) is positive and knee flexion (θ_K) is negative.

(continued)

EXAMPLE 11.2 *(CONTINUED)*

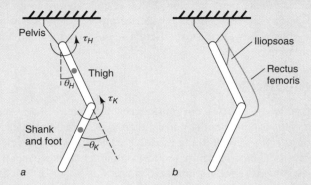

▲ **Figure 11.12** Simple model of the human lower limb during the swing phase of gait.

Table 11.2

	Thigh (T) or hip (H)	Shank (S) or knee (K)
L (m)	0.41	0.44
ρ (m)	0.18	0.21
m (kg)	8.7	4.7
I (kg m^2)	0.135	0.120
θ (°)	20	−50
τ (N·m)	Varies, see text.	Varies, see text.

L = segment length; ρ = distance from proximal joint to segment center of mass; m = segment mass; I = segment moment of inertia about center of mass; θ = joint angle; τ = joint moment.

The equations of motion for this system can be written in matrix-vector form as

$$M\ddot{\theta} = V + G + T \tag{11.13}$$

where M is the mass or inertia matrix, $\ddot{\theta}$ is a vector of joint angular accelerations, V is a vector of velocity-dependent moments, G is a vector of gravitational moments, and T is a vector of muscle moments. The equations of motion in explicit form are

$$\begin{bmatrix} -I_S - I_T - m_T\rho_T^2 - m_S\left(L_T^2 + \rho_S^2 + 2L_T\rho_S\cos\theta_K\right) & -I_S - m_S\rho_S\left(\rho_S + L_T\cos\theta_K\right) \\ -I_S - m_S\rho_S\left(\rho_S + L_T\cos\theta_K\right) & -I_S - m_S\rho_S^2 \end{bmatrix}\begin{bmatrix} \ddot{\theta}_H \\ \ddot{\theta}_K \end{bmatrix} =$$

$$\begin{bmatrix} L_T m_S\rho_S\sin\theta_K\left[\dot{\theta}_H^2 - \left(\dot{\theta}_H + \dot{\theta}_K\right)^2\right] \\ L_T m_S\rho_S\left(\sin\theta_K\right)\dot{\theta}_H^2 \end{bmatrix} + \begin{bmatrix} m_T g\rho_T\sin\theta_H + m_S g\left[L_T\sin\theta_H + \rho_S\sin\left(\theta_H + \theta_K\right)\right] \\ m_S g\rho_S\sin\left(\theta_H + \theta_K\right) \end{bmatrix} - \begin{bmatrix} \tau_H \\ \tau_K \end{bmatrix}$$

$$\tag{11.14}$$

where g is the acceleration due to gravity, and the symbols with overdots represent joint angular velocities $(\dot{\theta})$ and accelerations $(\ddot{\theta})$. Note that the values of most of the elements of the mass matrix are not constant but rather depend on θ_K. The mass matrix is a function of θ_K because the knee joint angle affects the mass distribution of the system relative to the suspension point at the hip. The contributions to the joint angular accelerations of any of the terms in the vectors on the right-hand side of equation 11.13 can be determined by matrix multiplying the vector by the inverse of the mass matrix. For example, the hip and knee joint angular accelerations induced by the muscle moments can be computed as

$$\ddot{\theta}_\tau = M^{-1} T \tag{11.15}$$

The inverse of the mass matrix is

$$M^{-1} = \begin{bmatrix} \dfrac{I_S + m_S \rho_S^2}{D} & \dfrac{-\left[I_S + m_S \rho_S \left(\rho_S + L_T \cos\theta_K\right)\right]}{D} \\ \dfrac{-\left[I_S + m_S \rho_S \left(\rho_S + L_T \cos\theta_K\right)\right]}{D} & \dfrac{I_S + I_T + m_T \rho_T^2 + m_S \left(L_T^2 + \rho_S^2 + 2L_T \rho_S \cos\theta_K\right)}{D} \end{bmatrix} \tag{11.16}$$

where D is defined as

$$D = \left(I_S + m_S \rho_S \left(\rho_S + L_T \cos\theta_K\right)\right)^2 - \left(I_S + m_S \rho_S^2\right)\left(I_S + I_T + m_T \rho_T^2 + m_S \left(L_T^2 + \rho_S^2 + 2L_T \rho_S \cos\theta_K\right)\right)$$

If the matrix multiplication in equation 11.15 is carried out, then the exact expressions for the hip and knee joint angular accelerations induced by the muscle moments are

$$\ddot{\theta}_{H\tau} = -M_{11}^{-1} \tau_H - M_{12}^{-1} \tau_K$$
$$\ddot{\theta}_{K\tau} = -M_{21}^{-1} \tau_H - M_{22}^{-1} \tau_K \tag{11.17}$$

where M_{ij}^{-1} indicates the entry in the ith row and jth column of the inverse of the mass matrix (equation 11.16). None of the elements of M^{-1} are zero, which will be true for most biomechanical systems; therefore, equation 11.17 indicates the important fact that the hip and knee joint angular accelerations are each determined by *both* the hip and knee joint muscle moments. The magnitudes of the elements of M^{-1}, τ_H, and τ_K can vary considerably over the course of a movement, which means that, for instance, the hip joint angular acceleration could potentially be influenced more by the knee joint moment than by the hip joint moment.

An example of a counterintuitive result can be provided by comparing the accelerations induced by the uniarticular and biarticular muscles acting on the lower limb model. Using the data from table 11.2 and the configuration shown in figure 11.12, we will reasonably assume that both the iliopsoas and rectus femoris have hip flexor moment arms of 0.03 m, and the rectus femoris has a knee extensor moment arm of 0.04 m. For these numerical values, the elements of M^{-1} are computed to be

$$M^{-1} = \begin{bmatrix} -0.999 & 1.794 \\ 1.794 & -6.276 \end{bmatrix}$$

If iliopsoas generates 500 N of force, this will result in muscle moments of 15 N·m and 0 N·m for τ_H and τ_K, respectively. If we enter these values into equation 11.17, we get

$$\ddot{\theta}_{H\tau} = 14.99°/s^2 \text{ and } \ddot{\theta}_{K\tau} = -26.92°/s^2$$

which means that iliopsoas accelerates both the hip joint and knee joint into flexion for this set of conditions (note the greater magnitude for knee flexion than for hip flexion). The finding that a uniarticular hip flexor muscle generates a hip flexion angular acceleration is to be expected. However, why does iliopsoas generate a knee flexion acceleration? This result can be understood as follows: The force applied by iliopsoas to the thigh segment causes the thigh to apply an anteriorly directed force on the shank segment at the knee joint. This causes a clockwise angular acceleration of the shank segment (referring to figure 11.12) of sufficient magnitude to result in a knee flexion acceleration.

(continued)

EXAMPLE 11.2 (CONTINUED)

Now consider the case where rectus femoris generates 500 N of force. The hip muscle moment, τ_H, is still 15 N·m, but the knee muscle moment, τ_K, is now 20 N·m. Substitution into equation 11.17 yields

$$\ddot{\theta}_{H\tau} = -20.89 \,{}^\circ\!/\mathrm{s}^2 \text{ and } \ddot{\theta}_{K\tau} = 98.61 \,{}^\circ\!/\mathrm{s}^2$$

The result for the knee is perhaps not surprising, because the rectus femoris induces a large knee extension angular acceleration. However, the result for the hip is counterintuitive. Despite generating exactly the same hip flexion moment as iliopsoas, the rectus femoris induces a hip extension angular acceleration. Similar to the prior example, this occurs because contracting the rectus femoris muscle gives rise to reaction forces that act on the thigh segment. This is not as easy to visualize as is the case in which the uniarticular iliopsoas induces knee flexion acceleration, which underlines the need to derive the complete equations of motion for the entire system being studied. Zajac and Gordon (1989) noted that the effects of biarticular muscles are highly dependent on the moment arm ratio of the two joints that the muscle spans. As a case in point, if the rectus femoris moment arm at the knee joint was actually 0.01 m, rather than 0.04 m (i.e., much smaller than the hip moment arm, rather than slightly larger), then the induced accelerations would be

$$\ddot{\theta}_{H\tau} = 6.02 \,{}^\circ\!/\mathrm{s}^2 \text{ and } \ddot{\theta}_{K\tau} = 4.46 \,{}^\circ\!/\mathrm{s}^2$$

Now the hip joint angular acceleration is again in the direction of flexion, and the knee extension angular acceleration is reduced in magnitude. Although these numerical results are specific to a confined set of joint angles and muscle forces, the general approach can readily be extended to examine the contributions of individual muscles to the motion of the limb segments over the full swing phase of walking (e.g., Piazza and Delp 1996) or to any other movement, given an appropriate model.

The preceding induced acceleration example focused exclusively on muscle-induced accelerations, but it is also possible to compute the accelerations induced by gravitational and velocity-dependent forces. The accelerations induced by gravitational forces for the preceding example would be given by

$$\ddot{\theta}_{Hg} = M_{11}^{-1}\, g_1 + M_{12}^{-1}\, g_2$$
$$\ddot{\theta}_{Kg} = M_{21}^{-1}\, g_1 + M_{22}^{-1}\, g_2 \tag{11.18}$$

where g_i indicates the entry in the ith row of the G vector in equation 11.13 (which only has one column). Likewise, the accelerations induced by velocity-dependent forces would be computed as

$$\ddot{\theta}_{Hv} = M_{11}^{-1}\, v_1 + M_{12}^{-1}\, v_2$$
$$\ddot{\theta}_{Kv} = M_{21}^{-1}\, v_1 + M_{22}^{-1}\, v_2 \tag{11.19}$$

where v_i indicates the entry in the ith row of the V vector in equation 11.13. This complete decomposition of the system accelerations can provide insight on whether a particular movement is dominated by muscular forces (T vector in equation 11.13) or nonmuscular forces (G and/or V vectors in equation 11.13).

Power Analyses

Several types of power analyses can be conducted with musculoskeletal models. The process of generating a simulation of movement generally yields muscle forces and velocities over the full movement, which allows muscle powers to be computed directly. This forms the basis for determining the mechanical work of locomotion at a finer level of detail than is possible using center of mass or inverse dynamics–based methods described in chapter 6. The ability to quantify individual muscle powers has also been used to better understand the dynamics of muscle function during complex movements. For example, Bobbert and colleagues (1986b) quantified the extent to which elastic recoil of tendinous tissues and transfer of power (from the knee to ankle) by the biarticular gastrocnemius each contribute to augmentation of ankle joint power in vertical jumping. Note that the energy transfer mechanism was computed in the context of a joint power analysis and reflects the ability of biarticular muscles to transfer mechanical energy between adjacent joints. However, a different power analysis based on induced accelerations can provide information on the flow of mechanical energy between body segments rather than between joints. The induced acceleration techniques described earlier can be extended to quantify how each muscle contributes to the instantaneous power of every body segment (Fregly and Zajac 1996). This type of analysis has been referred to by several names: *state-space energy analysis, induced power analysis,* and *segmental power analysis.* This latter term should not be confused with the segment-based energy and power techniques discussed in chapter 6.

Such induced power analyses have the potential to provide greater insight into muscle function than can be gained from induced accelerations alone. In a study of seated pedaling, Neptune and colleagues (2000) showed that much of the positive work done by the gluteus maximus during the downstroke generates energy to the leg segments rather than contributing directly to rotation of the crank. This can be seen in figure 11.13*a*, in which the net power has a large positive magnitude during the downstroke (0°-180°), simultaneous with a large positive magnitude for the limb segments and a low magnitude for the crank segment. This result occurs because the gluteus maximus does not generate a large tangential crank force to contribute to crank rotation during the downstroke. Thus, positive work by gluteus maximus serves primarily to increase the mechanical energy level of the limb segments. In contrast, the soleus generates a large tangential crank force but does relatively little positive work, as seen by the low net power in figure 11.13*b*. However, the soleus does transfer a substantial amount of energy from the leg segments to the crank, shown by the coincident negative limb power and positive crank power profiles that are nearly mirror images of each other. The amount of energy delivered by the soleus to the crank in this example exceeded what the muscle alone could generate, given its limited amount of shortening associated with the restricted ankle joint motion in pedaling. This form of power analysis was able to identify an important synergy, whereby the gluteus maximus generates mechanical energy and the soleus transfers that energy to the crank to produce a smooth and coordinated pedaling motion. This type of analysis represents the next level of detail beyond what can be inferred from joint powers derived from an inverse dynamics analysis (chapter 6).

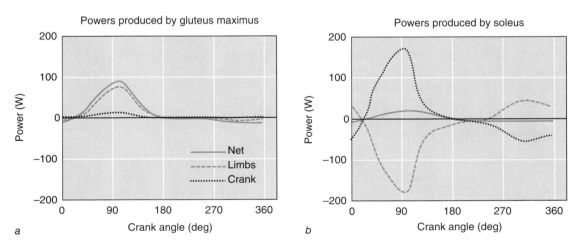

▲**Figure 11.13** Graphs showing the net, limb, and crank powers generated by the *(a)* gluteus maximus and *(b)* soleus muscles across the crank cycle during seated pedaling.

Adapted from *Journal of Biomechanics*, Vol. 33, R.R. Neptune, S.A. Kautz, and F.E. Zajac, "Muscle contributions to specific biomechanical functions do not change in forward versus backward pedaling," pgs. 155-64, copyright 2000, with permission of Elsevier.

Sensitivity Analysis

A typical musculoskeletal model can have tens to hundreds of parameters, and their numerical values will influence the model results to varying degrees. Sensitivity analysis is a technique commonly used to investigate how the overall model response is influenced by the values assigned to each of these many parameters. Two main types of sensitivity analyses have been used with musculoskeletal models. In the first, the goal is to determine how a specific factor influences human movement capacity. For example, a sport scientist working with elite athletes might be interested to know which muscle characteristic (e.g., strength, fiber type, shortening speed) is the primary determinant of sprinting speed. Given an appropriate model capable of simulating maximum-speed sprinting, specific muscle model parameters could be varied to determine which one most strongly affects sprinting speed (e.g., Miller et al. 2012a, 2012b). Note that this analysis would likely be done within a dynamic optimization framework, and the maximum-speed sprinting problem would need to be re-solved each time a muscle model parameter value was changed. The results of such a sensitivity analysis could provide useful fundamental and practical information on musculoskeletal function.

A second type of sensitivity analysis involves a systematic study of model responses to standardized changes in values of the model parameters. The goal here is to gain insight about the model itself; thus, this process is often part of the model development and evaluation phase. This form of analysis can also be useful, following a modeling study, in ascertaining the level of confidence that should be placed in the results. In this type of sensitivity analysis, the model parameter values are systematically modified to determine the effects on specific model outputs. For example, one might wish to determine the sensitivity of predicted joint moments to changes in the values of muscle and joint parameters. This type of analysis requires running an additional simulation for each parameter change but generally will not require re-solving the entire optimization problem. This type of analysis can also be used to evaluate the sensitivity of more global variables, such as how predicted ground reaction forces are altered by changes in foot and ground viscoelastic parameters in a walking model. Again, running a new simulation after each parameter change will be necessary. Depending on the exact nature of the perturbations and the specific focus of the sensitivity analysis, reoptimization of the movement sequence may or may not be required.

Usually, the results obtained with a musculoskeletal model are more sensitive to some parameters than to others. For example, several studies have shown that musculoskeletal model performance is highly sensitive to the values used for the rest length of the series elastic component in Hill-type muscle models but is less sensitive to the values used for other parameters such as the Hill dynamic constants (e.g., Scovil and Ronsky 2006). However, generalizations should be made cautiously, because sensitivities could easily vary between different musculoskeletal models or for the same model used to simulate different movement tasks. For example, maximum vertical jump height should be quite sensitive to the values assumed for peak isometric force in individual muscles, whereas the results obtained for submaximal pedaling should not depend as heavily on these values. Regardless, if a model output is shown to be highly sensitive to a particular model parameter, great care must be exercised both in determining model parameter values and in interpreting model results.

Model Validation

Model validation is an important but sometimes overlooked task. Surprisingly, the process of "validating" a model is not very well defined and can be context specific. One definition of model validation involves demonstrating that the model is strong and powerful for the task for which it was designed (Nigg 1999). Thus, validating a model would require demonstrating good agreement between model predictions and corresponding empirical measurements. The definition of "good agreement" may vary but might reasonably be based on the variability in the measured data (e.g., good predictions should fall within one standard deviation of the experimental mean). Although this process sounds straightforward, in practice it presents some difficulties. Consider a forward dynamics musculoskeletal model used to simulate walking. A valid model should predict gross kinematics and kinetics, as well as muscle and joint forces, that closely approximate measured values. Comparing empirical and model joint angles or ground reaction forces is readily accomplished, but opportunities to compare individual muscle or joint forces against measured values are far fewer. Even with easily measured variables, model predictions can only be compared with a limited set of experimental data. Thus, a model can only ever be validated for a restricted set of conditions, and extrapolations to nonobserved conditions involve a degree of uncertainty.

There have been some limited opportunities to compare model-predicted muscle or joint forces with directly measured values, but such primary forms of validation are relatively rare. Direct validation of predicted muscle forces is usually based on forces measured in nonhuman animals in a limited number of muscles (e.g., Prilutsky et al. 1997). More commonly, predicted muscle forces from musculoskeletal models are compared with measured EMG data. Considering the relations between EMG and muscle force (chapter 8), one would expect reasonable temporal agreement between EMG and muscle force profiles. Under carefully controlled circumstances, there may even be a limited basis for comparing relative EMG and muscle force amplitudes. However, in most movement scenarios, the phasing and amplitude of the EMG signals will vary from corresponding muscle force profiles. Therefore, at best such comparisons represent an indirect means of validating the predictions made with musculoskeletal models. Although surface EMG provides a noninvasive characterization of muscle actions, there is no analog for the noninvasive evaluation of predicted joint contact forces. Most opportunities to validate predicted joint forces have come from a limited set of instrumented joint replacements (e.g., Kim et al. 2009). For a detailed listing of modeling studies and forms of validation, the reader is referred to the review by Erdemir and colleagues (2007).

Many factors discussed in this chapter contribute to the validity of a model. Nearly every decision associated with the development of a model (e.g., number of DOF, number of muscles) and the use of a model (e.g., inverse versus forward dynamics, optimization cost function used) will affect model validity. The effort spent developing or selecting a model that is properly aligned with the research goals of a project will contribute substantially to the validity of the model. This involves both the model itself and the way in which it is used. Even a perfectly sound model will be of little value if it is combined with an inappropriate optimization cost function. Likewise, a highly biorealistic control algorithm will yield limited insight if it is linked to a poorly designed musculoskeletal model. When considering model validity, we must also remember that all musculoskeletal models are simplifications of reality and no single model will correctly predict all known phenomena in all conditions. Furthermore, good agreement between model outcomes and experimental data does not necessarily ensure that the model itself is valid. Particularly in complex models, there is the risk that "two wrongs make a right," whereby offsetting errors yield what appears to be a correct result. For example, good agreement with experimental joint kinetics may be a necessary condition for model validity, but it does not, by itself, guarantee that predicted muscle forces are also correct. In a multimuscle model, offsetting errors in moment arms for two synergistic muscles could result in incorrect force predictions in both muscles yet still yield the correct joint moment. Such errors can only be eliminated through the availability of better moment arm data for the two muscles in question and in fact would only be detected if a set of measured muscle forces were available for comparison.

The impact of errors on any particular modeling study is hard to predict and will depend heavily on the nature of the research question. A sensitivity analysis focusing on the key outcome variables for a particular study is useful in this regard. Researchers can ultimately maximize model validity by matching their model to their research question and by investigating the importance of parameters during model development and evaluation. Even if model validation must be largely indirect, useful insight can be gained from carefully designed and implemented models. For instance, a model used to simulate human walking that moves like experimental subjects, has ground reaction forces and joint moments that match those of the subjects, and predicts muscle forces that are temporally consistent with experimental EMG is likely to have captured many of the fundamental aspects of the mechanics of locomotion.

SUMMARY

This chapter introduced the basic concepts of musculoskeletal modeling. The major components of a musculoskeletal model were described, as well as techniques used to control musculoskeletal models and a range of possible analysis techniques. However, many of the details, by necessity, were excluded.

FROM THE SCIENTIFIC LITERATURE

Anderson, F.C., and M.G. Pandy. 2003. Individual muscle contributions to support in normal walking. *Gait and Posture* 17:159-69.

Neptune R.R., F.E. Zajac, and S.A. Kautz. 2004. Muscle force redistributes segmental power for body progression during walking. *Gait and Posture* 19:194-205.

Previously, we discussed dynamic optimization studies of vertical jumping. Similar techniques have been used to understand how muscles function in walking. Anderson and Pandy (2003) and Neptune and colleagues (2004) used forward dynamics simulations to identify how muscles and other forces (e.g., gravity) contribute to supporting the body during walking. The support offered by a particular force was based on its contribution to the vertical ground reaction force. Both groups reported that muscles in the ipsilateral limb were the primary contributors to the vertical ground reaction force and that gravity made the only other meaningful contribution. Other factors such as contralateral limb muscles and centrifugal forces made considerably smaller contributions. The contribution of gravity to the ground reaction force in this context reflects the passive bracing of the skeletal system.

The conclusions drawn between the two studies had a number of similarities. Both groups indicated that the vasti and gluteals made the largest contributions to the first peak in the vertical ground reaction force, whereas the plantar flexors made the greatest contributions to the second peak. Both groups also found that the hamstrings and rectus femoris made relatively small contributions to the vertical ground reaction force. However, the results from these two studies were not in full agreement. Even where there was general concurrence, the relative contributions of specific muscles to the ground reaction force were not the same. Furthermore, Anderson and Pandy (2003) reported that gluteus medius made the greatest muscular contribution to the vertical ground reaction force around the middle of the stance phase, whereas Neptune and colleagues (2004) reported that the primary contributors were the plantar flexors.

Some differences in findings between the two studies are not surprising, given the methodological differences. Anderson and Pandy (2003) used a 3-D model with 10 segments, 23 DOF, and 54 muscles. Walking was generated using dynamic optimization to find the muscle stimulation patterns that minimized the total metabolic energy consumed, divided by the distance traveled. Neptune and colleagues (2004) used a 2-D model with 7 segments, 9 DOF, and 30 muscles. Walking was generated using dynamic optimization to find the muscle stimulation patterns that optimally tracked a set of experimental gait data. The two studies also differed in terms of how each contribution to the ground reaction force was determined, particularly in terms of how the foot-ground interface was handled. Given these methodological differences, the broad similarities between studies are perhaps striking and point to the generality of several of the main conclusions. However, the differences between the two models also underlie one of the primary discrepancies in the results. The 2-D model used by Neptune and colleagues (2004) did not include a DOF for hip abduction and thus did not explicitly include hip abductor muscles. Therefore, it was not possible to predict the contribution of gluteus medius to support of the body during the middle of the stance phase. This again points to the need for researchers to understand how the characteristics of their model influence the results that are obtained. These two studies represent some of the earliest attempts to use advanced musculoskeletal modeling techniques to understand the mechanics of human walking. Several subsequent studies have further clarified the function of muscles in locomotion.

Musculoskeletal modeling is a young and rapidly expanding field that has witnessed many exciting advancements over the past few years.

In addition to the suggested readings listed at the end of the chapter, there are numerous other resources that interested readers may consult to build upon what they have learned in this chapter. At the time of this writing, there are currently two large, multiyear projects on biological modeling and simulation that include a major focus on musculoskeletal modeling:

Simbios (http://simbios.stanford.edu)

Virtual Physiological Human (www.vph-noe.eu)

These projects have national and international funding, suggesting strong support for the further advancement of the musculoskeletal modeling field. A more specific topic that is important in musculoskeletal modeling, but was only touched upon here, is the derivation of equations of motion (see example 11.2). For even the simplest of models, this process can be difficult and error-prone if pursued entirely by hand. Fortunately, several software packages are available to aid the modeler in this process:

Simbody (https://simtk.org/home/simbody)

MotionGenesis (http://motiongenesis.com)

SD/FAST (www.sdfast.com)

MSC Adams (www.mscsoftware.com)

SimMechanics (www.mathworks.com/products/simmechanics)

Open Dynamics Engine (http://opende.sourceforge.net)

These packages reflect a mix of open-source and commercial products; some are stand-alone applications (e.g., MotionGenesis) and others are integrated into other software packages (e.g., SimMechanics). Finally, the field of musculoskeletal modeling has advanced to the stage where there are now several purpose-built software packages designed to aid in development and use of musculoskeletal models:

OpenSim (http://simtk.org/home/opensim)

SIMM (www.musculographics.com)

AnyBody (www.anybodytech.com)

LifeMOD (www.lifemodeler.com)

GaitSym (www.animalsimulation.org)

MSMS (http://mddf.usc.edu)

These software packages also reflect both open-source and commercial options with a range of intended applications. Their mere presence and sheer numbers indicate the current value of musculoskeletal modeling as a research approach as well as the promise these techniques hold for the future.

SUGGESTED READINGS

Blemker, S.S., D.S. Asakawa, G.E. Gold, and S.L. 2007. Image-based musculoskeletal modeling: Applications, advances, and future opportunities. *Journal of Magnetic Resonance Imaging* 25:441-51.

Buchanan, T.S., D.G. Lloyd, K. Manal, and T.F. Besier. 2004. Neuromusculoskeletal modeling: Estimation of muscle forces and joint moments and movements from measurements of neural command. *Journal of Applied Biomechanics* 20:367-95.

Crowninshield, R.D., and R.A. Brand. 1981. The prediction of forces in joint structures: Distribution of intersegmental resultants. *Exercise and Sport Sciences Reviews* 9:159-81.

Erdemir, A., S. McLean, W. Herzog, and A.J. van den Bogert. 2007. Model-based estimation of muscle forces exerted during movements. *Clinical Biomechanics* 22:131-54.

Pandy, M.G. 2001. Computer modeling and simulation of human movement. *Annual Review of Biomedical Engineering* 3:245-73.

Piazza, S.J. 2006. Muscle-driven forward dynamic simulations for the study of normal and pathological gait. *Journal of Neuroengineering and Rehabilitation* 3:5-11.

Viceconti, M., D. Testi, F. Taddei, S. Martelli, G.J. Clapworthy, and S. Van Sint Jan. 2006. Biomechanics modeling of the musculoskeletal apparatus: Status and key issues. *Proceedings of the IEEE* 94:725-39.

Yamaguchi, T.G. 2001. *Dynamic Modeling of Musculoskeletal Motion: A Vectorized Approach for Biomechanical Analysis in Three Dimensions.* Boston: Kluwer.

Zajac, F.E., and M.E. Gordon. 1989. Determining muscle's force and action in multi-articular movement. *Exercise and Sport Sciences Reviews* 17:187-230.

Zajac, F.E., R.R. Neptune, and S.A. Kautz. 2002. Biomechanics and muscle coordination of human walking: Part I: Introduction to concepts, power transfer, dynamics and simulations. *Gait and Posture* 16:215-32.

Zajac, F.E., R.R. Neptune, and S.A. Kautz. 2003. Biomechanics and muscle coordination of human walking: Part II: Lessons from dynamical simulations and clinical implications. *Gait and Posture* 17:1-17.

PART IV

FURTHER ANALYTICAL PROCEDURES

Biomechanical analysis has some unique problems that other disciplines do not encounter. One of these problems concerns taking multiple derivatives of noisy data. Chapter 12 explains techniques for identifying and reducing the effects of noise in biomechanical signals, especially marker trajectories. It also outlines the frequency analysis of signals, which helps to characterize signals and determine how to evaluate the effectiveness of data smoothing techniques. Part IV also explores further analytical procedures that can be applied to biomechanical data. Chapter 13 outlines the theories and analytic methods used to investigate movement in complex systems with many degrees of freedom by using dynamical systems methods. This chapter focuses on how we assess and measure coordination and stability in changing movement patterns, and it examines the role of movement variability in health and disease. Chapter 14 explains several statistical tools for identifying the essential characteristics of any human movement. Biomechanists are faced with the sometimes daunting task of determining which variables from thousands of possibilities (linear and angular kinematics, linear and angular kinetics) best characterize a particular motion. Techniques in chapter 14 provide a means for selecting the best combination of these factors.

Chapter 12

Signal Processing

Timothy R. Derrick

A **signal** is a time- or space-varying quantity that conveys information. A **waveform** is a time- or space-varying quantity that contains either information, unwanted data called **noise,** or both. A waveform may take the form of a sound wave, voltage, current, magnetic field, displacement, or a host of other physical quantities. These are examples of continuous signals (meaning that the signal is present at all instances of time or space), also called **analog signals.** For convenience and to enable manipulation by computer software, we often convert continuous signals using **analog-to-digital** converters into a series of discrete values by sampling the phenomena at specific time intervals to create an equivalent digital signal (figure 12.1). The situation can also be reversed by **digital-to-analog** converters so that a digital signal can be viewed electronically. In this chapter, we

▶ define how to characterize a signal or waveform,

▶ examine the Fourier analysis of waveforms,

▶ explain the sampling theorem and Nyquist frequency,

▶ discuss how to ensure cyclic continuity, and

▶ review various data-smoothing techniques for attenuating noise from waveforms.

CHARACTERISTICS OF A SIGNAL

A sinusoidal time-varying signal has four characteristics: **frequency** *(f)*, **amplitude** *(a)*, offset *(a_0)*, and **phase angle** *(θ)*. These characteristics are depicted in the schematics in figure 12.2. The frequency represents how rapidly the signal oscillates; it is usually measured in cycles per second or hertz. One hertz is equal to 1 cycle per second. For example, the second hand of a clock completes one cycle every 60 s. Its frequency is one cycle per 60 seconds or 1/60 Hz. The frequency of a signal (figure 12.2*a*) is

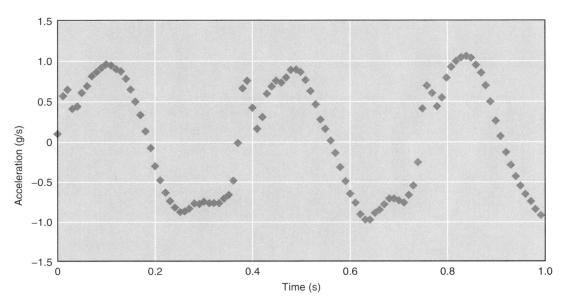

▲ **Figure 12.1** The digitized discrete representation of the acceleration of the head while a subject is running. The signal was sampled at 100 Hz (100 samples per second).

easy to determine in a single sine wave but more difficult to visualize in noncyclic signals with multiple frequencies. The amplitude of a signal (figure 12.2b) quantifies the magnitude of the oscillations. The offset (or direct current [DC] offset or DC bias) (figure 12.2c) represents the average value of the signal. The phase angle (or **phase shift**) in the signal (figure 12.2d) is the amount of time the signal may be delayed or time shifted.

Any time-varying signal or waveform, $w(t)$, is made up of these four characteristics. The following equation incorporates each of the four variables:

$$w(t) = a_0 + a \sin(2\pi ft + \theta) \tag{12.1}$$

Alternately, the equation can be written using the equivalent angular frequency, ω, where $\omega = 2\pi f$ (because f is in cycles per second or hertz, ω is in radians per second,

and there are 2π radians in a cycle). Thus another way to write this relationship is

$$w(t) = a_0 + a \sin(\omega t + \theta) \tag{12.2}$$

The time (t) is a discrete time value that depends on how frequently the signal is to be sampled. If the sampling frequency is 100 Hz (100 samples per second), then the sampling interval is the inverse (1/100th or 0.01). This means that there will be a sample or datum registered every 0.01 s. So, t is one of the discrete values in the set (0, 0.01, 0.02, 0.03, ..., T). The variable T represents the duration of the digitized signal. For example, by adding the equations for a 2 Hz sine wave to a 20 Hz sine wave, we created the following waveform (as illustrated in figure 12.3):

$$w(t) = \sin(2\pi 2t) + \sin(2\pi 20t) \tag{12.3}$$

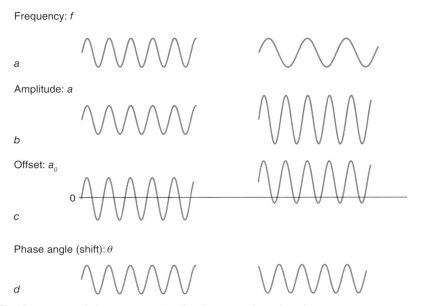

Frequency: f

a

Amplitude: a

b

Offset: a_0

c

Phase angle (shift): θ

d

▲ **Figure 12.2** The four essential components of a time-varying signal.

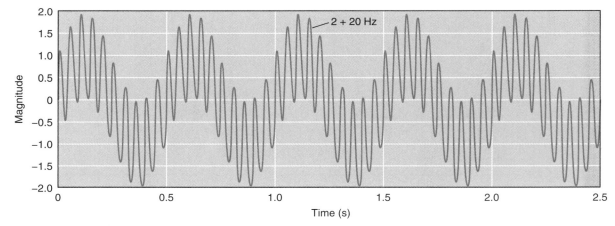

▲ **Figure 12.3** A 2 Hz and a 20 Hz sine wave summed over a 2.5 s period. The offset (a_0) and angle (θ) are zero for both waves, and the amplitudes are 1.

FOURIER TRANSFORM

Any time-varying signal can be represented by successively adding the individual frequencies present in the signal (Winter 1990). The a_n and θ_n values may be different for each frequency (f_n) and may be zero for any given frequency:

$$w(t) = a_0 + \sum a_n \sin(2\pi f_n t + \theta_n) \quad (12.4)$$

By using the cosine and sine functions, this series can be rewritten without the phase variable as

$$w(t) = a_0 + \sum [b_n \sin(2\pi f_n t) + c_n \cos(2\pi f_n t)] \quad (12.5)$$

This series is referred to as the *Fourier series*. The b_n and c_n coefficients are called the *Fourier coefficients*. They can be calculated using the following formulas:

$$a_0 = \frac{1}{T} \int_0^T w(t)\,dt \quad (12.6)$$

$$b_n = \frac{2}{T} \int_0^T w(t)\sin(2\pi f_n t)\,dt \quad (12.7)$$

$$c_n = \frac{2}{T} \int_0^T w(t)\cos(2\pi f_n t)\,dt \quad (12.8)$$

Here is another way of looking at it: If you want to know how much of a certain frequency (f_n) is present in a signal $w(t)$, multiply your signal by the sine wave $[\sin(2\pi f_n t)]$, take the mean value, and multiply it by 2. Repeat this process for the cosine wave and then add the squares of the two together, to get how much of the signal is composed of the frequency f_n. This is called the *power* at frequency f_n. Mathematically, the process is as follows:

$$a_0 = \text{mean}\,[w(t)] \quad (12.9)$$

$$b_n = 2 \times \text{mean}\,[w(t) \times \sin(2\pi f_n t)] \quad (12.10)$$

$$c_n = 2 \times \text{mean}\,[w(t) \times \cos(2\pi f_n t)] \quad (12.11)$$

$$\text{Power}(f_n) = b_n^2 + c_n^2 \quad (12.12)$$

The Fourier coefficients can be calculated from the equally spaced time-varying points with the use of a discrete Fourier transformation (DFT) algorithm (appendix H). Given the Fourier coefficients, we can construct the original signal using an inverse DFT algorithm. The DFT is a calculation-intensive algorithm. Faster and more commonly used are the fast Fourier transformations (FFTs). An FFT requires that the number of original data points be a power of 2 (. . . , 64, 128, 256, 512, 1024, 2048, . . .). If this is not the case, the usual method of obtaining a "power of 2" number of samples is to pad the data with zeros (add zeros until the number of points is a power of two). This creates two problems:

■ Padding reduces the power of the signal. Parseval's theorem implies that the power in the time domain must equal the power in the frequency domain (Proakis and Manolakis 1988). When you pad with zeros, you reduce the power (a straight line at zero has no power). You can restore the original power by multiplying the power at each frequency by $(N + L)/N$, where N is the number of nonzero values and L is the number of padded zeros.

■ Padding can introduce discontinuities between the data and the padded zero values if the signal does not end at zero. This discontinuity shows up in the resulting spectrum as increased power in the higher frequencies. To ensure that your data start and end at zero, you can apply a windowing function or subtract the trend line before performing the transformation. Windowing functions begin at zero, rise to 1, and then return to zero again. By multiplying your signal by a windowing function, you reduce the endpoints to zero in a gradual manner. Windowing should not be performed on data unless there are multiple cycles. Subtracting a trend line that connects the first point to the last point can be used as an alternative.

Most software packages give the result of an FFT in terms of a real portion and an imaginary portion. For a real discrete signal, the real portion corresponds to the cosine coefficient and the imaginary portion corresponds to the sine coefficient of the Fourier series equation. An FFT results in as many coefficients as there are data points (N), but half of these coefficients are a reflection of the other half. Therefore, the $N/2$ points represent frequencies from zero to one-half of the sampling frequency $(f_s/2)$. Each frequency bin has a width of f_s/N Hz. By increasing the number of data points (by padding with zeros or collecting for a longer period of time), you can decrease the bin width. This does not increase the resolution of the FFT; rather, it is analogous to interpolating more points from a curve.

Researchers often adjust the bin width so that each bin is 1 Hz wide. This is referred to as *normalizing the spectrum*. Adjusting the bin width changes the magnitude because the sum of the power frequency bins must equal the power in the time domain. Normalizing the spectrum allows data of different durations or sampling rates to be compared. The magnitude of a normalized spectrum is in units of (original units)^2/Hz.

A plot of the power at each frequency is referred to as the *power spectral density* (PSD) plot or simply the *power spectrum*. A PSD curve contains the same information as its time-domain counterpart, but it is rearranged to emphasize the frequencies that contain the greatest power rather than the point in time in the cycle at which the most power occurs. Figure 12.4 shows a leg acceleration curve along with its PSD.

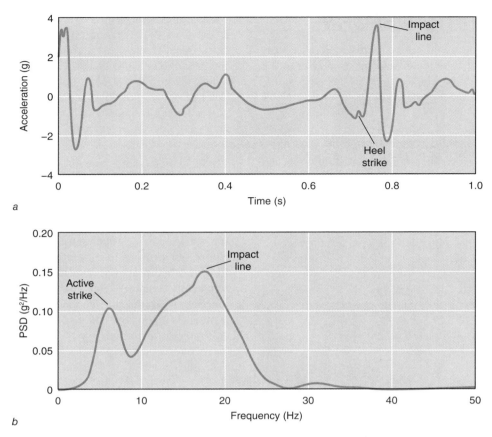

▲ **Figure 12.4** Leg acceleration during running in the *(a)* time and *(b)* frequency domains. The time domain graph shows two ground impacts, whereas the frequency domain graph is for a single stance phase.

TIME-DEPENDENT FOURIER TRANSFORM

The discrete Fourier transform (DFT) has the advantage that frequencies can be separated no matter when they occur in the signal. Even frequencies that occur at the same time can be separated and quantified. A major disadvantage is that we do not know when those frequencies are present. We could overcome this difficulty by separating the signal into sections and applying the DFT to each section. We would then have a better idea of when a particular frequency occurred in the signal. This process is called a time-dependent Fourier transform.

Because we are already able to separate frequencies, we will use that technique to build an intuitive feeling for how this transform works. If we take a signal that contains frequencies from 0 to 100 Hz, the first step is to separate the frequencies into two portions, 50 Hz and below and 50 Hz and above. Next, we take these two sections and separate them into two portions each. We now have sections of 0 to 50, 50 to 100, 0 to 25, 25 to 50, 50 to 75, and 75 to 100

Hz. This procedure—called *decomposition*—continues to a predefined level. At this point, we have several time-series representations of the original signal, each containing different frequencies. We can plot these representations on a three-dimensional (3-D) graph with time on one axis, frequency on a second axis, and magnitude on the third. Figure 12.5 shows a two-dimensional version of this type of graph with the contour lines illustrating the magnitudes at each frequency and time.

SAMPLING THEOREM

The process signal must be sampled at a frequency greater than twice as high as the highest frequency present in the signal itself. This minimum sampling rate is called the *Nyquist sampling frequency* (f_N). In human locomotion, the highest voluntary frequency is less than 10 Hz, so a 20 Hz sampling rate should be satisfactory; however, in reality, biomechanists usually sample at 5 to 10 times the highest frequency in the signal. This ensures that the signal is accurately portrayed in the time domain without missing peak values.

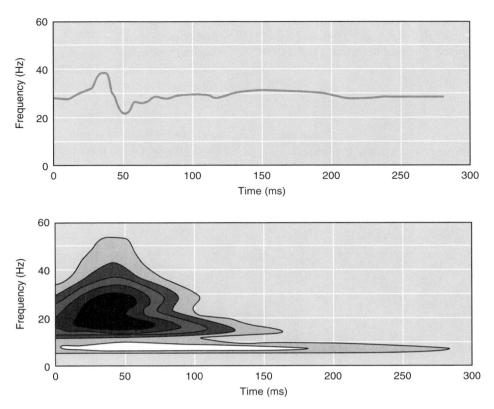

▲ **Figure 12.5** A 3-D contour map of the frequency by time values of a leg acceleration curve during running. The time domain curve is superimposed on the contour. There are two peaks in this curve. The high-frequency peak (approximately 20 Hz) occurs between 20 and 60 ms. The lower-frequency peak (approximately 8 Hz) occurs between 0 and 180 ms.

FROM THE SCIENTIFIC LITERATURE

Shorten, M.R., and D.S. Winslow. 1992. Spectral analysis of impact shock during running. *International Journal of Sport Biomechanics* 8:288-304.

The purpose of this study was to determine the effects of increasing impact shock levels on the spectral characteristics of impact shock and impact shock wave attenuation in the body during treadmill running. Three frequency ranges were identified in leg acceleration curves collected during the stance phase of running. The lowest frequencies (4-8 Hz) were identified as the active region as a result of muscular activity. The midrange frequencies (12-20 Hz) resulted from the impact between the foot and ground. There was also a high-frequency component (60-90 Hz) resulting from the resonance of the accelerometer attachment. Because these frequencies all occurred at the same time, it was impossible to separately analyze them in the time domain. Head accelerations were also calculated so that impact attenuation could be calculated from the transfer functions (TFs). TFs were calculated from the power spectral densities at the head (PSD_{head}) and the leg (PSD_{leg}) using the following formula:

$$TF = 10 \log_{10} (PSD_{head}/PSD_{leg}) \qquad (12.13)$$

This formula resulted in positive values when there was a gain in the signal from the leg to the head and negative values when there was an attenuation of the signal from the leg to the head. The results indicated that the leg impact frequency increased as running speed increased. There was also an increase in the impact attenuation so that head impact frequencies remained relatively constant.

The sampling theorem holds that if the signal is sampled at greater than twice the highest frequency, then the signal is completely specified by the data. In fact, the original signal *(w)* is given explicitly by the following formula (also see appendix I):

$$w(t) = \Delta \sum w_n \left[\frac{\sin\left[2\pi f_c(t - n\Delta)\right]}{\pi(t - n\Delta)} \right] \quad (12.14)$$

where Δ is the sample period (1/sampling frequency), f_c is $1/(2\Delta)$, w_n is the nth sampled datum, and t is the time. By using this formula (Shannon's reconstruction formula; Hamill et al. 1997), we can collect data at slightly greater than twice the highest frequency and then apply the reconstruction formula to "resample" the data at a higher rate (Marks 1993). Figure 12.6 illustrates the use of the resampling formula to reconstruct a running vertical GRF curve. The signal was originally sampled at 1000 Hz, and the impact peak was measured at 1345 N. Every 20th point was then extracted to simulate data sampled at 50 Hz. The peak value occurred between samples, with the nearest data point at 1316 N. This also changed the time of occurrence of the impact peak. After the reconstruction formula was applied to the 50 Hz data, the peak was restored to 1344 N with the same time of occurrence as the originally sampled data. With modern computers, there is little reason to undersample

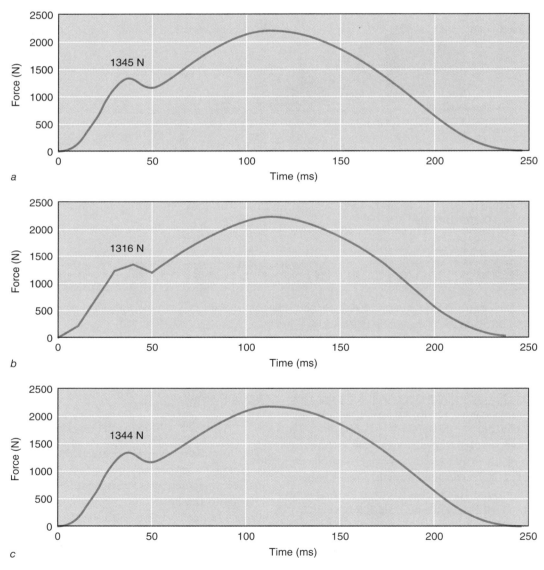

▲ **Figure 12.6** A running vertical GRF curve sampled at 1000 Hz (top), sampled at 50 Hz (middle), and sampled at 50 Hz but then reconstructed at 1000 Hz (bottom). The magnitude of the impact peak is identified in each graph. Reconstructing the signal results in a peak value very close to the original.

a signal unless the hardware is somehow limited, as is often the case when collecting kinematic data from video cameras with a sampling rate limited to 60 or 120 Hz. In many data collection systems it is necessary to collect **analog** data such as force or EMG data at sampling rates that are integral multiples of the sampling rate of the marker trajectories. For example, force and EMG may be collected at 1000 Hz when motion-capture data are collected at 200 frames per second (i.e., 5 times the video capture rate).

ENSURING CIRCULAR CONTINUITY

For the resampling formula to work correctly, the data must have circular continuity. To understand what circular continuity is, draw a curve from end to end on a piece of paper and then form a tube with the curve on the outside by rolling the paper across the curve. Circular continuity exists if there is no "discontinuity" where the start of the curve meets the end of the curve. This means that the first point on the curve must be equal to the last point. Nevertheless, the principle of circular continuity goes further: The slope of the curve at the start must equal the slope of the curve at the end. The slope of the slopes (the second derivative) must also be continuous. If you do not have circular continuity and you apply Shannon's reconstruction algorithm, you may be violating the assumption that only frequencies of less than half of the sampling frequency are present in the data. Discontinuities are by definition high-frequency changes in the data. If this occurs, you will see in the reconstructed data oscillations that have high amplitudes at the endpoints of the curve. These oscillations become smaller (damped) the farther you get from the endpoints, and they become much more evident if derivatives are calculated.

When data lack circular continuity, use the following steps (Derrick 1998) to approximate circular continuity (see figure 12.7):

1. Split the data into two halves.

2. Copy the first half of the data in reverse order and invert them. Attach this segment to the front of the original data.

3. Copy the second half of the data in reverse order and invert them. Attach this segment to the back of the original data.

4. Subtract the trend line from the first data point to the last data point.

Reversal of the first or second half of data is a procedure by which the first data point becomes the last data point of the segment, the second data point becomes the second

to last, and so on. Inversion is a procedure that flips the magnitudes about a pivot point. The pivot point is the point closest in proximity to the original data. Figure 12.7 shows a schematic diagram of the data after we sum the front and back segments and before we subtract the trend line.

Step 2 ensures circular continuity at the start of the original data set. Step 3 ensures circular continuity at the end of the original data set. Steps 2 and 3 together ensure that the slopes at the start and end of the new data set are continuous, but it is still possible to have a gap between the magnitude of the first point and the magnitude of the last point of the new data set. Step 4 removes this gap by calculating the difference between the trend line and each data point. Thus, the first and last points will be equal to zero.

If you fail to heed the sampling theorem, not only will you lose the higher frequencies, but the frequencies above the $1/2f_N$ (half the Nyquist frequency) actually fold back into the spectrum. In the time domain, this is referred to as *aliasing*. An antialiasing low-pass filter with a cutoff frequency (defined in the section titled Digital Filtering) greater than $1/2f_N$ applied to a signal before processing ensures no aliasing.

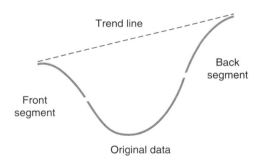

▲ **Figure 12.7** Schematic diagram of the procedure used to ensure that a signal has the property of circular continuity.

SMOOTHING DATA

Errors associated with the measurement of a biological signal may be the result of skin movement, incorrect digitization, electrical interference, artifacts from moving wires, or other factors. These errors, or "noise," often have characteristics that are different from the signal. **Noise** is any unwanted portion of a waveform. It is typically nondeterministic, lower in amplitude, and often in a frequency range different from that of the signal. For instance, errors associated with the digitizing process are generally higher in frequency than human movement. Noise that has a frequency different from those in the signal can be removed. If you were to plot the signal and the signal plus noise, it would look like figure 12.8. The

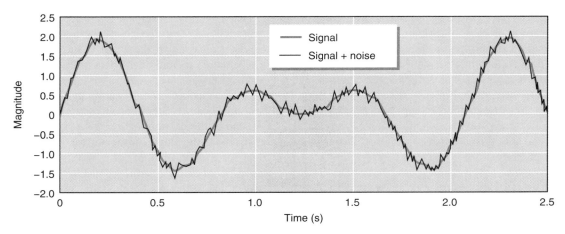

▲ **Figure 12.8** A biological signal with and without noise.

goal of smoothing is to eliminate the noise but leave the signal unaffected.

There are many techniques for smoothing data to remove the influence of noise. Outlined next are a number of the most popular. Each has its own strengths and weaknesses, and none is best for every situation. Researchers must be aware of how each method affects both the signal and the noise components of a waveform. Ideally, the signal would be unaffected by the smoothing process used to remove the noise, but most smoothing techniques affect the signal component to some extent.

Polynomial Smoothing

Any n data points can be fitted with a polynomial of degree $n - 1$ of the following form:

$$x(t) = a_0 + a_1 t + a_2 t^2 + a_3 t^3 + \ldots + a_{n-1} t^{n-1} \quad (12.15)$$

This polynomial will go through each of the n data points, so no smoothing has been accomplished. Smoothing occurs by eliminating the higher-order terms. This restricts the polynomial to lower-frequency changes and thus it will not be able to go through all of the data points. Most human movements can be described by polynomials of the ninth order or less. Polynomials produce a single set of coefficients that represent the entire data set, resulting in large savings in computer storage space. The polynomial also has the advantages of allowing you to interpolate points at different time intervals and of making the calculation of derivatives relatively easy. Unfortunately, they can distort a signal's true shape; see the article by Pezzack and colleagues (1977)—reviewed in chapter 1—for an example of this technique. In practice, avoid using polynomial fitting unless the signal is a known polynomial. For example, fitting a second-order (parabolic) polyno-

mial to the vertical motion of the center of gravity of an airborne body is appropriate. The path of the center of gravity during walking follows no known polynomial function, however.

Splines

A spline function (deBoor 2001) consists of a number of low-order polynomials that are pieced together in such a way that they form a smooth line. Cubic (third-order) and quintic (fifth-order) splines are the most popular for biomechanics applications (Wood 1982, Vaughan 1982). Splines are particularly useful if there are missing data in the data stream that need interpolation. In many motion capture systems, the "gap filling" procedures are done by fitting splines across the gaps in trajectories when two or more cameras were unable to view a marker and therefore the 3-D trajectory could not be reconstructed. Many techniques, such as **digital** filtering (discussed subsequently), require equally time-spaced data. Splines do not have this requirement.

Fourier Smoothing

Fourier smoothing consists of transforming the data into the frequency domain, eliminating the unwanted frequency coefficients, and then performing an inverse transformation to reconstruct the original data without the noise. Hatze (1981) outlined how to apply this method for smoothing displacement data.

Moving Average

A three-point moving average is accomplished by replacing each data point *(n)* by the average of $n - 1$, n, and $n + 1$. A five-point moving average uses the data points $n - 2$, $n - 1$, n, $n + 1$, and $n + 2$ and results in more smoothing than does a three-point moving average. Note that there will be undefined values at the start and end of the series. This method is extremely easy to implement but is incapable of distinguishing signals from noise. It will **attenuate** valid signal components and may not affect invalid noise components. A better choice is the digital filter.

Digital Filtering

A digital filter is a type of weighted moving average. The points that are averaged are weighted by coefficients in such a manner that a cutoff frequency can be determined. The **cutoff frequency** is the frequent at which a filter reduces the frequency's power by one half (half-power point) or equivalently reduces the amplitude to $\sqrt{2}/2$ or by −3 **decibels** or to approximately 0.707 of its original amplitude. In the case of a low-pass filter, frequencies below the cutoff are attenuated whereas frequencies above the cutoff are unaffected. The type of digital filter is determined by the frequencies that are passed through without attenuation. The following digital filters are all implemented in the same manner, but the coefficients are adjusted for a particular cutoff frequency. Signals that are band-passed or band-reject filtered are run through the filter with both low-pass and high-pass cutoff frequencies (figure 12.9).

Type of Filter

Filters can be constructed to attenuate different parts of the frequency spectrum. One or sometimes two cutoff frequencies are necessary to define which part of the frequency spectrum is attenuated and which part is left "as is," or passed unattenuated.

- ▶ Low-pass: The cutoff is selected so that low frequencies are unchanged but higher frequencies are attenuated. This is the most common filter type. It is often used to remove high frequencies from digitized kinematic data and as a digital antialiasing filter.
- ▶ High-pass: The cutoff is selected so that high frequencies are unchanged but lower frequencies are attenuated. It is used as a component in band-pass and band-reject filters or to remove low-frequency movement artifacts from low-voltage signals in wires that are attached to the body (e.g., electromyographic [EMG] signals).
- ▶ Band-pass: The frequencies between two cutoff frequencies are passed unattenuated. Frequencies below the lower cutoff and frequencies above the higher cutoff are attenuated. Such a filter is often used in EMG when there is movement artifact in the low-frequency range and noise in the high-frequency range.

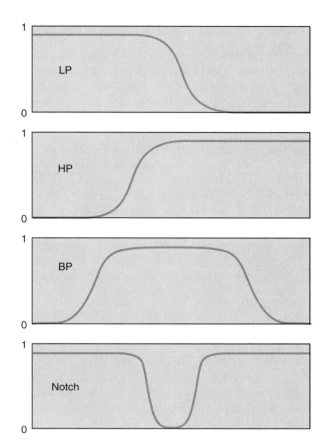

▲ **Figure 12.9** Frequency responses of different types of digital filters. The digital filter is implemented in the time domain, but it can be visualized in the frequency domain. The frequency response function is multiplied by the signal in the frequency domain and then transformed back into the time domain. LP = low-pass; HP = high-pass; BP = band-pass.

- ▶ Band-stop or band-reject: The frequencies between the two cutoff frequencies are attenuated. The frequencies below the lower cutoff and above the higher cutoff are passed unattenuated. This filter has little use in biomechanics.
- ▶ Notch: A narrow band or single frequency is attenuated. It is used to remove power-line noise (60 or 50 Hz) or other specific frequencies from a signal. This filter generally is not recommended for EMG signals, because they have significant power at 50 and 60 Hz (for additional information, see chapter 8).

Filter Roll-Off

Roll-off is the rate of attenuation above the cutoff frequency. The higher the order (the more coefficients) or the greater the number of times the signal is passed through the filter, the sharper the roll-off.

Recursive and Nonrecursive Filters

Recursive filters use both raw data and data that were already filtered to calculate each new data point. They sometimes are called infinite impulse response (IIR) filters. Nonrecursive filters use only raw data points and are called finite impulse response (FIR) filters. It is theoretically possible that a recursive filter will show oscillations in some data sets, but they will have sharper roll-offs. Data that are smoothed using a recursive filter will have a phase lag, which can be removed by putting the data through the filter twice—once in the forward direction and once in reverse. The filter is considered a *zero lag* filter if the net phase shift is zero. Digital filters distort the data at the beginning and end of a signal. To minimize these **distortions**, extra data should be collected before and after the portion that will be analyzed.

Optimizing the Cutoff

The selection of a cutoff frequency is very important when filtering data. This is a somewhat subjective determination based on your knowledge of the signal and the noise. A number of algorithms are used in an attempt to find a more objective criterion for determining the cutoff frequency (Jackson 1979). These optimizing algorithms typically are based on an analysis of the residuals, which are what is left over when you subtract the filtered data from the raw data. As long as only noise is being filtered, some of these values should be greater than zero and some less than zero. The sum of all of the residuals should equal zero (or at least be close). When the filter starts affecting the signal, the sum of the residuals will no longer equal zero. Some optimization routines use this fact to determine which frequency best distinguishes the signal from the noise

(figure 12.10). These algorithms are not completely objective, however, because you still must determine how close to zero the sum of the residuals is before selecting the optimal cutoff frequency.

Steps for Designing a Digital Filter

The following steps create a Butterworth low-pass, recursive digital filter. Modifications for a critically-damped and a high-pass filter are also discussed. Butterworth filters are said to be optimally flat in the pass band, making them highly desirable for biomechanical variables. This means that the amplitudes of the frequencies that are to be passed are relatively unaffected by the filter. Some filters, such as the Chebyshev or elliptic, have better roll-offs than the Butterworth filter, but they distort the amplitudes of the frequencies in the pass band.

1. Convert the cutoff frequency *(f_c)* from Hz to rad/s.

$$\omega_c = 2\pi f_c \qquad (12.16)$$

2. Adjust the cutoff frequency to reduce "warping" that resulted from the bilinear transformation.

$$\Omega_c = \tan\left[\frac{\omega_c}{2 \times \text{Sample rate}}\right] \qquad (12.17)$$

3. Adjust the cutoff frequency for the number of passes *(P)*. A pass occurs each time the data are passed through the filter. For every pass through the data, a second pass must be made in the reverse direction to correct for the phase shift. Increasing the number of passes increases the sharpness of the roll-off.

$$\Omega_N = \frac{\Omega_c}{\sqrt[4]{2^{\left(\frac{1}{P}\right)} - 1}} \qquad (12.18)$$

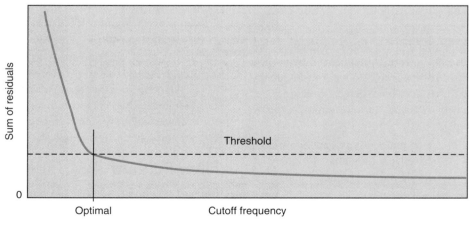

▲ **Figure 12.10** Selection of an optimal cutoff frequency using residual analysis.

4. Calculate the coefficients.

$$C_1 = \frac{\Omega_N^2}{\left(1 + \sqrt{2}\,\Omega_N + \Omega_N^2\right)}$$

$$C_2 = \frac{2\Omega_N^2}{\left(1 + \sqrt{2}\,\Omega_N + \Omega_N^2\right)} = 2C_1$$

$$C_3 = \frac{\Omega_N^2}{\left(1 + \sqrt{2}\,\Omega_N + \Omega_N^2\right)} = C_1$$

$$C_4 = \frac{2\left(1 - \Omega_N\right)^2}{\left(1 + \sqrt{2}\,\Omega_N + \Omega_N^2\right)}$$

$$C_5 = \frac{\left(2\Omega_N - \Omega_N^2 - 1\right)}{\left(1 + \sqrt{2}\,\Omega_N + \Omega_N^2\right)}$$

$$= 1 - \left(C_1 + C_2 + C_3 + C_4\right) \qquad (12.19)$$

5. Apply the coefficients to the data to implement the weighted moving average. Y_n values are filtered data and X_n values are unfiltered data. The filter is recursive because previously filtered data—Y_{n-1} and Y_{n-2}—are used to calculate the filtered data point, Y_n:

$$Y_n = C_1 X_{n-2} + C_2 X_{n-1} + C_3 X_n + C_4 Y_{n-1} + C_5 Y_{n-2} \qquad (12.20)$$

This filter is underdamped (damping ratio = 0.707) and will therefore "overshoot" and "undershoot" the true signal whenever there is a rapid change in the data stream. For more information, consult the article by Robertson and Dowling (2003). A critically damped filter can be designed (damping ratio = 1) by changing to 2 in each equation in step 4. The warping function must also be altered as follows:

$$\Omega_N = \frac{\Omega_c}{\sqrt{2^{\left(\frac{1}{2P}\right)} - 1}} \qquad (12.21)$$

In practice, there is little difference between the underdamped and critically damped filter. The distinction can be seen in response to a step input (a function that transitions from 0 to 1 in a single step). The Butterworth filter will produce an artificial minimum before the step and an artificial maximum after the step (Robertson and Dowling 2003). The Butterworth filter, however, has a better roll-off and therefore is better for filtering data that will be double differentiated, such as marker trajectories. In contrast, the critically damped filter is better for filtering rectified EMG data that will be used to determine onset times because this filter responds more rapidly and in the correct direction to quickly recruited muscles.

Furthermore, it is possible to calculate the coefficients so that the filter becomes a high-pass filter instead of a low-pass filter (Murphy and Robertson 1994). The first step is to adjust the cutoff frequency by the following:

$$f_c = \frac{f_s}{2} - f_{c-old} \qquad (12.22)$$

where f_c is the new cutoff frequency, f_{c-old} is the old cutoff frequency, and f_s is the sampling frequency. The coefficients (C_1 through C_5) are then calculated in the same way that they were for the low-pass filter and then the following adjustments are made:

$$c_1 = C_1,\; c_2 = -C_2,\; c_3 = C_3,$$
$$c_4 = -C_4,\; \text{and}\; c_5 = C_5 \qquad (12.23)$$

where c_1 through c_5 are the coefficients for the high-pass filter. The data can now be passed through the filter, forward and backward, just as in the low-pass filter case outlined previously.

Generalized Cross-Validation

Another commonly used data-smoothing technique designed by Herman Woltring (see From the Scientific Literature) called *generalized cross-validation* (GCV), occasionally described as the Woltring filter, behaves similarly to a bidirectional (zero-lag) Butterworth low-pass filter. One difference is that instead of specifying a cutoff frequency, the user may allow the software to determine its own characteristics or the user may enter a value corresponding to the mean square error (MSE) of the motion-capture system. This MSE is equivalent to the residual error computed during the calibration process of the motion capture system. In many gait laboratories this value could be 1 mm or less. Using too high a value results in oversmoothing the trajectories. This technique also has the advantage that it can fill gaps in the data stream because it does not require equidistantly sampled data (i.e., constant sampling rate).

SUMMARY

This chapter outlined the basic principles and rules for characterizing and processing signals acquired from any data collection system. Special emphasis was given to the frequency or Fourier analysis of signals and data smoothing. Data-smoothing techniques are of particular interest to biomechanists because of the frequent need to perform double time differentiation of movement patterns to obtain accelerations. As illustrated in chapter 1, small errors in digitizing create high-frequency noise that, after double differentiating, dominates true data history. Removing high-frequency noise prior to differentiation prevents this problem. The biomechanist, however, should be aware of how these smoothing processes affect data so that an appropriate method can be applied without distorting the true signal.

FROM THE SCIENTIFIC LITERATURE

Woltring, H.J. 1986. A FORTRAN package for generalized, cross-validatory spline smoothing and differentiation. *Advances in Engineering Software* 8(2):104-13.

The purpose of this technical report was to describe a software package for smoothing noisy data such as encountered by motion-capture systems. The report includes software written in FORTRAN that when given a set of data and various options computes natural B-spline functions that remove the high-frequency noise contained in the data stream. Dr. Woltring provided several means of creating the suitable smoothing function based on either a priori knowledge of the noise level of the data-capture system, the effective number of degrees of freedom in the smoothing function, or a generalized cross-validation or mean-squared prediction error criteria as described by Craven and Wahba (1979). His software assumes that the noise is uncorrelated (i.e., random) and additive and that the underlying signal is smoothly varying (i.e., has no rapid transients). Furthermore, the data stream does not need to be equidistant in time as is the case for digital filtering software. Although difficult to implement, the software is unique in that it does not require different cutoff frequencies for each marker trajectory. Several commercial software manufacturers (e.g., C-Motion, Vicon) have therefore included it as an alternative to Butterworth low-pass filters.

SUGGESTED READINGS

Antoniou, A. 2005. *Digital Signal Processing. Systems, Signals, and Filters.* Toronto: McGraw Hill Professional.

Press, W.H., B.P. Flannery, S.A. Teukolsky, and W.T. Vetterling. 1989. *Numerical Recipes in Pascal*, 422-97. New York: Cambridge University Press.

Smith, S.W. 1997. *The Scientist and Engineer's Guide to Digital Signal Processing.* San Diego, CA: California Technical Publishing.

Transnational College of LEX. 2012. *Who Is Fourier? A Mathematical Adventure.* 2nd ed. Trans. by Alan Gleason. Belmont, MA: Language Research Foundation.

Chapter 13

Dynamical Systems Analysis of Coordination

Richard E.A. van Emmerik, Ross H. Miller, and Joseph Hamill

Traditional analysis methods in biomechanics focus on data from a single joint or segment and typically derive discrete measures such as peak displacement, velocity, and force from individual time series data. Recent analysis methods from a dynamical systems approach instead focus more on coordination of joint or segmental angular rotations in which patterns of relative motion are important indicators of performance. In this chapter, we

- discuss the importance of a dynamical systems–based approach with an emphasis on coordination and coordination variability for biomechanical analysis;

- present the basic concepts of a dynamical systems–based approach;

- present well-established analysis techniques for assessing movement coordination, including relative phase and vector coding methods;

- present methods for assessing variability in human movement based on the coordination analysis methods; and

- discuss the benefits and limitations of each technique and the conditions under which these should and should not be used.

MOVEMENT COORDINATION

The human body contains almost 800 muscles that combine to bring changes in the action of the joints, segments, and overall movement of the body in space. For smooth, goal-directed movements to occur, the different degrees of freedom at each spatiotemporal scale (e.g., motor units, muscles, segments, joints) must be integrated into functional units. Coordination involves bringing the multiple degrees of freedom at each level into proper relations (Turvey 1990). Furthermore, this coordination must occur over multiple scales of space and time. This may be seen in the integration between, for example, motor unit firing patterns and force production at the muscle level or the coordination between respiratory and locomotor systems. One central question in the control of movement is how the wide range of degrees of freedom of the human body become coordinated.

Degrees of Freedom

The neurobiologist Paul Weiss (1941) defined *coordination* as the selective activation of degrees of freedom in such combinations that their united action will result in organized motor activity. Many years ago, the Russian movement scientist Nicolai Bernstein elegantly described the redundancy in the numbers of available components or degrees of freedom to carry out a movement task; our action system with its multiple degrees of freedom enables different solutions to a particular task. Bernstein defined coordination as mastering redundant degrees of freedom involved in a particular movement, thereby turning joints, segments, muscles, and motor units into controllable systems (Bernstein 1967; Turvey 1990).

The redundancy of degrees of freedom as explained by Bernstein is present even in basic movements involved in locomotion. Locomotion at different speeds entails coordinating the many degrees of freedom, where control activity is involved in recruiting physiological mechanisms with high degrees of freedom into mechanical movement "templates" or models with low degrees of freedom (Full and Koditschek 1999). These lower-dimensional mechanics could involve walking by vaulting, using the dynamics of the inverted pendulum, or running by bouncing, using mass-spring damper dynamics. In this neuromechanical perspective, the biomechanical properties of the body, in addition to the neural system, play a significant role in controlling the degrees of freedom in the human body in order to create coordinated locomotor patterns. Force-length and force-velocity relations in individual muscles, as well as biarticular muscles that couple two joints, can help stabilize movement and offer lower-level contributions to the degrees of freedom problem in human motor control (Van Ingen Schenau and Van Soest 1996).

Other researchers picked up on Bernstein's views on coordination and redundancy of degrees of freedom and observed that especially in the early phases of acquisition of new movement repertoires, reducing degrees of freedom by locking joints or simplifying relations between joints and segments is a strategy to allow controllability of the redundant degrees of freedom in the action system (e.g., Vereijken et al. 1992). Later in the learning process, degrees of freedom may be released, incorporated into the movement, or optimized. More recent research in the area of coordination and degrees of freedom is moving away from defining coordination as overcoming excessive or redundant degrees of freedom and proposes instead the concept of "abundancy," where all degrees of freedom contribute to the task, offering both stability and flexibility (e.g., Latash et al. 2002).

Oscillatory Movement Dynamics

Bernstein (1996, posthumous publication) identifies four levels of the construction of movement: the first level is muscle tone and focuses primarily on postural muscles and the maintenance of body orientation; the second level, muscular-articular links or synergies, subserves all possible patterns of locomotion in the terrestrial environment; levels three and four, space and actions, incorporate aspects of perception and intention in movements. Coordination at the level of synergies is most clearly revealed through the study and modeling of oscillatory movements (Turvey and Carello 1996). A major impetus for the development of such models was provided through the work of the physiologist Von Holst (1939/1973). Von Holst distinguished between two different forms of coordination, termed *absolute* and *relative coordination*. Two oscillators are in absolute coordination when they are phase and frequency locked. In contrast, oscillators in relative coordination maintain tendencies to particular phase couplings but exhibit a much wider array of observed frequency and phase relationships (see figure 13.1).

The phase relation between two oscillators has been proposed as a fundamental variable through which principles of coordination at the level of synergies can be revealed. This approach is represented by proponents of the synergetics (e.g., Haken et al. 1985; Kelso 1995) and the natural-physical approaches to movement coordination (Kugler and Turvey 1987; Turvey 1990). In the synergetics approach, a distinction is made between collective or "order" parameters and "control" parameters. Order parameters are variables that identify the macroscopic or collective behavior of a system. For example, in quadrupedal locomotion of animals, the different gait patterns (e.g., walk, trot, canter, and gallop) are characterized by changes in the phase relations between the legs. In this case the phase relation is the order parameter, which changes between the different gait modes but remains

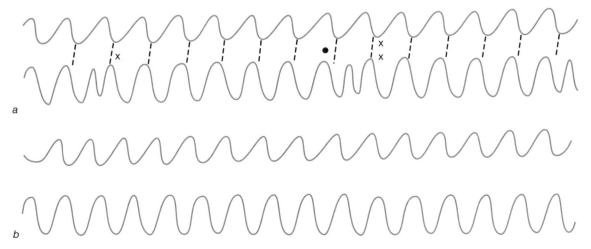

▲ **Figure 13.1** Rhythmic oscillations of the pectoral and dorsal fins of *Labrus* in the experiments of Von Holst. *(a)* Both fins in relative coordination. *(b)* Both fins in absolute coordination.

Reprinted from E. Von Holst, 1973, *The behavioral physiology of animals and man: Selected papers of E. Von Holst*, Vol. 1. (Coral Gables: University of Miami Press). Translated in 1979.

invariant within the particular gait mode. Phase relations between body segments or limbs are considered order parameters because of their fundamental reflection of cooperativity between components in the system.

The stability and transition dynamics of a synergy or movement pattern can be revealed by the systematic manipulation of a nonspecific control parameter. Control parameters can be used to reveal regions of stability and instability in coordination. In the quadrupedal locomotion example, gait speed can function as a control parameter, because the systematic manipulation of speed will reveal the different phase relations between the legs in the different gait modes. In the next section we discuss the application of relative phase techniques to the identification of stable coordination modes as well as transitions between these modes in human movement.

Movement Stability and Transitions

How a coordination analysis based on relative phase can help our understanding of movement stability and change is nicely illustrated in experiments by Kelso (1995) and his colleagues. Kelso asked participants to oscillate both index fingers in the transverse or sagittal planes, either in an in-phase coordination pattern (both fingers moving inward and then outward) or in an antiphase pattern (both fingers moving to the right and then to the left; similar to the motions of the fins in the experiments by Von Holst in figure 13.1). When a participant starts in the antiphase pattern, a systematic increase in the oscillatory frequency of both fingers will at first keep the coordination in the antiphase pattern, but at some critical frequency a sudden and abrupt transition to the in-phase coordination pattern occurs. An important feature of this transition process is the increase in variability of the relative phase between the fingers before the transition. Remarkably, when subjects subsequently reduce their frequency of oscillation, no return to the antiphase mode occurs. No transition is observed when you start with the in-phase pattern and increase the frequency.

Further analysis and modeling of these finger movement transitions revealed that the only stable finger coordination modes are in-phase and antiphase. These stable states are revealed by the dynamics of the relative phase between the fingers:

$$\dot{\phi} = \Delta\omega - a\sin(\phi) - 2b\sin(2\phi) + N \qquad (13.1)$$

where ϕ is the relative phase, $\dot{\phi}$ is the change in relative phase, the ratio of b/a is inversely related to the frequency of oscillation, and N represents random noise or fluctuations in the system. $\Delta\omega$ is the difference in natural frequency between the oscillators. When the symmetry between the oscillators is large (as is the case

in the finger oscillation experiment) and the frequency of oscillation is low, there are two stable coordination patterns or modes: in-phase ($\phi = 0°$ relative phase) and antiphase ($\phi = \pm\pi$) finger movements (see figure 13.2). These stable coordination patterns are called *attractors*. The depths of the potential wells in figure 13.2 are an indication of the strength (or stability) of the attractor. The different sizes of the potential wells help explain why there is a transition from antiphase to in-phase but not from in-phase to antiphase when frequency is increased; the potential well for the in-phase pattern is much deeper than the well for the antiphase pattern (figure 13.2a). Changes in coordination (also called *bifurcations* or *phase transitions*) emerge when, for critical values of the control parameter, the layout or the stability of the attractor changes. This can be seen in the different layouts of the attractors in figure 13.2a and b for the low- and high-frequency finger oscillation conditions when the symmetry in oscillator characteristics is high. At higher frequencies, the potential well for the antiphase mode disappears and there is only one stable mode at $\phi = 0°$ representing in-phase coordination.

When the symmetry between the oscillators is low (figure 13.2c and d), the number of stable coordinative states is also reduced. This can be observed, for example, when we try to coordinate leg movements and arm movements. Experiments show that coordination patterns are less stable compared with the patterns observed in the finger oscillation experiments. Fluctuations play a key role in these qualitative changes in pattern. These fluctuations can be visualized by movements of the "black ball" in the potential well. The deeper the well, the less likely it will be that the system will change state. However, changes in state are more likely to occur when the potential wells become shallow, as is the case in the low-symmetry example in figure 13.2.

Variability and Coordination

The multiple degrees of freedom involved in the coordination and control of human movement are a potential source of variability. In biomechanics and motor control, variability is traditionally equated with noise, is considered detrimental to system performance, and is typically eliminated from data as a source of error. In the assessment of coordination changes due to learning, aging, and disease, the presence of variability is still regarded as one of the most powerful indicators of poor performance. An increasing body of literature in the biological and physical sciences, however, stresses the beneficial and adaptive aspects of variability in system function. In the traditional perspective, higher levels of skill, competence, and health are associated with decreased variability. Current dynamical systems perspectives, however, have shown that the path to

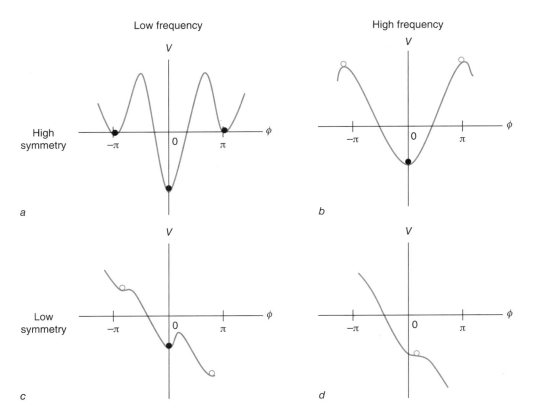

▲ **Figure 13.2** Potential landscapes of the stable and unstable pattern dynamics derived from the Haken, Kelso, and Bunz (HKB) model of rhythmic coordination. Filled circles show stable states. Open circles show unstable states. ϕ is relative phase, with $\phi = 0°$ representing in-phase coordination and $\phi = \pm\pi$ antiphase coordination. V is the potential function derived from equation 13.1.

Figure adapted from Kelso 1995.

frailty is identified by a loss of variability, reflecting fewer interactions between the degrees of freedom of the body (Lipsitz 2002). This hypothesis about loss of complexity can also be applied to neurological disease or orthopedic injuries (figure 13.3). Over time, reductions in effective degrees of freedom, interacting components, and synergies involved in the control of the biological system may become associated with a loss of variability. When these reductions in degrees of freedom and variability reach a critical threshold, injury or disease may emerge.

Increased variability or fluctuations are also essential features of abrupt changes (or phase transitions) in movement patterns. A signature feature of a phase transition is the appearance of instabilities in the coordination pattern. These instabilities are characterized by (1) a strong enhancement of fluctuations ("critical fluctuations") in the coordination before the transition and (2) a large increase in the time it takes to recover from a disturbance and return to steady state ("critical slowing down"). Critical fluctuations in complex systems arise due to the influences of elements at a more micro-

scopic level compared with the level of interest. These microscopic influences, together with environmental fluctuations, "pull" the system away from its current stable state or attractor. Fluctuations around the transition point from antiphase to in-phase finger movements have been observed during transitions in bimanual coordination (Kelso et al. 1986), during gait in the transition between walking and running (Diedrich and Warren 1995; Lamoth et al. 2009), and in coordination changes in the upper body and between arms and legs in human walking (Van Emmerik and Wagenaar, 1996a, 1996b; Wagenaar and Van Emmerik 2000).

Critical slowing down around the transition point is best shown by the changes and deformations in the potential landscape around the antiphase attractor (figure 13.2). As the control parameter is increased or the symmetry is low, the potential well becomes more and more shallow and a disturbance away from the fixed point will result in a slow relaxation back to the potential minimum (i.e., the convergence will be longer compared with the lower frequencies when the potential well is steep). Empirical evidence for this critical slowing down

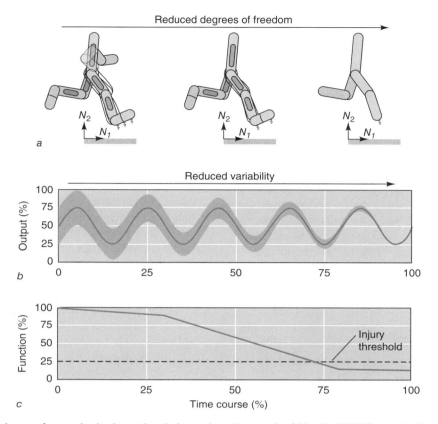

▲**Figure 13.3** Loss of complexity hypothesis based on the work of Lipsitz (2002) applied to injury or disease. Over time, reductions in effective degrees of freedom, interacting components, and synergies *(a)* become associated with a loss of variability in the system *(b)*. When these reductions in degrees of freedom and variability reach a critical threshold, injury or disease may emerge *(c)*.

in bimanual coordination has been provided by Scholz and colleagues (1987).

Research in biomechanics is beginning to explore the role of variability in movement. An important emphasis is the role that variability plays in the biomechanics of sport, injuries, and disease (Davids et al. 2003; Hamill et al. 1999, 2006; Wheat and Glazier 2006). When a movement is performed repetitively, the motions of the body's segments will vary somewhat, even for a cyclical motion like running. As mentioned previously, the traditional view is that variability decreases with the level of skilled performance and increases with the level of injury or disease. A common assumption in many studies on locomotion is that increased variability in traditional gait parameters (such as stride length and stride frequency) is associated with instability and increased risk of falls. Although increased variability in these spatiotemporal patterns of footfalls may indicate potential gait problems, an accurate understanding of the mechanisms that underlie instability and lead to falls requires insight into the dynamics of segmental coordination in the upper and lower body. As indicated earlier, in multiple degrees of freedom systems,

variability in performance is an inevitable and necessary condition for optimality and adaptability. Variability patterns in traditional gait parameters, therefore, might not reflect variability patterns in segmental coordination of upper-body and lower-extremity segments, as has been shown in research on Parkinson's disease (Van Emmerik et al. 1999).

In biomechanical research on running injuries, several studies have demonstrated an association between reduced coordination variability and orthopedic disorders (Hamill et al. 2006 for review). Coordination variability refers to the range of coordinative patterns that the organism exhibits while performing a movement. Coordination variability is often quantified as the between-trial or between–gait cycle standard deviation of the movement. Coordination variability is often assessed through relative phase and vector coding analysis methods. Several studies have shown that a certain amount of variability appears to be a signature of healthy, pain-free movement (e.g., Hamill et al. 1999; Heiderscheit et al. 2002; Miller et al. 2008). Hamill and colleagues (1999) and Heiderscheit and colleagues (2002) found that runners with unilateral patellofemoral pain were less variable in segment couplings involving the knee joint than were healthy runners. The authors suggest that this finding is indicative of a narrow range of coordination patterns that allowed for pain-free running. Miller and colleagues (2008) reached a similar conclusion on runners with and without a history of iliotibial band syndrome. Pollard and colleagues (2005) found that women were less variable than men in several lower-limb couplings during a high-speed unanticipated cutting maneuver, and the investigators suggested that these gender differences could be related to the prevalence of anterior cruciate ligament injuries in women. However, because all of these studies were retrospective in nature, the authors could not infer a causal relationship between variability and injury. Prospective studies on coordination variability and injury development are needed to assess this relationship.

In summary, from a dynamical systems perspective, variability is not inherently good or bad but rather indicates the range of coordination patterns used to complete the motor task. This offers a different view in comparison to the more traditional perspective that variability is undesirable. In contrast, there is a functional role for variability that expresses the range of possible patterns and transitions between patterns that a movement system can accomplish. Both abnormally low and high levels of variability may be detrimental to the functioning of the system.

FOUNDATIONS FOR COORDINATION ANALYSIS

In the next sections we present the basic concepts of dynamical systems that form the foundations of the analysis of coordination and coordination variability. These concepts include the notion of the state space and the different types of attractors, or preferred regions, in these state spaces that are instrumental in assessing stability and change in movement patterns.

The State Space

Biomechanical data can be represented graphically in a variety of different ways. A traditional and very informative type of graph is to display the changes in kinetic or kinematic parameters over time. Figure 13.4 presents an example of the changes in the left and right thigh angular displacements over a time period of 10 s while a subject is walking on a treadmill. Information about peak flexion and extension angles can be obtained from these graphs and quantified. A closer inspection of the graphs makes it quite evident that the coordination between these segments is primarily antiphase. However, especially during the stance phase there are more subtle changes in the patterns of coordination that are harder to obtain from these plots.

In the dynamical systems approach, the reconstruction of the so-called state space is essential in identifying the important features of the behavior of a system, such as its stability and ability to change and adapt to different environmental and task constraints. The state space is a representation of the relevant variables that will help identify these features. To help us understand and quantify coordination between joints or segments, it can be very useful to represent the system in a state space that is based on an angle-angle relative motion plot. Figure 13.5 presents a state space of the relative motion of the time series of the thigh and leg shown earlier in figure 13.4. This angle-angle plot can reveal

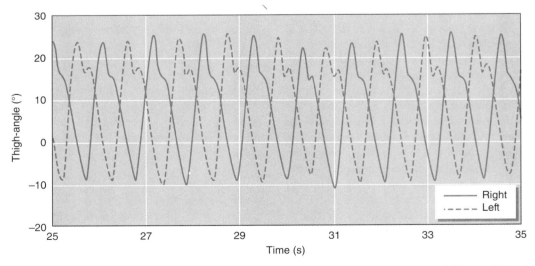

▲ **Figure 13.4** Individual time series of thigh (left and right) angular displacements while a subject is walking on a treadmill at a speed of 1.2 m/s.

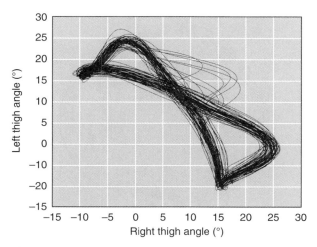

▲ Figure 13.5 Angle-angle plot of the thigh segmental angular time series presented in figure 13.4.

regions where coordination changes take place as well as parts of the gait cycle where there is relative invariance in coordination patterns. These coordinative changes in the angle-angle plots can be further quantified by vector coding techniques that are discussed later.

Another form of state space is where the position and velocity of a joint or segment are plotted relative to each other. This state-space representation is also often referred to as the *phase plane*. An example of a position-velocity phase-plane plot is shown in figure 13.6, where the angular displacement of the thigh is plotted against the angular velocity. This is a higher-dimensional state space because the time derivative of position is used to identify the pattern. The phase-plane representation is a

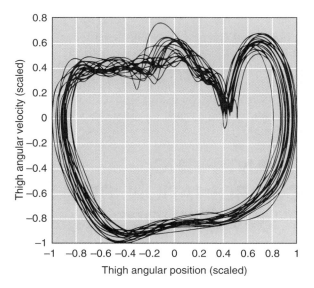

▲ Figure 13.6 Phase-plane plot of thigh angular displacement versus velocity.

first and critical step in the quantification of coordination using continuous relative phase techniques.

Attractors in State Space

In most forms of human movement, the dynamics in state space are limited to distinct regions, as can be seen in the phase-plane plot in figure 13.6, where the pattern in state space is limited to a fairly narrow, cyclical band. In dynamical systems, preferred regions in state space onto which the dynamics tend to settle are called *attractors*. Attractors can come in a variety of different forms. A seemingly simple kind of attractor is the point attractor. In this case, the dynamics in the system tend to converge onto one relatively fixed value in the state space. Figure 13.7, *c* and *d*, provides an example of a point attractor. The point attractor dynamics are shown in figure 13.7*c* by a relatively consistent value of discrete relative phase throughout the gait cycle. The discrete relative phase is based on the occurrence of peak flexion in the left and right thigh angular displacements, and the time series in 13.7*c* show that there is a consistent antiphase or 180° coordination pattern. This antiphase coordination can already be observed by comparing the individual time series in figure 13.4 but is more objectively quantified by the state-space plot in figure 13.7*d*, where so-called return maps (plotting the coordination at x_n vs. x_{n+1}, where *n* is the cycle number) identify a fixed-point attractor in state space. The relative phase dynamics are of the fixed-point type in this case, as a perturbation (sudden change in treadmill speed and return to the original speed; figure 13.7) results in a return to the original antiphase dynamics with a relative short latency. The perturbation here consisted of a brief (5 s) increase in treadmill speed from 1.2 m/s to 2.0 m/s, with a subsequent return to 1.2 m/s.

Whereas the coordination between the right and left legs can be identified in the form of a point attractor with a relatively fixed value of coordination from cycle to cycle, the dynamics of the individual limb segments show a very different pattern. These are clearly cyclical, as can be seen in the phase planes in figure 13.7, *a* and *b*. Attractors of this form are called *limit-cycle attractors*, and the dynamics converge onto a cyclical region in state space. These limit-cycle attractors are typically identified on the basis of a very narrow, overlapping band of the trajectories in the state space. However, the existence of the narrow band is a necessary but not sufficient condition to characterize the dynamics as a limit-cycle. An essential feature of the limit-cycle attractor is stability with respect to perturbation; to classify as a limit-cycle attractor, the system should also show resistance to perturbations. An example of this is given in figure 13.7, *a* and *b*. The regular cyclic pattern in the phase plane (solid line) represents the steady-state gait patterns while a subject was walking at a speed of 1.2

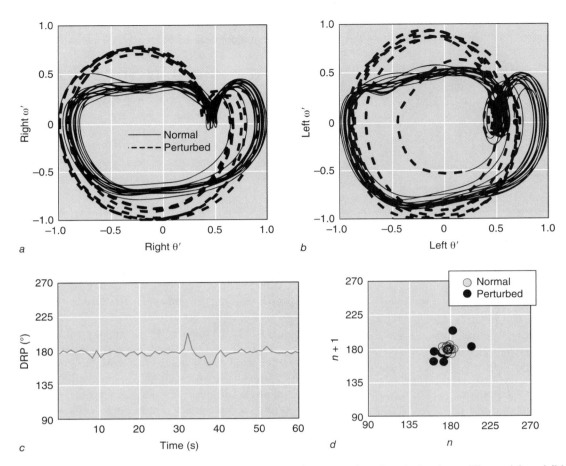

▲**Figure 13.7** Plots of point and limit cycle attractors under normal and perturbed conditions. *(a)* and *(b)* Thigh segmental angular position versus velocity phase planes while a subject was walking at 1.2 m/s. Speed was suddenly increased to 2.0 m/s after 30 s for a duration of 5 s, then returned to 1.2 m/s. *(c)* Time series of discrete relative phase (DRP), and *(d)* return map plot of current relative phase value versus the next relative phase value (DRP$_n$ vs. DRP$_{n+1}$). Open circles represent the stable coordination pattern (point attractor) around 180°, and the closed circles show the deviation from the attractor as a result of a sudden speed perturbation.

m/s. The dashed trajectories represent the perturbation phase of the trial when speed was suddenly increased to 2.0 m/s and the consequent return back to the original pattern. This return to the preperturbation dynamic is an essential feature of the limit-cycle attractor. The patterns in the phase plane can also serve as an energy plot. The convergence and divergence of trajectories in the phase plane can identify the loss and gain of energy in the system.

Higher-dimensional state spaces (three dimensions and up) can also reveal more complex types of attractors: quasiperiodic and chaotic attractors. The chaotic attractor demonstrates both stable attraction to a region in state space and variability. Figure 13.8 shows an example of the Lorenz attractor, a well-known chaotic attractor that emerges from dynamic interactions in fluid and air flow systems (Strogatz 1994). These dual features of stability and adaptability can be associated with a higher pattern

complexity that is now commonly regarded as reflective of healthy and expert systems (see figure 13.3). As an example, increased heart rate variability is considered an important indicator of healthy heart function, reflecting a degree of complexity in organization in which disruptions can be compensated for more easily (Glass 2001; Lipsitz 2002).

QUANTIFYING COORDINATION: RELATIVE PHASE METHODS

In the next section we present in more detail the basic procedures and analysis methods that are now commonly used to quantify coordination in biomechanical systems. All of the coordination analysis methods presented find their basis in the concepts of the state space and the

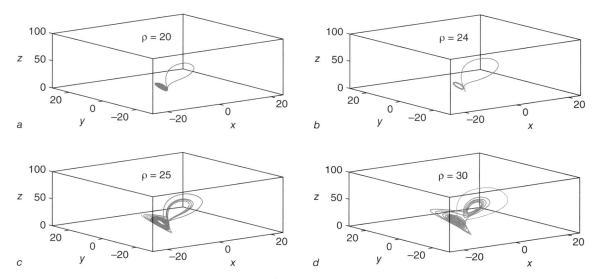

▲**Figure 13.8**   The Lorenz attractor. The state space dimensions *(x, y, z)* represent the dynamics of fluid flow. The control parameter ρ, the Rayleigh number that affects heat transfer, is increased from panels *a* to *d*. For small values of ρ, the system is stable (showing convection) and evolves to one of two fixed-points attractors. At a critical value of $\rho = \sim25$, the fixed points become repelling instead of attracting and complex chaotic trajectories and turbulent flows emerge (Strogatz 1994).

different attractor dynamics in these state spaces. We first discuss relative phase techniques (discrete relative phase, continuous relative phase, Fourier phase, and the Hilbert transform), followed by vector coding procedures for quantifying angle-angle diagrams. The benefits and limitations of each technique are discussed.

Discrete Relative Phase

Discrete relative phase (DRP) analysis allows for an assessment of coordination and changes in coordination based on relevant events in the time series of interest. In the examples of the thigh flexion and extension time series in figure 13.4, the relevant event may be peak forward flexion. DRP analysis derives the relative timing of corresponding peaks from the two different time series (such as the left and right thigh angular motions in figure 13.4), scaled to the overall cycle duration of one of the time series (figure 13.9):

$$DRP = \left(\frac{\Delta t}{T}\right) \times 360° \qquad (13.2)$$

where Δt represents the timing difference in the peak events in the two time series and T the overall cycle duration for one of the time series. Scaling to the cycle time is necessary to correct for differences in absolute duration of the cycle, which could influence the relative timing and the coordination. This would occur, for example, at different gait speeds, where stride cycle duration decreases with increases in speed. The DRP

is typically calculated for each cycle and then averaged across multiple cycles.

The advantage of the DRP analysis is its relative simplicity in that no further derivation (such as velocity, acceleration, and state space reconstruction) is needed. This allows the researcher to use the original signal such as the segmental angular time series of the thigh motion. Limitations of the DRP analysis are that in most cases only one data point of the time series is used in the assessment

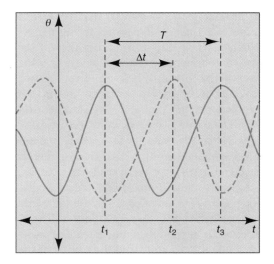

▲**Figure 13.9**   Discrete relative phase analysis procedure. θ represents angular excursion of the segmental or joint time series of interest (dashed and solid lines).

of coordination. This is usually not a problem in very regular signals in which the coordination does not change within the movement cycle, such as in the finger oscillation experiments discussed earlier. However, in signals in which the coordination can change within the cycle, the DRP analysis may be more problematic. Possible solutions can include choosing more points in the gait cycle for the DRP analysis or using vector coding or continuous relative phase methods (discussed later in this chapter). Another possible problem with the DRP analysis is that the peaks in the signal need to be well defined. It is not recommended that DRP analysis be conducted on time series with multiple peaks that are relatively inconsistent and may vary from cycle to cycle. This not only could provide incorrect coordination results but may artificially increase coordination variability measures derived from DRP analysis.

Discrete Relative Phase and Multifrequency Coordination

The discrete relative phase analysis outlined in the previous section works well for signals in which the peak events occur at similar frequencies within the cycle. In the example of left thigh versus right thigh coordination, the peak flexion happens once per stride cycle for each leg. In multifrequency coordination, as for example in the frequency coupling between the arms and the legs at lower walking speeds (a 2:1 coupling), the DRP analysis presented in the previous section is harder to implement. This is especially the case for systems that can display a much wider variety of frequency couplings, especially if these couplings can also change throughout the movement.

An important example in which the coordination is not limited to a 1:1 frequency coupling can be found in coupling between the respiratory and the locomotor systems. The coordination between the locomotor and respiratory rhythms is important for stable and efficient gait patterns. This coordination typically can occur in a variety of different frequency couplings that may change as a function of walking speed, fatigue, or training status.

Much of the research on locomotor-respiratory coordination (LRC) has focused on *entrainment* between these biological rhythms, which is defined as a strong coupling or "locking" of the frequency and phase relations of the two oscillations. Several mechanisms can act to entrain respiration and locomotor rhythms in various

FROM THE SCIENTIFIC LITERATURE

Diedrich, F.J., and W.H. Warren. 1995. Why change gaits? Dynamics of the walk to run transition. *Journal of Experimental Psychology: Human Perception and Performance* 21:183-201.

As adults, humans use two distinct gait modes (walking at slow speeds; running at fast speeds) with an abrupt transition between modes when speed is manipulated as a control variable. Diedrich and Warren (1995) examined intralimb coordination dynamics during the walk-run and run-walk transitions. In experiment 1, subjects performed continuous walk-run (or run-walk) transitions while the treadmill belt speed was continuously increased (or decreased) from 0.95 to 3.60 m/s over 30 s. Kinematic data (joint angles) were measured in the sagittal plane. Data were also collected while subjects walked at a range of steady belt speeds for 30 s. In experiment 2, the continuous transition condition of experiment 1 was repeated at a range of fixed stride frequencies around the subject's preferred stride frequency. The discrete relative phases (DRPs) of peaks in the joint angle time series were calculated. Specifically, the time between two successive hip extension or knee extension peaks was defined as 0° to 360°, and the timing of the ankle plantar flexion peak within this range was identified as the DRP. The walk-run transition speed (2.09 m/s) was slightly faster than the run-walk transition speed (2.05

m/s). In the continuous transition trials of experiment 1, a sudden decrease in the ankle-hip and ankle-knee DRP of about 50° occurred during the last walking step prior to running, where running was defined by the appearance of a flight phase. DRP increases by a similar magnitude immediately before the run-walk transition. From the steady speed condition, variability in DRP (quantified as the within-trial standard deviation) increased near and within the gait transition region (Froude number ≈ 0.5) but did not decrease following the transition. Similar results were seen in the variability of the stride frequency. As evidence that speed rather than another stride parameter was the control variable behind the gait transitions, in experiment 2 subjects consistently transitioned from walking to running at a speed of 2.2 m/s regardless of the imposed stride frequency. The authors interpreted the overall results as evidence that human locomotion shares many characteristics of a nonlinear dynamical system, with preferred gaits characterized by stable phase relations and hysteresis, and while phase transitions are preceded by an increase in variability and loss of stability of the current gait mode.

species of animals. For example, horses have an almost fixed breathing to stride ratio of 1:1 due to the constraints put on the thoracic region from the repeated impact loading of the forelimbs striking the ground (Bramble and Carrier 1983). In addition, the visceral mass dynamics (visceral "piston") may constrain the coupling between these systems. Human locomotion, in contrast, has a much wider range of frequency couplings between locomotion and breathing rhythms. Researchers have observed couplings of limb movements to breaths that include 1:1, 2:1, 3:1, 3:2, 4:1, and 5:2, with 2:1 being used most often (Bramble and Carrier 1983). These frequency relationships may also change during running and walking and as a function of running experience.

To assess different frequency couplings and their impact on coordination, a modification to the DRP analysis technique is needed. This technique was developed by McDermott and colleagues (2003) in their work on locomotor-respiratory coordination, but it can be applied to multifrequency coupling generally. The technique first generates a relative phase time series between end-inspiration (EI) of the breathing rhythm and heel-strikes of the locomotor rhythm from the consecutive gait cycles on the basis of the following formula:

$$\text{DRP}_{mf} = \left(\frac{t + nT}{T}\right) \times 360° \qquad (13.3)$$

where DRP_{mf} represents discrete relative phase in multifrequency systems, n is the number of complete stride cycles between each heel-strike and the subsequent EI, T is the time duration of the stride in which EI occurred, and t is the time lag from the beginning of the stride in which EI occurred to the subsequent EI. To illustrate this calculation of relative phase, simulated breath and heel-strike signals are plotted in figure 13.10. The heel-strike

signal in this example is a pulse wave with a period $T = 100$ ms, and the breath signal in this example is a sine wave with a period of 300 ms with a 75 ms phase shift. The resulting DRP_{mf} values for the three heel-strikes in breath 1 (with n equal to 2, 1, and 0) are therefore 990°, 630°, and 270°, respectively. Note that if there are 3 heel-strikes within one breath as in this example, the relative phase value of each heel-strike will fall within the specific ranges:

$$1080° \geq \text{DRP}_{mf1} > 720°$$

$$720° \geq \text{DRP}_{mf2} > 360°$$

$$360° \geq \text{DRP}_{mf3} > 0°$$

According to this relationship, the relative phase value of the first heel-strike (DRP_{mf1}) in the breath cycle will fall in the highest range and the last (DRP_{mf3} in this example) will fall in the lowest range (between 0° and 360°). Sample time series of relative phase from an individual walking at preferred speed at different stride frequencies are shown in figure 13.11a. The left part before the solid vertical line represents walking at 20% below preferred stride frequency (PSF), whereas the right part shows the transition to walking at 20% above PSF. The periodicity of these time series is then assessed by creating return maps (plotting the relative phase time series against time-lagged versions of the same time series; see figure 13.11, b and c). Frequency couplings are identified from these maps by applying specific range criteria to the appropriate map. Convergence to the line of identity in the return maps reveals the periodicity in the coupling (see McDermott et al. 2003 for a detailed overview of the methods). For example, at PSF − 20% (left column), there is convergence in the third-order return map (x_n vs. x_{n+3}) showing a dominant 3:1 coupling. For walking at

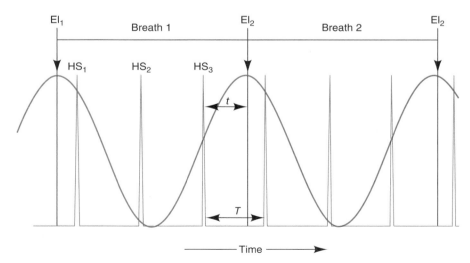

▲ **Figure 13.10** Derivation of multifrequency relative phase in equation 13.3.

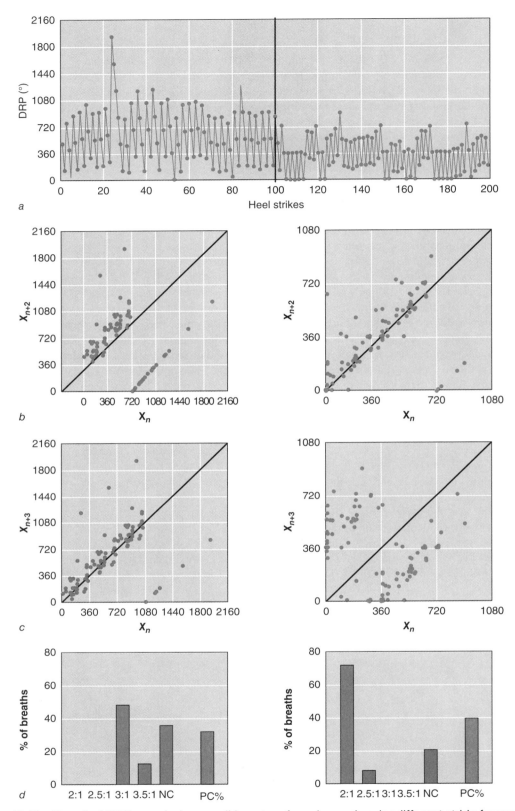

▲ Figure 13.11 Result of DRP$_{mf}$ analysis on walking at preferred speed under different stride frequency conditions from a representative subject. Left column: 20% below preferred stride frequency; right column: 20% above preferred stride frequency. *(a)* DRP time series, where the vertical bold line represents transition from −20% to +20% of preferred stride frequency. *(b)* and *(c)* Return maps based on these data. *(d)* Summary statistics of frequency and phase couplings (PC). NC = noncouplings.

PSF + 20%, there is convergence in the second-order return map (x_n vs. x_{n+2}), identifying a dominant 2:1 coupling at higher stride frequencies.

The next step in the LRC analysis is to establish a distribution of frequency couplings and noncouplings for each trial, expressed as a percentage of the total number of breaths. The frequency coupling that occurs most often is defined as the dominant coupling and is an indicator of coupling strength. The second dominant coupling and noncouplings are used as measures of variability in frequency coupling. Histograms of the percentage of occurrence of the different frequency couplings are shown in figure 13.11d for the low (left bottom panel) and high (right bottom panel) stride frequency conditions.

The analysis process so far quantifies the varying frequency coupling patterns that are observed both within and across the sample data sets independent of the phase relations that occur. Because phase is often dependent on the frequency relation, particularly for half-integer couplings, the next step in the LRC analysis is to quantify the phase coupling based on the frequency couplings identified within each time series. To accomplish this, phase coupling (PC) is assessed by the dispersion of points (noncouplings not included) from the line of identity in the lowest range of each of the return maps (DRP_{mf} between 0 and 360). This analysis represents the phase of the heel-strike preceding the EI (figure 13.10) of the corresponding breath cycle. Perfect phase coupling (no variability) occurs when all points fall on the line of identity. Variability in phase coupling is measured in terms of deviations from this line. Therefore, phase coupling is quantified by first calculating the Euclidian distance of each point (d_n) from the line of identity and then summing the weighted distances (wd):

$$wd_n = \begin{cases} 1 - \dfrac{|d_n|}{40\cos(45)}, & d_n \leq 40 \\ 0, & d_n > 40 \end{cases} \qquad (13.4)$$

$$\text{PC} = \frac{\sum_{n=1}^{m} wd_n}{m} \times 100\% \qquad (13.5)$$

where m is the total number of points in the lowest range of each of the return maps. This measure systematically weights points with distances from the line of identity that are greater than or equal to 40° by 0, and it weights those with distances less than 40° according to their distance and expresses their sum as a percentage of highest possible sum. Perfect phase coupling is shown by PC = 100%, with variability in the phase coupling increasing with further decrements from 100%.

The use of discrete relative phase in conjunction with systematic time lags and return maps allows us to quantify both frequency and phase coupling in a system that is highly variable and where multiple frequency couplings are present. These patterns could then be assessed in light of changes in a relevant parameter such as locomotor speed or frequency to reveal different transition dynamics. McDermott and colleagues (2003) used the LRC analysis technique (equation 13.3) to assess entrainment in the coupling between locomotion and respiration. The underlying assumption in much of the research on locomotion-respiration coupling is that a higher degree of entrainment is associated with more efficient energy utilization during movement. McDermott and colleagues found that entrainment alone is not appropriate for studying the highly variable interaction between locomotor and breathing rhythms. Results showed that greater locomotor experience is not associated with higher levels of entrainment but is reflected in differences in the adaptability to imposed speed or movement frequency conditions. This multifrequency DRP analysis may also be useful in quantifying coordination in general and with applications to training, disease, and aging.

Continuous Relative Phase

Continuous relative phase (CRP) is often considered a higher-order measure of the coordination between two segments or two joints. This higher order emerges from the derivation of CRP from the movement dynamics in the phase plane of the two joints or segments. Continuous relative phase analysis has been used to identify stability and transitions in the dynamics of bimanual coordination (see Kelso, 1995). Others have used CRP analysis to characterize joint or segmental coordination during gait (Hamill et al. 1999; Van Emmerik et al. 1999).

Although CRP may seem relatively easy to implement, several key concepts regarding the method and the interpretation must be addressed. In the original finger coordination experiments (see figure 13.2), CRP was used as a higher-resolution form of DRP. In these experiments, both the DRP and the CRP essentially yielded the same results: they demonstrated the change from antiphase coordination to in-phase coordination. The finger oscillations are characterized by sinusoidal movements, which allows for an interpretation of the DRP as representative of the instantaneous phase relation of two oscillators at multiple points in a cycle. However, this is not the case for oscillations that deviate from sinusoidal or even in sine waves whose frequencies are other than $0.5/\pi$ Hz (Peters et al. 2003). As CRP expands from predominantly sinusoidal signals into nonsinusoidal signals and partial oscillations, it is necessary to take great care in the calculation of CRP.

The CRP quantifies the coordination between two oscillators based on the difference in their phase plane angles. In gait studies, these oscillators are usually either two segments or two joints (i.e., thigh and leg segments or hip and knee joints). Note that although we model the motion of the segments and joints as a physical oscillator in the phase plane, these segments and joints have properties more complex than these more elementary physical oscillators. The coordination relationship between the oscillators in the human body is often described as coupling when we are quantifying the relation between the kinematics of two body segments that are linked anatomically and mechanically.

Construction of the Phase Plane

The first step in calculating CRP is to construct the phase planes of the two oscillators under investigation (figure 13.12). To reiterate, the phase plane is usually constructed from the position-velocity (or angular position–angular velocity) histories of the segment or joint in question (figure 13.6). However, the reconstruction of higher-order information for the assessment of coordination does not have to be limited to the position-velocity phase plane. We could also construct higher-order state spaces from velocity-acceleration, moment-angle, or moment-moment time series.

When one is constructing the plane using angular position and velocity, the angular position is typically generated from kinematic data collected via a motion-capture system. Angular velocity is then calculated with an appropriate differentiation method. Great care must be taken in the calculation of angular velocity depending on whether the angles have been calculated via a 2-D or 3-D analysis. In a 2-D analysis, angular velocity can be

calculated by using a central difference approximation to determine the time derivative of angular position. However, with a 3-D analysis, angular velocity cannot be calculated by simply differentiating the angular position (see chapter 2 for a detailed explanation for the calculation of angular velocity in 3-D).

Normalization of State Variables

A particularly important step in the CRP procedure involves normalizing the angular position and angular velocity profiles (θ' and ω', respectively, in figure 13.12). Normalization of the two signals that make up the phase plane is necessary to account for the amplitude and frequency differences in the signals. Normalization of amplitude is necessary because the amplitude of the velocity data is usually much greater than the amplitude of the position data. Velocity will then have a disproportionate impact on the phase angle. Comparing two signals that differ substantially in angular excursions will lead to the same problems. When the differences in frequencies between the two signals are not taken into account, artifacts in the final CRP measure appear in the form of a low-frequency oscillation. For a complete description of the necessity of normalizing these signals, see Peters and colleagues (2003).

Two methods for normalization have been used in the literature. In one case, the angular position and angular velocity can be normalized to a unit circle using the maximum and minimum values (Van Emmerik and Wagenaar 1996a). By using this method, however, we lose information regarding zero velocity (i.e., the zero on the vertical axis of the phase plot does not correspond to an actual zero velocity). In the second case (Burgess-Limerick et al. 1993), the angular velocity parameter is allowed to "float" below or above either −1 or +1 on the vertical axis. The latter method of normalization appears to be the most often used in the biomechanics literature. For this method, the normalized angular position in the range from −1 to +1 is

$$\theta'_i = \frac{2 \times \left[\theta_i - \min(\theta_i)\right]}{\max(\theta_i) - \min(\theta_i)_{-1}} \qquad (13.6)$$

where θ' is the normalized angular position, θ is the original angular position, and i is a data point in the cycle. A similar procedure is done with the angular velocity normalization, but this one will keep zero velocity at the origin of the phase plane:

$$\omega'_i = \frac{\omega_i}{\max\left[\max(\omega_i), \max(-\omega_i)\right]} \qquad (13.7)$$

where ω' is the normalized angular velocity, ω is the original angular velocity, and i is a data point in the cycle. Keeping the zero velocity at the origin allows the

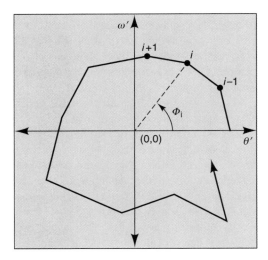

▲ **Figure 13.12** Construction of phase plane for continuous relative phase analysis. θ' represents normalized angular position, ω' normalized angular velocity, and $\phi(i)$ the phase angle at frame i.

researcher to use convergence and divergence in the phase plane to assess the energy changes in the system. Another way to normalize the phase plane is to divide the velocity by the mean oscillation frequency obtained over multiple cycles. This method has been used in research on the dynamics of interlimb coordination (Sternad et al. 1999).

In gait studies in which a number of cycles or strides are to be used, the decision on the maximum value must be determined. Normalization could be accomplished on a cycle-by-cycle basis or it could be done on the maximum value over multiple cycles. It has been demonstrated that from a qualitative point of view, the differences in these procedures are minimal. However, normalizing to the maximum value of multiple strides will better maintain the true spatial properties among cycles (Hamill et al. 2000). An example of normalizing to the maximum of a number of gait cycles is shown in figure 13.7. A potential problem with this method can arise when the maximum value is obtained from an aberrant cycle in the data. Careful inspection of the data is needed to avoid such outliers that may artificially change the maximum value used in the normalization procedure.

Once the normalization procedures have been accomplished, the next step is to scale the angular position and angular velocity profiles to an equivalent number of data points in each of the oscillators. In many cases, this is done by scaling the period of the cycle to 100%.

Calculation of the Phase Angle and Relative Phase

Finally, the phase plane is constructed by plotting the angular position versus angular velocity (figures 13.12 and 13.13). This phase plane is constructed for each of the oscillators. For each of the oscillators, the phase angle, $\phi(t)$, is obtained by calculating the four-quadrant arctangent angle relative to the right horizontal at each instant in the cycle by

$$\phi(i) = \tan^{-1}\left(\frac{\omega'(i)}{\theta'(i)}\right) \qquad (13.8)$$

where ϕ is the phase angle, ω' is the normalized angular velocity, and θ' is the normalized angular position (figure 13.12). The CRP angle of the coupling of the two oscillators (i.e., the two segments or joints) is then calculated as

$$CRP(i) = \phi_A(i) - \phi_B(i) \qquad (13.9)$$

where $\phi_A(i)$ is the phase angle of one oscillator and $\phi_B(i)$ is the phase angle of the other oscillator at data point i in the cycle. Phase angle (ϕ) values can range from 0° to 360° at any instant in time throughout the movement cycle (or from 0° to +180° and 0° to −180° with the discontinuity between quadrants 2 and 3, as used in figure 13.13).

This polar distribution results in potential discontinuities when we calculate the phase difference as is done in the CRP. To eliminate these discontinuities one can either apply circular statistical methods (Batchelet 1981; Fisher 1993; also see later sections) or limit the resulting CRP to the range 0° to 180°. Phase angle differences greater than 180° are then subtracted from 360°. This procedure preserves the full distribution of phase angles in the polar plot and allows for CRP measures without discontinuities. Simply folding the phase angles into two or even one quadrant can significantly affect the interpretation of the CRP data (Wheat et al. 2002). Maintaining the full polar plot for the assessment of CRP is essential for a correct interpretation of this coordination measure.

The CRP angles can be represented as a function of the period of the cycle (see figure 13.13). Given the assumption that the motions of the two oscillators are relatively sinusoidal (although they may not be) and are one-to-one in frequency ratio, information about their phasing can be derived. When the CRP angle is 0°, the two oscillators are perfectly in-phase. An example of in-phase coordination between two oscillators is the movement of the two windshield wipers on a car. A CRP angle of 180° indicates that the oscillators are perfectly antiphase. In the case of the windshield wipers, antiphase movement would result when both wipers rotate to the center of the windshield at the same time. Any CRP angle between 0° and 180° indicates that the oscillators are out of phase but could be relatively in-phase (closer to 0°) or antiphase (closer to 180°). Note that in the example of the windshield wipers, both the position and velocity are matched. This may not be the case in many of the human movement examples discussed before.

It is often tempting to use the CRP angle to discuss which oscillator is leading and which is lagging relative to the other oscillator. Because the phase angle of one oscillator is subtracted from the phase angle of another, the lead-lag interpretation is often assumed. However, the calculation of CRP described previously does not allow for such an interpretation. The lack of a lead-lag interpretation results from the fact that CRP is derived from two phase angles that in turn are obtained from angular position and velocity data. In addition, the normalization procedures used to calculate CRP make lead-lag interpretations difficult. Therefore, obtaining information on lead-lag information should not be the primary objective of CRP analysis. Excellent tools such as auto-correlation and cross-correlation analysis can provide insights into lead and lag aspects. The CRP is an assessment of coordination, and the phase plane is a higher-order representation of the dynamics. Any difference in orientation in phase angles reflects coordination in this phase plane. Assessment of lead-lag information, a single-dimensional temporal variable, should not be an

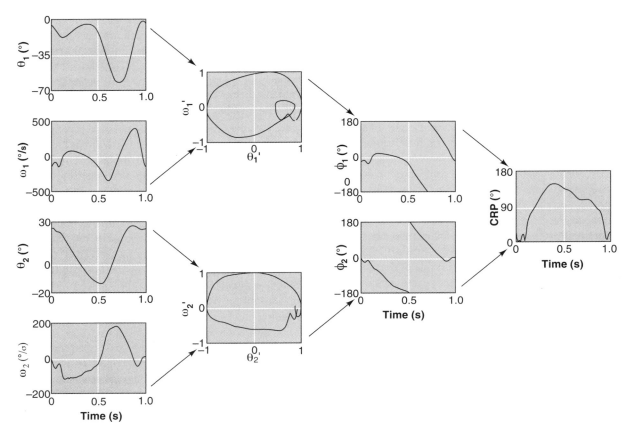

▲ **Figure 13.13** Procedures involved in continuous relative phase analysis. From left to right, the first column represents the original time series of angular position and velocity for the knee and hip joints during walking. The second column represents the normalized phase plane for each segment. The phase angle derived from each phase plane is depicted in the third column and the continuous relative phase (CRP; the difference between the two phase angles) in the fourth column.

objective of CRP analysis and is better performed with traditional cross-correlational methods.

Dependent Measures Derived From a CRP Analysis

There is little agreement in the literature on the types of dependent measures that are derived from a CRP analysis. It should be clear that an analysis of each coupling separately must be undertaken. The simplest measures to derive are the CRP angles at discrete points in the cycle. For example, in gait, we may define the cycle as a complete stride (i.e., from an event in one limb to the next occurrence of the same event in the same limb). We can then use the CRP angle at the initiation of the cycle or at any key (possibly functional) point throughout the cycle.

The most common measures derived from CRP data are the averages over either a discrete portion of the cycle period or a well-defined functional unit of the cycle. In the first case, the angles could be averaged over each 10% of the cycle. If we divide the cycle into functional units and use the right limb as an example, the first functional

unit may be from right foot contact to maximum knee flexion of the right limb, the second from maximum knee flexion to right toe-off, and the third from right toe-off to right foot contact. Another method of presenting CRP angles was introduced by Van Emmerik and Wagenaar (1996b). These authors analyzed the distribution of the relative phase across the stride cycle through CRP histogram analysis. This method of binning and the formation of histograms is explained in more detail later in the section on vector coding.

Assessment of Coordination Variability Based on CRP Analysis

The CRP time series obtained from equations 13.8 and 13.9 can be used to obtain a measure of coordination variability between two joints or segments. For a proper assessment of coordination variability, the following two key aspects need to be addressed: (1) Average variability measures should not be obtained directly from CRP time series that vary systematically throughout the movement (stride) cycle, and (2) variability measures can only be obtained from data that do not contain discontinuities. These two aspects are discussed in more detail.

To obtain a measure of variability, we typically calculate the standard deviation with respect to the average CRP in the data. However, this procedure can only be applied to a cycle in which the CRP maintains a fairly constant value throughout the cycle. This is observed in the finger oscillation experiments discussed earlier. As long as the coordination mode is stable (i.e., antiphase throughout the cycle), the standard deviation will present a measure of variability of coordination. However, in the case of human walking and running, for example, CRP fluctuates systematically throughout the gait cycle (figure 13.13). In this case, using the standard deviation around the mean CRP will provide spurious data, because the standard deviation also reflects the systematic changes in CRP across the gait cycle.

A better approach is to obtain a coordination variability measure that reflects the cycle-to-cycle variability. The procedure for this between-cycle variability is shown in figure 13.14. In figure 13.14a, the dashed lines represent the individual movement cycles and the solid line the average cycle. For n trials (each scaled to 100% of gait cycle), we obtain the between-cycle standard deviation of the CRP across the n trials for each percentage point of the gait cycle. The result is a

new CRP variability time series of the between-cycle standard deviation, as shown in figure 13.14b. Various summary dependent measures can then be obtained from this new CRP variability time series, including an overall average across the time series, variability across stance and swing phases, or variability across other identified functional phases of the gait cycle. The quantification of the between-cycle or between-trial coordination variability based on CRP will only be accurate when no discontinuities in the original CRP data exist. Procedures to avoid these discontinuities were discussed earlier in the assessment of the CRP analysis procedures and are addressed in the section on vector coding.

Alternative Measures of Relative Phase

This section presents some alternative methods for calculating the relative phase between two signals. These methods are the relative Fourier phase analysis and the Hilbert transform. A key aspect of both of these methods is that they are not as sensitive as CRP to oscillations in time series that deviate significantly from sinusoidal patterns. The Hilbert transform can also deal with nonstationary signals.

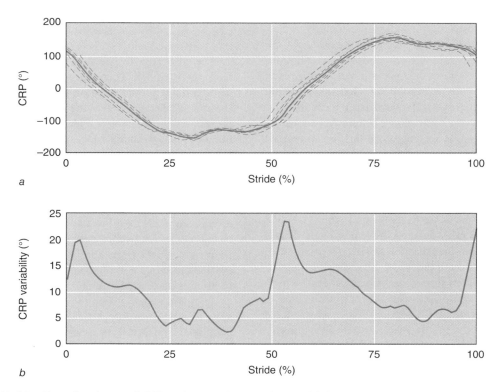

a

b

▲ **Figure 13.14** Coordination variability plots and procedures. *(a)* Multiple continuous relative phase (CRP) cycles superimposed. *(b)* Between-cycle variability at each percentage point of stride for the entire stride cycle.

FROM THE SCIENTIFIC LITERATURE

Miller, R.H., S.A. Meardon, T.R. Derrick, and J.C. Gillette. 2008. Continuous relative phase variability during an exhaustive run in runners with a history of iliotibial band syndrome. *Journal of Applied Biomechanics* 24:262-70.

When dynamical systems techniques are applied to the analysis of cyclical human movement, a certain degree of variability in coordination is generally indicative of a healthy state, whereas a relative lack of variability has been associated with an injured or pathological state. In addition, the onset of fatigue has been suggested as an injury factor in long-distance running. This study compared coordination variability between healthy runners with and without a history of iliotibial band syndrome (ITBS), a common overuse injury of the knee that often recurs. Kinematic data were collected throughout a treadmill run to volitional exhaustion. The authors assessed lower-extremity coordination using continuous relative phase (CRP) and quantified variability as the standard deviation of CRP between strides. The ITBS runners were less variable in coordination patterns during the swing phase that included segment motions previously

associated with ITBS and knee pain. They were also less variable in their coordination at heel-strike but were not less variable during stance. There was no main effect for fatigue and no interaction for fatigue and group. The authors concluded that a history of ITBS is associated with a relative lack of coordination variability in couplings that would be expected to place strain on the iliotibial band. This finding was reconciled with a modeling analysis from a previous study that found that the ITBS group exhibited greater peak strain and strain rate of the iliotibial band than did the control group. The authors speculated that the similar degrees of variability between the ITBS and control groups during stance indicated an adaptation made by the ITBS group to recover from their injuries. The authors suggested that a prospective cohort would demonstrate less variability during stance prior to injury.

Relative Fourier Phase

The previously noted method for calculating CRP involves some restrictions that are particularly important when interpretations are made on the basis of CRP analysis. First, deviations from sinusoidal oscillations in the segmental or joint movement patterns may result in spurious oscillations in the CRP. Second, to obtain within-trial variability in relative phase to measure coordinative stability, the CRP should be relatively constant during a movement cycle. Therefore, only between-trial or between–gait cycle variability can be assessed with the CRP methods described previously.

With locomotor kinematics, typically one or both of these assumptions are violated. Positional signals generally do not oscillate as perfect sinusoids, and their CRP fluctuates considerably during a movement cycle. This is especially the case when one is assessing coordination of joint angular motion in the lower extremity during gait. In these cases, when signals deviate substantially from a sinusoidal pattern, CRP assessments of changes in coordination and variability of coordination may become problematic and alternative methods may be required.

If inferences on variability and coordination modes are desired under these conditions, an alternative approach that depends less critically on these assumptions is to use the relative Fourier phase (RFP) method introduced by Lamoth and colleagues (2002). RFP

assumes that multiple harmonics in the original positional signals contribute to their temporal evolution. A particular relevant harmonic is extracted, and the relative phase is assessed based on characteristics of this harmonic.

The first step for performing RFP is to define a harmonicity index. There are no hard rules for this procedure, but it should result in the definition of a scalar that quantifies the power of the various harmonics in the positional signals. Lamoth and colleagues (2002) defined the index of harmonicity (IH) as

$$IH = \frac{P_0}{\sum_{i=0}^{5} P_i} \qquad (13.10)$$

where P_i is the power spectral density of the $(i+1)$th harmonic, and P_0 is the power spectral density of the first harmonic, which oscillates at the signal's fundamental frequency. An IH = 1 means that there are no oscillations outside the signal's fundamental frequency and that the rotation is perfectly harmonic.

The next step is to define a window width (i.e., a span of time) over which the frequency content of the positional signals will be assessed. The window should be much shorter than the total time span of the signals, but again there are no hard rules for defining its width. Previous research has used a window width four times the period of the fundamental frequency of the signal with the greatest IH (Lamoth et al. 2002).

With the harmonicity index and the window width defined, the RFP angle (ϕ_{rfp}) can be calculated. This measure is analogous to the CRP angle (ϕ_{crp}) from the traditional CRP approach. The window begins at the earlier possible time span (with the left edge of the window at the first time point), and the portions of both signals within this window are transformed into the frequency domain. The phase shift of each signal at a common frequency is recorded. Lamoth and colleagues (2002) defined this frequency as the fundamental frequency of the signal with the greatest harmonicity index. This process is repeated until the window has spanned the entire time series.

The resulting sets of phase shifts are reconstructed in the time domain. These new time series are defined as the Fourier phase angles. The ϕ_{rfp} is then calculated at each time point as the difference between the two Fourier phases (representing each joint or segment).

Because the sinusoidal oscillator assumption has not been violated, conclusions can be drawn on a coordination mode between the original positional signals. For example, $\phi_{rfp} = 0°$ indicates in-phase coordination between the positional signals, whereas $\phi_{rfp} = 180°$ indicates antiphase coordination. However, these coordination modes apply only to oscillations at the fundamental frequency.

The Hilbert Transform

The state space is usually constructed from time-delayed copies of a signal, or from derivatives of the signal with respect to time. However, these are not the only suitable methods. For relative phase analysis of two signals, one alternative is to construct the phase plane using the Hilbert transform (Rosenblum and Kurths 1998). The Hilbert transform is an analytic signal processing approach that allows for unambiguous assessment of the phase difference of two arbitrary, nonstationary signals that are also nonsinusoidal.

Consider an arbitrary time-varying signal $s(t)$, and define $\tilde{s}(t)$ as the Hilbert transform of this signal:

$$\tilde{s}(t) = \frac{1}{\pi} \int_{-\infty}^{\infty} \frac{s(t)}{t - \tau} d\tau \qquad (13.11)$$

Since this integral is improper when $t = \tau$, it is taken according to the Cauchy principal value. The relative phase between two time-varying signals $s_1(t)$ and $s_2(t)$ is then

$$\phi_{hrp} = \phi_1 - \phi_2 = \tan^{-1}\left(\frac{\tilde{s}_1 s_2 - s_1 \tilde{s}_2}{s_1 s_2 + \tilde{s}_1 \tilde{s}_2}\right) \qquad (13.12)$$

where ϕ_{hrp} quantifies the phase relationship based on the Hilbert transform between the signals.

The Hilbert transform approach can be applied to any arbitrary time-varying signal and is not limited to sinusoidal oscillators. It thus may be more appropriate for analyzing data such as joint angular excursions during gait, which typically do not oscillate sinusoidally. Because no derivatives are performed in constructing the phase plane, the Hilbert transform approach avoids magnifying noise in the original signals and may be more appropriate for analyzing noisy or raw data.

Benefits and Limitations of Relative Phase Analysis

The relative phase analysis techniques that have been presented offer the researcher various tools to assess movement coordination in biomechanical systems. Each technique presented has distinct advantages and disadvantages. The discrete relative phase analysis (DRP) is a relatively simple method requiring no further derivation or reconstruction of other time series (such as velocity or acceleration). A potential drawback of DRP is that this is mostly based on one salient event in the time series. This is no problem if prior work exists that has clearly identified this particular event as critical. However, in signals with multiple events or in cases in which peaks or valleys occur inconsistently during different times in the cycle, the DRP may not be the best analysis method. Continuous relative phase or vector coding techniques may be more appropriate under these conditions.

Continuous relative phase (CRP) allows for a quantification of coordination across an entire movement cycle, such as the stride cycle during walking and running. This may provide the researcher with better insights into the overall coordination patterns. Another advantage of CRP is that it reflects the coordination dynamics in the phase plane, incorporating both the displacement and velocity of the joint or segment. This can provide more detailed information regarding changes in coordination dynamics. CRP analysis works best on signals that are close to sinusoidal. When the signal deviates strongly from a sinusoidal pattern, spurious oscillations in CRP may occur. Normalization will control for some of these, but other techniques such as the Hilbert transform may be better when one is working with strongly nonsinusoidal signals.

The relative Fourier phase analysis described earlier does adequately address the issues raised in coordination analyses involving nonsinusoidal signals. In this regard, relative Fourier phase would be better suited, for example, to assess coordination of joint angles during gait, because these typically show nonsinusoidal oscillations. However, one of the drawbacks to this method is the loss of the additional oscillations in the original component time series that results when the analysis is limited to the dominant frequency only. By focusing on the fundamental frequency in the component time series and using this as the basis for the coordination measure, this method bears more similarities to the DRP methods described earlier and may not address the full complexity of the original signals.

CRP provides higher-order information about coordination by incorporating the dynamics in the position-velocity phase plane. For more complex signals, CRP reflects the differences in the phase angles that are derived from the phase plane. It is often difficult or not possible to refer CRP findings to the purely spatial interpretations of phasing. In rehabilitation procedures in clinical practice or during motor learning, the attention and focus often may be on improving the spatial relationships between joints or segments. In those instances, vector coding procedures may provide a better means of assessing coordination changes.

QUANTIFYING COORDINATION: VECTOR CODING

The relative motion between the angular time series of two joints or segments has been used to distinguish normal from disordered gait patterns, symmetrical and asymmetrical gait patterns in motor development,

and changes in coordination in sport as a function of expertise (see Wheat and Glazier 2006 for an overview). Examples of relative motion plots representing the coordination of the thigh versus leg segmental angular displacements for walking and running are presented in figure 13.15. As mentioned before, these relative motion plots are also known as *angle-angle diagrams*, because they usually depict the changes in the angular rotations of the joints or segments of the body.

Various techniques have been developed over time to quantify the relative motion patterns in angle-angle diagrams. This started with the chain encoding method developed by Freeman (see Whiting and Zernicke 1982). This technique involves the superposition of a grid on the relative motion plot that transforms the original data into discrete numerical codes (1-7), quantifying the directions of change in the relative motion plot. This chain encoding technique, however, may lose important information in the subtle changes of the relative motion patterns. Sparrow and colleagues (1987) addressed this limitation by introducing a vector-based coding scheme in which the angles between consecutive data points in the relative motion plot were calculated. Tepavac and Field-Fote (2001) used the coding technique developed by Sparrow and colleagues (1987) to quantify the variability or similarity in relative motion plots across multiple gait

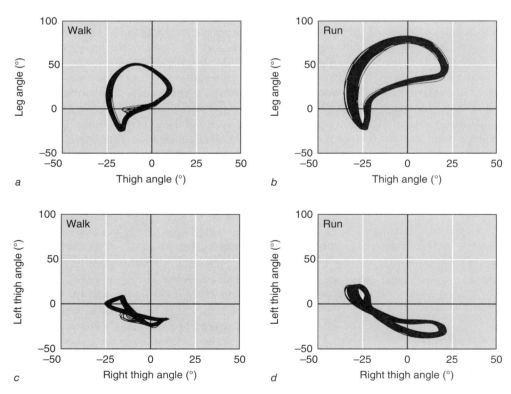

▲ **Figure 13.15** Angle-angle diagrams from walking and running gait trials. *(a, b)* Intralimb hip-knee coordination for walking and running. *(c, d)* Interlimb coordination for walking and running.

cycles. In the following sections we discuss the detailed procedures, the quantification of dependent measures, and the benefits and limitations of vector coding analysis.

Vector Coding Procedures

This section presents the procedures involved in the quantification of relative motion patterns between two joints or segments. To illustrate, we use data from the angle-angle diagram of the intralimb thigh-leg angular relative motion in the sagittal plane during walking and running (figure 13.15). The relative motion between the two segments can be quantified with the coupling angle (γ), an angle subtended from a vector adjoining two successive time points relative to the right horizontal (figure 13.16):

$$\gamma_{j,i} = \tan^{-1}\left(\frac{y_{j,i+1} - y_{j,i}}{x_{j,i+1} - x_{j,i}}\right) \qquad (13.13)$$

where $0° \leq \gamma \leq 360°$, i represents consecutive data points in a cycle (here expressed as a percent stance), and j identifies the multiple gait cycles. Since these angles are directional and obtained from polar distributions ($0°$-$360°$), taking the arithmetic mean of a series of angles can result in errors in the average value not representing the true orientation of the vectors. Therefore, mean coupling angles ($\overline{\gamma}_i$) must be computed using circular statistics (Batschelet 1981; Fisher 1993). Within a subject and then across the group, $\overline{\gamma}_i$ is calculated from the mean horizontal ($\overline{x}_i$) and vertical ($\overline{y}_i$) components across the multiple gait cycles (j) for each percentage (i) of stance:

$$\overline{x}_i = \frac{1}{n}\sum_{j=1}^{n}\left(\cos\gamma_{j,i}\right) \qquad (13.14)$$

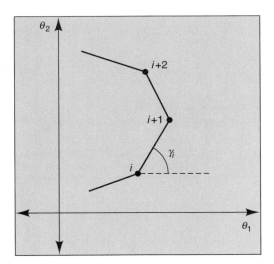

▲ **Figure 13.16** Derivation of the coupling angle *(γ)* for vector coding analysis.

$$\overline{y}_i = \frac{1}{n}\sum_{j=1}^{n}\left(\sin\gamma_{j,i}\right) \qquad (13.15)$$

The length of the mean vector is then defined as

$$r_i = \left(\overline{x}_i^2 + \overline{y}_i^2\right)^{1/2} \qquad (13.16)$$

and has a well-defined angle versus the positive horizontal axis. This angle is referred to as the *coupling angle γ* (see figures 13.16 and 13.17). The mean coupling angle across the different gait cycles is again defined for each percentage *(i)* of the cycle:

$$\overline{\gamma}_i = \begin{cases} \arctan\left(\overline{y}_i / \overline{x}_i\right) \; if \; \overline{x}_i > 0 \\ 180 + \arctan\left(\overline{y}_i / \overline{x}_i\right) \; if \; \overline{x}_i < 0 \end{cases} \qquad (13.17)$$

The vector coding analysis can provide a measure of coordination variability, very similar to the variability obtained through DRP or CRP analysis. As in the relative phase techniques, variability should not be obtained across an entire movement (e.g., gait) cycle, because the coordination likely changes in most cases. Unless the coordination remains constant, variability measures directly taken across the gait cycle are not recommended. The correct method is to first establish the cycle-to-cycle variation for each point in the (gait) cycle (after scaling each trial to 100%) across multiple gait cycles. Coordination variability measures can then be obtained as averages across the gait cycle of this between-cycle variation (a global variability measure), or more locally at salient points or intervals across the cycle (such as early stance, midstance, or swing).

Further Quantification of the Coupling Angle Time Series

Further analysis of the time series of the coupling angle is needed to obtain quantitative measures of coordination across the gait cycle. The average of the coupling angle across the cycle could represent the coordination, but this will only apply when the coordination does not systematically change across the cycle. As can be seen from the pattern in figure 13.17, this may not be the case. One can obtain averages across different phases of the gait cycle or obtain histograms of the distribution of the coupling angles.

Another approach is to divide the coupling angles into specific "bins" that may reveal different coordination tendencies. This method was developed by Chang and colleagues in 2008 in an analysis of forefoot-rearfoot coordination. In the current example of thigh-leg coordination, four unique coordination patterns can be identified (see figure 13.18): antiphase, in-phase, thigh segment phase, and leg segment phase. The four

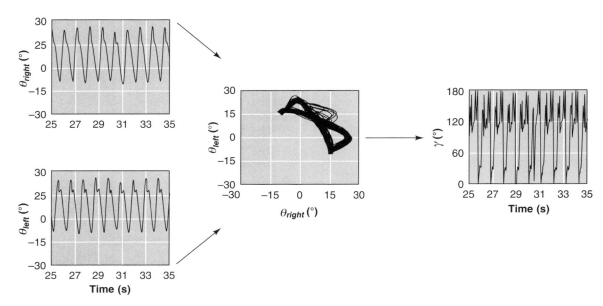

▲ **Figure 13.17** Procedures involved in vector coding analysis. From left to right, the first column depicts the left and right thigh segmental angular rotations. The second column represents the angle-angle plot of these segmental rotations, and the third column the time series of the resulting coupling angle (γ).

coordination patterns emerge from the classification scheme that divides the polar distribution into bins of 45° each. These four patterns are based on the vertical, horizontal, and diagonal (positive and negative) line segments. In-phase couples (centered on the 45° and 225° positive diagonal) rotate in the same direction; an example is concurrent thigh and leg flexion or extension. Antiphase couples (centered on the 135° and 315° negative diagonal) rotate in opposite directions; this occurs when thigh flexion is countered by leg extension. Phase couples dominated by a single segment occur when the coupling angles parallel the horizontal ($\gamma =$

0° or 180°: a thigh phase) or parallel the vertical ($\gamma = 90°$ or 270°: leg phase).

The coupling angle time series over the entire gait cycle (equation 13.17) can also be used to assess the degree of similarity between two different angle-angle plots. Sparrow and colleagues (1987) developed a pattern similarity method that cross-correlates the coupling angle time series of two different relative motion plots. This pattern similarity method was later extended by Tepavac and Field-Fote (2001) to allow for comparisons of similarity of vector coding time series derived from multiple angle-angle plots.

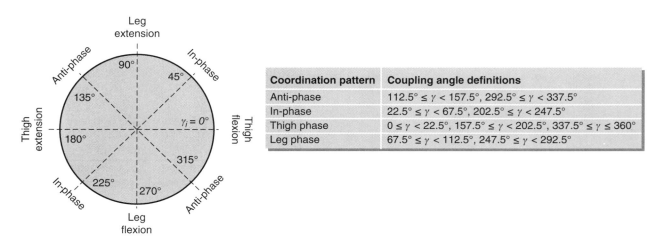

Coordination pattern	Coupling angle definitions
Anti-phase	$112.5° \leq \gamma < 157.5°$, $292.5° \leq \gamma < 337.5°$
In-phase	$22.5° \leq \gamma < 67.5°$, $202.5° \leq \gamma < 247.5°$
Thigh phase	$0 \leq \gamma < 22.5°$, $157.5° \leq \gamma < 202.5°$, $337.5° \leq \gamma \leq 360°$
Leg phase	$67.5° \leq \gamma < 112.5°$, $247.5° \leq \gamma < 292.5°$

▲ **Figure 13.18** Classification of coordination patterns based on vector coding analysis.

Chang, R., R.E.A. van Emmerik, and J. Hamill. 2008. Quantifying rearfoot-forefoot coordination in human walking. *Journal of Biomechanics* 41:3101-5.

The literature on the mechanics of the foot during gait has characterized the stable foot by a decreasing medial longitudinal arch angle that is accompanied by forefoot pronation and concurrent rearfoot supination—in other words, antiphase motion. This article presents an implementation of the vector coding technique as discussed in the previous sections to facilitate the quantification and interpretation of forefoot and rearfoot coordination. The following coordinative patterns between the rearfoot segment and forefoot segment were quantified: in-phase, antiphase, rearfoot phase, and forefoot phase. Nine skin markers were placed on the rearfoot and forefoot segments according to a multisegment foot model. Healthy young subjects performed standing calibration and walking trials (1.35 m/s) while a three-dimensional motion-capture system acquired their kinematics. Rearfoot-forefoot joint angles were derived, and the arch angle was inferred from the sagittal plane. Changes in coupling angles were categorized into one of the four coordination patterns. Arch kinematics were consistent with the literature; in stance, the arch angle reached peak dorsiflexion and then decreased rapidly. However, antiphase coordination was not the predominant pattern during push-off (see figure 13.19). The data from this study suggest that the coordinative interactions between the rearfoot and forefoot are more complicated than previously described. The technique offers a new perspective on coordination of the foot during gait and may provide insight into deformations of underlying tissues, such as the plantar fascia.

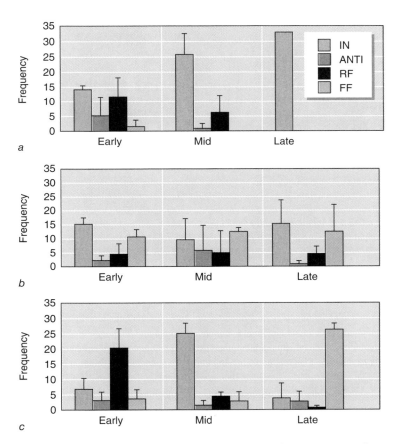

▲ **Figure 13.19** Histograms of rearfoot-forefoot coordination derived from vector coding analysis during walking for the early, mid and late stance phase. *(a)* Sagittal plane coordination, *(b)* frontal plane coordination, and *(c)* transverse plane coordination. IN = in-phase coordination; ANTI = antiphase coordination; RF = dominant rearfoot motion; FF =dominant forefoot motion.

Benefits and Limitations of Vector Coding

Using vector coding procedures to assess coordination has advantages: (1) The absence of velocity data will make the appearance of additional oscillations in the quantification of coordination less likely in case of time series that are not sinusoidal; (2) the coupling angle provides direct information about the movement patterns without the need to derive higher-order variables; and (3) vector coding procedures may have less stringent requirements regarding the normalization of data. Especially in clinical practice it may be important to have coordination measures that are readily interpretable in terms of the movement patterns that need to be changed. Higher-order variables, such as those used in the calculation of CRP, may be harder to interpret and implement into clinical practice. However, these higher-order variables may be more sensitive in detecting and diagnosing the more subtle changes in movement patterns in the progression of disease or recovery from injury.

OVERVIEW OF COORDINATION ANALYSIS TECHNIQUES

The implementation of the different coordination and variability analysis techniques discussed in this chapter depends first and foremost on the nature of the research question but is also to a large degree determined by the benefits and limitations inherent in each technique. An overview of the pros and cons of the coordination analyses discussed is presented in Table 13.1. We must take several factors into account in deciding what technique to use: (1) at what "level" an understanding of coordination is desired (spatial vs. higher order); (2) the nature of the signals, that is, to what degree the time series are sinusoidal or deviate from sinusoidal; (3) whether the coordination involves 1:1 frequency or multifrequency relations; and (4) whether single events in the gait or movement cycle or analysis of the entire phase or cycle is needed. Assessing these factors and knowing the benefits and limitations of each technique will help us choose the

Table 13.1 Summary of the Benefits and Limitations of the Various Coordination Analysis Techniques

Coordination analysis technique	Pros	Cons
DRP	Provides relative simplicity in that no reconstruction of higher-dimensional state space is required.	Provides coordination assessment based on single event in time series; is unreliable when peaks in time series are not well defined or change; this may affect variability.
Multifrequency DRP	Allows coordination assessment for signals with different frequencies.	Entails a more elaborate technique than regular DRP, involving construction of return maps.
CRP	Allows for coordination assessment across entire movement (gait) cycle; includes higher-order dynamics based on phase plane; these higher-order dynamics may be more sensitive in detecting performance changes.	Requires normalization to address frequency and amplitude differences between signals; is not applicable to signals that deviate substantially from a sinusoidal pattern; the coordination results are difficult to reflect back to a spatial joint-segment motion interpretation only.
Fourier phase	Can be applied to nonsinusoidal signals through comparison of phase relation of fundamental frequency only; is easier to apply than DRP in case peaks are not well defined.	Possibly removes information regarding coordination between the two time series; may not provide different information compared with DRP.
Hilbert transform	Can be applied to both sinusoidal and nonsinusoidal signals; does not require numerical derivatives; avoids the magnification of noise in original signals.	Involves a more elaborate technique that requires application of analytical techniques to derive the state space.
Vector coding	Is applicable to both sinusoidal and nonsinusoidal data; has less stringent normalization requirements; when applied to angle-angle plots may be easier to use in clinical applications and interpretations.	Entails loss of higher-order information compared with CRP; may reduce sensitivity.

most effective coordination analysis technique to study a particular research question.

SUMMARY

This chapter presented an overview of the conceptual basis of analysis techniques used to assess coordination in human movement. The conceptual basis of these coordination analysis techniques has its foundation in dynamical systems approaches to human movement. The dynamical systems approach has offered new insights and tools to assess stability and change in movement patterns. The procedures involved in the application of different coordination analysis methods were presented, and the benefits and limitations of these different techniques were discussed. This overview shows that a number of powerful and varied techniques are available to identify coordination changes in the analysis of human movement. Careful consideration should be given to the benefits and limitations of each coordination analysis method before it is applied to the question of interest.

SUGGESTED READINGS

Miller, R.H., R. Chang, J.L. Baird, R.E.A. van Emmerik, and J. Hamill. 2010. Variability in kinematic coupling assessed by vector coding and continuous relative phase. *Journal of Biomechanics* 43:2554-60.

Hamill, J., R.E.A. van Emmerik, B.C. Heiderscheit, and L. Li. (1999). A dynamical systems approach to lower extremity running injuries. *Clinical Biomechanics* 14:297-308.

Tepavac, D., and E.C. Field-Fote. 2001. Vector coding: A technique for the quantification of intersegmental in multicyclic behaviors. *Journal of Applied Biomechanics* 17:259-70.

Van Emmerik, R.E.A., M.T. Rosenstein, W.J. McDermott, and J. Hamill J. 2004. A nonlinear dynamics approach to human movement. *Journal of Applied Biomechanics* 20:396-420.

Wheat, J.S., and P.S. Glazier. 2006. Measuring coordination and variability in coordination. In *Movement System Variability*, eds. K. Davids, S.J. Bennett, and K.M. Newell, 167-81. Champaign, IL: Human Kinetics

Chapter 14

Analysis of Biomechanical Waveform Data

Kevin J. Deluzio, Andrew J. Harrison, Norma Coffey, and Graham E. Caldwell

Several chapters in this text are concerned with the measurement and calculation of biomechanical kinematic and kinetic data related to human motion. For example, in chapter 2 on 3-D kinematics we learned how to measure the joint angles of flexion-extension, adduction-abduction, and axial rotation. The development of new technologies has allowed biomechanists to use more sophisticated methods and construct much more complex experiments. As a result, biomechanics studies often generate large quantities of data. These data may present in the form of times series (e.g., kinematic or kinetic parameters with respect to time) or coordination structures such as angle-angle diagrams, coupling angle curves, continuous relative phase, or phase-plane plots, as seen in chapter 13. The common features of all these data are that they are high-dimensional and can be represented as curves or groups of curves.

Analysis of such data is challenging, and although various procedures have been explored, no standard technique for analysis of curve data has been established. Experimental studies often aim to determine differences in such curve data sets between several groups of individuals, for example, differences between injured and control subjects, differences between athletes grouped according to skill level, differences between children at various developmental levels, or changes in movement patterns in response to some form of intervention. The goal of the analysis depends on the research question that the data were collected to address, but most analyses require an initial process of data reduction in which the raw data are transformed into a smaller, more useful form to be used in downstream analysis techniques such as statistical hypothesis testing.

In this chapter we consider the analysis of biomechanical data. Advanced methods of detecting patterns in multidimensional signals are fairly well established in the sciences but are not yet commonly used in biomechanics research. Principal component analysis (PCA) and functional data analysis (FDA) are two such approaches. In this chapter, we

- ▶ address the particular challenges associated with the analysis of waveform or time series data,
- ▶ explain how PCA and FDA can be used to detect and interpret differences in the shape and amplitude of waveform data,
- ▶ demonstrate how PCA and FDA output can be used to statistically test differences between groups or conditions, and
- ▶ compare the differences between PCA and FDA.

BIOMECHANICAL WAVEFORM DATA

Most biomechanical data characterizing human movement appear as temporal waveforms or time series representing specific joint measures such as angles, moments, or forces measured over time. For example, the data in figure 14.1 illustrate the knee flexion angle measured for one complete gait cycle on 50 subjects with knee osteoarthritis and 60 subjects without. These data have a few striking characteristics. First, there are a lot of data. If we represent each of the 110 waveforms at each percentage of the gait cycle (from 0% to 100%), we have $110 \times 101 = 11,110$ data points. For this reason the data are often described as multidimensional. Second, there is a general shape or underlying pattern to the data that the waveforms generally follow. For a given waveform, a specific value is related to neighboring values in the same waveform and also to the values of other waveforms. The strength of the relation between waveform values can be described as collinearity and can be referred to as the correlation structure in the data. Third, there is substantial variability in the data. Part of this variation is within each group and is related to differences in knee joint motion between subjects (so-called subject-to-subject variation). Another part of the variation is due to the differences in knee joint motion between the two groups (so-called between-group variation) and it is this variation that we are usually interested in because it relates to an objective of biomechanical data analysis: to detect and interpret differences in the waveform data between the subject groups of interest.

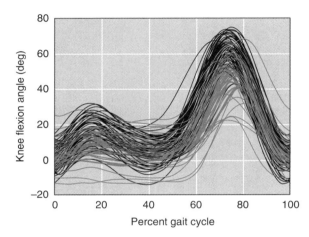

▲ **Figure 14.1**   Knee flexion angle waveform data for 50 subjects with advanced knee osteoarthritis (OA) (green lines) and 60 subjects without OA (black lines).

How do we meet this objective in biomechanical studies that generate large amounts of temporal waveform data? The challenge is in finding the most salient features of the data and retaining the most important features while not losing important discriminatory information. Traditionally, this has been accomplished by extracting discrete parameters from the waveform data (e.g., peak or minimal values). Although this results in a small set of parameters that can be compared across subjects, it comes at the cost of losing much of the temporal information in the waveform. Such approaches lead to severe reductions in data and much important information is discarded, thus rendering these approaches unsatisfactory (Donoghue et al. 2008; Donà et al. 2009). Also, the fact that different studies use different definitions (i.e., average across phase, or value at specific temporal event) for discrete parameters has been suggested as a reason for inconsistent conclusions in the biomechanics literature (O'Connor and Bottum 2009). Furthermore, the definition of parameters can be difficult, particularly in the study of pathological motion. In some cases, individual subjects may not

exhibit a particular waveform feature. For example, consider the four different knee adduction moment waveforms, measured during gait and presented in figure 14.2, starting with the classic double-peak adduction moment waveform shape that is commonly reported (figure 14.2a). The other three waveforms are atypical waveforms with no definitive peaks (figure 14.2b), no definitive second peak (figure 14.2c), and no definitive first peak (figure 14.2d). Even though these waveforms are atypical, they are present to some degree in various populations. In a study evaluating the adduction moment in asymptomatic and osteoarthritic subjects, Hurwitz and colleagues (2002) acknowledged that 52% of the subjects with knee osteoarthritis, compared with 29% of their asymptomatic counterparts, did not have a definitive second peak. In an effort to capture information about the shape of biomechanical waveform data, researchers have turned to a variety of techniques that retain the temporal information within the data.

These techniques that consider the entire waveform are inherently more complex than the selection and analysis of discrete waveform parameters because it is more difficult to compare the many features of a set of waveforms than a set of discrete parameters. Although a peak or minimal value in a time series may be evident and easily quantified, how do you quantify two complex waveforms that vary in magnitude and shape? Several techniques have been developed to do so, including multivariate statistical techniques such as principal component analysis, factor analysis, and correspondence analysis, as well as other mathematical methods such as Fourier analysis, fuzzy analysis, fractal analysis, wavelet transforms, and neural networks. A very good comparison of these techniques applied to gait analysis was presented by Chau (2001a, 2001b). Our objective in the present chapter is to describe the use of two of these techniques, principal component analysis (PCA) and functional data analysis (FDA), in some detail.

We begin with principal component analysis, which is an orthogonal decomposition technique, meaning that the resulting individual principal components are independent of each other. Mathematically, PCA is an orthogonal transformation that converts a number of correlated variables into a smaller number of uncorrelated, independent variables called *principal components*. PCA is ideally suited to data reduction and interpretation and has been used effectively in biomechanics research to analyze temporal waveform data in several varied applications including gait (Landry et al. 2007; Muniz and Nadal 2009), balance (Pinter et al. 2008), ergonomics (Wrigley et al. 2006), and surface electromyography (Hubley-Kozey et al. 2006; Perez and Nussbaum 2003; Wooten et al. 1990). Similar solutions to the problem of feature selection are often associated with slightly different names. The techniques referred

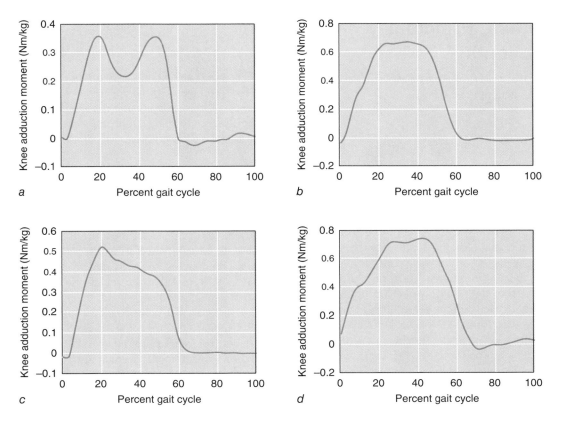

▲**Figure 14.2** Adduction moment waveform data for four subjects. *(a)* Typical waveform, *(b)* waveform with no definitive peaks, *(c)* waveform with no definitive second peak, *(d)* waveform with no definitive first peak.

to as *singular value decomposition* (SVD) or *Karhunen-Loève* (KL) expansion can be considered equivalent to PCA (Gerbrands 1981). These solutions are statistically driven techniques for pattern recognition that avoid any assumptions about the structure of the data or the relative importance of particular events or portions of the signal. We follow the PCA section with a description of the related but distinct technique known as functional data analysis (FDA).

PRINCIPAL COMPONENT ANALYSIS

In chapter 12 we introduced the concept of Fourier analysis, in which a time-varying signal is represented as a combination of sinusoids at various frequencies and phase shifts. There is a direct analogy between Fourier analysis and principal component analysis of waveform data. In Fourier analysis, the set of underlying sinusoids used to represent the time series waveform can be thought of as the *basis functions*, and a particular waveform is associated with a specific set of coefficients of the basis functions, known in that case as the Fourier coefficients. In biomechanics there are many uses for

Fourier analysis, and waveform sinusoidal features can be extracted using Fourier analysis (Chao et al. 1983).

However, time series data can be represented by other basis functions too, not just those found with Fourier analysis. In many cases, sinusoids are not a natural fit to the raw waveform data, and a better set of basis functions may come from the waveform data themselves. When applied to time series data, PCA computes and extracts a unique set of basis functions from the waveforms based on the variation that is present in the waveform data. These basis functions are known as principal components and are related to the shape of the waveform and, in particular, to modes of variation within the data. As with Fourier analysis, each waveform in the original data set is associated with a set of unique scores: the coefficients of the basis functions. The three major strengths of PCA are that (1) the principal components are independent of each other, (2) only a few of the principal components are required to adequately represent the original waveform data, and (3) the scores can be used in downstream analyses (e.g., as the dependent variables in hypothesis testing, discriminant analysis, and cluster analysis) to detect and interpret differences in waveform shape between subjects. We will see later in the chapter that unique basis functions are also the underpinnings of functional data analysis.

Calculating Principal Components

In the case of a set of time series waveform data, we can represent the data in matrix form,

$$X = \begin{bmatrix} x_{11} & x_{12} & \cdots & x_{1p} \\ x_{21} & x_{22} & \cdots & x_{2p} \\ \vdots & \vdots & \ddots & \vdots \\ x_{n1} & x_{n2} & \cdots & x_{np} \end{bmatrix} \quad (14.1)$$

where each row represents a complete time series waveform for one subject (n = number of time series and rows), and each column represents the values at one particular instant of the waveform for all subjects (p = the number of time points). Thus, each knee flexion angle curve from figure 14.1 would occupy a row in the matrix, with each column representing a time point of the angle waveforms. For these 110 subjects, each with waveforms sampled at 101 time points (0%-100% gait cycle), $n = 110$ and $p = 101$. Principal component analysis is performed on the columns of X so that the correlated variables in this case are the p time samples each observed on n subjects.

We are interested in the variation in these data, in terms of both how the waveform changes over time and how one subject varies from another. One way of expressing the variance structure within the data is the covariance[1] matrix of the columns of X, here signified as S.

$$S = \begin{bmatrix} s_{11} & s_{12} & \cdots & s_{1p} \\ s_{21} & s_{22} & \cdots & s_{2p} \\ \vdots & \vdots & \ddots & \vdots \\ s_{p1} & s_{p2} & \cdots & s_{pp} \end{bmatrix} \quad (14.2)$$

The diagonal elements, s_{ii}, represent the variance at each instant of the waveform, found by calculating the mean of the ith column of X and then the average squared distance between this mean and all n waveform values at that particular time instant:

$$s_{ii} = \frac{\sum_{k=1}^{n}(x_{ki} - \bar{x}_i)^2}{n-1} \quad (14.3)$$

where i is the column and n is the number of rows (subjects). The off-diagonal elements, s_{ij}, represent the covariance between each pair of time instants,

$$s_{ij} = \frac{\sum_{k=1}^{n}(x_{ki} - \bar{x}_i)(x_{kj} - \bar{x}_j)}{n-1} \quad (14.4)$$

where i and j are two of the columns and n is the number of rows (subjects). If the covariances are not zero, then there is a linear relationship between the two variables. The strength of this relationship can be represented by the correlation coefficient

$$r_{ij} = \frac{s_{ij}}{s_{ii}s_{jj}} \quad (14.5)$$

The covariance matrix S contains the variability structure from the original data, and the off-diagonal elements are nonzero in general, meaning that the columns of the original data waveforms are correlated. The principal components are extracted from this covariance matrix S. Recall that the set of principal components that we seek are uncorrelated, meaning that they will be associated with a covariance matrix that has all the off-diagonal elements equal to zero. The transformation process from the original data covariance matrix S to the principal components covariance matrix D is known as *diagonalization*, or orthogonal decomposition from linear algebra, and can be written as

$$U'SU = D \quad (14.6)$$

The matrix U can be thought of as an orthogonal transformation matrix that realigns the original data into a new coordinate system. The new coordinates are the principal components, and they are aligned with the directions of variation in the data. The columns of U are the eigenvectors of S and are often called principal component *loading vectors*. D is a diagonal covariance matrix whose elements, λ_i, are the eigenvalues of S, and each eigenvalue is a measure of the variation associated with each principal component. The number of nonzero diagonal elements of D is the maximum number of principal components. This is equal to the lesser of the number of subjects n, or the length of the waveform p, which corresponds to the rank r of S. In our example of the knee flexion angle waveforms, the rank of S = 101, so the maximum number of principal components is 101. We shall see later that in practice we use only a small fraction of the maximum number of principal components. The final step is to use the matrix U to transform the original data[2] into the new uncorrelated principal components (Z),

$$\underset{(nxr)}{Z} = \left[\underset{(nxp)}{X - \bar{X}}\right]\underset{(pxr)}{U} \quad (14.7)$$

[1]Another option is to use the correlation matrix of the columns of X. This is equivalent to the covariance matrix of columns of X after they have been scaled to unit variance and zero mean. Such scaling is recommended when the variables are measured in different units.

[2]The covariance matrix is independent of the mean of X, so the mean is removed from the data to avoid any ambiguity.

Each column of Z is a principal component, and the elements of the column are referred to as principal component (PC) scores. After calculation, the principal components are ordered according to the amount of variance that each explains in the original data, so that the first explains the maximum amount of variance, the second has maximum variance subject to being orthogonal to the first, and so on. The variance of each principal component is given by the eigenvalues, λ_i, which are the diagonal elements of matrix D. PCA is a variance-preserving transformation so that the total variation in the original raw data is captured by the principal components. The most commonly used measure of the total data variation is the sum of the variances of each variable, which is equal to the sum of the diagonal elements of S. The sum of the diagonal elements of a matrix is known as the trace *(tr)* of a matrix, so that

$$tr(S) = tr(D) \qquad (14.8)$$

In this way we can quantify the portion of the total variation explained by each principal component,

$$\text{Variation Explained by PCi} = \frac{\lambda_i}{tr(S)} = \frac{\lambda_i}{\sum \lambda} \qquad (14.9)$$

In the case of the knee flexion waveform data (figure 14.1), we see that more than 90% of the variation in the data is captured by the first 3 PCs and almost 99% of the variation is explained by the first 5. The breakdown by PC is shown in table 14.1.

Table 14.1 Variation Explained by Each PC of the Flexion Angle Data

PC	Variation explained (%)	Cumulative variation explained (%)
PC1	61.5	61.5
PC2	19.9	81.4
PC3	12.5	93.9
PC4	2.8	96.7
PC5	2.0	98.7

PC Scores

To recap, the transformed variables known as *principal components* are represented by the columns of the matrix Z and are just linear combinations of the original data. The individual transformed observations are referred to as *PC scores* and are represented by the individual elements of each column of Z. The first four PC scores for the knee flexion waveforms of

figure 14.1 are shown in the scatterplots of figure 14.3. Each data point represents a specific subject; the subjects with knee OA are shown as open black squares, whereas the subjects without OA are shown as open green diamonds. Note that group separation in the PC scores is more apparent than in the raw waveform data. Furthermore, this separation is primarily seen in the plot of PC1 versus PC2. Therefore, we can determine that PC2 and PC1 are capturing variation related to group differences, whereas PC3 and PC4 are capturing variation that is not related to differences between the groups. We can test for these differences objectively by performing statistical analysis on the PC score data as shown in figure 14.4. In this example the analysis was restricted to simple *t*-tests; however, as we describe later in this chapter, the PC scores can be used in more complex statistical models.

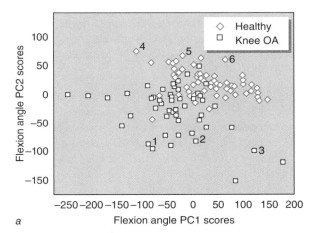

a

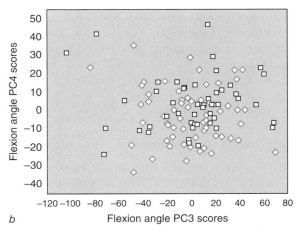

b

▲ **Figure 14.3** PC score values for both the subjects with OA (open black squares) and the subjects without OA (open green diamonds). *(a)* PC1 versus PC2 scores. The numbers are associated with the waveforms for those subjects shown in figure 14.8e. *(b)* PC3 versus PC4 scores.

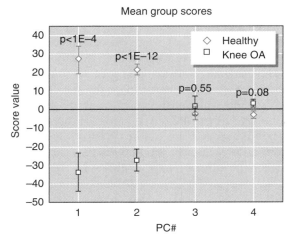

▲**Figure 14.4** PC score comparisons. The mean ± SEM for both the subjects with OA (open black squares) and the subjects without OA (open green diamonds) are shown for each PC. Unpaired Student's *t*-test with a Holm-Sidak correction for multiple comparisons was performed to compare the difference in means for each PC. The adjusted *p*-values are given above each comparison.

PC Loading Vectors

The eigenvectors, or columns of U, are referred to as loading vectors. The loading vectors associated with the first three principal components of the knee flexion angle data are shown in Figure 14.5. These are the unique set of basis vectors, referred to earlier, that PCA extracts from the waveform data. The loading vectors are themselves waveforms and are the same length as the original waveform data. Therefore, in figure 14.5, the loading vectors for the knee flexion angle data are plotted as a function of the gait cycle. The vertical axis measures the individual coefficients of each loading vector. It is these coefficients that are combined with the original waveform data to produce the PC scores. Explicitly, the first PC score for subject *i* would be calculated as

$$z_{1i} = \left(x_{i1} - \overline{x}_1\right)u_{11} + \left(x_{i2} - \overline{x}_2\right)u_{21}$$
$$+\cdots+\left(x_{1p} - \overline{x}_p\right)u_{p1} \qquad (14.10)$$

The PC score, z_{1i}, is just a linear combination of each time sample (mean corrected) of that particular subject's waveform and the PC loading vector coefficients. If the coefficients are close to zero for particular time samples, then these times samples contribute very little to the PC score. Therefore, we can examine the shape of the loading vectors to gain insight into the variation that each PC captures. In our knee flexion angle waveform example, the time samples refer to the portion of the gait cycle, and the portions of the gait cycle associated with larger magnitudes of the loading vector are more important to

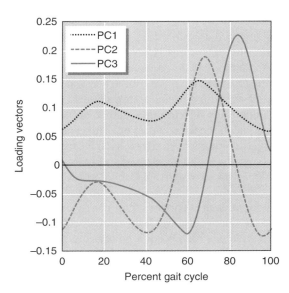

▲**Figure 14.5** The loading vector coefficients are shown for the first three principal components extracted from the knee flexion waveform data.

that particular principal component. For example, PC2 emphasizes the portions of the gait cycle around late stance (~40% gait cycle) and early swing (~70% gait cycle) more than early stance (~15% gait cycle). We address the important topic of principal component interpretation in a later section. However, first we need to determine how many principal components we should retain.

Selecting the Number of Principal Components to Retain

The real strength of PCA lies in the fact that the majority of the variation is usually explained by the first few principal components, and these usually contain the most relevant information from the original data. The remaining components can often be disregarded without loss of important information. Indeed, we saw that only 3 PCs were needed to explain almost 95% of the variation in the knee flexion waveform data (Table 14.1). How many principal components need be considered in a given analysis? There are a variety of stopping rules that attempt to find the cutoff between informative variation and noise. The most simple of these is to define the cutoff based on the percentage of variation explained by the retained PCs, typically 90% to 95%. The scree test is a graphical technique that is based on plotting the eigenvalues in order from largest to smallest. The graph usually resembles a cliff, and the cutoff point is the elbow between the large and small eigenvalues (figure 14.6). The word *scree* is a geological term referring to

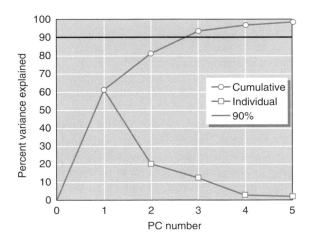

▲ Figure 14.6 Scree plot. The eigenvalues associated with each PC are scaled to percentage of variation explained, and both the percentage of variation explained and the cumulative variation explained are plotted versus the PCs.

the rubble at the bottom of a cliff. A full discussion of various techniques for choosing the number of PCs can be found in Jackson (1991).

We can also consider the problem of choosing the number of principal components in terms of how well the retained principal components represent the original waveform data. We introduced PCA as a means of transforming the original data into a set of PC scores and loading vectors, but it is also possible to use the principal component scores and the loading vectors to estimate the original waveform data.

Estimating the Waveform Data From Principal Components

We can rearrange equation 14.7 to express the original waveforms in terms of a linear combination of the loading vectors, U, and the PC scores, Z. This is done by multiplying both sides by the inverse of matrix U, but because U is an orthogonal matrix, the inverse is just the transpose of the matrix (indicated by the superscript t) below.

$$X = ZU^t + \bar{X} \qquad (14.11)$$

In this way the data are reconstructed in terms of the individual contributions from each loading vector. The PC score indicates the level of contribution from each of the loading vectors. We can see this explicitly if we expand the preceding equation for an individual waveform, say waveform i:

$$\underset{(1 \times p)}{\vec{x}_i} = \underset{(1 \times 1)}{z_{i1}} \underset{(1 \times p)}{\vec{u}_1} + \underset{(1 \times 1)}{z_{i2}} \underset{(1 \times p)}{\vec{u}_2} + \cdots + \underset{(1 \times 1)}{z_{ir}} \underset{(1 \times p)}{\vec{u}_r} + \underset{(1 \times p)}{\bar{x}} \quad (14.12)$$

If we use all of the principal components as indicated in this equation, then we end up with an exact reproduction of the original data. However, it is possible to obtain a reasonable estimate of the waveform data using only a subset of the first few PCs, in the same way that time series data can be reconstructed with just a few Fourier harmonics (see chapter 12). This process is illustrated in figure 14.7 for a subject without knee osteoarthritis (figure 14.7a) and a subject with knee osteoarthritis (figure 14.7b). In each case the solid green line is the original waveform data, whereas the dotted black line is an estimate of those data using only the first PC, computed using the first term of equation 14.12:

$$\vec{x}_i^* = z_{i1}\vec{u}_1 + \bar{x} \qquad (14.13)$$

where $\vec{x}_i^*$ is the estimate of the individual waveform, z_{i1} is the PC1 score for that particular subject, $\vec{u}_1$ is the loading vector for the first PC, and $\bar{x}$ is the average of all the waveforms. Estimates for the subject with OA and the subject without OA therefore only differ in the value for their particular PC1 score z_{i1}. We can see that the estimate is better for the subject without OA than it is for the subject with OA. However, the estimates for both subjects improve with the addition of more PCs and

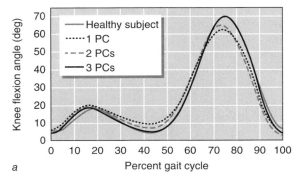

a

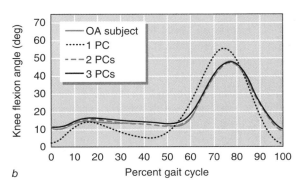

b

▲ Figure 14.7 PC estimation of original waveform data. These plots reveal how an individual waveform can be estimated using a subset of PCs for *(a)* a subject without OA and *(b)* a subject with OA.

are very good for both subjects with three PCs. It can be useful to examine the residuals between the estimate and the original data, and indeed the sums of squares of these differences (*SSres*) can be used as measure of how well the PCs represent the original data.

$$SSres = \sum \left(\vec{x} - \vec{x}^* \right)^2 \qquad (14.14)$$

Interpreting the Principal Components

What do the computed principal components tell us about our data? The process of relating the principal components to the original biomechanical variable is important because we are usually interested in interpreting the differences in principal component scores that we observe. In the case of the flexion angle data, the subjects with OA were very different from subjects without OA with respect to the first two principal component (figures 14.3 and 14.4). But what features of the knee flexion angle curves are associated with these two principal components? Therefore, we would like to identify and interpret the feature of the knee flexion angle that is associated with these principal components.

The PCA method extracts features of variation from the original waveform data. These features can be interpreted based on the fact that each principal component is associated with a particular shape change in the waveforms. Interpretation is accomplished by examining the shape of the loading vector and individual waveforms corresponding to high and low values of the PC score. It can be helpful to examine the loading vectors simultaneously with waveforms that represent extreme values of each principal component (figure 14.8). The first three loading vectors from the knee flexion angle data are shown in the top row of figure 14.8. A zero-line is helpful in identifying whether a positive *z*-score applied to the loading vector will have a positive (additive) or negative (subtractive) effect on the mean waveform. For the first PC, the loading vector has all positive coefficients; therefore, the PC1 scores will reflect a weighted average of the knee flexion angle waveform data. PC1 scores will tend to be high if the waveform is on average greater than the mean waveform and low if the waveform is lower than the mean waveform. This effect can be visualized by examining individual subject waveforms corresponding to high and low PC scores. One option is to select the waveforms corresponding to the 5th and 95th percentile PC scores (Deluzio and Astephen 2007). However, the two selected waveforms may also differ with respect to PCs other than the one of interest. It can be helpful to select the extreme PC scores using the scatterplots of the PC scores (figure 14.3).

The middle row of plots in figure 14.8 contains individual subject waveforms corresponding to extreme values

of the PC scores. For example the knee flexion angle data for subjects with high (solid lines) and low (dashed) PC1 scores are shown in figure 14.8*d*. Subject waveforms corresponding to extreme values of PC2 and PC3 are shown in figure 14.8, *e* and *f*. Note that the individual waveforms in figure 14.8*e* are numbered so they can be linked to the PC score values in figure 14.3*a*. Selecting individual waveforms in this way can help us understand the effect of changes in PC scores. However, it can be challenging to select waveforms that differ only in the PC that we are trying to interpret. For example, comparing waveforms 3 and 4 reveals differences in both PC1 and PC2. To isolate the features of the waveform corresponding to only the PC of interest, we can also consider reconstructing waveform data using only one PC.

We can create two waveforms $\vec{x}_H$ and $\vec{x}_L$, representing waveforms corresponding to a high and a low value of the PC, by adding and subtracting a scalar multiple of the loading vector, $\vec{u}_i$, to the average waveform, $\overline{x}$. A convenient scalar multiple is one standard deviation (SD) of the corresponding PC scores, $SD(\vec{z}_i)$:

$$\vec{x}_H = \overline{x} + SD\left(\vec{z}_i\right) \times \vec{u}_i$$
$$\vec{x}_L = \overline{x} - SD\left(\vec{z}_i\right) \times \vec{u}_i \qquad (14.15)$$

These reconstructions are shown in figure 14.8, *g* through *i*. Comparing these two waveforms in figure 14.8*g* reveals that differences in PC1 scores are associated with a vertical shift in the knee flexion waveform data. Therefore, PC1 can be interpreted as a measure of the overall average, or the magnitude, of the waveform. It is very often the case that the largest source of variation, and hence the first principal component, is associated with the overall average magnitude of the waveform data.

Interpretation of PC2 follows a similar procedure of examining the loading vector (figure 14.8*b*) and reconstructed waveforms associated with high and low PC scores (figure 14.8*h*). In this case, the loading vector has both positive and negative coefficients, and peaks in the loading vector correspond with peaks in the original waveform data. Large PC2 scores are associated with waveforms that have high flexion angles in swing (60%-80% gait cycle) and low flexion angles at late stance (40% gait cycle) and late swing–early stance (90%-10% gait cycle), whereas lower PC2 scores are associated with a flatter profile waveform. Therefore, PC2 can be interpreted as a measure of the overall range of motion throughout the gait cycle. Subjects with large PC2 have a large range of motion. Recall that the largest difference between the subjects with and without OA was with respect to PC2 (figure 14.3). In other words, the subjects with OA had significantly reduced range of motion during gait.

The third principal component also has positive and negative coefficients, but the peaks do not coincide with the peaks of the original waveforms (figure 14.8*c*). The

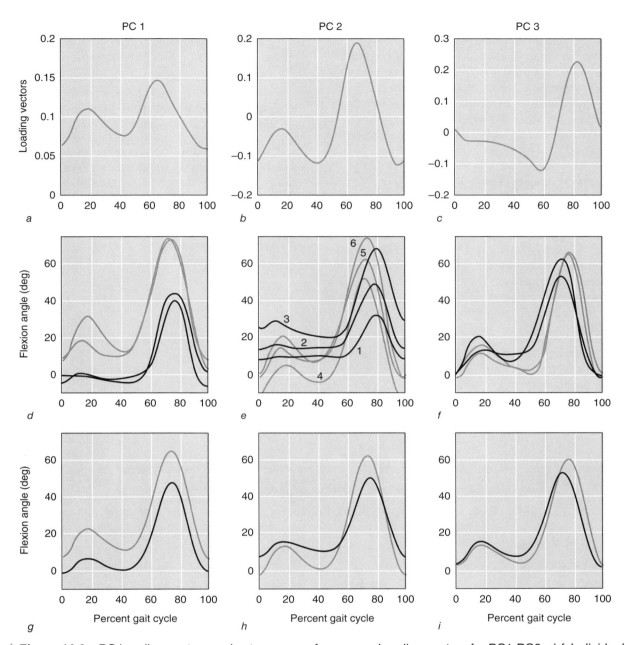

▲ **Figure 14.8** PC Loading vectors and extreme waveforms. *a-c*: Loading vectors for PC1-PC3. *d-f*: Individual subject waveforms associated with low (black lines) and high (green lines) scores. The numbers on the waveforms in figure 14.8*e* identify the same PC scores shown in figure 14.3*a*. *g-i*: Waveform reconstructions based on the mean waveform ±1 standard deviation (SD) of the PC scores times the loading vector for each PC. The green line represents the high (+1 SD) value of each PC, and the black line represents the low (−1 SD) value of the PC.

two reconstructed waveforms differ from each other in that there is a phase shift between them (figure 14.8*i*). The waveform associated with the low PC3 score leads the other. These kinds of temporal shifts in waveform data are common, and they can be isolated through PCA. Recall that there were no differences between the groups with respect to PC3 (figures 14.3 and 14.4) and that the differences in PC1 and PC2 exist independent of the effect of

this phase shift. Testing for differences in the other PCs is equivalent to removing the phase shifts in the data.

In summary, the plots of waveform data corresponding to high and low PC scores for a given loading vector are essential to the interpretation. Differences between the high and low PC score waveforms can be attributed to the features captured by the loading vector. Plots of these extremes are highly valuable because they characterize the effect of a loading vector in the original units and demonstrate the range of observations in the sample. Waveform reconstruction using a range of PC score values (i.e., ±SD) is the best way to isolate the waveform shape changes captured by each PC. Examining individual waveforms corresponding to extreme PC scores can also aid interpretation, but one has to be mindful that changes seen in the selected waveforms are usually affected by the variance of other PCs.

Hypothesis Testing Using Principal Components

In addition to detecting and interpreting differences in the shape of waveform data, PCA transforms the data from p waveform values to a small number of PC scores for each subject. These PC scores can be used for hypothesis testing of differences in the waveform data. There are three properties of PCA that make it well suited to this task:

▶ The principal components, or features extracted from the waveform data, are optimal in the sense that they explain a maximal amount of variance in the original data. Therefore, differences in the waveform data due to different experimental conditions should be revealed as features of variation.

▶ The principal components are orthogonal to each other, so that hypothesis testing of the PC scores amounts to testing independent features of the data.

▶ The PC scores tend to be "well behaved" statistically; that is, they are usually normally distributed. Therefore, parametric statistical techniques can be applied to perform hypothesis tests.

The starting point for hypothesis testing is a data matrix that contains the waveform data from all subjects, under all the experimental conditions under investigation. Consider the case where the researcher is interested in determining the effect of two experimental conditions on some biomechanical waveform measure. The data can be represented as follows:

$$\begin{bmatrix} x1_{11} & x1_{12} & \cdots & x1_{1p} & z1_{11} & \cdots & z1_{1k} \\ x1_{21} & x1_{22} & \cdots & x1_{2p} & z1_{21} & \cdots & z1_{2k} \\ \vdots & \vdots & \ddots & \vdots & \vdots & \vdots & \vdots \\ x1_{n1} & x1_{n2} & \cdots & x1_{np} & z1_{n1} & \cdots & z1_{nk} \\ x2_{11} & x2_{12} & \cdots & x2_{1p} & z2_{11} & \cdots & z2_{13} \\ x2_{21} & x2_{22} & \cdots & x2_{2p} & z2_{21} & \cdots & z2_{23} \\ \vdots & \vdots & \ddots & \vdots & \vdots & \vdots \\ x2_{m1} & x2_{m2} & \cdots & x2_{mp} & z2_{m1} & \cdots & z2_{mk} \end{bmatrix}$$

FROM THE SCIENTIFIC LITERATURE

Smith, A.J., D.G. Lloyd, and D.J. Wood. 2004. Pre-surgery knee joint loading patterns during walking predict the presence and severity of anterior knee pain after total knee arthroplasty. *Journal of Orthopaedic Research* 22:260-6.

In the treatment of knee joint arthritis with total knee arthroplasty, it has been observed that some patients have anterior knee pain even after the surgery. Gait patterns have been shown to be predictive of surgical outcome after high tibial osteotomy (Prodromos et al. 1985; Wang et al. 1990) and of component migration after total knee arthroplasty (Hilding et al. 1999). Smith and colleagues (2004) hypothesized that preoperative gait patterns would be related to clinical outcome following total knee arthroplasty. Their study focused on the sagittal plane knee flexion moment and examined whether the preoperative flexion moment was related to postoperative anterior knee pain.

One of the most interesting findings was related to the pattern of the knee flexion moment waveform. The authors extracted the first four principal components (PCs) from the knee flexion moment waveforms. A cluster analysis was then performed on the principal component scores to identify three patterns of knee flexion moment waveforms, which were classified as either biphasic, flexor,

and extensor. Two important findings were that the vast majority (95%) of the control subjects had the biphasic pattern, and that the presurgery pattern was related to the postsurgery pattern. This confirmed that abnormal postsurgery gait patterns could be explained in part by the presence of these patterns before surgery.

Perhaps more important, these authors related the preoperative knee flexion moment data to the presence and severity of knee pain after total knee arthroplasty. They used logistic regression to predict presence of knee pain and multiple linear regression to predict severity of knee pain. The best predictor in both cases was the PC2 scores. The second principal component was interpreted as the magnitude of the early midstance flexion moment. It is interesting to note that the PC2 scores were better predictors than were typically chosen waveform parameters such as the peak value during early midstance. This was the first prospective study to relate presurgery knee loading to the presence and severity of knee pain after total knee arthroplasty.

where x1 and x2 represent the p-dimensional waveform data for condition 1 and condition 2, respectively, and z1 and z2 represent the k PC scores resulting from the PCA. Each PC score of interest (columns on the right-hand partition) is then used in Student's t-tests to determine whether there are differences between the two groups. For example, in figure 14.4, we can see that there is a significant difference between the OA and the control groups in the average joint angle (PC1) and the range of motion in knee flexion angle (PC2). The basic approach is the same for more complicated statistical analysis, such as repeated-measures analysis of variance (Landry et al. 2007) or discriminant analysis (Deluzio and Astephen 2007).

FUNCTIONAL DATA ANALYSIS

With the PCA approach, data are described in the form of waveforms composed of individual time samples. In contrast, the key concept in functional data analysis (FDA) is that the entire sequence of measurements for a movement or condition is viewed as a function or a single entity rather than a series of individual data points (Ryan et al. 2006). The term *functional data analysis* was introduced by Ramsay and Dalzell (1991), who outlined several practical reasons for considering data analysis from such a functional perspective. Use of this term refers our attention to the intrinsic structure of the collected data we seek to analyze.

Biomechanical data are usually obtained over a number of discrete time points and assumed to be generated by some underlying function denoted by $y_i(t)$; the data points are a series of "snapshots" of that function at various points in time. As with PCA, FDA does not require that the data be time series in nature; rather, FDA can be applied to curves of various forms (e.g., phase-plane plots). We can also assume that the data should display a certain degree of smoothness, and much research has been directed toward the development of techniques that condition raw data so that they exhibit smooth characteristics (see chapter 12 on signal processing). Associated with smoothness is the assumption that adjacent data values are linked by the underlying function and, therefore, it is unlikely that adjacent values will vary greatly. Finally, we also assume that functional data sets possess a number of derivatives that are also smooth. Ramsay and Silverman (2005) provide a comprehensive reference to the main concerns and theoretical developments in FDA over the last decade.

As with PCA, FDA can incorporate several different procedures as a means of achieving the stated goals, but the following steps are typically used when FDA is applied to the analysis of biomechanical data:

1. Data representation: derivation of smooth functions
2. Registration of data, that is, time normalization or landmark registration (optional)
3. Functional principal component analysis
4. Presentation and analysis of results

Here, the procedures involved in these steps are described with reference to typical applications in the biomechanics literature.

Step 1: Derivation of Smooth Functions

Because biomechanical data are typically obtained by sampling data points at regular intervals, our functional data are usually observed at discrete time points t_{ij}, where i denotes the ith subject and there are $i = 1, \ldots, N$ subjects in the sample, and $j = 1, \ldots, n_i$, where n_i is the number of observations in the record for subject i. Because the records for different subjects in a sample can have different lengths and because measurements can be taken at different times for each subject, the index j is used to indicate the particular times at which a value was measured for a subject. Letting j range from 1 to n_i indicates that the total number of values measured can be different for each of the i subjects in the sample. Therefore, if 20 values are measured for subject 1, there are 20 distinct time points $(t_{11}, t_{12}, \ldots, t_{120})$ at which a value was taken, where t_{11} denotes the first time that a value was measured for subject 1, t_{12} denotes the second time that a value was measured for subject 1, and so on. For our purposes, we will assume that $n_i = n$; that is, the number of measurements taken for each subject is the same. However, it should be noted that this is not a requirement of FDA and all results still apply when the number of measurements taken for each subject is different. Also note that even though we assume that the number of measurements taken for each subject is the same, the time points at which those measurements were taken can still vary from subject to subject.

Often the collected data will contain a true signal mixed with measurement error or noise that can be represented mathematically as:

$$y_{ij} = \underbrace{y_i\left(t_{ij}\right)}_{\text{Signal}} + \underbrace{\varepsilon_{ij}}_{\text{Noise}} \qquad (14.16)$$

where y_{ij} is the raw data for subject i, $y_i(t)$ is a smooth function to be estimated, and $j = 1, \ldots, n$ data points.

Various techniques have been used to attenuate the noise in raw data and thereby improve the accuracy of our analyzed data, including polynomial smoothing (Miller and Nelson 1973), digital filtering (Winter 2009), and more recently spline basis smoothing approaches such as the application of general cross-validatory (GCV) splines (Craven and Wahba 1979; Woltring 1986). The application of basis function expansions to remove noise and smooth biomechanical data is widespread, as these algorithms provide effective smoothing and are more flexible than previous methods. However, spline basis functions also provide inherent advantages for analyzing the functional nature of biomechanical data and for examining functional changes in data sets.

Smoothing Using Basis Functions

In the earlier description of PCA, we discussed the concept of basis functions, an underlying set of mathematical functions (e.g., Fourier series) that can be used to describe any given waveform. Ramsay and Silverman (2005) provide a detailed description of the application of basis functions in FDA. Briefly, assume we wish to obtain a mathematical function for a biomechanical data curve $y_i(t)$. We do not know in advance the nature of any particular function, so we need a system that can be applied to many types of data. Therefore, we need to define a set of basic mathematical building blocks that can be combined to describe the functional nature of our data. If we start with a known set of basis functions $[\phi_1(t), \ldots, \phi_K(t)]$, we can approximate our unknown function using a linear combination of a sufficiently large number (K) of these functions. The more basis functions used (i.e., the larger the value of K), the closer we get to exact interpolation of the data. Fewer basis functions lead to smoother estimated data curves, but the residual difference between the smooth function and the original noisy data may increase as the number of basis functions is reduced. There are many choices of possible basis functions, including polynomial basis functions, Fourier basis functions, B-spline basis functions, and wavelet basis functions. The choice of basis function depends on the characteristic behavior of the data being analyzed, and no single basis is suitable for all data types. Once the basis functions have been chosen, the smooth function $y_i(t)$ can then be expressed as a linear combination of those basis functions:

$$y_i(t) = \sum_{k=1}^{K} c_{ik} \phi_k(t) \quad (14.17)$$

where $\phi_k(t)$ is the kth basis function (evaluated at time t) with weight c_{ik} and K is the total number of basis functions. The entire data sequence for subject i can be written in matrix notation as:

$$y_i = \phi c_i \quad (14.18)$$

where c_i is a vector of length K containing the basis function coefficients for subject i and Φ is a matrix with n rows and K columns containing the K basis functions evaluated at times t_{ij}. Then the task of functional data smoothing is to find the vector of coefficients c_i. For a

sample of N individuals we need to estimate N coefficient vectors c_i, $i = 1, \ldots, N$, such that:

$$Y = \Phi C \quad (14.19)$$

where Y is an $n \times N$ matrix of functional observations with each column containing the raw data for a particular subject, C is a $K \times N$ matrix of basis function coefficients, and Φ is the $n \times K$ basis function matrix.

For ease of presentation, let us assume that we just have a single data record for a single subject i with $j = 1, \ldots, n$ data points. If we choose the number of basis functions K and minimize the residual sum of squares error (SSE), the K-length vector of coefficients c_i can be estimated via least squares by minimizing:

$$\begin{aligned} SSE &= \sum_{j=1}^{n} \left[y_{ij} - y_i(t_{ij}) \right]^2 \\ &= \sum_{j=1}^{n} \left[y_{ij} - \sum_{k=1}^{K} c_{ik} \phi_k(t_{ij}) \right]^2 \\ &= (y_i - \Phi c_i)'(y_i - \Phi c_i) \quad (14.20) \end{aligned}$$

where y_i, c_i, and Φ are as defined previously. If $K = n$, this implies we can choose c_i such that $y_i(t_{ij}) = y_{ij}$; that is, the data are interpolated without any smoothing. If $K < n$, this implies that the data will be smoothed by some degree. The choice for optimizing K can be quite complex. It is also difficult to control the amount of smoothing since different levels of smoothing are achieved by simply adding or removing some basis functions.

Smoothing splines provides an alternative approach that allows more control over the smoothing process and alleviates the need to choose a value for K. When using smoothing splines, we set $K = n$, which implies that the data are interpolated and the resulting function estimates will not be very smooth; that is, the estimates will be "rough." Although we no longer need to specify K, we need an estimation that represents the raw data well but also ensures that the resulting function estimates are smooth. Therefore, we need to control for the overfitting (or roughness) induced by setting $K = n$. This can be achieved with a penalty term that penalizes the curvature (a measure of the roughness) of the estimated function. A standard mathematical method of measuring the curvature of a twice-differentiable curve $y_i(t)$ is to calculate its integrated squared second derivative (Green and Silverman 1994). The influence of the roughness penalty term is controlled by a smoothing parameter λ, ensuring that the fitting of a particular curve is determined not only by its goodness of fit (least squares fit) but also by its roughness. Note that the use of λ here to represent the smoothing parameter should not be confused with the unrelated term λ_i, which represents the eigenvalues in PCA.

The addition of the roughness penalty term results in a modified optimization criterion as given by

$$
\begin{aligned}
\text{PENSSE} &= \sum_{j=1}^{n} \left[y_{ij} - y_i \left(t_{ij} \right) \right]^2 + \lambda \int D^2 y_i(t)\, dt \\
&= \sum_{j=1}^{n} \left[y_{ij} - \sum_{k=1}^{K} c_{ik} \phi_k \left(t_{ij} \right) \right]^2 + \lambda \int D^2 y_i(t)\, dt \\
&= \left(y_i - \Phi c_i \right)' \left(y_i - \Phi c_i \right) + \lambda \int D^2 y_i(t)\, dt \quad (14.21)
\end{aligned}
$$

where PENSSE represents the penalized residual sum of squares error, $D^2 y_i(t)$ denotes the second derivative of $y_i(t)$, and λ is the smoothing parameter. The first term on the right-hand side (least squares) of the preceding equation controls fit to the data, whereas the second term (roughness penalty) controls the smoothness of the resulting function estimate. The choice of smoothing parameter is important because as λ increases, more emphasis is placed on smoothness and less on fit to the data, whereas as λ decreases more emphasis is placed on fit to the data and less on smoothness. $\lambda = 0$ gives the least squares fit.

A value for λ can be determined by subjective or objective means. The aim of the smoothing process is to provide curves that are stable and interpretable but also faithfully represent the raw data. If we vary the smoothing parameter, different features of the data can be explored and a subjective choice can be made for λ. An alternative approach is an objective automatic method that allows the smoothing parameter to be chosen by the data. Cross-validation and generalized cross-validation (Craven and Wahba 1979) are popular approaches for choosing an appropriate value for λ but have been criticized for finding values of λ that are "too small" or that undersmooth the data (Hastie and Tibshirani 1990). The generalized maximum likelihood criterion proposed by Wahba (1985) partially remedies this problem. In general, it is recommended to use automatic methods such as cross-validation only as a guideline or starting point before making a subjective choice for λ (Ramsay and Silverman 2002). Figures 14.9 show the effect of changing λ on the smoothness of the resulting estimates. In all cases, λ is very small and changing λ by only a minimal amount can yield much smoother results.

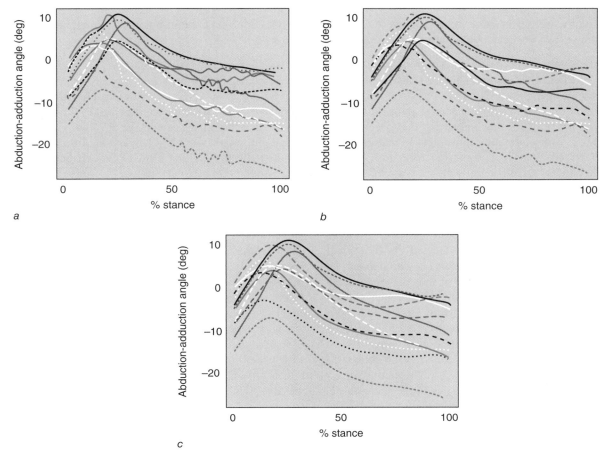

▲ Figure 14.9 Adduction-abduction angle of ankle joint complex during running. *(a)* Data are smoothed with λ set to 1×10^{-9}. *(b)* Data are smoothed with λ set to 1×10^{-6}. *(c)* Data are smoothed with λ set to 1×10^{-4}.

Sometimes the number of observations recorded for each record can be very large, such as when motion-capture equipment provides hundreds or thousands of measurements per second. In such cases, setting $K = n$ can cause computational difficulties. It is usually possible to use a smaller number of basis functions and still use the roughness penalty approach to approximate the curves.

Choosing Basis Functions

Basis functions should be chosen to match the features of the data being analyzed. For example, if the observed waveforms are periodic, a Fourier basis may be appropriate. Alternatively, if the observed functions are locally smooth and nonperiodic, B-splines may be appropriate; if the observed data are noisy but contain informative "spikes" that need to avoid the effect of severe smoothing, a wavelet basis may be appropriate. The ultimate choice of basis functions should provide the best possible approximation using a relatively small number of functions K.

B-splines are very useful basis functions for smoothing kinematic data because their structure is designed to provide the smooth function with the capacity to accommodate changing local behavior. B-splines consist of polynomial pieces joined at certain values of x, called *knots*. Eilers and Marx (1996) outlined the general properties of a B-spline basis. Once the knots are known, it is relatively easy to compute the B-splines using the recursive algorithm of de Boor (1978).

Step 2: Registration of Data

An intermediate optional step often included in FDA is landmark registration. In many cases we find that smoothed curves may follow a similar overall pattern but the timing or position on the curve of certain important features (e.g., global maximum or minimum, zero crossings) may differ from one record (i.e., curve) to another. Landmark registration identifies the location of a number of visible features or landmarks and shifts each

FROM THE SCIENTIFIC LITERATURE

Ryan, W., A.J. Harrison, and K. Hayes. 2006. Functional data analysis in biomechanics: A case study of knee joint vertical jump kinematics. *Sports Biomechanics* 5:121-38.

This study provided a detailed description of the application of FDA to lower-limb angular kinematic data on children performing the vertical jump. The aim of the study was to examine differences in joint kinematics with respect to developmental stages that describe the progression from immature movement to mature form: stage 1, initial; stage 2, elementary; and stage 3, mature. Although these stages generally progress as children get older, they are not precisely linked to age. Comparisons were made between landmark-registered and unregistered knee-joint angle data from countermovement vertical jumping. Following landmark registration, step 3 of FDA was then implemented, which derived the functional principal components (FPCs) on the registered data. The landmark-registered functional principal components localized the timing of the bottom of the crouch. Consequently, the first FPC on landmark-registered data more accurately represented the extent of the variability in the knee-joint angle at the bottom of the crouch irrespective of when this event occurred. The second FPC on landmark-registered data accounted for 16.8% of the total variation and had very similar characteristics to the second FPC on unregistered data, but it is important to note that the plus and minus graphs swapped positions. This occasionally happens in FDA, and it is important to consider the shape of the high and low scoring graphs rather than their respective signs. Both registered and unregistered second

FPCs were associated with the range of knee-joint flexion. The percentage of variation explained by the second FPC increased with landmark registration. The third FPC on registered data was similar to the third FPC on unregistered data, although the percentage of variation explained was slightly higher at 7.9%. Inspection of the third FPC graphs indicated that the third FPC described the rate, range, and smoothness of knee-joint-flexion action during the countermovement.

The study also demonstrated that functional principal component scores could be used to determine group differences or trends. Results of the developmental stage analysis showed that the range of scores on the first FPC progressively decreased from developmental stage 1 to stage 3. This suggested that the knee-joint actions of stage 1 subjects were highly variable in relation to the mean curve, whereas stage 3 subjects used a knee-joint action that was closer to the ensemble mean curve.

Stage analysis of second FPC scores showed no clearly identifiable trend. However, analysis of the third FPC scores showed a very clear stage-wise increase in the third FPC score from stage 1 to stage 3. Ryan and colleagues (2006) suggested that high scorers on the third FPC were likely to use the stretch-shortening cycle more effectively in performing the vertical jump, and the authors concluded that developmentally mature jumpers made more effective use of the stretch-shortening cycle.

curve accordingly so that these features occur together, permitting a more intuitive comparison between records. Landmark registration can provide a much more meaningful cross-sectional average for curves that contain an obvious feature such as a minimum. The point-wise variation may also be reduced considerably. However, landmark registration is not always suitable to every analysis process because landmarks may be missing from certain curves or the timing of landmarks may be ambiguous. Therefore, this step is not always included in FDA, and the decision whether to register the data should be made on a case-by-case basis.

The importance of landmark registration can be illustrated when one considers how individual factors produce slight alterations of curve shape in a series of data. Finding an ensemble average of time points in a series of curves may eliminate an important source of variability if the curves were not properly aligned beforehand. Godwin and colleagues (2009) examined moment curves during a manual lifting task and registered them according to two well-defined inflection points on the graph that corresponded to the common events of box pick-up and box release. These authors used double registration to eliminate temporal variation between subjects so that the timing of features on individual curves became identical to the timing of features found in a reference curve. There are times when landmark registration may not be appropriate and can produce unintended effects (Clarkson et al. 2005); therefore, landmark registration should be used only when it is justified in removing unwanted temporal or spatial variation.

Step 3: Functional Principal Component Analysis

Functional principal component analysis (FPCA) is an extension of the classic multivariate technique to the functional domain. In this case, eigenfunctions rather than eigenvectors are used to represent the principal components. A major advantage of FPCA is that it produces principal components that are functions defined in the same domain as the original functional observations and, consequently, the functional principal components (FPCs) extracted in the analysis have a definite biomechanical interpretation. Figure 14.10 shows three FPCs for the leg abduction-adduction (Leg ABD) angle during stance phase in running (Coffey et al. 2011). The FPCs are presented as functions over the time series (i.e., expressed in the same domain as the original function), and a multiple of each FPC can be added to (or subtracted from) the overall group mean curve to demonstrate the exact abduction-adduction movement characteristics of subjects that score high (or low) on each FPC.

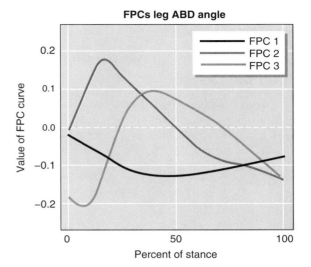

▲ Figure 14.10 The functional principal components for leg abduction-adduction movement (Coffey et al. 2008).

Reprinted from *Human Movement Science*, Vol. 30(1); N. Coffey et al., "Common functional principal component analysis: A new approach to analyzing human movement data," pgs. 1144-1166. Copyright 2011, with permission of Elsevier.

As with the traditional PCA, each FPC will account for a certain proportion of variance, and it is necessary to determine how many FPCs are required to complete a meaningful analysis. This can be achieved by inspection of the scree plot as shown in figure 14.11. This shows that as the number of FPCs increases, the amount of variance accounted will approach 100% and each succeeding FPC accounts for progressively less and less variation. In figure 14.11 it is clear that only two or three FPCs account for more than 95% of the variation in the sample,

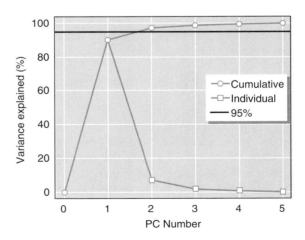

▲ Figure 14.11 Scree plot of FPCs. In this example it is clear that the first 3 FPCs account for more than 95% of the variation in the sample.

and on this basis there is little point in considering more than three FPCs for this data set.

Calculating Functional Principal Components

Functional principal components analysis results in eigenfunctions, and therefore the rth FPC, denoted as $\xi_r(t)$, is now represented as a function describing a particular pattern of behavior over the whole time interval. Let

$$\bar{y}(t) = \sum_{i=1}^{N} y_i(t) \qquad (14.22)$$

denote the mean of the functional data set and

$$v(s,t) = N^{-1} \sum_{i=1}^{N} \left[y_i(s) - \bar{y}(s) \right] \left[y_i(t) - \bar{y}(t) \right] \quad (14.23)$$

denote the covariance function of the functional data set, where s and t are time points. Calculating the FPCs involves the orthogonal decomposition of the covariance function (rather than covariance matrix as in multivariate PCA) to determine the dominant modes of variation in the data. Each FPC is estimated by solving

$$\int v(s,t)\xi_r(t)dt = \rho_r \xi_r(s) \qquad (14.24)$$

where ρ_r is an appropriate eigenvalue. The proportion of variation accounted for by the rth FPC is

$$\frac{\rho_r}{\sum \rho_r} \qquad (14.25)$$

For further analysis, each subject can be weighted by each of the FPCs extracted, resulting in scalars referred to as *FPC scores*. That is, for each subject, one FPC score is calculated for each FPC extracted. The score on the rth FPC for the ith subject is calculated as

$$f_{ir} = \int \xi_r(t) \left[y_i(t) - \bar{y}(t) \right] dt \qquad (14.26)$$

These FPC scores can then be used in further statistical analysis to ascertain group trends.

Smoothing of Functional Principal Components

Step 1 of the FDA process emphasized the assumption that our raw data may contain noise derived from measurement, but we always assume that the raw data represent some underlying smooth function. In functional principal component analysis, we can extract the FPCs from the data after they have been smoothed with the smoothing parameter λ as earlier described. However, this does not guarantee that the resulting FPCs will be smooth. An alternative approach is to work with the raw data and ensure smoothness of the resulting FPCs

by incorporating smoothing into the extraction of the FPCs. Figure 14.12 shows the effect of deriving FPCs on raw functional data and specifying various levels of the smoothing parameter for the FPCs.

When B-splines are used, it doesn't usually make much difference whether smoothing is applied to the raw data or the FPCs. If the raw data contain a lot of measurement error or noise, then this may be an important consideration. The general philosophy in functional data analysis is to leave the smoothing step until the final output of the analysis has been computed. For example, if you are interested in the actual subject and group mean curves, it may be best to smooth the raw data. If your ultimate interest is the FPCs, then it is normal to smooth the extracted FPCs.

Step 4: Presentation and Analysis of Results

After we complete functional principal component analysis, the next step is to present the results in ways that allow insightful interpretation. Ramsay and Silverman (2005) recommend the use of graphs presenting the ensemble mean curve of the original $y_i(t)$ data, designated $\bar{y}(t)$, and the functions obtained by adding and subtracting a suitable multiple of each FPC function, for example, $\bar{y}(t) \pm c \times \xi_1(t)$. Figure 14.13 shows this form of presentation for the first three FPCs on the leg abduction-adduction angle data sets of Coffey and colleagues (2011). The interpretations of these graphs are relatively simple. Figure 14.13a shows the first three FPCs plotted with respect to percentage of stance. Figure 14.13b shows that high scorers in the first functional principal component (FPC1), illustrated by the plus (+) signs, are characterized by a leg abduction-adduction angle that is lower than the mean function throughout stance. Conversely, low scorers illustrated by the minus (−) signs are characterised by a leg abduction-adduction angle that is higher than the ensemble mean angle throughout stance. Figure 14.13c shows the effect of FPC2 relative to the ensemble mean function, with high positive scorers tending to display smaller leg ABD angles at heel strike and high negative scorers tending to display leg ABD angles that are greater than average at heel-strike. Therefore FPC2 describes the precise leg abduction-adduction movements associated with heel strike. Figure 14.13d shows FPC3 for the same data set, and it is clear that high positive scorers tend to display increased leg abduction-adduction range of motion and negative scorers tend to display decreased leg abduction-adduction range of motion. Therefore FPC3 is representative of leg abduction-adduction range of motion throughout the stance phase.

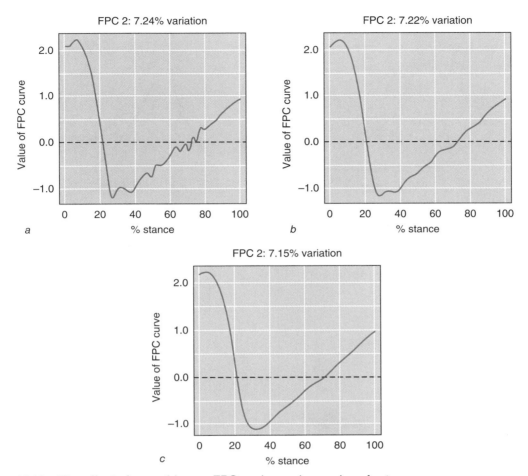

▲ **Figure 14.12** The effect of smoothing on FPCs using various values for λ.

Analyzing Functional Principal Component Scores

It is often helpful, in addition to presenting the FPCs relative to the ensemble mean function, to present an analysis of the FPC scores to allow group or treatment comparisons. Donoghue and colleagues (2008) showed that FPC scores could be subjected to group analysis using analysis of variance (ANOVA) techniques and that effect size statistics could be applied to the FPC scores. Figure 14.14 shows the analysis of the first principal component FPC1 on ankle dorsiflexion during stance phase in running and the accompanying group analysis of FPC1 scores on subjects with Achilles tendonitis who were wearing customized orthotics, the same subjects without orthotics, and the uninjured limbs from the same subjects (control). The ANOVA on FPC1 scores showed a significant difference with a large effect size between subjects without orthotics and the control condition.

In other studies, Godwin and colleagues (2009) described the use of a modified functional ANOVA technique that could be applied to FPC scores, and Epi-fanio and colleagues (2008) showed that FPCA could be used to differentiate between normal and pathological patterns in a sit-to-stand movement. A further advantage of FPCA is that the results of this approach provide outcomes that relate directly to the behavior of subjects. This is illustrated in figure 14.15, which shows that the functional movement pattern of a high scorer (subject 17) mimics the pattern of movement predicted by the high scoring FPC, which is depicted by the + line in the graph. Similarly, the lowest scoring subject on this FPC (subject 28) has an adduction-abduction angle function that mimics that low scoring pattern on this FPC.

Analyzing Coordination Using Bivariate Functional Principal Components

A key area in biomechanics and motor control is the analysis of coordination. Analysis of coordination patterns requires an examination of curves that capture two or more parameters simultaneously. Therefore, analysis of coordination requires the simultaneous analysis of

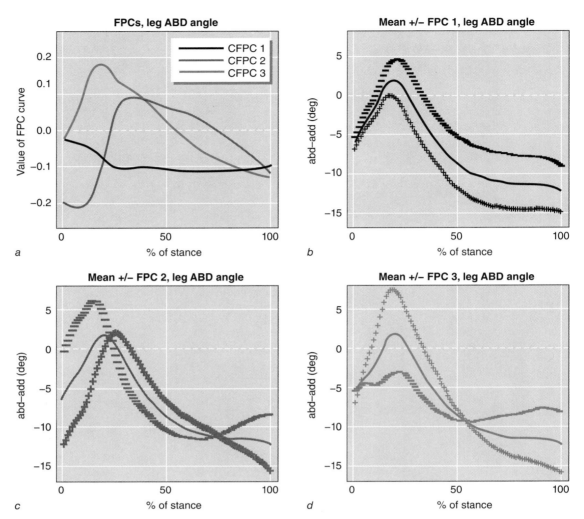

Figure 14.13 Visualizing functional principal components (FPCs). In *a*, the first three FPCs are shown for leg abduction-adduction angle during stance phase in running. The effect of high (+) or low (-) scores for each of the three FPCs is shown in panels *b*, FPC 1; *c*, FPC2, and *d*, FPC 3. See text for details on interpretation.

Reprinted from *Human Movement Science*, Vol. 30(1); N. Coffey et al., "Common functional principal component analysis: A new approach to analyzing human movement data," pgs. 1144-1166. Copyright 2011, with permission of Elsevier.

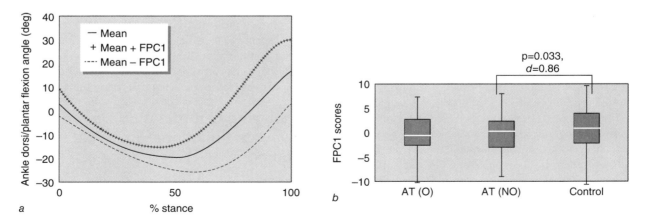

Figure 14.14 Functional principal component for ankle dorsiflexion on stance phase in running and analysis of FPC scores for injured subjects wearing orthotics, AT(O); injured subjects without orthotics, AT(NO); and the uninjured limb (control).

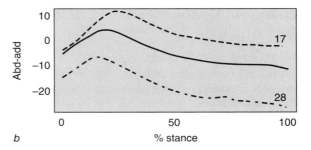

▲Figure 14.15 *(a)* The mean abduction-adduction angle function plus or minus the first FPC on abduction-adduction angle. *(b)* The mean abduction-adduction angle function and abduction-adduction angle functions of the subjects with the highest positive score (subject 17) and lowest negative score (subject 28) on FPC1. The similarity of subject 17's abduction-adduction angle function to the mean + FPC1 and the similarity of subject 28's abduction-adduction angle function to the mean – FPC1 is obvious.

more than one function. The procedures of FPCA can be extended from a single variable to multiple variables. In an extension of the work by Ryan and colleagues (2006), Harrison and colleagues (2007) illustrated that bivariate FPCA could be used to analyze coordination using hip-knee angle-angle patterns in the vertical jump. Although standard FPCA extracts FPCs from a single variable $[\xi_r(t)]$, in bivariate FPCA each component consists of a vector of FPCs such that $\xi_r(t) = [\xi_r^{HIP}(t), \xi_r^{KNEE}(t)]$, where *HIP* and *KNEE* represent the hip and knee angle data sets during the jump. We can extract these bivariate

FPCs using FDA methods. To interpret the bivariate FPCs, we plot the mean functions across the time domain, point by point $[\bar{y}^{HIP}(t), \bar{y}^{KNEE}(t)]$. Combining the data points at each time point forms the ensemble mean hip-knee graph. At each point on the curve, we construct an arrow representing the FPC for each angle, $[\bar{y}^{HIP}(t) + c \times \xi_r^{HIP}(t), \bar{y}^{KNEE}(t) + c \times \xi_r^{KNEE}(t)]$, where ξ_r^{HIP} is the FPC of the hip and *c* is a weighting factor chosen to allow effective display of the FPC. Figure 14.16 shows the coordination analysis on vertical jump using bivariate FPCA.

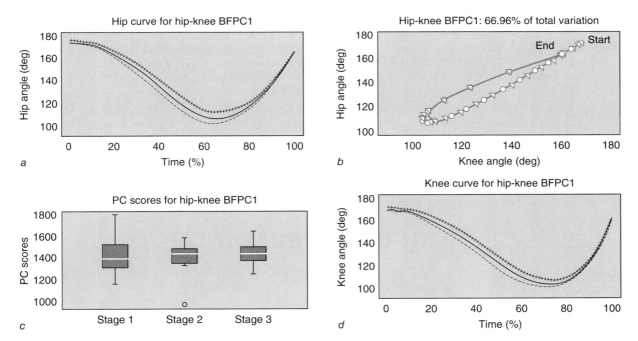

▲Figure 14.16 Functional principal component analysis of the first bivariate functional principal component for hip–knee joint coordination during vertical jumping (Harrison et al. 2007). *(b)* The mean hip-knee angle-angle plot with bivariate functional principal component scores at each time point represented as vector arrows. Hip and knee functional principal component (FPC) plots are presented in *(a)* and *(d)* and placed adjacent to the hip and knee axes, respectively. *(c)* The distribution of bivariate functional principal component scores with respect to developmental stage.

Additional Resources for Functional Data Analysis

In their text, Ramsay and Silverman (2005) considered several case studies illustrating how FDA ideas work in practice in a diverse range of subject areas including biomechanics. Further illustration of FDA with practical examples can be found on the FDA website: www.psych.mcgill.ca/misc/fda. Additional resources for implementing FDA procedures in MATLAB and the freeware statistics application R are available. Ramsay and colleagues (2009) have produced a useful text on the implementation of FDA techniques with both R and Matlab, and Clarkson and colleagues (2005) demonstrate how these techniques can be implemented in S+.

COMPARISON OF PCA AND FDA

Both principal component analysis of waveform data and functional data analysis have the same objective of extracting features from waveform data that reveal how the waveforms, or curves, differ from each other. Whereas PCA is essentially a data reduction technique that follows some initial treatment of the data (e.g., smoothing, filtering, averaging, registration or time normalizing), FDA is essentially a collection of statistical techniques that encompass the estimation and analysis of waveform data as well as the data reduction of principal component analysis.

The key difference between PCA and FDA is the data representation before the calculation of principal components. The input data for PCA are either actual observations or interpolated observations, where the number of observations is consistent for each subject. In the FDA approach, each waveform is represented by a function, that is, a specific set of basis function coefficients. This basic difference leads to features of data collection and analysis unique to each method. In PCA, all input data waveforms must have exactly the same number of data points (same number of matrix columns), with each data column synchronized in time; this is most easily accomplished through synchronized data collection. In FDA, each waveform is represented by a function; therefore, measurements can be taken at different time points and the number of points in each waveform can vary.

Another important difference is that FDA incorporates data-processing steps such as smoothing and curve registration, whereas these would have to be done as preliminary analyses prior to PCA. The smoothness of the functions generated by the FDA means that the consideration of time derivatives, continuous phase-plane plots, or bivariate functions can easily be incorporated within the process. Although FDA provides a more flexible analysis, it is less accessible to many users because it requires some level of programming skill using Matlab, S-Plus, or R. By contrast, PCA can be carried out with widely available statistical software applications such as SPSS or Minitab.

SUMMARY

Biomechanical data sets are generally high-dimensional. Analysis of these data in ways that allow effective interpretation without severe loss of data is challenging. Here we have presented two related methods, PCA and FDA, which both provide data reduction while retaining important information from the entire data waveform or time series. Both PCA and FDA are particularly well suited to the demands of analyzing high-dimensional biomechanical data. An important feature is their ability to distinguish between groups of subjects based on differences in shape or pattern of the waveform data. The application of standard statistical procedures to the resulting PCA or FPCA scores provides an objective way to test for differences in these patterns. We have shown how the results of the analysis can be presented in ways to allow insightful interpretation of the principal components in terms of the characteristic patterns in biomechanics.

SUGGESTED READINGS

PCA

Cappellini, G., Y.P. Ivanenko, R.E. Poppele, and F. Lacquaniti. 2006. Motor patterns in human walking and running. *J Neurophysiol.* 95:3426-37.

Chau, T. 2001. A review of analytical techniques for gait data. Part 1: Fuzzy, statistical and fractal methods. *Gait Posture* 13:49-66.

Daffertshofer, A., C.J.C. Lamoth, O.G. Meijer, and P.J. Beek. 2004. PCA in studying coordination and variability: A tutorial. *Clinical Biomechanics* 19:415-28.

Deluzio, K.J., and J.A. Astephen. 2007. Biomechanical features of gait waveform data associated with knee osteoarthritis: An application of principal component analysis. *Gait and Posture* 25:86-93.

Jackson, J.E. 1991. *A User's Guide to Principal Components.* New York: Wiley.

O'Connor, K.M., and M.C. Bottum. 2009. Differences in cutting knee mechanics based on principal components analysis. *Medicine and Science in Sports and Exercise* 41(4):867-78.

Sadeghi, H., F. Prince, S. Sadeghi, and H. Labelle. 2000. Principal component analysis of the power developed in the flexion/extension muscles of the hip in able-bodied gait. *Medical Engineering and Physics* 22(10):703-10.

FDA

Clarkson, D.B., C. Fraley, C. Gu, and J.O. Ramsay. 2005. *S+ Functional Data Analysis*. New York: Springer.

Coffey, N., O. Donoghue, A.J. Harrison, and K. Hayes. 2011. Common functional principal components analysis—A new approach to analyzing human movement data. *Human Movement Science* 30:1144-66.

Donà, G., E. Preatoni, C. Cobelli, R. Rodano, and A.J. Harrison. 2009. Application of functional principal component analysis in race walking: An emerging methodology. *Sports Biomechanics* 8(4):284-301.

Donoghue, O., A.J. Harrison, N. Coffey, and K. Hayes. 2008. Functional data analysis of running kinematics in chronic Achilles tendon injury. *Medicine and Science in Sports and Exercise* 40(7):1323-35.

Functional Data Analysis website. www.psych.mcgill.ca /misc/fda.

Hooker, G. List of publications and resources. Personal website. www.bscb.cornell.edu/~hooker.

Ramsay, J., G. Hooker, and S. Graves. 2012. *Functional Data Analysis With R and MATLAB*. New York: Springer.

Ramsay, J.O., and B.W. Silverman. 2002. *Applied Functional Data Analysis*. New York: Springer-Verlag.

Ramsay, J.O., and B.W. Silverman. 2005. *Functional Data Analysis*. 2nd ed. New York: Springer.

Simonoff, J.S. 1996. *Smoothing Methods in Statistics*. New York: Springer.

International System of Units (Système International, SI)

SOME RULES FOR REPORTING SI UNITS

■ Do not use a period after the abbreviated versions of metric units unless the unit appears at the end of a sentence. For example, 35.6 N (short for newtons), 3.00 kg (short for kilograms), or 0.500 s (short for seconds) are correct forms; but 40.0 m., 20.5 kPa., or 20.5 sec. are incorrect forms.

■ A centered dot (·) is used to separate abbreviated SI units involving combined quantities, such as N·s for newton seconds or kg·m² for kilogram meters squared. However, a decimal point (.) is also acceptable and is usually much easier to use.

■ Do not capitalize a unit derived from a proper name when spelling out the unit even though the unit's abbreviation is a capital letter. Examples of such units include the watt (W), the newton (N), the hertz (Hz), the pascal (Pa), and the joule (J).

■ Use a slash (/) to indicate an arithmetic division of units, such as m/s for meters per second or N/m² for newtons per square meter.

■ Do not mix abbreviations and unabbreviated forms in an expression. For instance, the following are incorrect forms: newtons per m, kg.meters, N.seconds, and watts/kg.

■ The prefixes hecto, deca, deci, and centi should be avoided, except for the measurements of area, volume, and length, such as hectare, deciliter, and centimeter.

■ When pronouncing metric units that have a prefix, always place the accent on the complete prefix, that is, kilo'-meter versus ki-lo'-meter or kilo-meter'.

■ Always type a space between the numeric part and the number except for °C, ° (angle), and %. Examples: 76.4 W, 20.4°C, 13.45%, 678 N·m, and 45.2°.

■ When writing out numbers with more than four digits on either side of the decimal point, use a space instead of a comma to separate digits into groups of three as, for example, 23 400 m or 0.002 63 m. This is because the comma is used in many countries as a decimal point. It is permissible to omit the blank in four-digit numbers, such as 1002, 9980, and 0.1234.

Quantity*	Name	Symbol	Formula
KINEMATIC DOMAIN			
Length *(l, r, s, x, y, z)*	**meter**	**m**	
Area *(A)*	square meter	m²	
	hectare	ha	hm² =10 000 m²
Volume *(V)*	cubic meter	m³	
	liter	L	1 dm³
Linear velocity, speed *(v)*	meter per second	m/s	
Linear acceleration *(a)*	meter per second squared	m/s²	
Linear jerk *(j)*	meter per second cubed	m/s³	
Plane angle *(θ, ψ, α, β, γ)*	radian	rad	m/m = 1
	degree	deg, °	π/180 rad
	minute	'	1/60°
	second	"	1/360°
	revolution	r	2π rad, 360°
Angular velocity *(ω)*	radian per second	rad/s	
Angular acceleration *(α)*	radian per second squared	rad/s²	
Solid angle *(Ω)*	steradian	sr	
INERTIAL PROPERTY DOMAIN			
Mass *(m)*	**kilogram**	**kg**	
	metric ton or tonne	t	1 Mg = 1000 kg
Moment of inertia *(I, J)*	kilogram meter squared		kg·m²
Density *(ρ)*	kilogram per cubic meter		kg/m³
Viscosity *(η)*	pascal second		Pa·s
TIME (TEMPORAL) DOMAIN			
Time *(t)*	**second**	**s**	
	minute	min	60 s
	hour	h	3600 s
	day	d	86 400 s
	year	a	31.536 Ms
Frequency *(f)*	hertz	Hz	1/s
KINETIC DOMAIN			
Force *(F)*	newton	N	kg·m/s²
Moment of force *(M)*, torque *(t)*	newton meter		N·m
Pressure *(P)*	pascal	Pa	N/m²
	millibar	mbar	1 mbar = 100 Pa
Stress *(σ or τ)*	pascal	Pa	N/m²
Energy *(E)*, work *(W)*	joule	J	kg·m²/s²
Power *(P)*	watt	W	J/s
Linear impulse	newton second		N·s or kg·m/s

KINETIC DOMAIN			
Linear momentum *(p)*	kilogram meter per second		kg·m/s or N·s
Angular impulse	newton meter second		N·m·s or kg·m²/s
Angular momentum *(L)*	kilogram meter squared per second		kg·m²/s or N·m·s
ELECTRICAL DOMAIN			
Current *(I)*	**ampere**	**A**	W/A
Voltage *(V)*	volt	V	s·A
Charge *(Q)*	coulomb	C	J/s
Power *(P)*	watt	W	
Resistance *(R)*, impedance *(Z)*	ohm	Ω	V/A
Capacitance *(C)*	farad	F	C/V
Magnetic flux *(Φ)*	weber	Wb	V·s
Magnetic flux density *(B)*	tesla	T	Wb/m²
Inductance *(L)*	henry	H	Wb/A
Conductance *(G)*	siemens	S	A/V
Electric energy *(E)*	joule	J	W·s
TEMPERATURE DOMAIN			
Temperature (T)	**kelvin**	**K**	
	degree Celsius	°C	
CHEMICAL DOMAIN			
Amount of substance *(n)*	**mole**	**mol**	
Concentration *(c)*	mole per cubic meter	mol/m³	
LIGHT DOMAIN			
Luminous intensity *(I)*	**candela**	**cd**	
Luminous flux *(Φ)*	lumen	lm	cd·sr
Illuminance *(E)*	lux	lx	lm/m²
Radiant intensity *(I)*	watt per steradian		W/sr
Radiance *(L)*	watt square meter per steradian		W·m²/sr

*Base units are in **bold.**

SI Prefixes

Multiplication factor	Prefix	Symbol
1 000 000 000 000 000 000 000 000 = 10^{24}	yotta	Y
1 000 000 000 000 000 000 000 = 10^{21}	zetta	Z
1 000 000 000 000 000 000 = 10^{18}	exa	E
1 000 000 000 000 000 = 10^{15}	peta	P
1 000 000 000 000 = 10^{12}	tera	T
1 000 000 000 = 10^{9}	giga	G
1 000 000 = 10^{6}	mega	M
1000 = 10^{3}	kilo	k
100 = 10^{2}	hecto*	H
10 = 10^{1}	deca* or deka*	da
0.1 = 10^{-1}	deci*	d
0.01 = 10^{-2}	centi*	c
0.001 = 10^{-3}	milli	m
0.000 001 = 10^{-6}	micro	μ
0.000 000 001 = 10^{-9}	nano	n
0.000 000 000 001 = 10^{-12}	pico	p
0.000 000 000 000 001 = 10^{-15}	femto	f
0.000 000 000 000 000 001 = 10^{-18}	atto	a
0.000 000 000 000 000 000 001 = 10^{-21}	zepto	z
0.000 000 000 000 000 000 000 001 = 10^{-24}	yocto	y

*These prefixes normally are not used except for special quantities, such as the hectare, the centimeter, and the decibel.

Selected Factors for Converting Between Units of Measure

Unit	Conversion factor
AREA	
1 acre	= 0.405 ha
1 square inch	= **645.16** mm²
ENERGY, WORK	
1 calorie	= **4.1868** J
1 Calorie (dietetic-1 kcal)	= 4.1855 kJ
1 erg	= **0.1** μJ
1 foot pound-force	= 1.356 J
FORCE, WEIGHT	
1 dyne	= **10** μN
1 kilopond (kg force)	= **9.806 65** N
1 pound-force	= 4.448 N
LENGTH	
1 foot	= **30.48** cm = **0.3048** m
1 inch	= **2.54** cm
1 yard	= **0.9144** m
1 mile	= **1.609 344** km
MASS	
1 ounce (avoirdupois)	= 28.35 g
1 pound (avoirdupois)	= 0.4536 kg
1 slug	= 14.59 kg
1 stone (14 lb, UK)	= 6.350 kg
1 ton (long, 2240 lb, UK)	= 1.016 Mg
1 ton (short, 2000 lb)	= 0.907 Mg

(continued)

(continued)

Unit	Conversion factor
POWER	
1 British thermal unit (BTU) per hour	= 0.293 W
1 horsepower (electric)	= **746** W
PRESSURE, STRESS	
1 atmosphere (standard)	= **101.325** Pa
1 mmHg (0°C)	= 133.3 Pa
1 pound per square inch (psi) (lb/in^2)	= 6.895 kPa
TEMPERATURE	
1 Fahrenheit degree	= **5/9** K**
VELOCITY	
1 mph	= **0.447 04** m/s = **1.609 344** km/h
VOLUME	
1 cubic foot	= 0.028 32 m^3
1 cubic inch	= 16.39 cm^3
1 gallon (imperial)	= 4.546 L
1 gallon (US)	= 3.785 L

All values in **bold** are exact conversions.

*The large calorie commonly used by dietitians to refer to the amount of energy in foods is defined by an international agreement.

**Add 32° after converting from Celsius to Fahrenheit or subtract 32° before converting from Fahrenheit to Celsius. Note that a Celsius degree is equal to a kelvin (K).

Appendix C

Basic Electronics

This appendix gives a brief overview of the elementary concepts of electronic circuits that are relevant to the collection of human movement data. The topics discussed include basic electronic components, Ohm's law, circuit diagrams, and the functions of several common laboratory instruments, such as amplifiers and electrogoniometers. Students interested in further detail or sample problems on any particular topic should refer to textbooks on electronics or linear circuits (such as *Schaum's Outline* [O'Malley 1992] or Winter and Patla 1997). The focus here is on simple concepts about steady-state circuits and how they apply to common measurements of human movement.

Electronics notation and symbols are standardized across different fields. Here, we use the notations and symbols given in this table:

Basic SI Electrical Units

Quantity	Symbol	SI unit	SI abbreviation
Current	I	ampere	A
Voltage	V	volt	V
Resistance	R	ohm	Ω
Capacitance	C	farad	F
Power	P	watt	W

CIRCUIT DIAGRAMS

A **circuit diagram** is a formal means of representing an electric circuit. We use these diagrams in this appendix to illustrate different examples. Circuit diagrams have many conventions, the most common of which are these:

► Components are represented by standard icons with their sizes noted.

► Wires are represented by straight lines for zero resistance.

► Wires are drawn only in north-south-east-west directions.

► A connection of two wires is indicated by a solid dot.

► One wire passing over another is indicated by a short loop.

► Interface points are indicated by open dots and labeled.

Circuit diagrams can, of course, become very complicated. The conventions just listed are displayed in figure C.1. This diagram shows the symbols for various components: a 9-volt (V) battery and its ground, a 100-ohm (Ω) resistor, a 10 Ω variable resistor, and a 1-microfarad (μF) capacitor. A discussion on these electrical components and several principles of electricity follows.

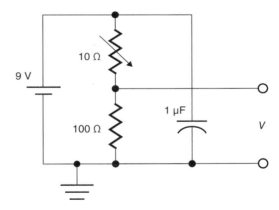

▲ **Figure C.1** Circuit diagram of a 9 V battery powering two resistors and a capacitor. The lower side of the battery is grounded. The voltage *(V)* is the quantity that we measure. The 10 Ω resistor has a variable resistance with a maximum of 10 Ω.

ELECTRIC CHARGE, CURRENT, AND VOLTAGE

Electric charge can be either positive or negative, depending on whether we are dealing with protons or electrons. Electricity is the flow of electrons through some medium, whether through a wire in a house or lightning through the air. The basic SI unit of electric charge is the **coulomb** (C). It represents about 6.25×10^{18} electrons. The rate of flow of electricity, or **current,** has units of **amperes,** or amps (A); 1 A is a flow rate of 1 C/s. As practical examples, consider that a handheld calculator requires a few microamperes (μA) to operate, a D-cell battery supplies about 100 mA, a car battery offers a maximum of about 2 A, and a typical house circuit provides 20 A. Current flow occurs when there is a difference between the electrical potential energy at two sites. This potential difference is called a **voltage.** One volt is defined as 1 joule (J) of energy per 1 C of charge. A D-cell battery offers 1.25 V, a car battery offers 12 V, and house electricity averages 110 V. A human electromyograph (EMG), in contrast, is on the order of μV.

A point of zero voltage is called a **ground**. This is never an absolute quantity but rather a defined reference point in a circuit. Thus, two circuits can have their own grounding references, but there may be a potential difference between the grounds of the two circuits. For example, in a small battery-powered circuit such as a clock or flashlight, ground is typically defined as the negative terminal of the battery powering the circuit. In house applications, ground is defined as the potential of the surrounding soil. This is accomplished by connecting the circuit to a metal rod driven into the earth. This house ground is different from the ground in any battery-powered circuit unless a connection is made between them. As another example, jump-starting a car is dangerous because potential differences can exist between two cars; even though the battery in each car is 12 V, their tires insulate them from the road (which is the ground). In human movement, we often see these principles applied in EMG recording because different voltage potentials can exist over the skin surface of the body depending on what muscles are active. We often record EMG with a separate grounding plate on a bony landmark away from the musculature.

Voltage and current are related (as is discussed later in this appendix), and this is often a source of confusion. The principles, stated previously, must be remembered: Current is the flow of electrons, and voltage is a potential energy difference that can cause electron flow. If current is flowing between two sites, then there must be a voltage difference between them. However, there can be a voltage difference without the flow of current; in that case, there is no complete circuit for the current to flow through. For example, there is a voltage difference between the terminals of a wall outlet, regardless of whether an appliance is connected to it. Current only flows between the terminals when an appliance is connected to them and turned on. An extreme example is that birds can land on a high-voltage overhead power line without being harmed. The same principle applies

to people who work on electrical lines: As long as workers are highly insulated from the ground, it is possible for them to touch the wire with their bare hands. When contact is made, a person is thousands of volts higher than the ground, but because virtually no current can flow through the insulation, the worker is unharmed. However, when a power line is broken in a storm and one end falls to the ground, touching the wire can be fatal because making contact with the wire connects a circuit to the ground.

Circuits are often difficult to conceptualize because they cannot be visualized directly. A measurement instrument, such as a voltmeter, oscilloscope, or computer, must be used to establish the state of a circuit. This is an abstract task, and it can be helpful to use the flow of a fluid through a pipe system as an analogy. Electric current (amperage) is analogous to the rate of fluid flow through the pipe (i.e., liters per second). Voltage is analogous to the pressure in the pipe system. Thus, if water is flowing through a hose, there must be a pressure difference between the ends of the hose; however, we can have a closed, pressurized container with no water leaking out of it. Flow implies that a potential energy difference exists. The fact that a potential energy difference exists, however, does not imply that something is flowing. Other fluid examples are offered throughout this appendix to illustrate key points.

We most often think of voltage as the strength of a power supply. However, it is also an important quantity that we measure. In biophysical systems, we almost always measure a voltage, not a current. This is primarily a matter of ease of use and the relative durability of voltmeters compared with ammeters. When we speak of a biophysical **signal**, we are referring to a time-varying voltage produced by a human subject or some device attached to it.

RESISTORS

Electrical **resistivity** is a fundamental material property: As electrons pass through a material, energy is dissipated as heat. **Resistance** is a measure of this effect in a specific object. Resistance is measured in units of **ohms** (Ω), and thus resistivity has units of ohms per meter (Ω/m). In other words, the resistance of an object is a function of the resistivity of its material as well as the object's dimensions. In particular, resistance is directly proportional to the length of the material. Regarding fluid flow, resistivity is analogous to the friction that exists between a fluid and the pipe through which it flows; resistance is analogous to the total frictional force of the pipe system. The total resistance of a pipe depends on its frictional characteristics as well as its length. Electrical resistivities of materials vary over many orders of magnitude. For example, copper wire has a resistivity of about 10^{-4} Ω/m; human skin, 20 to 50 kΩ/m; semiconductors such as silicon, around 10^5 Ω/m; and wood, about 10^{13} Ω/m.

EXAMPLE C.1

Estimate the resistance of 1 cm of copper wire using the resistivity just given, 10^{-4} Ω/m.

See answer C.1 on page 386.

A **resistor** is a device that resists electricity. Typical resistor sizes vary from around 1 Ω to 1 MΩ. Knowledge of the resistances within a circuit is critical to understanding its behavior. Indeed, we typically use our knowledge of resistors to manipulate the flow of current and perform the desired function. When studying the human movement, we often need to be aware of the resistances in both our instruments and the human body.

Some resistors have variable resistances. A common type of variable resistor is the **potentiometer**, often called a *pot*. Some potentiometers can be adjusted by turning them (a rotary pot), whereas others slide linearly. Volume controls on radios can take both forms, as can dimmer switches for indoor lighting.

Most circuits include multiple resistances. Thus, it is important to understand how resistors act when connected together. The two basic manners of connecting are in a series and parallel. In a **series** connection, there is one path. One resistor follows the other, and all current flowing through one resistor must also flow through the other (figure C.2). The total resistance of two resistors in series is equal to the sum of the resistances, that is,

$$R_s = R_1 + R_2 \tag{C.1}$$

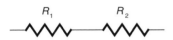

▲ **Figure C.2** Circuit diagram of two resistors in series.

If more resistors are added to the series, the total resistance is equal to the sum of each resistance:

$$R_s = R_1 + R_2 + R_3 + \ldots + R_n \tag{C.2}$$

In a **parallel** connection, there is branching (figure C.3). The total current flowing through the system is divided between two or more resistors. The total resistance R_p of two or more resistors in parallel is given by

$$\frac{1}{R} = \frac{1}{R_1} + \frac{1}{R_2} + \frac{1}{R_3} + \cdots + \frac{1}{R_n} \tag{C.3}$$

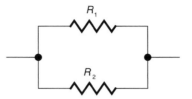

▲ **Figure C.3** Circuit diagram of two resistors in parallel.

For the case of two resistors in parallel, this reduces to

$$R = \frac{R_1 R_2}{R_1 + R_2} \tag{C.4}$$

EXAMPLE C.2

a. What is the total resistance of two 10 Ω resistors in series? In parallel?

See answer C.2a on page 386.

b. What is the total resistance of a 10 Ω resistor and a 1 Ω resistor in series? In parallel?

See answer C.2b on page 386.

CAPACITORS

A **capacitor** is a device that stores electric charge; in our analogy to fluid flow, a capacitor is equivalent to a tank or a bucket that holds water. Its behavior is very different from that of a resistor and is not discussed in detail here. The important thing about capacitance is that it is a common physical property that we often must account for. It typically attenuates the voltage that we try to measure, and its effects can be noticeable on certain data. For example, high-speed devices such as telephones and computer networks have very thin cables because the capacitance of thicker cables would essentially absorb the small amounts of electricity being sent through them. This is analogous to the fact that a garden hose holds water: Water does not come out of the hose for a few seconds after the faucet is turned on because the water must first fill the hose to capacity. It is for this reason that some accelerometers have extremely thin cables. Similarly, EMG electrodes are preamplified to provide a stronger source of electricity that can overcome the capacitance of the wires.

Note that capacitance is not a *bad* factor but simply a factor that must be taken into account. We in fact exploit the behavior of capacitors so that radios can be tuned to different stations. Capacitors can also be used to filter signals in the same way as the digital filters introduced in chapter 1 and detailed in chapter 11. Readers interested in relevant examples may again refer to any linear-circuits text.

Along with capacitors, **impedance** is also important. Impedance, denoted Z, is a more general term for all of the factors that limit electrical flow through a circuit. Thus, impedance includes the net effects of all resistors and capacitors in the circuit.

The symbol for a capacitor—two lines—represents the two plates that hold the electric charge. Sometimes the plates are drawn parallel, but at other times, the plate with the lower voltage is denoted with a curved shape.

OHM'S LAW

Ohm's law is perhaps the most fundamental law in all of electronics. It states that the voltage across a resistor equals the product of its resistance and the current flowing through it:

$$V = IR \qquad (C.5)$$

where V is the voltage across the resistor, I is the current through the resistor, and R is the magnitude of the resistance. There are different ways to express this law. If we increase the voltage in a circuit, the current increases in proportion, and if we increase the size of a resistor, the current decreases proportionately. When plotted, this function is a straight line, as shown in figure C.4. This is a linear function. It remains true regardless of the magnitude of the current or how it changes over time. We express this mathematically as

$$V(t) = I(t)\, R \qquad (C.6)$$

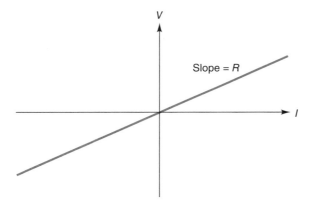

▲ **Figure C.4**   Graphical representation of Ohm's law.

Because of this linearity, resistor circuits are the most straightforward to analyze, although they, too, can get complicated.

Ohm's law is analogous to fluid flow. The electrical resistance R corresponds to the resistance of the piping, the current I corresponds to the volume rate of fluid flow, and the voltage V corresponds to pressure. If we increase the voltage, more current flows, in the same way that if we increase the pressure of water in a pipe, more water flows. If we increase the resistance in a circuit, less current flows—again, just like water.

EXAMPLE C.3

a. What is the current flowing in this circuit?

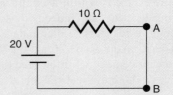

See answer C.3a on page 387.

b. In this circuit, what is the size of the resistor R?

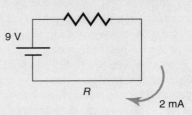

See answer C.3b on page 387.

c. Suppose a 110 V house outlet is wired to a 15 A circuit breaker. What is the minimum resistance that can be applied to this outlet?

See answer C.3c on page 387.

In practice, we are not concerned with currents flowing through loops, as these examples have illustrated. Instead, we often speak of the **voltage drop** across a resistor, which has to do with the manner in which you measure voltage. Because voltage is a potential difference between two points, a **voltmeter** measures the difference with two probes. For example, in the circuit in example C.3b, if we placed one probe before the 9 V battery and one probe after it, the voltmeter would register a voltage gain of 9 V. If we placed the probes across the resistor, the voltmeter would measure a voltage drop of 9 V (i.e., it would read −9 V). If you placed the probes across points A and B, the voltmeter would register 0 V because there is no resistance between these points. In measuring any electrical device, there is a specific component across which the changes in voltage are measured.

Earlier in this appendix, we discussed impedance. When measuring impedance, the formula is analogous to Ohm's law:

$$Z = \frac{V}{I} \qquad (C.7)$$

where Z is the impedance, V is the voltage across the circuit, and I is the current through it. If a circuit is made up entirely of resistors, then the impedance is equal to the resistance. However, for reasons beyond the scope of our discussion, if the voltage varies with time, we will observe the effects of the capacitance of the circuit.

POWER LAWS

Sliding friction between two objects generates heat. In a similar manner, electrical resistance generates heat. This is simply a matter of energetics: If the electrical potential between sites is different and current flows between them, the energy must be dissipated in some manner, whether through a motor, a light bulb, or a heating element. The power dissipated by a resistor is given by

$$P = IV \qquad (C.8)$$

where P is the power dissipated by the resistor, I is the current through it, and V is the voltage across it. That is, the power dissipated by a resistor as heat is given by the product of the current flowing through and the voltage across it. Power, as in mechanical applications, has units of watts (W). With Ohm's law, we can also derive two other forms of the power law,

$$P = I^2 R = \frac{V^2}{R} \qquad (C.9)$$

where R is the magnitude of the resistance. These equations demonstrate that heating devices such as ovens and hair dryers work by having low resistances. A heating element (a coil) is simply a resistor; as current flows through it, energy dissipates as heat. Using the equation on the left, we see that the power increases as the square of the current. Therefore, a decrease in the resistance of the heating element causes a proportional increase in the current.

EXAMPLE C.4

a. What is the resistance of the heating coils of a 1200 W toaster that runs on 110 V house circuitry?

See answer C.4a on page 387.

b. In an earlier example, we had a 110 V house outlet on a 15 A circuit breaker. What is the maximum wattage appliance you can plug into this outlet?

See answer C.4b on page 387.

MEASUREMENT OF PHYSICAL SYSTEMS

Having discussed the basic behaviors of simple circuit components, we now turn to the way we use these components in the laboratory. We begin with a discussion of how we convert human movements into electrical signals that our computers can measure.

Transducers

In the vast majority of cases in which we measure physical quantities electrically, we measure changes in voltage. This is a fundamental principle that cannot be overemphasized. A *0* or a *1* in a computer is represented by a voltage of 0 or 5 V, respectively. When sound is transmitted through a wire to a speaker, the changes in voltage are interpreted as sound. When radio signals are transmitted to a satellite, these, too, are registered by the voltages they impart on the receiver. This is also true for the measurement of EMG activity, force, and even the reflections of body markers to a camera's lens.

The process of converting a physical dimension into a voltage is called **transduction**. A device that performs this function is a **transducer**. Some of the types of transducers are force, pressure, linear displacement, rotary displacement, and acceleration transducers. The common principle in all of these devices is that the quantity being measured causes the resistance of the transducer to change. For example, a force transducer (used in a force platform) has tiny resistors that deform slightly when force is applied. An electrogoniometer has a rotary resistor that changes as it is rotated. When these resistances change, then, in accordance with Ohm's law, a constant current through a transducer causes the voltage to change proportionately.

EXAMPLE C.5

Suppose we have a blood pressure transducer connected to a 10 mA current supply. As the pressure changes from 80 to 120 mmHg, the transducer's resistance changes from 1000 to 1200 Ω. What will the voltage outputs be at these two pressures?

See answer C.5 on page 387.

Voltage Dividers

How would we measure a sensor with a variable resistance? This is slightly more complicated than the blood pressure example, because most electrical supplies have a constant voltage, not a constant current. Suppose we have a variable resistor and connect a voltage source across it as shown in figure C.5. The standard nomenclature is to label the source voltage V_{in} and the measured voltage V_{out}. For a simple circuit like this, no matter how much the variable resistance R_V changes, V_{out} will always equal V_{in}, so this circuit is useless for measuring changes in R_V.

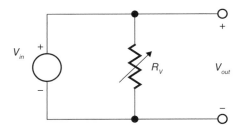

▲ **Figure C.5** Circuit diagram of a variable resistor connected to a voltage source.

In a modification of this circuit (figure C.6), a resistor R is in series with the variable resistance. We want to know the voltage V_{out}. To do this, we can determine using Ohm's law that the current, I, is

$$I = \frac{V_{in}}{R + R_V} \tag{C.10}$$

Because current flows through both resistors, we can substitute it in Ohm's law for R_V and V_{out}:

$$V_{out} = \frac{V_{in} R_V}{R + R_V} \tag{C.11}$$

This circuit is referred to as a **voltage divider**. V_{out} for this circuit varies over an easily measurable range when R and R_V are of similar magnitudes. The circuit is commonly used for a simple potentiometer.

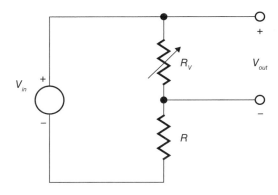

▲ **Figure C.6** Circuit diagram of a resistor in series with a variable resistance.

EXAMPLE C.6

In the voltage divider, what is V_{out} for the following cases?

1. $V_{in} = 15$ V, $RV = 100\ \Omega$, and $R = 100\ \Omega$
2. $V_{in} = 15$ V, $RV = 110\ \Omega$, and $R = 100\ \Omega$
3. $V_{in} = 15$ V, $RV = 100\ \Omega$, and $R = 10\ \Omega$
4. $V_{in} = 15$ V, $RV = 110\ \Omega$, and $R = 10\ \Omega$

See answer C.6 on page 387.

Wheatstone Bridges

Voltage dividers have two problems. In many sensors, the variability of resistance is small, often less than 5%. Also, we often "zero" a sensor rather than subtracting a constant voltage to establish zero for the quantity we are measuring. These difficulties are overcome with a **Wheatstone bridge** (figure C.7), a circuit of two parallel voltage dividers. In its neutral state, it has four equivalent resistances. When the variable resistance changes, we can compare the amount by which the variable resistance has changed from its neutral state by measuring V_{out}.

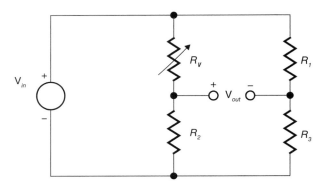

▲ **Figure C.7** Circuit diagram of a Wheatstone bridge.

EXAMPLE C.7

In the Wheatstone bridge, what is the formula for V_{out}? Let $R_1 = R_2 = R_3$.

See answer C.7 on page 387.

COMMON LABORATORY INSTRUMENTS

Many common laboratory instruments use the principles discussed in this appendix. These instruments include

(1) the linear variable differential transducer (LVDT), (2) the electrogoniometer, (3) strain-gauge force transducers, and (4) amplifiers.

Linear Variable Differential Transducer

The LVDT (figure C.8) is a common instrument that measures a linear movement over a short range of motion, typically less than 30 cm. Its main cylinder contains a finely manufactured and calibrated **linear potentiometer**. Therefore, its resistance changes linearly as it is moved. LVDTs can measure to within fractions of a millimeter—a computer-controlled milling machine, for instance, measures to within 2.5 μm. Common laboratory applications include treadmill inclination adjustments, digital calipers, footwear impact testers, and knee arthrometers (for measuring joint laxity or stiffness).

▲ **Figure C.8** Linear variable differential transducer as used in a footwear impact tester.

Electrogoniometer

An electrogoniometer, as its name suggests, measures joint angles electronically. Its basic component is a **rotary potentiometer**. Its internal structure is diagrammed in figure C.9. The end terminals (A and C) are connected to the ends of the resistive material. The middle terminal (B) is connected to a rotating slider. As the knob of this slider is turned, the middle contact moves across the resistive material. Because resistance is a function of material length, we observe the change in resistance. For example, if we have a 10 kΩ potentiometer, the resistance from A to C will measure 10 kΩ. As the rotating slider is moved from A to C, we will measure a resistance across A and B that changes from 0 to 10 kΩ, while the resistance from B to C changes from 10 kΩ to 0 (figure C.10).

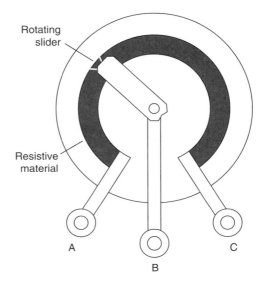

▲ **Figure C.9** Schematic of a rotary potentiometer.

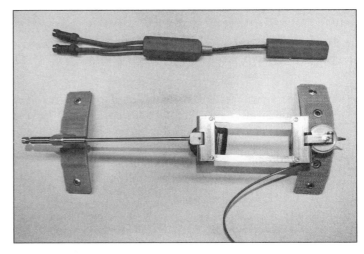

▲ **Figure C.10** The top instrument is a Biometrics electrogoniometer that measures angles about two axes. The bottom instrument is a uniaxial electrogoniometer with a four-bar linkage (potentiometer is on the right).

Strain-Gauge Force Transducers

When force is applied to a material, it deforms. This is called mechanical **strain**. Because resistance is a function of material length, we observe a change in resistance when a material is deformed. This is the basic principle of a **strain gauge**. If we have a resistor with a precisely known resistance glued to a deformable object, we can measure the object's change in resistance as it deforms. The gauges themselves are usually much smaller in area than a postage stamp, but equally as thin (see figure C.11). Once glued to the surface of a structure, they bend with the material without altering the structural properties. Strain gauges are usually placed in a Wheatstone bridge circuit.

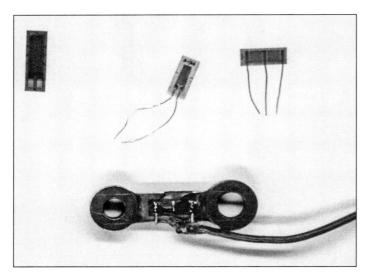

▲ **Figure C.11** Three types of strain gauges and a strain-gauged link (bottom) for measuring axial loads.

Strain gauges are commonly used to measure forces in human movement with such devices as floor-mounted force plates, tension transducers, pressure transducers, and even accelerometers. It is also common in biomedical research to mount gauges to orthoses and prostheses as well as to cadaver samples of bone, cartilage, and tendon.

Amplifiers

An **amplifier** is a device that increases the voltage of a signal. Figure C.12 shows how the idealized amplifier is designated in circuit diagrams.

The most common type of amplifier is the **operational amplifier**. These are commonly installed on

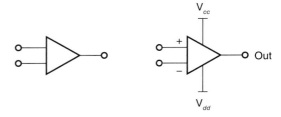

▲ **Figure C.12** Symbols for an operational amplifier. The detailed form, on the right, labels the inputs and the power supply for the amplifier V_{cc} and V_{dd}.

silicon chips. Unlike resistors and capacitors, *op-amps* are active circuit elements and therefore require current to power them. Op-amps have many different uses and implementations. Two common connections are the inverting and noninverting configurations (figure C.13). The noninverting op-amp circuit increases the magnitude of the voltage it measures. The inverting op-amp circuit increases the incoming voltage and **inverts** it (i.e., takes the negative of it).

The performance of an amplifier is called the **gain**. Gain is the ratio of the incoming voltage to the amplified voltage, $\frac{V_{out}}{V_{in}}$. In op-amp circuits, gain is controlled by altering the ratio of the resistors R_F and R. For noninverting configurations, the gain is given by $1+\frac{R_F}{R}$, and for inverting configurations the gain is $-\frac{R_F}{R}$. For a variable gain, a potentiometer may be substituted for either R_F or R.

Applications for amplifiers are too numerous to mention. Their most common use is for turning weak electromagnetic waves into audible sound in radios and cell phones. In human movement science, we use them to measure tiny EMG, electrocardiographic, and electroencephalographic signals; they also are used in force plates and other force transducers and in accelerometers. They can be used to construct **analog** filters, integrators, and differentiators.

An important characteristic of op-amps and other active circuits is their **input impedance**. This is a measure of the sensitivity of the op-amp: High input impedance means, in effect, that the op-amp needs to draw very little current from the measured quantity to function. This is very important in human movement, because most biophysical signals are extremely small. Ideally, the input impedance of an EMG amplifier, for example, would be infinite. Typically, amplifiers have input impedances of 1 MΩ, but EMG or bioamplifiers have input impedances of 10 MΩ or more. EMG amplifiers require higher impedances because skin can have resistances of about 20 to 100 kΩ and, when unprepared, as high as 2 MΩ or more.

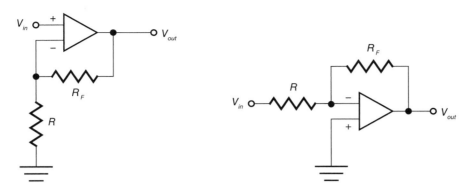

▲ **Figure C.13** Circuit diagrams for noninverting (left) and inverting (right) operational amplifiers.

SUGGESTED READINGS

Bobrow, L.S. 1987. *Elementary Linear Circuit Analysis.* 2nd ed. Oxford, UK: Oxford University Press.

Cathey, J.J. 2002. *Schaum's Outline of Electronic Devices and Circuits.* 2nd ed. New York: McGraw-Hill.

Cobbold, R.S.C. 1974. *Transducers for Biomedical Measurements: Principles and Applications.* Toronto: Wiley.

Horowitz, P., and W. Hill. 1989. *The Art of Electronics.* 2nd ed. Cambridge, MA: Cambridge University Press.

Ohanian, H.C. 1994. Electric force and electric charge. In *Principles of Physics.* 2nd ed. New York: Norton.

O'Malley, J. 1992. *Schaum's Outline of Basic Circuit Analysis.* 2nd ed. New York: McGraw-Hill.

Winter, D.A., and A.E. Patla. 1997. *Signal Processing and Linear Systems for Movement Sciences.* Waterloo, ON: Waterloo Biomechanics.

Vectors and Scalars

A quantity that is completely specified by its magnitude is referred to as a scalar. Examples of scalars are mass, density, work, energy, and volume. They are treated mathematically as real numbers and, as such, obey all of the usual rules of algebra.

For any analysis in 3-D, however, the use of a physical quantity that expresses direction is important. The location of any vector in 3-D space relative to an origin is referred to as a position vector. A **vector** is a quantity that requires both direction and magnitude for its complete specification and must add according to the parallelogram law. In this text, a vector is designated by an arrow over the vector name (e.g., $\vec{A}$).

The method most often used for defining a position in 3-D space is the Cartesian coordinate system, in which a position vector has three coordinates that uniquely distinguish the distance from the origin of the coordinate system to a point in space. These coordinates are mutually orthogonal. Therefore, any location in this 3-D space can be defined using Cartesian coordinates or the projection of each component onto the reference frame. For example, vector $\vec{A}$ in figure D.1 is located at distances A_x from the **Y-Z** plane, A_y from the **X-Z** plane, and A_z from the **X-Y** plane. The location of this position

is defined by the coordinates (A_x, A_y, A_z). Two vectors are considered equal if all of the respective components of the vectors are equal. That is, vectors $\vec{A}$ and $\vec{B}$ are equal if $A_x = B_x$, $A_y = B_y$, and $A_z = B_z$.

Another method of representing the components of point A is by using unit vectors. Unit vectors are defined as vectors of unit length along each of the axes of the coordinate system and are specified as $\vec{i}$, $\vec{j}$, and $\vec{k}$ along the X, Y, and Z axes, respectively (figure D.2). Specifically, a unit vector, $\vec{e}_A$ is defined as

$$\vec{e}_A = \frac{\vec{A}}{\|\vec{A}\|}, \text{ where } \|\vec{e}_A\| = 1 \quad (D.1)$$

or a vector with unit length and in the direction of $\vec{A}$. The double bars mean the norm or magnitude of the vector; this operation is defined later in this appendix. The unit vectors associated with each of the axes of the coordinate system are defined as

$$\text{in the } x\text{-direction, } \vec{i} = (1,0,0); \quad (D.2)$$

$$\text{in the } y\text{-direction, } \vec{j} = (0,1,0); \text{ and} \quad (D.3)$$

$$\text{in the } z\text{-direction, } \vec{k} = (0,0,1). \quad (D.4)$$

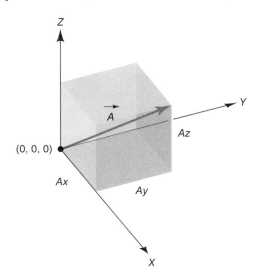

▲ **Figure D.1**　A 3-D coordinate system.

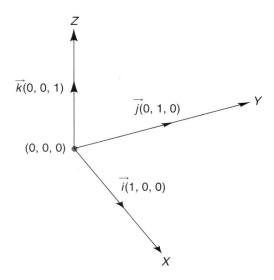

▲ **Figure D.2**　Orientation of the unit vector coordinates in a coordinate system.

Any vector can then be represented using its Cartesian coordinates or as the vector sum of its unit vectors. For example, the vector $\vec{A}$ can be represented as

$$\vec{A} = \left(A_x, A_y, A_z \right) \tag{D.5}$$

or in unit vector form by

$$\vec{A} = A_x \vec{i} + A_y \vec{j} + A_z \vec{k} \tag{D.6}$$

A vector can be changed to a unit vector by dividing each component by the norm or magnitude of the vector. That is, for a vector $\vec{A}$, the unit vector in the direction of $\vec{A}$ is

$$\vec{e}_A = \frac{A_x}{\|\vec{A}\|} \vec{i} + \frac{A_y}{\|\vec{A}\|} \vec{j} + \frac{A_z}{\|\vec{A}\|} \vec{k} \tag{D.7}$$

VECTOR OPERATIONS

To illustrate the following operations, two vectors $\vec{A}$ = (1, 5, 2) and $\vec{B}$ = (6, 1, 3) are used in the examples throughout this appendix.

Magnitude or Norm of a Vector

The magnitude or norm of a vector represents the length of the vector. For a vector $\vec{A}$, with components A_x, A_y, and A_z representing distances from the Y-Z, X-Z, and X-Y planes, respectively, the norm of the vector is calculated using the Pythagorean relationship as follows:

$$\|\vec{A}\| = \sqrt{A_x^2 + A_y^2 + A_z^2} \tag{D.8}$$

EXAMPLE

For A = (1, 5, 2), the norm or the magnitude of A is

$$\|\vec{A}\| = \sqrt{1^2 + 5^2 + 2^2}$$

$$\|\vec{A}\| = \sqrt{30} = 5.48.$$

Addition of Vectors

If vectors $\vec{A}$ and $\vec{B}$ are designated as $\vec{A}$ = (A_x, A_y, A_z) and $\vec{B}$ = (B_x, B_y, B_z), then

$$\vec{A} + \vec{B} = \left(A_x + B_x, A_y + B_y, A_z + B_z \right) \tag{D.9}$$

Using unit vector notation, the sum of A and B is

$$\vec{A} + \vec{B} = \left(A_x + B_x \right) \vec{i} + \left(A_y + B_y \right) \vec{j} + \left(A_z + B_z \right) \vec{k} \tag{D.10}$$

EXAMPLE

For $\vec{A} = (1, 5, 2)$ and $\vec{B} = (6, 1, 3)$, the vector $\vec{A} + \vec{B}$ is

$$\vec{A} + \vec{B} = (1 + 6) \vec{i} + (5 + 1) \vec{j} + (2 + 3) \vec{k}$$

$$\vec{A} + \vec{B} = 7\vec{i} + 6\vec{j} + 5\vec{k} = (7, 6, 5).$$

Subtraction of Vectors

Technically, there is no operation in which vectors subtract; however, the negative of one vector may be added to another vector. Thus, if we wished to find the difference between vectors $\vec{A}$ and $\vec{B}$, then $\vec{A} - \vec{B}$ is really $\vec{A} + \left(-\vec{B} \right)$. As a result, if vectors $\vec{A}$ = A_x, A_y, A_z and $\vec{B}$ = B_x, B_y, B_z, then the difference between two vectors A and B is

$$\vec{A} + \left(-\vec{B} \right) = \left(A_x - B_x \right) \vec{i} + \left(A_y - B_y \right) \vec{j} + \left(A_z - B_z \right) \vec{k} \tag{D.11}$$

EXAMPLE

For $\vec{A} = (1, 5, 2)$ and $\vec{B} = (6, 1, 3)$, the vector $\vec{A} - \vec{B}$ is

$$\vec{A} - \vec{B} = (1 - 6) \vec{i} + (5 - 1) \vec{j} + (2 - 3) \vec{k}$$

$$\vec{A} - \vec{B} = -5\vec{i} + 4\vec{j} - 1\vec{k} \text{ or } (-5, 4, -1).$$

Multiplication of a Vector by a Scalar

Vectors can be multiplied by a scalar by multiplying each component by the scalar. Thus, for a vector $\vec{A}$ and scalar c,

$$c\vec{A} = cA_x \vec{i} + cA_y \vec{j} + cA_z \vec{k} \tag{D.12}$$

EXAMPLE

For $\vec{A}$ = (1, 5, 2) and scalar c = 2, the vector $c\vec{A}$ is

$$c\vec{A} = 2(1)\vec{i} + 2(5)\vec{j} + 2(2)\vec{k}$$

$$c\vec{A} = 2\vec{i} + 10\vec{j} + 4\vec{k} = (2, 10, 4).$$

Dot or Scalar Product

Given two vectors $\vec{A}$ and $\vec{B}$, whose components are A_x, A_y, A_z and B_x, B_y, B_z, respectively, the dot or scalar product, $\vec{A} \cdot \vec{B}$, is defined by the equation

$$\vec{A} \cdot \vec{B} = A_x B_x + A_y B_y + A_z B_z \quad \text{(D.13)}$$

The result of the dot product calculation is always a scalar and hence the alternative name "scalar product." Because the result is a quantity with only magnitude and no direction, we can state that the dot product operation is commutative (i.e., $\vec{A} \cdot \vec{B} = \vec{B} \cdot \vec{A}$).

EXAMPLE

For $\vec{A} = (1, 5, 2)$ and $\vec{B} = (6, 1, 3)$, $\vec{A} \cdot \vec{B}$ is

$$\vec{A} \cdot \vec{B} = (1 \times 6) + (5 \times 1) + (2 \times 3)$$

$$\vec{A} \cdot \vec{B} = 6 + 5 + 6$$

$$\vec{A} \cdot \vec{B} = 17.$$

In analytical geometry, the cosine of the angle between two line segments is given by the following formula:

$$\cos \theta = \frac{A_x B_x + A_y B_y + A_z B_z}{|\vec{A}||\vec{B}|} \quad \text{(D.14)}$$

We can rewrite this equation using our previous definition of the dot product to be

$$\cos \theta = \frac{\vec{A} \cdot \vec{B}}{|\vec{A}||\vec{B}|} \quad \text{(D.15)}$$

or

$$\vec{A} \cdot \vec{B} = |\vec{A}||\vec{B}| \cos \theta \quad \text{(D.16)}$$

Geometrically, $\vec{A} \cdot \vec{B}$ is equal to the length of the projection of $\vec{A}$ on $\vec{B}$, times the magnitude of $\vec{B}$. If the angle between these two vectors is 90° and the cosine of the angle is zero, then the dot product is equal to zero. That is, $\vec{A}$ is perpendicular to $\vec{B}$ provided that neither is equal to (0, 0, 0). From the definitions of unit vector coordinates, $\vec{i}$, $\vec{j}$, and $\vec{k}$, the following relationships hold:

$$\vec{i} \cdot \vec{i} = 1 \quad \text{(D.17)}$$

$$\vec{j} \cdot \vec{j} = 1 \quad \text{(D.18)}$$

$$\vec{k} \cdot \vec{k} = 1 \quad \text{(D.19)}$$

because the angle and the cosine of the angle between the pairs of coordinates is zero. Since the unit vector coordinates are perpendicular to each other, the cosine of 90° is zero. Thus,

$$\vec{i} \cdot \vec{j} = 0 \quad \text{(D.20)}$$

$$\vec{i} \cdot \vec{k} = 0 \quad \text{(D.21)}$$

$$\vec{j} \cdot \vec{k} = 0 \quad \text{(D.22)}$$

When the vectors $\vec{A}$ and $\vec{B}$ are unit vectors, then the dot product of the two vectors is the cosine of the angle between them:

$$\vec{A} \cdot \vec{B} = \cos \theta \quad \text{(D.23)}$$

Cross or Vector Product

For two vectors, $\vec{A}$ and $\vec{B}$, the cross product, $\vec{A} \times \vec{B}$, is defined by the following equation:

$$\vec{A} \times \vec{B} = \left(A_y B_z - A_z B_y, \ A_z B_x - A_x B_z, \ A_x B_y - A_y B_x \right) \quad \text{(D.24)}$$

The result of this calculation is always a vector, hence the alternative name, *vector product*. Geometrically, the resulting vector, $\vec{C}$, of the cross product of two vectors, $\vec{A}$ and $\vec{B}$, is perpendicular to the plane formed by both $\vec{A}$ and $\vec{B}$ (Figure C.3). The direction of the vector $\vec{C}$ is determined by the right-hand rule. That is, if the fingers of the right hand are placed along the vector $\vec{A}$ and rotated to the vector $\vec{B}$, the right thumb would point in the direction of the resulting vector $\vec{C}$. It should be intuitive that

$$\vec{A} \times \vec{B} \neq \vec{B} \times \vec{A} \quad \text{(D.25)}$$

However,

$$\vec{A} \times \vec{B} = -\vec{B} \times \vec{A} \quad \text{(D.26)}$$

Therefore, we say that the cross product operation is not commutative.

It is clear that the vector $\vec{C} = \vec{A} \times \vec{B}$ is determined from a right-hand triad of the three vectors, $\vec{A}$, $\vec{B}$, and

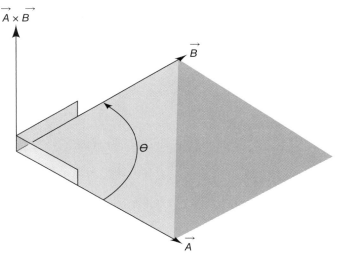

▲ **Figure D.3** Geometric representation of the cross or vector product of $\vec{A} \times \vec{B}$.

$\vec{C}$. Therefore, since $\vec{C}$ is perpendicular to the plane of $\vec{A}$ and $\vec{B}$, the magnitude of the cross product can be written as

$$\left|\vec{A} \times \vec{B}\right| = \left|\vec{A}\right|\left|\vec{B}\right|\sin\theta \tag{D.27}$$

where $\left|\vec{A}\right|$ and $\left|\vec{B}\right|$ are the norms of the two vectors and θ is the angle between $\vec{A}$ and $\vec{B}$. In terms of unit vector coordinates, it can be shown that

$$\vec{i} \times \vec{i} = 0 \tag{D.28}$$

$$\vec{j} \times \vec{j} = 0 \tag{D.29}$$

$$\vec{k} \times \vec{k} = 0 \tag{D.30}$$

because the angle between each of these pairs of unit vectors is zero and the sine of $0°$ is zero. Also,

$$\vec{i} \times \vec{j} = \vec{k} \tag{D.31}$$

$$\vec{j} \times \vec{k} = \vec{i} \tag{D.32}$$

$$\vec{k} \times \vec{i} = \vec{j} \tag{D.33}$$

Rather than trying to remember the formula for a cross product, it is probably easier to determine the result as the expansion of a determinant (see appendix E for more details about computing determinants). If a 2×2 matrix, [M], is defined as

$$[M] = \begin{bmatrix} A & B \\ C & D \end{bmatrix} \tag{D.34}$$

then the determinant of this matrix, $\left|M\right|$, is

$$\left|M\right| = AD - BC \tag{D.35}$$

We can use this calculation to determine the resulting vector from a cross product. The cross product can be written as follows:

$$\vec{A} \times \vec{B} = \begin{vmatrix} \vec{i} & \vec{j} & \vec{k} \\ A_x & A_y & A_z \\ B_x & B_y & B_z \end{vmatrix} \tag{D.36}$$

Using the concept of the determinant and that $\vec{i} \times \vec{j} = \vec{k}$, $\vec{j} \times \vec{k} = \vec{i}$, and $\vec{k} \times \vec{i} = \vec{j}$, we can reconfigure the structure above to be a series of determinants as follows:

$$\vec{A} \times \vec{B} = \begin{vmatrix} A_y & A_z \\ B_y & B_z \end{vmatrix}\vec{i} + \begin{vmatrix} A_z & A_x \\ B_z & B_x \end{vmatrix}\vec{j} + \begin{vmatrix} A_x & A_y \\ B_x & B_y \end{vmatrix}\vec{k} \tag{D.37}$$

Calculating the determinant of each 2×2 matrix gives us

$$\vec{A} \times \vec{B} = \left(A_y B_z - A_z B_y\right)\vec{i} + \left(A_z B_x - A_x B_z\right)\vec{j} + \left(A_x B_y - A_y B_x\right)\vec{k} \tag{D.38}$$

or

$$\vec{A} \times \vec{B} = \left[A_y B_z - A_z B_y, \; A_z B_x - A_x B_z, \; A_x B_y - A_y B_x\right] \tag{D.39}$$

or

$$\vec{A} \times \vec{B} = \left[A_y B_z - A_z B_y, \; -\left(A_x B_z - A_z B_x\right), \; A_x B_y - A_y B_x\right] \tag{D.40}$$

EXAMPLE

For $\vec{A} = (1, 5, 2)$ and $\vec{B} = (6, 1, 3)$, $\vec{A} \times \vec{B}$ is

$$\vec{A} \times \vec{B} = \begin{vmatrix} \vec{i} & \vec{j} & \vec{k} \\ 1 & 5 & 2 \\ 6 & 1 & 3 \end{vmatrix}$$

$$\vec{A} \times \vec{B} = \begin{vmatrix} 5 & 2 \\ 1 & 3 \end{vmatrix} \vec{i} + \begin{vmatrix} 2 & 1 \\ 3 & 6 \end{vmatrix} \vec{j} + \begin{vmatrix} 1 & 5 \\ 6 & 1 \end{vmatrix} \vec{k}$$

$$\vec{A} \times \vec{B} = \left[(5 \times 3) - (2 \times 1)\right]\vec{i} + \left[(2 \times 6) - (1 \times 3)\right]\vec{j} + \left[(1 \times 1) - (5 \times 6)\right]\vec{k}$$

$$\vec{A} \times \vec{B} = [15 - 2]\vec{i} + [12 - 3]\vec{j} + [1 - 30]\vec{k}$$

$$\vec{A} \times \vec{B} = 13\vec{i} + 9\vec{j} - 29\vec{k}.$$

To conduct a 3-D analysis, you must be very familiar with vectors and scalars. For a more in-depth exposure, we suggest that you refer to a textbook such as those listed below that deals particularly with vectors and vector operations.

SUGGESTED READINGS

Beezer, R.A. *A First Course in Linear Algebra.* http://linear.ups.edu/download.html (this is a free book online).

Friedburg, S.H., A.J. Insel, and L.E. Spence. *Linear Algebra.* Englewood Cliffs, NJ: Prentice-Hall.

Matrices and Matrix Operations

Many of the computations done in 3-D kinematics can be accomplished relatively efficiently using *matrix algebra*. A **matrix** is any rectangular array of numbers. Each number in the array is an element. If the array has m rows and n columns, the matrix is referred to as an $m \times n$ matrix. Each element of the matrix can then be identified by its row and column positions. The number of rows and columns in a matrix denotes the order of the matrix. For example, the 3×3 matrix [A] can be written as

$$[A] = \begin{bmatrix} a_{11} & a_{12} & a_{13} \\ a_{21} & a_{22} & a_{23} \\ a_{31} & a_{32} & a_{33} \end{bmatrix} \tag{E.1}$$

The element a_{23} is the element in the second row and the third column, or the element in the (2, 3) place. Generally, each element in matrix [A] can be referred to as a_{ij}, where i is the row number and j is the column number.

TYPES OF MATRICES

A matrix having only one row with several columns is a *row matrix*. For example,

$$[A] = \begin{bmatrix} a_{11} & a_{12} & a_{13} \end{bmatrix} \tag{E.2}$$

Conversely, a matrix can have several rows but only one column. This type of matrix is referred to as a *column matrix*. For example,

$$[A] = \begin{bmatrix} a_{11} \\ a_{21} \\ a_{31} \end{bmatrix} \tag{E.3}$$

We will encounter a number of special matrices in this chapter. A *square matrix* has the same number of rows and columns. A **diagonal matrix** is a square matrix in which the elements on the diagonal ($a_{11}, a_{22}, \ldots, a_{nn}$) are nonzero and the others are zero. This matrix [A] is a square, diagonal matrix:

$$[A] = \begin{bmatrix} a_{11} & 0 & 0 \\ 0 & a_{22} & 0 \\ 0 & 0 & a_{33} \end{bmatrix} \tag{E.4}$$

An *identity matrix* is a square matrix whose diagonal elements equal 1. Designated as [I] matrices, they are written like this:

$$[I] = \begin{bmatrix} 1 & 0 & 0 \\ 0 & 1 & 0 \\ 0 & 0 & 1 \end{bmatrix} \tag{E.5}$$

The *transpose* of matrix [A] is designated as $[A]^T$ and is one in which the rows and columns are interchanged.

$$[A] = \begin{bmatrix} a_{11} & a_{12} & a_{13} \\ a_{21} & a_{22} & a_{23} \\ a_{31} & a_{32} & a_{33} \end{bmatrix}, \text{ then } [A]^T = \begin{bmatrix} a_{11} & a_{21} & a_{31} \\ a_{12} & a_{22} & a_{32} \\ a_{13} & a_{23} & a_{33} \end{bmatrix} \tag{E.6}$$

MATRIX OPERATIONS

A 3-D analysis involves many operations concerning matrices with which you must become familiar. As with vector operations, readers are urged to refer to more detailed presentations of the concepts in other sources.

To illustrate the following operations, a matrix [A] will be used throughout the examples.

$$[A] = \begin{bmatrix} 6 & 1 & 3 \\ -1 & 1 & 2 \\ 4 & 1 & 3 \end{bmatrix} \tag{E.7}$$

Determinant of a 3 × 3 Matrix

In Appendix D, we demonstrated the method of calculating the determinant of a 2 × 2 matrix. We now show how to calculate the determinant of a 3 × 3 matrix. Given the matrix [A], we copy the first two columns to the right of the matrix. We first draw three arrows to the right beginning with the column farthest to the left. The elements of each arrow are multiplied together. For example, for the first arrow, the product of the elements is $(a_{11} \times a_{22} \times a_{33})$. We then draw three arrows to the left beginning with the column farthest to the right. The elements of each arrow are also multiplied together. The determinant is calculated by adding the products of the arrows to the right and subtracting each of the products of the arrows to the left. This technique is illustrated next.

$$\det[A] = \begin{vmatrix} a_{11} & a_{12} & a_{13} \\ a_{21} & a_{22} & a_{23} \\ a_{31} & a_{32} & a_{33} \end{vmatrix} \begin{matrix} a_{11} & a_{12} \\ a_{21} & a_{22} \\ a_{31} & a_{32} \end{matrix}$$

$$= (a_{11} \times a_{22} \times a_{33}) + (a_{12} \times a_{23} \times a_{31}) + (a_{13} \times a_{21} \times a_{32})$$

$$- (a_{12} \times a_{21} \times a_{33}) - (a_{11} \times a_{23} \times a_{32}) - (a_{13} \times a_{22} \times a_{31})$$

EXAMPLE

For $[A] = \begin{bmatrix} 6 & 1 & 3 \\ -1 & 1 & 2 \\ 4 & 1 & 3 \end{bmatrix}$, the determinant of $[A]$ is

$$\det[A] = \begin{vmatrix} 6 & 1 & 3 \\ -1 & 1 & 2 \\ 4 & 1 & 3 \end{vmatrix}$$

$$\det[A] = (6 \times 1 \times 3) + (1 \times 2 \times 4) + (3 \times -1 \times 1)$$

$$- (1 \times -1 \times 3) - (6 \times 2 \times 1) - (3 \times 1 \times 4)$$

$$\det[A] = 18 + 8 - 3 - (-3) - 12 - 12$$

$$\det[A] = 2.$$

Cofactor Matrix

A further matrix operation that is important in 3-D kinematics is the calculation of the *cofactor matrix*. This operation is used in the calculation of the inverse of a matrix. The formula for calculating the cofactor matrix $[A^c]$ of a 3×3 matrix $[A]$ is

$$[A^c] = \begin{vmatrix} (a_{22} \times a_{33} - a_{23} \times a_{32}) & -(a_{21} \times a_{33} - a_{23} \times a_{31}) & (a_{21} \times a_{32} - a_{22} \times a_{31}) \\ -(a_{12} \times a_{33} - a_{13} \times a_{32}) & (a_{11} \times a_{33} - a_{13} \times a_{31}) & -(a_{11} \times a_{32} - a_{12} \times a_{31}) \\ (a_{12} \times a_{23} - a_{13} \times a_{22}) & -(a_{11} \times a_{23} - a_{13} \times a_{21}) & (a_{11} \times a_{22} - a_{12} \times a_{21}) \end{vmatrix} \quad (E.9)$$

EXAMPLE

For $[A] = \begin{bmatrix} 6 & 1 & 3 \\ -1 & 1 & 2 \\ 4 & 1 & 3 \end{bmatrix}$, $[A^c]$ is

$$[A^c] = \begin{vmatrix} (1 \times 3 - 2 \times 1) & -((-1) \times 3 - 2 \times 4) & (-1 \times 1 - 1 \times 4) \\ -(1 \times 3 - 3 \times 1) & (6 \times 3 - 3 \times 4) & -(6 \times 1 - 1 \times 4) \\ (1 \times 2 - 3 \times 1) & -(6 \times 2 - 3 \times (-1)) & (6 \times 1 - 1 \times (-1)) \end{vmatrix}$$

$$[A^c] = \begin{vmatrix} 1 & 11 & -5 \\ 0 & 6 & -2 \\ -1 & -15 & 7 \end{vmatrix} .$$

Inverting a Matrix

The *inverse* of a matrix [D] is designated as [D]$^{-1}$ and is one that satisfies the following expression:

$$[A]^{-1}[A] = [I] \qquad (E.10)$$

where [A] ≠ 0 (i.e., not singular) and [I] is the identity matrix.

The inverse of a matrix is calculated in several steps: (1) calculate the determinant of the matrix; (2) calculate the cofactor matrix; (3) take the transpose of the cofactor matrix; and (4) multiply the transpose of the cofactor matrix by the inverse of the determinant. To invert matrix [A], the formula then is

$$[A]^{-1} = \frac{1}{\det[A]}[A^c]^T \qquad (E.11)$$

EXAMPLE

For $[A] = \begin{bmatrix} 6 & 1 & 3 \\ -1 & 1 & 2 \\ 4 & 1 & 3 \end{bmatrix}$, the inverse matrix of [A] is

$$[A]^{-1} = \frac{1}{\det[A]}\left[\text{cofactor} \begin{vmatrix} 1 & -11 & -5 \\ 0 & 6 & -2 \\ -1 & -15 & 7 \end{vmatrix}\right]^T$$

$$[A]^{-1} = \frac{1}{2}\left[\text{cofactor} \begin{vmatrix} 1 & 0 & -1 \\ -11 & 6 & -15 \\ -5 & -2 & 7 \end{vmatrix}\right]$$

$$[A]^{-1} = \begin{vmatrix} 0.5 & 0 & -0.5 \\ -5.5 & 3 & -7.5 \\ -2.5 & -1 & 3.5 \end{vmatrix}.$$

To understand 3-D kinematics, it is important to be fully cognizant of the mathematical manipulations that must be performed with matrices. For a more in-depth exposure to matrices and matrix algebra, we suggest that you refer to a textbook that deals particularly with vectors and matrices.

SUGGESTED READINGS

Beezer, R.A. *A First Course in Linear Algebra*. http://linear.ups.edu/download.html. (this is a free book online).

Friedburg, S.H., A.J. Insel, and L.E. Spence. *Linear Algebra*. Englewood Cliffs, NJ: Prentice-Hall.

Numerical Integration of Double Pendulum Equations

A fourth-order Runge-Kutta method is more difficult to implement in computer code than an Euler method, and it is described here in a format that can be readily adapted to a computer language. The first step is to implement functions that return the angular accelerations of each segment given the angular positions and velocities of each segment, that is,

$$\alpha_T = f(\theta_T,\ \theta_L,\ \omega_T,\ \omega_L) \text{ and } \alpha_L = f(\theta_T,\ \theta_L,\ \omega_T,\ \omega_L).$$

These functions implement the result of simultaneously solving equations G1 and G2 in appendix G (equations of motion for the thigh and leg/foot) for α_T and α_L, that is,

$$\alpha_T = \frac{C_1 I_2 - C_2 A}{B}$$

$$\alpha_L = \frac{C_2 \left(I_T + m_L L_T^2\right) - C_1 A}{B}$$

where

$A = m_L L_T d_L \cos(\theta_L - \theta_T)$

$B = I_L \left(I_T + m_L L_T^2\right) - A^2$

$C_1 = m_L L_T d_L \omega_L^2 \sin(\theta_L - \theta_T) + [a_{Hx}\cos \theta_T + (a_{Hy} + g) \sin\theta_T] (m_T d_T + m_L L_T)$

$C_2 = - m_L L_T d_L \omega_T^2 \sin(\theta_L - \theta_T) + [a_{Hx} \cos \theta_L + (a_{Hx}y + g) \sin\theta_L] m_L d_L$

The next step is to implement functions that compute the change in velocity,

$$\Delta\omega_T = f(\theta_T,\ \theta_L,\ \omega_T,\ \omega_L) \text{ and } \Delta\omega_L = f(\theta_T,\ \theta_L,\ \omega_T,\ \omega_L)$$

which are implemented by using an Euler method applied to equations G1 and G2:

$$\Delta\omega_T = \alpha_T (\theta_T,\ \theta_L,\ \omega_T,\ \omega_L) \Delta t \text{ and}$$

$$\Delta\omega_L = \alpha_L (\theta_T,\ \theta_L,\ \omega_T,\ \omega_L) \Delta t$$

where Δt is the time increment of the numerical integration procedure.

Two functions are needed to compute the change in angular positions using the Euler method:

$$\Delta\theta_T = \omega_T \Delta t + \frac{1}{2}\alpha_T\left(\theta_T,\theta_L,\omega_T,\omega_L\right)\Delta t^2$$

and

$$\Delta\theta_L = \omega_L \Delta t + \frac{1}{2}\alpha_L\left(\theta_T,\theta_L,\omega_T,\omega_L\right)\Delta t^2.$$

In the functions α, $\Delta\omega$, and $\Delta\theta$, the values for θ_T, θ_L, ω_T, and ω_L are temporary function parameters and therefore should be given different names when used in computer code. Using these functions, a fourth-order Runge-Kutta method can be implemented. Given values for θ_T, θ_L, ω_T, and ω_L at an instant in time, the values of these four parameters at the next instant in time are computed using the following procedure:

$$\delta\omega_{T1} = \Delta\omega_T\left(\theta_T,\ \theta_L,\ \omega_T,\ \omega_L\right)$$

$$\delta\omega_{L1} = \Delta\omega_L\left(\theta_T,\ \theta_L,\ \omega_T,\ \omega_L\right)$$

$$\delta\theta_{T1} = \Delta\theta_T\left(\theta_T,\ \theta_L,\ \omega_T,\ \omega_L\right)$$

$$\delta\theta_{L1} = \Delta\theta_L\left(\theta_T,\ \theta_L,\ \omega_T,\ \omega_L\right)$$

$$\delta\omega_{T2} = \Delta\omega_T\left(\theta_T + \frac{\partial\theta_{1T}}{2},\theta_L + \frac{\partial\theta_{1L}}{2},\omega_T + \frac{\partial\omega_{1T}}{2},\omega_L + \frac{\partial\omega_{1L}}{2}\right)$$

$$\delta\omega_{L2} = \Delta\omega_L\left(\theta_T + \frac{\partial\theta_{1T}}{2},\theta_L + \frac{\partial\theta_{1L}}{2},\omega_T + \frac{\partial\omega_{1T}}{2},\omega_L + \frac{\partial\omega_{1L}}{2}\right)$$

$$\delta\theta_{T2} = \Delta\theta_T\left(\theta_T + \frac{\partial\theta_{1T}}{2},\theta_L + \frac{\partial\theta_{1L}}{2},\omega_T + \frac{\partial\omega_{1T}}{2},\omega_L + \frac{\partial\omega_{1L}}{2}\right)$$

$$\delta\theta_{L2} = \Delta\theta_L\left(\theta_T + \frac{\partial\theta_{1T}}{2},\theta_L + \frac{\partial\theta_{1L}}{2},\omega_T + \frac{\partial\omega_{1T}}{2},\omega_L + \frac{\partial\omega_{1L}}{2}\right)$$

$$\delta\omega_{T3} = \Delta\omega_T\left(\theta_T + \frac{\partial\theta_{2T}}{2},\theta_L + \frac{\partial\theta_{2L}}{2},\omega_T + \frac{\partial\omega_{2T}}{2},\omega_L + \frac{\partial\omega_{2L}}{2}\right)$$

$$\delta\omega_{L3} = \Delta\omega_L\left(\theta_T + \frac{\partial\theta_{2T}}{2},\theta_L + \frac{\partial\theta_{2L}}{2},\omega_T + \frac{\partial\omega_{2T}}{2},\omega_L + \frac{\partial\omega_{2L}}{2}\right)$$

$$\delta\theta_{T3} = \Delta\theta_T\left(\theta_T + \frac{\partial\theta_{2T}}{2},\theta_L + \frac{\partial\theta_{2L}}{2},\omega_T + \frac{\partial\omega_{2T}}{2},\omega_L + \frac{\partial\omega_{2L}}{2}\right)$$

$$\delta\theta_{L3} = \Delta\theta_L\left(\theta_T + \frac{\partial\theta_{2T}}{2},\theta_L + \frac{\partial\theta_{2L}}{2},\omega_T + \frac{\partial\omega_{2T}}{2},\omega_L + \frac{\partial\omega_{2L}}{2}\right)$$

$$\delta\omega_{T4} = \Delta\omega_T\left(\theta_T + \delta\theta_{3T},\ \theta_L + \delta\theta_{3L},\ \omega_T + \delta\omega_{3T},\ \omega_L + \delta\omega_{3L}\right)$$

$$\delta\omega_{L4} = \Delta\omega_L\left(\theta_T + \delta\theta_{3T},\ \theta_L + \delta\theta_{3L},\ \omega_T + \delta\omega_{3T},\ \omega_L + \delta\omega_{3L}\right)$$

$$\delta\theta_{T4} = \Delta\theta_T\left(\theta_T + \delta\theta_{3T},\ \theta_L + \delta\theta_{3L},\ \omega_T + \delta\omega_{3T},\ \omega_L + \delta\omega_{3L}\right)$$

$$\delta\theta_{L4} = \Delta\theta_L\left(\theta_T + \delta\theta_{3T},\ \theta_L + \delta\theta_{3L},\ \omega_T + \delta\omega_{3T},\ \omega_L + \delta\omega_{3L}\right)$$

Next value for $\omega_T = \omega_T + \frac{1}{6}\left(\partial\omega_{T1} + \partial\omega_{T4}\right) + \frac{1}{3}\left(\partial\omega_{T2} + \partial\omega_{T3}\right).$

$$\text{Next value for } \omega_L = \omega_L + \frac{1}{6}\left(\partial\omega_{L1} + \partial\omega_{L4}\right) + \frac{1}{3}\left(\partial\omega_{L2} + \partial\omega L_{L3}\right).$$

$$\text{Next value for } \theta_T = \theta_T + \frac{1}{6}\left(\partial\theta_{T1} + \partial\theta_{T4}\right) + \frac{1}{3}\left(\partial\theta_{T2} + \partial\theta_{T3}\right).$$

$$\text{Next value for } \theta_L = \theta_L + \frac{1}{6}\left(\partial\theta_{L1} + \partial\theta_{L4}\right) + \frac{1}{3}\left(\partial\theta_{L2} + \partial\theta_{L3}\right).$$

The first step of this Runge-Kutta procedure is the Euler method. The remaining steps compute three more Euler-method estimates. The final values are weighted averages of the four estimates. This procedure repeats until the desired end conditions—such as an instant in time or a leg angle—are reached.

Appendix G

Derivation of Double Pendulum Equations

Table G.1   Notation

m	Segment mass
I_{cm}	Segment moment of inertia about mass center
I	Segment moment of inertia about proximal end
L	Segment length
d	Distance from segment mass center to proximal end
$\theta, \omega,$ and α	Segment angular position, velocity, and acceleration
a_H	Acceleration of the hip
g	Acceleration of gravity
v	Velocity of mass center

The subscripts T and L denote the thigh and leg/foot, respectively; the subscripts x and y denote anterior-posterior and vertical coordinate system directions, respectively.

The kinetic energy of the double pendulum is derived from the first principle that the kinetic energy of an object in general planar motion equals the sum of its rotational and translational kinetic energies. The rotational kinetic energy is calculated using the moment of inertia about the mass center, and the translational velocity equals the linear velocity of the mass center.

$$KE = \frac{1}{2}\left(I_{Tcm}\omega_T^2 + I_{Lcm}\omega_L^2 + m_T v_T^2 + m_L v_L^2\right)$$

where v_T is the linear velocity of the thigh center of mass, which is the vector sum of the velocities of the hip and the thigh mass center relative to the hip; that is,

$$v_T^2 = (v_{Hx} + \omega_T d_T \cos\theta_T)^2 + (v_{Hy} + \omega_T d_T \sin\theta_T)^2$$

and where v_L is the linear velocity of the leg/foot center of mass, which is the vector sum of the velocities of the hip, the knee relative to the hip, and the leg/foot mass center relative to the knee; that is,

$$v_L^2 = (v_{Hx} + \omega_T L_T \cos\theta_T + \omega_L d_L \cos\theta_L)^2 + (v_{Hy} + \omega_T L_T \sin\theta_T + \omega_L d_L \sin\theta_L)^2$$

The potential energy of the double pendulum is derived by determining the effects of the three degrees of freedom, namely the position of the hip and the angular positions of the thigh and leg, on the segment mass centers. As shown in figure G.1, the hip position changes both mass centers. The thigh angular position affects the locations of both thigh and leg/foot mass centers, whereas the leg angular position affects only the leg/foot center of mass.

Direct analysis of this geometry yields

$$PE = g\left[(m_T + m_L)\, y_H + (m_T d_T + m_L L_T)\,(1 - \cos\theta_T) + m_L\, d_L\,(1 - \cos\theta_L)\right].$$

Lagrange's equation of motion,

$$\frac{d}{dt}\frac{\partial KE}{\partial \omega_i} - \frac{\partial KE}{\partial \theta_i} + \frac{\partial PE}{\partial \theta_i} = M_i$$

is then implemented as follows:

Thigh

$$\frac{\partial KE}{\partial \omega_T} = I_T \omega_T + m_T d_T \left(v_{Hx}\cos\theta_T + v_{Hy}\sin\theta_T\right) +$$
$$m_L\left[(v_{Hx} + \omega_T L_T \cos_T + \omega_L\, d_L\, \cos\theta_L)(L_T \cos\theta_T) + (v_{Hy} + \omega_T L_T \sin\theta_T + \omega_L\, d_L\, \sin\theta_L)(L_T \sin\theta_T)\right]$$

$$\frac{d}{dt}\frac{\partial KE}{\partial \omega_T} = I_T \alpha_T + m_T d_T \left(a_{Hx}\cos\theta_T - v_{Hx}\theta_T\sin\theta_T + a_{Hy}\sin\theta_T + v_{Hy}\omega_T\cos\theta_T\right)$$
$$+ m_L\left[(v_{Hx} + \omega_T L_T \cos\theta_T + \omega_L\, d_L\, \cos\theta_L)(-L_T\, \omega_T\, \sin\theta_T)\right.$$
$$+ (a_{Hx} + \alpha_T L_T \cos\theta_T - \omega_T^2 L_T \sin\theta_T + \alpha_L\, d_L\, \cos\theta_L - \omega_L^2\, d_L\, \sin\theta_L)(L_T \cos\theta_T)$$
$$+ (v_{Hy} + \omega_T L_T \sin\theta_T + \omega_L\, d_L\, \sin\theta_L)(L_T\, \omega_T\, \cos\theta_T)$$
$$\left. + (a_{Hy} + \alpha_T L_T \sin\theta_T + \omega_T^2 L_T \cos\theta_T + \alpha_L\, d_L\, \sin\theta_L + \omega_L^2\, d_L\, \cos\theta_L)(L_T \sin\theta_T)\right]$$

$$= I_T\, \alpha_T + m_T\, d_T\, \left[(a_{Hx} + v_{Hy}\, \omega_T)\cos\theta_T + (a_{Hy} - v_{Hx}\, \omega_T)\sin\theta_T\right]$$
$$+ m_L\left[L_T^2\, \alpha_T + (v_{Hx} + \omega_L\, d_L\, \cos\theta_L)(-L_T\, \omega_T\, \sin\theta_T)\right.$$
$$+ (a_{Hx} + \alpha_L\, d_L\, \cos\theta_L - \omega_L^2\, d_L\, \sin\theta_L)(L_T \cos\theta_T) + (v_{Hy} + \omega_T^2\, d_L\, \sin\theta_L)(L_T\, \omega_T\, \cos\theta_T)$$
$$\left. + (a_{Hy} + \alpha_L\, d_L\, \sin\theta_L + \omega_L^2\, d_L\, \cos\theta_L)(L_T \sin\theta_T)\right]$$

$$= (I_T + m_L\, L_T^2)\, \alpha_T + m_T\, d_T\, \left[(a_{Hx} + v_{Hy}\, \omega_T)\cos\theta_T + (a_{Hy} - v_{Hx}\, \omega_T)\sin\theta_T\right]$$
$$+ m_L\left[L_T\, \omega_T\, (v_{Hy}\cos\theta_T - v_{Hx}\sin\theta_T) + L_T\, (a_{Hx}\cos\theta_T + a_{Hy}\sin\theta_T)\right.$$
$$\left. + L_T\, d_L\, \omega_T\, \omega_L\, \sin(\theta_L - \theta_T) + L_T\, d_L\, \alpha_L\, \cos(\theta_L - \theta_T) - L_T\, d_L\, \omega_L^2\, \sin(\theta_L - \theta_T)\right]$$

$$= (I_T + m_L\, L_T^2)\, \alpha_T + m_L\, L_T\, d_L\, \left[\alpha_L\, \cos(\theta_L - \theta_T) - \omega_L^2\, \sin(\theta_L - \theta_T) + \omega_T\, \omega_L\, \sin(\theta_L - \theta_T)\right]$$
$$+ (m_T d_T + m_L\, L_T)\, \left[(a_{Hx} + v_{Hy}\, \omega_T)\cos\theta_T + (a_{Hy} - v_{Hx}\, \omega_T)\sin\theta_T\right]$$

$$\frac{\partial KE}{\partial \theta_T} = m_T d_T \omega_T\left(-v_{Hx}\sin\theta_T + v_{Hy}\cos\theta_T\right)$$
$$+ m_L L_T\left[(v_{Hx} + \omega_T L_T \cos\theta_T + \omega_L\, d_L\, \cos\theta_L)(-\omega_T L_T \sin\theta_T)\right.$$
$$\left. + (v_{Hy} + \omega_T L_T \sin\theta_T + \omega_L d_L \sin\theta_L)(\omega_T L_T \cos\theta_T)\right]$$

$$= m_T d_T \omega_T (v_{Hy}\cos\theta_T - v_{Hx}\sin\theta_T) + m_L L_T \omega_T (v_{Hy}\cos\theta_T - v_{Hx}\sin\theta_T)$$
$$+ m_L L_T d_L \omega_T \omega_L (\cos\theta_T \sin\theta_L - \sin\theta_T \cos\theta_L)$$

$$= \omega_T (m_T d_T + m_L L_T)(v_{Hy}\cos\theta_T - v_{Hx}\sin\theta_T) + m_L L_T d_L \omega_T \omega_L \sin(\theta_L - \theta_T)$$

$$\frac{\partial PE}{\partial \theta_T} = g\left(m_T d_T + m_L L_T\right)\sin\theta_T$$

Leg/Foot

$$\frac{\partial KE}{\partial \omega_L} = I_L \omega_L + m_L\left[\left(v_{Hx} + \omega_T L_T \cos\theta_T + \omega_L d_L \cos\theta_L\right)\left(d_L \cos\theta_L\right)\right.$$
$$\left. + (v_{Hy} + \omega_T L_T \sin\theta_T + \omega_L\, d_L\, \sin\theta_L)(d_L \sin\theta_L)\right]$$

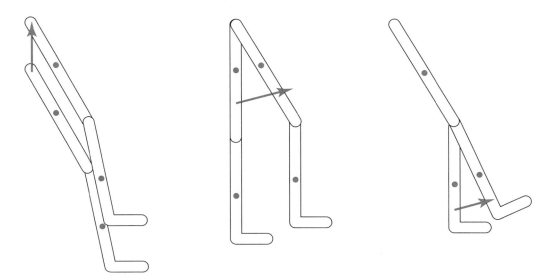

▲ **Figure G.1**   The hip position, thigh position, and leg/foot position each have different effects on the mass centers.

$$\frac{d}{dt}\frac{\partial KE}{\partial \omega_L} = I_L \alpha_L + m_L \left[\left(v_{Hx} + \omega_T L_T \cos\theta_T + \omega_L d_L \cos\theta_L \right)\left(-d_L \omega_L \sin\theta_L \right) \right]$$
$$+ (a_{Hx} + \alpha_T L_T \cos\theta_T - \omega_T^2 L_T \sin\theta_T + \alpha_L d_L \cos\theta_L - \omega_L^2 d_L \sin\theta_L)(d_L \cos\theta_L)$$
$$+ (v_{Hy} + \omega_T L_T \sin\theta_T + \omega_L d_L \sin\theta_L)(d_L \omega_L \cos\theta_L) + (a_{Hy} + \alpha_T L_T \sin\theta_T + \omega_T^2 L_T \cos\theta_T$$
$$+ \alpha_L d_L \sin\theta_L + \omega_L^2 d_L \cos\theta_L)(d_L \sin\theta_L)$$

$$= I_L \alpha_L + m_L d_L \left[(a_{Hx} + v_{Hy} \omega_L) \cos\theta_L + (a_{Hy} - v_{Hx} \omega_L) \sin\theta_L \right]$$
$$+ m_L \left[\alpha_T L_T d_L \cos(\theta_L - \theta_T) + \omega_T^2 L_T d_L \sin(\theta_L - \theta_T) + \alpha_L d_L^2 - L_T d_L \omega_T \omega_L \sin(\theta_L - \theta_T) \right]$$

$$= I_L \alpha_L + m_L d_L \left[(a_{Hx} + v_{Hy} \omega_L) \cos\theta_L + (a_{Hy} - v_{Hx} \omega_L) \sin\theta_L \right]$$
$$+ m_L L_T d_L \left[\alpha_T \cos(\theta_L - \theta_T) + \omega_T^2 \sin(\theta_L - \theta_T) - \omega_T \omega_L \sin(\theta_L - \theta_T) \right]$$

$$\frac{\partial KE}{\partial \theta_L} = m_L \left[\left(v_{Hx} + \omega_T L_T \cos\theta_T + \omega_L d_L \cos\theta_L \right)\left(-\omega_L d_L \sin\theta_L \right) \right]$$
$$+ (v_{Hy} + \omega_T L_T \sin\theta_T + \omega_L d_L \sin\theta_L)(\omega_L d_L \cos\theta_L) = m_L \left[d_L \omega_L (v_{Hy} \cos\theta_L - v_{Hx} \sin\theta_L) \right]$$
$$+ L_T d_L \omega_T \omega_L (\sin\theta_T \cos\theta_L - \cos\theta_T \sin\theta_L)] = m_L \left[d_L \omega_L (v_{Hy} \cos\theta_L - v_{Hx} \sin\theta_L) \right]$$
$$- L_T d_L \omega_T \omega_L \sin(\theta_T - \theta_L)]$$

$$\frac{\partial PE}{\partial \theta_L} = g m_L d_L \sin\theta_L$$

Thus, applying Lagrange's equation yields:

Thigh

$$(I_T + m_L L_T^2)\, \alpha_T + m_L L_T d_L \left[\alpha_L \cos(\theta_L - \theta_T) - \omega_L^2 \sin(\theta_L - \theta_T) + \omega_T \omega_L \sin(\theta_L - \theta_T) \right]$$
$$+ (m_T d_T + m_L L_T) \left[(a_{Hx} + v_{Hy} \omega_T) \cos\theta_T + (a_{Hy} - v_{Hx} \omega_T) \sin\theta_T \right]$$
$$- \left[\omega_T (m_T d_T + m_L L_T)(v_{Hy} \cos\theta_T - v_{Hx} \sin\theta_T) + m_L L_T d_L \omega_T \omega_L \sin(\theta_L - \theta_T) \right]$$
$$+ g(m_T d_T + m_L L_T) \sin\theta_T$$
$$\rightarrow (I_T + m_L L_T^2)\, \alpha_T + m_L L_T d_L \left[\alpha_L \cos(\theta_L - \theta_T) - \omega_L^2 \sin(\theta_L - \theta_T) \right]$$
$$+ (m_T d_T + m_L L_T)[a_{Hx} \cos\theta_T + (a_{Hy} + g) \sin\theta_T] = 0 \tag{G.1}$$

Leg/Foot

$$I_L a_L + m_L d_L \left[(a_{Hx} + v_{Hy} \omega_L) \cos\theta_L + (a_{Hy} - v_{Hx} \omega_L) \sin\theta_L \right]$$

$$+ m_L L_T d_L \left[\alpha_T \cos(\theta_L - \theta_T) + \omega_L^2 \sin(\theta_L - \theta_T) - \omega_T \omega_L \sin(\theta_L - \theta_T) \right]$$

$$- m_L \left[d_L \omega_L (v_{Hy} \cos\theta_L - v_{Hx} \sin\theta_L) + L_T d_L \omega_T \omega_L \sin(\theta_T - \theta_L) \right] + g\, m_L d_L \sin\theta_L = 0$$

$$\rightarrow I_L \alpha_L + m_L d_L \left[L_T \alpha_T \cos(\theta_L - \theta_T) + L_T \omega_T^2 \sin(\theta_L - \theta_T) + a_{Hx} \cos\theta_L + (a_{Hy} + g) \sin\theta_L \right] = 0 \quad \text{(G.2)}$$

Discrete Fourier Transform Subroutine

This is the BASIC code to calculate the power spectrum, power(), of a time series, h(numpnts), where numpnts is the number of data in h. Arrays s(), c() hold the coefficients of the sine and cosine functions. There will be only m values in the arrays s(), c(), and power(). The method used is slow but simple.

```
DIM s(numpnts), c(numpnts), h(numpnts), power(numpnts)
pi = 3.14159265
w = (2 * pi) / numpnts
m = numpnts / 2 + 1
FOR k = 1 TO m
k1w = (k - 1) * w
FOR j = 1 TO numpnts
alpha = k1w * (j - 1)
s(k) = s(k) + h( j) * SIN(alpha)
c(k) = c(k) + h( j) * COS(alpha)
NEXT j
s(k) = 2 * s(k)
c(k) = 2 * c(k)
power(k) = s(k)^2 + c(k)^2
NEXT k
```

Shannon's Reconstruction Subroutine

This is the BASIC code to implement Shannon's formula for reconstructing data at sampling rates above the Nyquist rate.

```
'olddelta = original sampling rate
'newdelta = new sampling rate
'samptime = duration of the trial
'fc = Nyquist frequency
'newpoints = number of reconstructed data points
'oldpoints = original number of points
'd!(n) = nth point of the original signal
's!(i) = ith point in the reconstructed signal
pi = 3.14159265
samptime = (oldpoints - 1) * olddelta
newpoints = samptime/newdelta
fc = 1/(2 * olddelta)
fc2 = 2 * fc
FOR i = 1 To newpoints
t = (i - 1) * newdelta
FOR n = 1 To oldpoints
IF (t - (n - 1) * olddelta) <> 0 Then
m = sin(fc2 * pi * (t - (n - 1) * olddelta))/(pi * t - (n - 1) * olddelta))
ELSE
m = 1/olddelta
END IF
newdata(i) = newdata(i) + olddata(n) * m
NEXT n
newdata(i) = newdata(i) * olddelta
NEXT I
```

Example Answers

CHAPTER 3

Example 3.1

$$m_{thigh} = P_{thigh}\, m_{total} = 0.100 \times 80.0 = 8.00 \text{ kg}$$

Example 3.2

$$x_{cg} = -12.80 + 0.433\,[7.3 - (-12.80)] = -4.10 \text{ cm}$$

$$y_{cg} = 83.3 + 0.433\,(46.8 - 83.3) = 67.5 \text{ cm}$$

Notice that the coordinates of the thigh's center of gravity (−4.10, 67.5) must fall between the two endpoints. Carefully preserve the signs of the coordinates during the computations. These coordinates were taken from frame 10 of table 1.3.

Example 3.3

$$l_{thigh} = \sqrt{\left[7.3 - (-12.80)\right]^2 + \left(46.8 - 83.3\right)^2} = 41.67 \text{ cm}$$

$$k_{cg} = K_{cg(thigh)} \times L_{thigh} = 0.323 \times 41.67$$
$$= 13.46 \text{ cm} = 0.1346 \text{ m}$$

In example 3.1, the thigh mass was calculated to be 8.00 kg.

$$I_{cg} = mk^2 = 18.00 \times 0.1346^2 = 0.326 \text{ kg·m}^2$$

Notice that the radius of gyration (k_{cg}) was calculated before the moment of inertia could be computed. This in turn required calculating the thigh segment's length (L_{thigh}). Also notice that the units of the radius of gyration were converted to meters before squaring.

To compute the moment of inertia about the proximal end of this thigh, we must apply the parallel axis theorem after first computing the distance from the thigh center to the proximal end. This distance is called $r_{proximal}$.

$$r_{proximal} = 0.433 \times L_{thigh} = 0.433 \times 41.67 =$$
$$18.04 \text{ cm} = 0.1804 \text{ m}$$

$$I_{proximal} = I_{cg} + m_{thigh}\, r_{proximal}^2 = 0.326 + 18.00$$
$$\times 0.1804^2 = 0.912 \text{ kg·m}^2$$

Notice that the moment of inertia about the proximal end is larger than that about the center of gravity.

The moment of inertia is always smallest about an axis through the center of gravity.

CHAPTER 4

Example 4.1

The issue with drawing the wind is that we are unable to locate its center of pressure. This is the same problem with FBDs of swimmers and cyclists.

CHAPTER 5

Example 5.1a

The FBD for this example is almost the same as figure 5.10d, except that in the FBD for this example we included the forces of gravity and the vertical GRF. Note that X- and Y-axes have been drawn to indicate positive direction. For the sake of example, note that the horizontal acceleration (−64 m/s²) is drawn with an arrow in the negative horizontal direction (to the left) with a positive 64 m/s². Similarly note that the angular acceleration is drawn in a negative (clockwise) direction with a positive 28 rad/s² value. In these two cases, it is also acceptable to draw them pointing in the opposite directions with negative values.

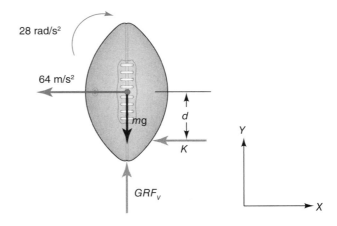

As indicated by the gray arrows, the unknowns in this diagram are the vertical GRF, the force K, and the distance d, between K's line of action and the center of the ball. We start by solving for the forces in the vertical direction because they are the simplest in this case.

$$\sum F_y = ma_y$$

$$GRF_y - mg = ma_y$$

$$\Rightarrow GRF_y - (0.25 \text{ kg})(9.81 \text{ m/s}^2) = 0$$

$$\Rightarrow GRF_y = 2.45 \text{ N}$$

This is the expected result when the ground is simply supporting the weight of the ball.

When solving for the horizontal force K, note that the FBD is used to determine that K should have a minus sign because it points to the left (−x). That we calculated a positive K means that the force acts in the direction drawn in the FBD, which means the force is directed in the negative x direction.

$$\sum F_x = ma_x$$

$$-K = ma_x$$

$$\Rightarrow -K = (0.25 \text{ kg})(-64 \text{ m/s}^2)$$

$$\Rightarrow K = 16.0 \text{ N}$$

As is standard in human-movement problems, we calculate moments about the ball's mass center. The FBD shows that there is only one force, K, which does not act through the mass center. Its moment equals the product Kd. We write this term down and then put a negative sign in front of it because this moment causes a *clockwise* (negative) effect. This somewhat tricky part requires visualization. Imagine the mass center as a fixed point and observe that the force K turns about the mass center in a clockwise fashion.

$$\sum M = I\alpha$$

$$-K(d) = I\alpha$$

$$\Rightarrow d \frac{(0.04 \text{ kg} \cdot \text{m}^2)(-28 \text{ rad/s}^2)}{16 \text{ N}} = -0.070 \text{ m}$$

The distance, d, to the force is negative, indicating that the ball was kicked below the mass center.

Example 5.1b

The FBD is almost the same as in the previous example, except for the addition of the horizontal tee force. Note that the tee force has a positive sign:

$$\sum F_x = ma_x$$

$$-K + F_T = ma_x$$

$$\Rightarrow -K + 4 \text{ N} = (0.25 \text{ kg})(-64 \text{ m/s}^2) = -16 \text{ N}$$

$$\Rightarrow K = 20.0 \text{ N}$$

This is the logical result: The kicking force was the sum of the tee force and the ball's reaction.

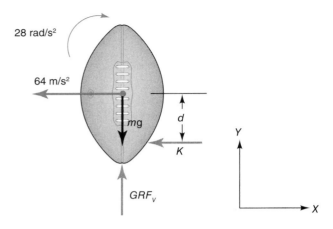

In summing moments about the mass center, note that the moment of the force K is negative, as before, but that the moment of the force F_T is positive:

$$\sum M = I\alpha$$

$$-K(d) + F_T(0.15 \text{ m}) = I\alpha$$

$$\Rightarrow d \frac{(0.04 \text{ kg} \cdot \text{m}^2)(-28 \text{ rad/s}^2) - 4(0.15 \text{ m})}{16 \text{ N}}$$

$$= -0.1075 \text{ m}$$

Thus, the force was lower on the ball in this instance.

Example 5.2

This is an apparently complex situation resolvable with an FBD; in fact, it is very similar to our football example. To draw this correctly, we should first identify the forces involved: body weight (gravity), the unknown GRFs, and the reaction of a body mass being accelerated. There is also an unknown ankle moment, M_A. In the FBD, they look like this:

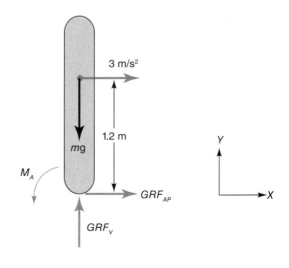

Because the commuter maintains a still posture, her body's angular acceleration is 0 rad/s. The unknown reactions GRF_x, GRF_y, and M_A are at the floor; they are drawn in positive coordinate system directions and determined with our three equations of motion:

$$\sum F_x = ma_x$$

$$GRF_x = ma_x$$

$$\Rightarrow GRF_x = (60 \text{ kg})(3 \text{ m/s}^2) = 180.0 \text{ N}$$

$$\sum F_y = ma_y$$

$$GRF_y - mg = ma_y$$

$$\Rightarrow GRF_y - (60 \text{ kg})(9.81 \text{ m/s}^2) = 0$$

$$\Rightarrow GRF_y = 589 \text{ N}$$

Both of these reactions are positive, indicating that they act in the directions drawn.

We will again sum moments about the mass center. Note that we have drawn the unknown moment M_A in a counterclockwise direction, so it is positive in the first equation below. The moment of the vertical GRF_y is zero and the moment of the horizontal GRF_x is positive because it points counterclockwise about the mass center.

$$\sum M = I\alpha$$

$$M_A + GRF_x(1.2 \text{ m}) = I\alpha$$

$$\Rightarrow M_A + (180 \text{ N})(1.2 \text{ m}) = (130 \text{ kg·m}^2)(0 \text{ rad/s}^2)$$

$$\Rightarrow M_A = -216 \text{ N·m}$$

This is a large ankle moment, which is why sudden subway starts usually cause people to either take a step or grab a handle.

Example 5.3

The FBD of the racket is quite simple. The only tricky part is that the hand has an unknown action. Therefore, we must draw unknown actions in each coordinate system direction:

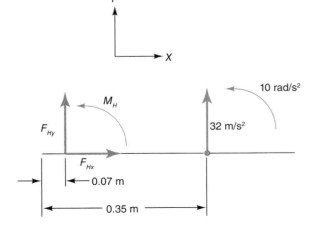

Solving our three equations of motion:

$$\sum F_x = ma_x$$

$$F_{Hx} = (0.5 \text{ kg})(0 \text{ m/s}^2)$$

$$\Rightarrow F_{Hx} = 0 \text{ N}$$

$$\sum F_y = ma_y$$

$$F_{Hy} = (0.5 \text{ kg})(32 \text{ m/s}^2) = 16 \text{ N}$$

$$\sum M = I\alpha$$

$$M_H - F_{Hy}(0.35 \text{ m} - 0.07 \text{ m}) = (0.1 \text{ kg·m}^2)(10 \text{ rad/s}^2)$$

$$\Rightarrow M_H = (16 \text{ N})(0.35 \text{ m} - 0.07 \text{ m})$$
$$+ (0.1 \text{ kg·m}^2)(10 \text{ rad/s}^2) = 5.5 \text{ N·m}$$

A moment is the result of a force couple, that is, two equal, opposite, noncollinear forces. We can assume that the hand area by the forefinger is pushing the racket handle forward and the area of the little finger is pulling it backward:

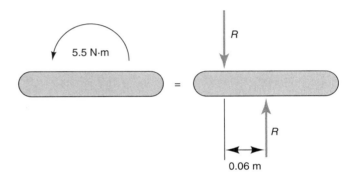

If these points are about 6 cm apart, we can estimate that each member of the force couple is about 91.7 N.

Example 5.4

We have measured the endpoints of the object with our camera and determined where the object's mass center is located. We start by redrawing our FBD with the distances in the X and Y directions between the forces and the mass center, about which we are going to calculate the moments of force:

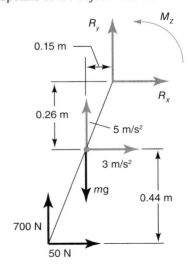

We then apply the three equations of 2-D motion:

$$\sum F_x = ma_x$$

$$R_x + 50\ \text{N} = (8\ \text{kg})(3\ \text{m/s}^2)$$

$$\Rightarrow R_x = -26.0\ \text{N}$$

Note that R_x is negative. This means that the force points in the direction opposite to that shown in the FBD.

$$\sum F_y = ma_y$$

$$R_y + 700\ \text{N} - mg = (8\ \text{kg})(5\ \text{m/s}^2)$$

$$\Rightarrow R_y = -700\ \text{N} + (8\ \text{kg})(9.81\ \text{m/s}^2) + (8\ \text{kg})(5\ \text{m/s}^2) = -581.5\ \text{N}$$

As with R_x, our calculated value for R_y is negative, indicating that it points downward, the opposite of what we drew in the FBD.

In writing our moment equation, we first write down all the terms and then decide whether each term is positive or negative according to the right-hand rule:

$$\sum M_z = I\alpha$$

$$M_z + (50\ \text{N})(0.44\ \text{m}) - (700\ \text{N})(0.25\ \text{m}) - R_x(0.26\ \text{m}) + R_y(0.15\ \text{m}) = (0.2\ \text{kg·m}^2)(10\ \text{rad/s}^2)$$

We then substitute the numerical values for R_x and R_y that we calculated previously. Note that we put these in parentheses with their negative signs to preserve the signs that we determined for the signs of the moments:

$$\Rightarrow M_z + (50\ \text{N})(0.44\ \text{m}) - (700\ \text{N})(0.25\ \text{m}) - (-26\ \text{N})(0.26\ \text{m}) + (-582\ \text{N})(0.15\ \text{m}) = (0.2\ \text{kg·m}^2)(10\ \text{rad/s}^2)$$

We then bring all terms but our unknown moment M_z to the right-hand side of the equation and solve:

$$\Rightarrow M_z = -(50\ \text{N})(0.44\ \text{m}) + (700\ \text{N})(0.25\ \text{m}) + (-26\ \text{N})(0.26\ \text{m}) - (-582\ \text{N})(0.15\ \text{m}) + (0.2\ \text{kg·m}^2)(10\ \text{rad/s}^2) = 235\ \text{N·m}$$

Example 5.5

The FBD is

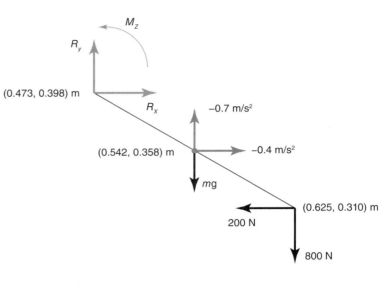

$$\sum F_x = ma_x$$

$$R_x - 200\ \text{N} = (0.1\ \text{kg})(-0.4\ \text{m/s}^2)$$

$$\Rightarrow R_x = 200\ \text{N}$$

The exact value is 199.96 N, but it rounds to 200 N. The crank is so light that its mass-acceleration product is negligible. The same thing happens in the y-direction:

$$\sum F_y = ma_y$$

$$R_y - 800\ \text{N} - (0.1\ \text{kg})(9.81\ \text{m/s}^2) = (0.1\ \text{kg})(-0.7\ \text{m/s}^2)$$

$$\Rightarrow R_y = 801\ \text{N}$$

Again, it is helpful when calculating moments to redraw the FBD with distances between points:

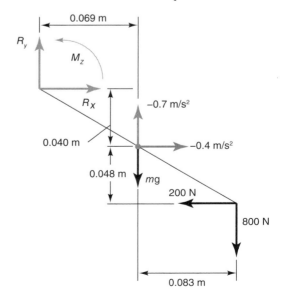

$$\sum_i M = I\alpha$$

$$M_Z - R_x(0.040 \text{ m}) - R_y(0.069 \text{ m})$$
$$- (200 \text{ N})(0.048 \text{ m}) - (800 \text{ N})(0.083 \text{ m})$$
$$= (0.003 \text{ kg·m}^2)(10 \text{ rad/s}^2)$$

$$\Rightarrow M_Z = (200 \text{ N})(0.040 \text{ m}) + (801 \text{ N})(0.069 \text{ m})$$
$$+ (200 \text{ N})(0.048 \text{ m})$$
$$+ (800 \text{ N})(0.083 \text{ m}) + (0.1 \text{ kg·m}_2)(10 \text{ rad/s}^2)$$

$$\Rightarrow M_Z = 139.3 \text{ N·m}$$

Example 5.6

In this example we need only to section the arm at the shoulder and analyze that one piece. Because this is a static case, the ma and $I\alpha$ terms are zero. To construct the FDB, we section the arm and draw the three unknowns at the shoulder joint. We also draw the weight of the upper arm, forearm, and hand:

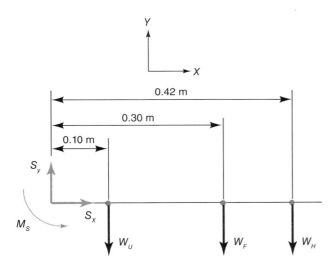

To solve for the three unknowns:

$$\sum_i F_x = ma_x$$

$$S_x = 0 \text{ N}$$

$$\sum_i F_y = ma_y$$

$$S_y - W_U - W_F - W_H = 0$$

$$\Rightarrow S_y = W_U + W_F + W_H$$

$$\Rightarrow S_y = (4 \text{ kg})(9.81 \text{ m/s}^2) + (3 \text{ kg})(9.81 \text{ m/s}^2)$$
$$+ (1 \text{ kg})(9.81 \text{ m/s}^2) = 78.5 \text{ N}$$

In this example, it is convenient to calculate moments about the shoulder. Once again, the procedure is first to write down the products of forces and distances and then establish the sign of each moment.

$$\sum_i M_z = 0$$

$$M_S - W_U(0.10 \text{ m}) - W_F(0.30 \text{ m}) - W_H(0.42 \text{ m}) = 0$$

$$\Rightarrow M_S = W_U(0.10 \text{ m}) + W_F(0.30 \text{ m}) + W_H(0.42 \text{ m})$$

$$\Rightarrow M_S = (4 \text{ kg})(9.81 \text{ m/s}^2)(0.10 \text{ m})$$
$$+ (3 \text{ kg})(9.81 \text{ m/s}^2)(0.30 \text{ m})$$
$$+ (1 \text{ kg})(9.81 \text{ m/s}^2)(0.42 \text{ m}) = 16.9 \text{ N·m}$$

Example 5.7

The FBD is almost identical to the previous example, except for the added weight W_1:

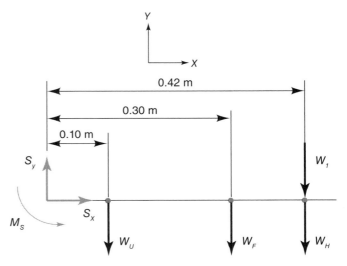

Solving for the three unknowns:

$$\sum_i F_x = ma_x$$

$$S_x = 0 \text{ N}$$

$$\sum_i F_y = ma_y$$

$$S_y - W_U - W_F - W_H - W_1 = 0$$

$$\Rightarrow S_y = W_U + W_F + W_H + W_1$$

$$\Rightarrow S_y = (4 \text{ kg})(9.81 \text{ m/s}^2) + (3 \text{ kg})(9.81 \text{ m/s}^2)$$
$$+ (1 \text{ kg})(9.81 \text{ m/s}^2)$$
$$+ (2 \text{ kg})(9.81 \text{ m/s}^2) = 98.1 \text{ N}$$

This is the expected change (compared with example 5.6) of about 20 N.

Once again, to sum the moments, we first write down each moment, then establish its sign:

$$\sum M = 0$$

$$M_S - W_U(0.10 \text{ m}) - W_F(0.30 \text{ m})$$
$$- (W_H + W_1)(0.42 \text{ m}) = 0$$

$$\Rightarrow M_S = W_U(0.10 \text{ m}) + W_F(0.30 \text{ m})$$
$$+ (W_H + W_1)(0.42 \text{ m})$$

$$\Rightarrow M_S = (4 \text{ kg})(9.81 \text{ m/s}^2)(0.10 \text{ m})$$
$$+ 3 \text{ kg} (9.81 \text{ m/s}^2)(0.30 \text{ m})$$
$$+ (1 \text{ kg} + 2 \text{ kg})(9.81 \text{ m/s}^2)(0.42 \text{ m}) = 25.1 \text{ N·m}$$

Example 5.8

The FBD of the forearm is again similar to the previous example. We draw the new unknowns at the elbow joint in a positive sense and calculate the distances from the forces to the elbow:

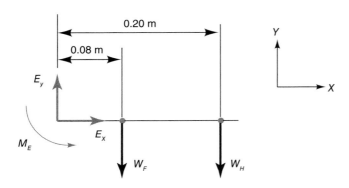

Solving for the three unknowns:

$$\sum F_x = ma_x$$

$$E_x = 0 \text{ N}$$

$$\sum F_y = ma_y$$

$$E_y - W_F - W_H = 0$$

$$\Rightarrow E_y = W_F + W_H$$

$$\Rightarrow E_y = (3 \text{ kg})(9.81 \text{ m/s}^2) + (1 \text{ kg})(9.81 \text{ m/s}^2) = 39.2 \text{ N}$$

It is convenient to calculate moments about the elbow:

$$\sum M = 0$$

$$M_E - W_F(0.08 \text{ m}) - W_H(0.20 \text{ m}) = 0$$

$$\Rightarrow M_E = W_F(0.08 \text{ m}) + W_H(0.20 \text{ m})$$

$$\Rightarrow M_E = (3 \text{ kg})(9.81 \text{ m/s}^2)(0.08 \text{ m})$$
$$+ (1 \text{ kg})(9.81 \text{ m/s}^2)(0.20 \text{ m}) = 4.3 \text{ N·m}$$

To solve for the upper arm, the method of sections requires that we draw the upper arm with forces and moments at the elbow having directions equal but opposite to those on the forearm:

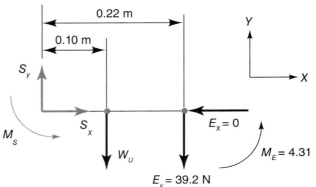

$$\sum F_x = ma$$

$$S_x - E_x = 0$$

$$\Rightarrow S_x = 0 \text{ N}$$

$$\sum F_Y = ma$$

$$S_y - W_U - E_y = 0$$

$$\Rightarrow S_Y = W_U + E_y$$

$$\Rightarrow S_Y = (4 \text{ kg})(9.81 \text{ m/s}^2) + 39.2 \text{ N} = 78.4 \text{ N}$$

This is the same value we had calculated earlier. Summing the moments about the shoulder:

$$\sum M = 0$$

$$M_S - M_E - W_U(0.10 \text{ m}) - E_y(0.22 \text{ m}) = 0$$

$$\Rightarrow M_S = M_E + W_U(0.10 \text{ m}) + E_y(0.22 \text{ m})$$

$$\Rightarrow M_S = 4.32 \text{ N·m} + (4 \text{ kg})(9.81 \text{ m/s}^2)(0.10 \text{ m})$$
$$+ (39.2 \text{ N})(0.22 \text{ m}) = 16.87 \text{ N·m}$$

This is also the same value that we calculated earlier.

Example 5.9

The FBD of the foot:

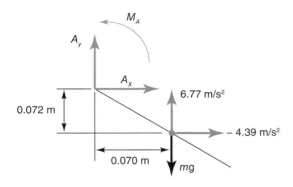

Solving for the reaction forces:

$$\sum F_x = ma_x$$

$$A_x = m_f a_f$$

$$\Rightarrow A_x = (1.2 \text{ kg})(-4.39 \text{ m/s}^2) = -5.3 \text{ N}$$

$$\sum F_y = ma_y$$

$$A_y - m_f g = m_f a_f$$

$$\Rightarrow A_y = m_f g + m_f a_f$$

$$\Rightarrow A_y = (1.2 \text{ kg})(9.81 \text{ m/s}^2)$$
$$+ (1.2 \text{ kg})(6.77 \text{ m/s}^2) = 19.9 \text{ N}$$

Because the foot mass is small, the reactions are small.

Solving for the ankle joint moment, we sum moments about the mass center. There are three moments; of these, the ankle moment M_A is positive because that is how it was drawn. The moments of the reaction forces are both negative because they turn clockwise about the mass center:

$$\sum M = I\alpha$$

$$M_A - A_x(0.072 \text{ m}) - A_y(0.070 \text{ m}) = I_f \alpha_f$$

$$\Rightarrow M_A = A_x(0.072 \text{ m}) + A_y(0.070 \text{ m}) + I_f \alpha_f.$$

We now substitute the numerical values of the reaction forces A_x and A_y in parentheses so that we preserve their signs:

$$\Rightarrow M_A = (-5.3 \text{ N})(0.072 \text{ m}) + (19.9 \text{ N})(0.070 \text{ m})$$
$$+ (0.011 \text{ kg·m}^2)(5.12 \text{ rad/s}^2)$$

$$\Rightarrow M_A = 1.01 \text{ N·m}$$

This is a very small joint moment. Although it is essentially zero, because it is positive it is a dorsiflexor moment, assuming the person is facing the right.

The FBD of the leg is as follows. Note that we placed the numerical values for the ankle reactions into this diagram and retained their original signs.

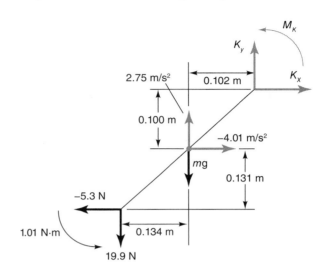

$$\sum F_x = ma_x$$

$$K_x - A_x = m_l a_l$$

$$\Rightarrow K_x = A_x + m_l a_l$$

$$\Rightarrow K_x = -5.3 \text{ N} + (2.4 \text{ kg})(-4.01 \text{ m/s}^2) = -14.9 \text{ N}$$

$$\sum F_y = ma_y$$

$$K_y - A_y - m_{lg} = m_l a_l$$

$$\Rightarrow K_y = A_y + m_{lg} + m_l a_l$$

$$\Rightarrow K_y = 19.9 \text{ N} + (2.4 \text{ kg})(9.81 \text{ m/s}^2)$$
$$+ (2.4 \text{ kg})(2.75 \text{ m/s}^2) = 50.0 \text{ N}$$

When summing leg moments, note that the moments of the vertical (y) forces are positive, whereas the moments of the horizontal (x) forces are negative.

$$\sum M = I\alpha$$

$$M_K - M_A - K_x(0.100 \text{ m}) + K_y(0.102 \text{ m})$$
$$- A_x(0.131 \text{ m}) + A_y(0.134 \text{ m}) = I_l \alpha_l$$

$$\Rightarrow M_K = M_A + K_x(0.100 \text{ m}) - K_y(0.102 \text{ m})$$
$$+ A_x(0.131 \text{ m})$$
$$- A_y(0.134 \text{ m}) + I_l \alpha_l$$

Note how we again substitute for the reaction forces the numerical values in parentheses to preserve their signs:

$$\Rightarrow M_K = 1.01 \text{ N·m} + (-14.9 \text{ N})(0.100 \text{ m})$$
$$- (50.0 \text{ N})(0.102 \text{ m}) + (-5.3 \text{ N})(0.131 \text{ m})$$
$$- (19.90 \text{ N})(0.134 \text{ m}) + (0.064 \text{ kg·m}^2)(-3.08 \text{ rad/s}^2)$$

$$\Rightarrow M_K = -9.1 \text{ N·m}$$

This is a small knee flexor moment, assuming the person is facing to the right.

This is the FBD of the thigh. Again note that we placed the numerical values for the knee reactions into this diagram with the same signs that they were calculated as having.

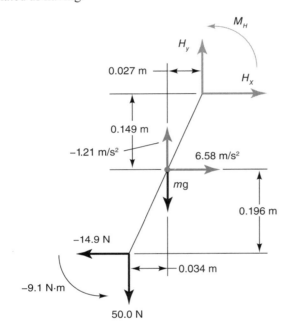

$$\sum F_x = ma_x$$

$$H_x - K_x = m_t a_t$$

$$\Rightarrow H_x = K_x + m_t a_t$$

$$\Rightarrow H_x = -14.9 \text{ N} + (6.0 \text{ kg})(6.58 \text{ m/s}^2) = 24.6 \text{ N}$$

$$\sum F_y = ma_y$$

$$H_y - K_y - m_t g = m_t a_t$$

$$\Rightarrow H_y = K_y + m_t g + m_t a_t$$

$$\Rightarrow H_y = 50.0 \text{ N} + (6.0 \text{ kg})(9.81 \text{ m/s}^2)$$
$$+ (6.0 \text{ kg})(-1.21 \text{ m/s}^2) = 101.6 \text{ N}$$

$$\sum M = I\alpha$$

$$M_H - M_K - H_x(0.149 \text{ m}) + H_y(0.027 \text{ m})$$
$$- K_x(0.196 \text{ m}) + K_y(0.034 \text{ m}) = I_t(\alpha_t)$$

$$\Rightarrow M_H = M_K + H_x(0.149 \text{ m}) - H_y(0.027 \text{ m})$$
$$+ K_x(0.196 \text{ m}) - K_y(0.034 \text{ m}) + I_t(\alpha_t)$$

$$\Rightarrow M_H = -9.1 \text{ N·m} + (24.6 \text{ N})(0.149 \text{ m})$$
$$- (101.6 \text{ N})(0.027 \text{ m}) + (-14.9 \text{ N})(0.196 \text{ m})$$
$$- (50.0 \text{ N})(0.034 \text{ m}) + (0.130 \text{ kg·m}^2)(8.62 \text{ rad/s}^2)$$

$$\Rightarrow M_H = -11.7 \text{ N·m}$$

Because this is a negative result, it is a hip extensor moment, assuming the person is facing to the right.

Example 5.10

The free-body diagram of the foot is the same as for the swing phase except for the ground reaction forces. We must be careful to place the ground reaction forces in the proper location. It is also helpful to draw them in the positive direction and put their values on, positive or negative. Solving for the reaction forces:

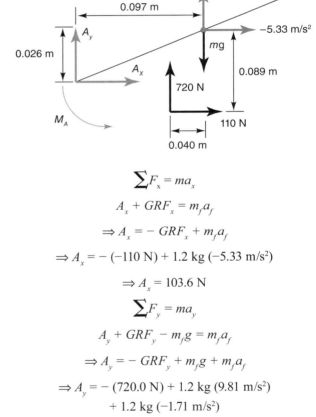

$$\sum F_x = ma_x$$

$$A_x + GRF_x = m_f a_f$$

$$\Rightarrow A_x = -GRF_x + m_f a_f$$

$$\Rightarrow A_x = -(-110 \text{ N}) + 1.2 \text{ kg}(-5.33 \text{ m/s}^2)$$

$$\Rightarrow A_x = 103.6 \text{ N}$$

$$\sum F_y = ma_y$$

$$A_y + GRF_y - m_f g = m_f a_f$$

$$\Rightarrow A_y = -GRF_y + m_f g + m_f a_f$$

$$\Rightarrow A_y = -(720.0 \text{ N}) + 1.2 \text{ kg}(9.81 \text{ m/s}^2)$$
$$+ 1.2 \text{ kg}(-1.71 \text{ m/s}^2)$$

$$\Rightarrow A_y = -710.3 \text{ N}$$

Because the foot mass is small, the ankle joint reaction forces are nearly equal and opposite to the ground reaction forces. Note that the vertical reaction A_y is negative; this means that the actual force is pushing down on the ankle joint, which is a logical result given that this joint is bearing body weight.

Solving for the ankle joint moment, we sum moments about the mass center. There are five moments; of these the ankle moment M_A is positive because of how it is drawn. The moment of the horizontal joint reaction force is positive because it runs counterclockwise around the mass center; however, the moment of vertical joint reaction force is negative because it turns clockwise about the mass center. Similarly, the moment of the horizontal ground reaction force is positive, and the moment of the vertical ground reaction force is negative. Note that the location of the center of pressure is critical in establishing the moment of the vertical GRF; also note that the moment arm of the moment of the horizontal GRF is the vertical position of the mass center because that force is located on the ground (i.e., at $y = 0.0$ m):

$$\sum M = I\alpha$$

$$M_A - A_x(0.026 \text{ m}) - A_y(0.097 \text{ m}) + GRF_x(0.089 \text{ m}) - GRF_y(0.040 \text{ m}) = I_f\alpha_f$$

$$\Rightarrow M_A = A_x(0.026 \text{ m}) + A_y(0.097 \text{ m}) - GRF_x(0.089 \text{ m}) + GRF_y(0.040 \text{ m}) + I_f\alpha_f$$

$$\Rightarrow M_A = (103.6 \text{ N})(0.026 \text{ m}) + (-710.3 \text{ N})(0.097 \text{ m}) - (-110 \text{ N})(0.089 \text{ m}) + (720 \text{ N})(0.040 \text{ m}) + (0.011 \text{ kg·m}^2)(-20.2 \text{ rad/s}^2)$$

$$\Rightarrow M_A = -33.2 \text{ N·m}$$

This is a plantar flexor action. The free-body diagram of the leg is shown here.

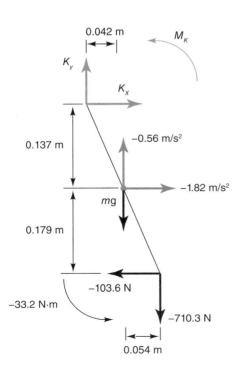

$$\sum F_x = ma_x$$

$$K_x - A_x = m_l a_l$$

$$\Rightarrow K_x = A_x + m_l a_l$$

$$\Rightarrow K_x = 103.6 \text{ N} + 2.4 \text{ kg}(-1.82 \text{ m/s}^2)$$

$$\Rightarrow K_x = 99.2 \text{ N}$$

$$\sum F_y = ma_y$$

$$K_y - A_y - mg = m_l a_l$$

$$\Rightarrow K_y = A_y + mg + m_l a_l$$

$$\Rightarrow K_y = -710.3 \text{ N} + 2.4 \text{ kg}(9.81 \text{ m/s}^2) + 2.4 \text{ kg}(-0.56 \text{ m/s}^2)$$

$$\Rightarrow K_y = -688.1 \text{ N}$$

When summing leg moments, note that the moments of both the vertical and horizontal forces are negative.

$$\sum M = I\alpha$$

$$M_K - M_A - K_x(0.137 \text{ m}) - K_y(0.042 \text{ m}) - A_x(0.179 \text{ m}) - A_y(0.054 \text{ m}) = I_l\alpha_l$$

$$\Rightarrow M_K = M_A + K_x(0.137 \text{ m}) + K_y(0.042 \text{ m}) + A_x(0.179 \text{ m}) + A_y(0.054 \text{ m}) + I_l\alpha_l$$

$$\Rightarrow M_K = -33.2 \text{ N·m} + (99.2 \text{ N})(0.137 \text{ m}) + (-688.1 \text{ N})(0.042 \text{ m}) + (103.6 \text{ N})(0.179 \text{ m}) + (-710.3 \text{ N})(0.054 \text{ m}) + (0.064 \text{ kg·m}^2)(-22.4 \text{ rad/s}^2)$$

$$\Rightarrow M_K = -69.8 \text{ N·m}$$

This is a knee flexor moment.

The free-body diagram of the thigh is as follows. Again note that we have placed the numerical values for the knee reactions into this diagram with the same signs that they were calculated as having:

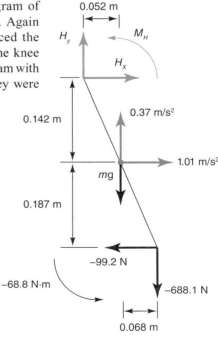

$$\sum_i F_x = ma_x$$

$$H_x - K_x = m_f a_t$$

$$\Rightarrow H_x = K_x + m_f a_t$$

$$\Rightarrow H_x = 99.2 \text{ N} + 6.0 \text{ kg}(1.01 \text{ m/s}^2)$$

$$\Rightarrow H_x = 105.3 \text{ N}$$

$$\sum_i F_y = ma_y$$

$$H_y - K_y - mg = m_f a_t$$

$$\Rightarrow H_y = K_y + mg + m_f a_t$$

$$\Rightarrow H_y = -688.1 \text{ N} + 6.0 \text{ kg}(9.81 \text{ m/s}^2)$$
$$+ 6.0 \text{ kg}(0.37 \text{ m/s}^2)$$

$$\Rightarrow H_y = -627.0 \text{ N}$$

As when we summed the leg moments, note that the moments of both the vertical and horizontal forces are negative:

$$\sum_i M = I\alpha$$

$$M_H - M_K - H_x(0.142 \text{ m}) - H_y(0.052 \text{ m})$$
$$- K_x(0.187 \text{ m}) - K_y(0.068 \text{ m}) = I_t \alpha_t$$

$$\Rightarrow MH = M_K + H_x(0.142 \text{ m}) + H_y(0.052 \text{ m})$$
$$+ K_x(0.187 \text{ m}) + K_y(0.068 \text{ m}) + I_t \alpha_t$$

$$\Rightarrow M_H = -69.8 \text{ N·m} + (105.3 \text{ N})(0.142 \text{ m})$$
$$+ (-627.0 \text{ N})(0.052 \text{ m}) + (99.2 \text{ N})(0.187 \text{ m})$$
$$+ (-688.1 \text{ N})(0.068 \text{ m}) + (0.130 \text{ kg·m}^2)(8.6 \text{ rad/s}^2)$$

$$\Rightarrow M_H = -114.6 \text{ N·m}$$

This is a hip extensor moment. Note how all three of these lower-extremity joint moments are much larger than their swing-phase counterparts.

CHAPTER 6

Example 6.1

First, convert the workload to newtons:

$$2.5 \times 9.81 = 24.53 \text{ N}$$

Second, calculate the number of revolutions of the crank:

$$R = 60 \times 20 = 1200$$

Finally, calculate the work:

$$W = 24.53 \times 1200 \times 6 = 176,616 \text{ J} = 176.6 \text{ kJ}$$

Notice, that the final answer was converted to kilojoules (kJ).

Example 6.2

$$W = \Delta E = E_{final} - E_{initial}$$

Assuming that the only change in energy results from a change in speed:

$$\text{Work} = 1/2 \, m \, v^2 - 0 = 1/2 \times 80.0 \times 6^2 = 1440 \text{ J}$$

$$\text{Power} = \text{Work/Duration} = 1440/4 = 360 \text{ W}$$

Example 6.3

First, compute the potential energy:

$$E_{gpe} = 18.0 \times 9.81 \times 1.20 = 212.9 \text{ J}$$

Second, calculate the translational kinetic energy:

$$E_{tke} = 1/2 \times 18.0 \times 8^2 = 576 \text{ J}$$

Third, calculate the rotational kinetic energy:

$$E_{rke} = 1/2 \times 0.50 \times 20.0^2 = 100.0 \text{ J}$$

Last, sum to obtain the total energy:

$$E_{tme} = 212.9 + 576 + 100 = 889 \text{ J}$$

APPENDIX C

Example C.1

Total resistance (R) is the product of the resistivity and the length:

$R = (10^{-4} \, \Omega/\text{m})(10^{-2} \text{ m})$, which equals $10^{-6} \, \Omega$.

Example C.2a

In series, the total resistance is $10 \, \Omega + 10 \, \Omega$, which equals $20 \, \Omega$. In parallel, the total resistance is

$$\frac{(10 \, \Omega)(10 \, \Omega)}{10 \, \Omega + 10 \, \Omega}$$

which equals $5 \, \Omega$.

Example C.2b

In series, the total resistance is $10 \, \Omega + 1 \, \Omega$, which equals $11 \, \Omega$. In parallel, the total resistance is

$$\frac{(10 \, \Omega)(1 \, \Omega)}{10 \, \Omega + 1 \, \Omega}$$

which equals $0.909 \, \Omega$.

Note that when resistors are in series, the total resistance must be greater than that of the largest resistor in the circuit. However, when resistors are connected in parallel, the total resistance must be less than that of the smallest resistor in the circuit.

Example C.3a

By Ohm's law,

$$I = \frac{V}{R} \Rightarrow I = \frac{20\ V}{10\ \Omega} \Rightarrow I = 2\ A$$

Example C.3b

By Ohm's law,

$$R = \frac{V}{I} \Rightarrow R = \frac{9\ V}{0.002\ A} \Rightarrow R = 4.54\ k\Omega$$

Example C.3c

By Ohm's law,

$$R = \frac{110\ V}{15\ A} \Rightarrow R = 7.3\ \Omega$$

Example C.4a

Rearranging the power law,

$$R = \frac{V^2}{P} \Rightarrow R = 10.1\ \Omega$$

Example C.4b

Using the power law,

$$P = (15\ A)(110\ V) \Rightarrow P = 1650\ W$$

Example C.5

The change in voltage will be given by Ohm's law:

$$V_{80} = 1000\ \Omega\ (10\ mA) = 10.0\ V$$

$$V_{120} = 1200\ \Omega\ (10\ mA) = 12.0\ V$$

This change is easily measured, because typical voltmeters measure to units at least as small as millivolts.

Example C.6

$$V_{out} = \frac{15\ V(100\ \Omega)}{100\ \Omega + 100\ \Omega} = 7.5\ V \qquad 1.$$

$$V_{out} = \frac{15\ V(110\ \Omega)}{100\ \Omega + 100\ \Omega} = 8.25\ V \qquad 2.$$

$$V_{out} = \frac{15\ V(100\ \Omega)}{10\ \Omega + 100\ \Omega} = 13.63\ V \qquad 3.$$

$$V_{out} = \frac{15\ V(110\ \Omega)}{10\ \Omega + 110\ \Omega} = 13.75\ V \qquad 4.$$

Voltage dividers are most sensitive to changes in one resistor when the two resistances are of similar magnitude. Note that between cases 1 and 2 and between cases 3 and 4, R_V changed by 10%. When R was 100 Ω, there was a 0.75 V (5%) change in V_{out} when R_V changed by 10%. However, when R was 10 Ω, the 10% change in R_V yielded only a 0.12 V (0.8%) change in V_{out}.

Example C.7

V_{in} is applied across each of two voltage dividers. From the previous example, we can write a formula for the voltage across R_2 as

$$V_2 = R_2 \frac{V_{in}}{R_V + R_2}$$

and we can write a formula for the voltage across R_3 as

$$V_3 = R_3 \frac{V_{in}}{R_1 + R_3}$$

V_{out} represents the difference in voltage between these two points, that is,

$$V_{out} = V_2 - V_3 = R_2 \frac{V_{in}}{R_V + R_2} - R_3 \frac{V_{in}}{R_1 + R_3}$$

Letting $R_1 = R_2 = R_3$ and doing some algebra yields

$$V_{out} = V_{in} \frac{R_1 - R_V}{2(R_1 + R_V)}$$

This is the classic formula for the output voltage of a Wheatstone bridge.

Glossary

acceleration—A rate of change of velocity and second time derivative of displacement; symbol is *a*. Measured in meters per second squared (m/s^2) and sometimes *g* (= 9.81 m/s^2).

accelerometer—An instrument that directly measures acceleration; used in impact testing and car crash studies.

accelerometry—Measurement of acceleration by an accelerometer.

action (force)—A force created when a reaction force is possible; see *law of reaction*.

ammeter—A device for measuring current or amperage.

ampere—An SI unit of electrical current; equals 1 coulomb of charge per second; symbol is A.

amplifier—A device for increasing the magnitude of an analog signal.

amplitude—Half of the peak-to-peak magnitude in a sinusoidal signal. The value *a* in the sinusoidal function, $W = a \sin(2\pi t + \theta)$. See also *frequency* and *phase angle*.

amplitude distortion—Any change in the true amplitude or magnitude of a signal.

analog—*(a)* A voltage-varying signal; *(b)* a continuous signal; *(c)* the opposite of digital, as in analog vs. digital signal, timepiece, or computer.

analog signal—A continuous electrical signal that has the same characteristics as another physical signal (e.g., force, pressure, acceleration).

analog-to-digital (A/D)—A process by which an analog signal is converted to a digital signal that is suitable to input into a digital computer or device.

analogue—A variant of *analog*.

angular impulse—The time integral of a resultant moment of force acting on a body.

angular momentum—A product of mass moment of inertia and angular velocity. Symbol is *L*.

anthropometry—The measurement of human physical dimensions and the relationships these measurements have with performance.

area moment of inertia—The second moment of a geometrical body.

attenuate—To reduce signal magnitude; opposite of *amplify*.

axis of rotation—The axis about which a body appears to rotate.

biomechanics—The science that studies the influence of forces on living bodies.

body segment parameters—The inertial or physical properties of body segments, especially mass, density, locations of center of mass and center of gravity, and moment of inertia.

capacitor—An electrical component that stores electrical charge; symbol is *C*.

center of gravity—The same as the *center of mass* when the body is near a large astronomical body (earth).

center of mass—The point at which any line passing through it divides the body's mass in half.

center of percussion—The point on a body at which a collision causes no pressure at the suspension point.

center of pressure—The point at which an equivalent single force causes the same effect on a rigid body as a distributed force.

central force—A force that is always directed at a single point in space; a force that acts through the center of gravity of a body.

centrifugal force—A pseudo-force that is "felt" when a body is following a curved path; the negative of the *centripetal force*.

centripetal force—A force that causes a body to follow a curved path; always directed toward the center of curvature of the path.

centroidal moment of inertia—The moment of inertia of a body about an axis through its center of mass.

cinematography—*(a)* Recording images on film; *(b)* the study of the factors that influence the quality of recording of film images.

circuit diagram—A formal means of representing an electric circuit, in which standard icons and straight lines are used.

coefficient of kinetic friction—The ratio of kinetic friction to normal force, $\mu_{kinetic}$.

coefficient of restitution—The ratio of changes in velocity of two undeformed bodies after and before colliding with each other.

coefficient of static friction—The ratio of maximum static friction to normal force, μ_{static}.

concentric (contraction)—A contraction in which the muscle force is directed toward the center of the muscle; that is, the muscle shortens while contracting.

contraction—The state a muscle is in when it has been induced internally (neurally) or externally (electrically) to shorten; requires neural or electrical stimulation and chemical energy, in the form of adenosine triphosphate or creatine phosphate, and produces an electromyographic signal and force in the tendon. Note that muscle can shorten (concentric contraction), lengthen (eccentric contraction), or remain the same length (isometric contraction) and still be in a state of contraction; however, a muscle can produce a force passively without being in contraction.

Coriolis force—A pseudo-force that appears when a body is moving within a moving frame of reference; for example, an airplane flying due north appears to follow a curved path with respect to the rotating earth.

coulomb—An SI unit of charge, corresponding to about 6.25 $\times$ 10^{18} electrons; its abbreviation is C.

crosstalk—Any distortion to a waveform caused by another nearby waveform. For example, one muscle's EMG may include a nearby muscle's EMG or a longitudinal force signal may be distorted by a torsional force signal.

current—The rate of flow of electrons or electricity; measured in amps.

cutoff frequency—The frequency at which a filter attenuates amplitude by –3 decibels (0.707 or $\sqrt{2}/2$).

decibel—One-hundredth of a bel, a unit for describing the ratio of two powers or intensities or for comparison to a reference power or intensity; abbreviated dB. Used in electronics and acoustics. For intensities, n dB = 20 $\log_{10}$(I1/I2). For powers, n dB = 10 $\log_{10}$(P1/P2). For example, an amplifier gain (intensity) of 1000 = 20 $\log_{10}$(1000/1) = 60 dB; a power gain of 20 dB is a gain of 100.

deformable body—A body that can deform under the influence of forces; an elastic body.

diagonal matrix—A matrix whose elements are all zero except elements that have the same row and column numbers.

digital—A numeric; can be represented by a number suitable for use by digital computers.

digital-to-analog (D/A)—A process by which a digital signal can be converted to an equivalent analog (voltage-varying) signal.

digitizer—A device for converting positional information to digital form, usually used to quantify motion from filmed or videotaped images.

direct dynamics—Derivation of kinematics from forces and moments of force.

displacement—The vector that quantifies change of linear position of a particle or a body's center of gravity; sometimes called *linear displacement*.

distortion—Any error introduced to a signal; see *amplitude, frequency,* and *phase distortion*.

dynamic response—The mechanical response of a physical system while under the influence of applied forces. Can also apply to analog systems or mathematical models; however, the applied forces become applied signals or mathematical functions.

dynamics—The mechanics of bodies in motion; see *direct dynamics* and *inverse dynamics*.

dynamometry—The measurement of mechanical forces, moments of force, and power.

eccentric (contraction)—A contraction in which the muscle force is directed away from the center of the muscle; that is, the muscle lengthens while contracting.

eccentric force or thrust—An impulsive force with a line of action that does not pass through the center of gravity of the body, causing angular acceleration.

electrocardiography (ECG)—A recording of the electrical potentials produced by the cardiac muscles.

electrogoniometer (elgon)—A goniometer that measures joint angles electronically; often consists of a potentiometer with two armatures.

electromyogram (EMG)—A recording from an electromyograph.

electromyograph—A device for measuring the electrical potentials produced by skeletal muscles; usually consists of a differential amplifier with high input impedance (10 MΩ) and high common mode rejection (>80 dB).

electromyography—The recording of electrical potentials produced by skeletal muscles.

energy—The ability to do work; can be potential or kinetic.

entropy—The loss of usable energy after any transformation of energy from one form to another.

ergometer—A device for measuring mechanical work or permitting human exercise (an exercise machine), such as bicycle or rowing ergometers.

ergometry—The measurement of mechanical work.

ergonomics—*(a)* The study of the factors influencing human work, especially in the workplace; *(b)* literally, work economics; *(c)* "fitting the task to the worker."

event—A unique instant in time, such as the heel-strike in walking, ball contact in striking or impacting activities, and the "catch" in rowing.

external force—Any environmental force that acts on a body.

force—The action of one body on another.

force couple—The turning effect of two parallel forces of equal magnitude but opposite direction; a free moment.

force platform (or plate)—An instrumented, rigid plate capable of quantifying forces applied to its surface.

free-body diagram (FBD)—A diagram of a body free from its environment but including all the external forces it experiences.

free moment—A moment of force caused by forces, especially force couples, at which the location of the axis of rotation is arbitrary.

frequency—The cyclic rate of a periodic signal in cycles per second or hertz (Hz). Can also mean an angular frequency (ω) in radians per second where $\omega = 2\pi f$. The value f in the sinusoidal function, $W = a \sin(2\pi ft + \theta)$. See also *amplitude* and *phase angle*.

frequency distortion—Any change in the frequency spectrum of a signal.

frequency response—The frequency spectrum of a system or device in response to a mechanical stimulus. Can also apply to analog systems or mathematical models.

friction—A force caused by the transverse (i.e., shearing) interactions of two surfaces.

gain—In an amplifier, the ratio of the original voltage to the amplified voltage.

g-force—A pseudo-force that occurs when a body is rapidly accelerated; results from the inertia of the body, that is, its reluctance to accelerate in response to the applied force.

goniometer—A device for measuring joint angles; see *electrogoniometer*.

ground—The electrical reference point assigned zero voltage.

ground reaction force (GRF)—A single equivalent force equal to the sum of a distribution of forces applied to a surface.

Hall-effect (transducer)—An effect due to the movement of electrons perpendicularly to a permanent magnet; used to quantify force.

hysteresis—The maximum difference between the loading and unloading curves of a *transducer*.

impedance—The sum of all effects on current flow, including resistance and capacitance.

impulse—The time integral of the resultant force acting on a body.

inertia—The reluctance of a body to change its state of rest or motion along a straight line; measured by mass and moment of inertia.

inertial force—A force equal to the negative of the resultant force; used with d'Alembert's principle (the sum of the resultant force and inertial force equals zero).

input impedance—The resistance between the input of a circuit and its ground.

internal force—A force whose action and reaction occur within the same body; a muscle force whose origin and insertion act within the same free body.

inverse dynamics—Computation of forces or moments of force from a body's kinematics and inertial properties.

invert—In electronics, to take the negative of a voltage.

isokinetic (contraction)—Contraction of a joint at which the joint angular speed is constant (compare *isovelocity contraction*). It can also mean a contraction that is produced by a constant moment of force.

isometric (contraction)—A constant-length contraction, meaning a muscle contraction in which the muscle has no appreciable change in its length.

isotonic (contraction)—*(a)* A contraction of an excised (in vitro) muscle in which the muscle contracts against a constant load; *(b)* weightlifting, that is, a whole muscle (and joint) contraction against a constant load, such as a weight, barbell, or dumbbell; *(c)* a contraction in which a muscle contracts against an artificially produced constant load.

isovelocity (contraction)—A muscle contraction in which the muscle shortens or lengthens at a constant velocity or speed or when a joint's angular velocity is constant (compare with *isokinetic contraction*).

kilopond—A force equal to a 1 kg mass on earth; equal to 9.81 N.

kinematics—The study of motion without regard to its causes or quantities of motion, such as velocity, speed, acceleration, and angular displacement.

kinetic friction—Dry friction that occurs when two contacting surfaces are in motion.

kinetics—The study of the causes of motion; the study of forces and moments of force and their characteristics, such as work, energy, impulse, momentum, and power.

law of acceleration—Newton's second law; the acceleration of an object is proportional to the sum of the external forces.

law of gravitation—See *universal law of gravitation*.

law of inertia—Newton's first law; in the absence of an external force, an object remains motionless or in constant speed along a straight line.

law of reaction—Newton's third law; for a force to exist, there must be a reaction force equal in magnitude but opposite in direction.

limiting static friction force—The maximum static friction force before two contacting surfaces "slip."

linearity—The ability of a transducer to produce an output signal that is directly proportional to the input amplitude; the closeness of the relationship between a transducer's input and output signals to a straight line as measured statically or at low frequency. Measured by Pearson's product moment correlation coefficient *(r)*, which is the same as a linear least-squares curve fit.

linear potentiometer—A resistor with a fixed connection on each end and a sliding connection between the two ends so that translation of the sliding connection alters the resistance between it and each end.

local angular momentum—The product of a body's centroidal mass moment of inertia and its angular velocity.

mass moment of inertia—The second moment of mass of a body about a particular axis.

matrix—Any rectangular array of numbers in which each number in the array is an element.

mechanics—A science that studies the influence of forces on bodies.

moment (arm)—*(a)* The perpendicular distance from a point to a line or surface; radius; the moment arm in the moment of force, moment of momentum, and moment of inertia; *(b)* short for moment of force.

moment of force—The turning effect of a force on a body, also called torque. Symbol is M or τ. Measured in newton meters (N·m).

moment of inertia—The reluctance of a body to change its rotational state. Symbol is I or $\bar{I}$. Measured in kilogram meters squared (kg·m^2).

moment of inertia tensor—A 3×3 matrix corresponding to the three-dimensional moment of inertia of a body.

moment of momentum—Vector product of position and linear momentum (i.e., $\vec{r} \times m\vec{v}$). Same as remote angular momentum.

momentum—A product of mass and linear velocity; Newton's quantity of motion. Symbol is p.

motion-analysis system—A system for collecting and processing the motion of sensors or markers attached to a body.

motion capture—Any process that records the motion of a body over time. Some systems require that the body have reflective (i.e., passive) markers; other systems require transmitters (i.e., active) to send location information; still others may simply record the motion visually for later processing to determine the body's trajectory.

natural frequency—The frequency at which an object resonates or oscillates most strongly to an applied force, also called the *resonant frequency*.

net force—A force equivalent to the sum of all forces acting across a joint.

net moment of force—A moment of force equivalent to the sum of all moments of force acting across a joint.

noise—Any unwanted random or systematic component in a waveform. Random noise that is uncorrelated with the true signal can be reduced by filtering or averaging. Systematic noise is caused by interference produced by external sources and can be reduced in a variety of ways, including removal of the source of the interference or shielding the electronics.

normal—*(a)* Perpendicular to a surface or line; *(b)* perpendicular to tangential.

normal force—The force component that is perpendicular to a surface (compare *tangential force).*

normalize—To perform a form of scaling involving division of a set of numbers by a factor such as body weight, cycle time, or maximum force.

ohm—An SI unit of electrical resistance; symbol is Ω.

Ohm's law—A law stating the linear relation between voltage and current in a linear circuit, $V = IR$.

operational amplifier—A specific type of electronic component that amplifies a voltage.

parallel—In electronics, a connection scheme in which the corresponding ends of two or more devices are connected so that the electrical current branches through one or another of the devices.

parallelogram law—A law that defines the addition of vectors; the resultant of two vectors is the diagonal of a parallelogram formed from the two vectors.

phase—A period of time, such as a swing phase or recovery phase.

phase angle—The amount of lead or lag of a sinusoidal waveform compared with a second sinusoidal waveform of the same frequency, used in a Fourier series; measured in degrees or radians. The value θ in the sinusoidal function, $W = a \sin(2\pi f t + \theta)$. See also frequency and phase angle. Measured in degrees (°) or radians (rad).

phase distortion—Any change in the phase angle or time delays in a signal.

piezoelectric effect—Occurs when certain crystals, such as quartz, are mechanically stressed, causing a voltage.

piezoelectric (transducer)—An effect from pressure exerted on certain crystals that causes them to produce a voltage; used in force and acceleration transducers and precision timepieces.

potentiometer—An electronic device that permits variable resistance; used in electrogoniometers, amplifier controls, volume controls, and other similar devices.

power—Rate of doing mechanical work, symbol is P. Measured in watts (W).

pressure—The force exerted over an area; symbol is p; units are pascals (Pa).

principal distance—The camera's distance setting, which should be set equal to the distance between the lens and the object being filmed.

principal mass moments of inertia—The diagonal elements of an inertia tensor.

products of inertia—The off-diagonal elements of an inertia tensor.

pseudo-force—An apparent force that exists as the result of a moving frame of reference; includes *centrifugal forces, g-forces,* and *Coriolis forces.*

radius of gyration—The radius of a point mass that has the same mass moment of inertia as a particular body.

reaction (force)—A force that occurs whenever an action force is created (see *law of reaction).*

refine—To scale or otherwise transform digitized motion-picture data to real units in a known frame of reference.

remote angular momentum—The same as *moment of momentum.*

resistance—In electronics, the effect of a particular device or component that is directly proportional to the voltage applied across it. Symbol is R. Measured in ohms (Ω).

resistivity—A material property that expresses the electrical resistance per unit length.

resistor—A device that limits current flow in direct proportion to the voltage across it.

resultant—The vector sum of two or more vectors; see *parallelogram law.*

resultant force—The sum of all forces acting on a body; also called *external force.*

Riemann integration—Integration by adding the elements and multiplying by the duration of the sample.

rigid body—A group of particles occupying fixed positions with respect to each other; a theoretical body that is not deformable and has fixed inertial properties.

rotary potentiometer—A resistor with a fixed connection on each end and a sliding connection between the two ends so that rotating the control alters the resistance between the sliding connection and each end.

scalar—A quantity that can be characterized by a magnitude alone (e.g., mass, distance, speed, time, energy, and power); compare *vector.*

scale—To alter the magnitude of a digital signal; to multiply by a constant; compare *normalize, refine.*

sensitivity—For transducers, this is the ratio of the input signal divided by the output signal, for example, 100 N/V.

series—In electronics, a connection between two or more components in which one follows the other so that all current passes through each component.

signal—The information content of a waveform; the opposite of noise.

Simpson's rule integration—A method of integration based on Simpson's rule approximation of a series of data.

spatial synchronization—The synchronization of two or more data sets in space.

static friction—Dry friction when there is no relative motion of two contacting surfaces.

statics—The mechanics of bodies at rest or in uniform (constant linear) motion.

strain—The change in length divided by resting length; normalized deformation.

strain gauge—A resistor-based device designed to be attached to the surface of a material so that its resistance changes as the material deforms.

stress—The loading force per cross-sectional area; measured in kilopascals (kPa); the normalized load force.

tangent—*(a)* A line that is perpendicular to the normal surface or curve; *(b)* the slope of a line or rise/run; *(c)* the tangent of an angle (tan θ); the ratio of opposite to adjacent sides of a right triangle.

tangential—The direction that is parallel to the tangent line of a curved path; perpendicular to the normal.

tangential force—The force component that is perpendicular to a *normal force* and parallel to a surface.

telemetry—The transmission of a signal over a distance, usually by radio transmission.

temporal—Relating to time; in the time domain.

temporal synchronization—Synchronization of two or more data sets in time.

tensor—A mathematical or physical quantity possessing a specified system of components for every coordinate system; a generalized vector with more than three components, each of which is a function of the coordinates of an arbitrary point in space of an appropriate number of dimensions.

thermodynamics—The branch of science concerned with transduction of heat and energy.

torque—See *moment of force*. Used especially when the moment of force is about the longitudinal axis of a body.

trajectory—The path that an object or point travels through space.

transducer—A device that is actuated by power from one system and supplies power, usually in another form, to a second system; a device that changes one form of energy to another. An input transducer converts a physical signal, such as force, temperature, or power, into an electrical signal, usually voltage. An output transducer converts an electrical signal into a physical quantity, as, for example, loudspeakers, oscillographs, and multimeters do.

transduction—The process of converting a physical dimension into a voltage.

trapezoidal integration—A method of integrating a series of numbers by adding adjacent trapezoids.

universal law of gravitation—A law stating that two objects have a force of attraction proportional to the product of their masses divided by the square of the distance between their centers of mass.

vector—A mathematical expression possessing magnitude and direction that adds according to the *parallelogram law*; examples are force, acceleration, and displacement, but not finite rotations; compare *scalar*.

velocity—A vector rate of change of displacement that includes the direction of motion; symbol is *v*. Measured in meters per second (m/s).

voltage—An SI unit of electrical potential, equal to 1 J of energy per coulomb of charge; symbol is *V* or *E*. Measured in volts (V).

voltage divider—A circuit of two or more resistors in series, with their object being to use or measure voltage at an intermediate point.

voltage drop—The voltage across a resistor.

voltmeter—A device that measures voltage.

waveform—Any continuously varying quantity consisting of signal or noise components.

weight—A force due to the gravitational attraction of a massive body such as the earth or moon.

Wheatstone bridge—An electrical circuit including two parallel pairs of series-connected resistors.

work—The change in energy of a body. Measured in joules (J).

References

Abdel-Aziz, Y.I., and H.M. Karara. 1971. Direct linear transformation from comparator coordinates into object space coordinates in close-range photogrammetry. In *American Society for Photogrammetry Symposium on Close Range Photogrammetry.* Falls Church, VA: American Society for Photogrammetry.

Albert, W.J., and D.I. Miller. 1996. Takeoff characteristics of single and double axel figure skating jumps. *Journal of Applied Biomechanics* 12:72-87.

Aleshinsky, S.Y. 1986a. An energy "sources" and "fractions" approach to the mechanical energy expenditure problem. I: Basic concepts, description of the model, analysis of a one-link system movement. *Journal of Biomechanics* 19:287-94.

Aleshinsky, S.Y. 1986b. An energy "sources" and "fractions" approach to the mechanical energy expenditure problem. IV: Criticism of the concept of "energy transfers within and between links." *Journal of Biomechanics* 19:307-9.

Aleshinsky, S.Y. 1986c. An energy "sources" and "fractions" approach to the mechanical energy expenditure problem. V: The mechanical energy expenditure deduction during motion of a multi-link system. *Journal of Biomechanics* 19:310-5.

Alexander, R.M. 1990. Optimum take-off techniques for high and long jumps. *Philosophical Transactions of the Royal Society of London B* 329:3-10.

Alexander, R.M. 1992. Simple models of walking and jumping. *Human Movement Science* 11:3-9.

Alexander, R.M., M.B. Bennett, and R.F. Ker. 1986. Mechanical properties and function of the paw pads of some mammals. *Journal of Zoology (London)* 209:405-19.

Allard, P., A. Cappozzo, A. Lundberg and C. Vaughan. 1998. *Three-Dimensional Analysis of Human Locomotion.* Chichester, UK: John Wiley & Sons.

Allard, P., R. Lachance, R. Aissaoui, and M. Duhaime. 1996. Simultaneous bilateral able-bodied gait. *Human Movement Science* 15:327-46.

Ambroz, C., A. Scott, A. Ambroz, and E.O. Talbott. 2000. Chronic low back pain assessment using surface electromyography. *Journal of Occupational and Environmental Medicine* 42:660-9.

American Society for Photogrammetry and Remote Sensing. 1980. *Manual of Photogrammetry,* 4th ed. Falls Church, VA: American Society for Photogrammetry and Remote Sensing.

An, K.-N., and E.Y.S. Chao. 1991. Kinematic analysis. In *Biomechanics of the Wrist Joint,* eds. K.-N. An, R.A. Berger, and W.P. Cooney. New York: Springer-Verlag.

An, K.N., K. Takahashi, T.P. Harrigan, and E.Y. Chao. 1984. Determination of muscle orientations and moment arms. *Journal of Biomechanical Engineering* 106:280-2.

Anderson, F.C., and M.G. Pandy. 2001. Dynamic optimization of human walking. *Journal of Biomechanical Engineering.* 123:381-390.

Anderson, F.C., and M.G. Pandy. 2003. Individual muscle contributions to support in normal walking. *Gait and Posture* 17:159-69.

Andersson, E.A., J. Nilsson, and A. Thorstensson. 1997. Intramuscular EMG from the hip flexor muscles during human locomotion. *Acta Physiologica Scandinavica* 161:361-70.

Andreassen, S., and L. Arendt-Nielsen. 1987. Muscle fibre conduction velocity in motor units of the human anterior tibial muscle: A new size principle parameter. *Journal of Physiology* 391:561-71.

Andreassen, S., and A. Rosenfalck. 1978. Recording from a single motor unit during strong effort. *IEEE Transactions on Biomedical Engineering* 25:501-8.

Andriacchi T.P., E.J. Alexander, M.K. Toney, C. Dyrby, and J. Sum. 1998. A point cluster method for in vivo motion analysis: Applied to a study of knee kinematics. *Journal of Biomechanical Engineering* 120:743-9.

Andriacchi, T.P., B.J. Andersson, R.W. Fermier, D. Stern, and J.O. Galante. 1980. A study of lower limb mechanics during stair climbing. *Journal of Bone and Joint Surgery* 62:749-57.

Aoki, F., H. Nagasaki, and R. Nakamura. 1986. The relation of integrated EMG of the triceps brachii to force in rapid elbow extension. *Tohoku Journal of Experimental Medicine* 149:287-91.

Arampatzis, A., S. Stafilidis, G. De Monte, K. Karamanidis, G. Morey-Klapsing, and G.P. Bruggemann. 2005. Strain and elongation of the human gastrocnemius tendon and aponeurosis during maximal plantar flexion effort. *Journal of Biomechanics* 38:833-41.

Arnold, A.S., S. Salinas, D.J. Asakawa, and S.L. Delp. 2010. Accuracy of muscle moment arms estimated from MRI-based musculoskeletal models of the lower extremity. *Computer Aided Surgery* 5:108-119.

Astephen, J.L., K.J. Deluzio, G.E. Caldwell, and M.J. Dunbar. 2008. Biomechanical changes at the hip, knee, and ankle joints during gait are associated with knee osteoarthritis severity. *Journal of Orthopedic Research* 26(3):332-41.

Audu, M.L., and D.T. Davy. 1985. The influence of muscle model complexity in musculoskeletal motion modeling. *Journal of Biomechanical Engineering* 107:147-57.

Audu, M.L., and D.T. Davy. 1988. A comparison of optimal control algorithms for complex bioengineering studies. *Optimal Control Applications and Methods* 9:101-6.

Bach, T.M. 1995. Optimizing mass and mass distribution in lower limb prostheses. *Prosthetics and Orthotics Australia* 10:29-34.

Bahler, A.S. 1967. Series elastic component of mammalian skeletal muscle. *American Journal of Physiology* 213:1560-4.

Baildon, R., and A.E. Chapman. 1983. A new approach to the human muscle model. *Journal of Biomechanics* 16:803-9.

Barkhaus, P.E., and S.D. Nandedkar. 1994. Recording characteristics of the surface EMG electrodes. *Muscle and Nerve* 17:1317-23.

Barter, J.T. 1957. *Estimation of the Mass of Body Segments.* WADC Technical Report 57-260. Wright-Patterson Air Force Base, OH.

Bartolo, A., R.R. Dzwonczyk, C. Roberts, and E. Goldman. 1996. Description and validation of a technique for the removal of ECG contamination from diaphragmatic EMG signal. *Medical and Biological Engineering and Computing* 34:76-81.

Basmajian, J.V. 1989. *Biofeedback: Principles and Practices for Clinicians.* Baltimore: Lippincott Williams & Wilkins.

Basmajian, J.V., and G. Stecko. 1962. A new bipolar electrode for electromyography. *Journal of Applied Physiology* 17:849.

Batschelet, E. 1981. *Circular Statistics in Biology.* London: Academic Press.

Beer, F.P., E.R. Johnston, Jr., D.F. Mazurek, P.J. Cornwell, and E.R. Eisenberg. 2010. *Vector Mechanics for Engineers; Statics and Dynamics.* 9th ed. Toronto: McGraw-Hill.

Bell, A.L. and R.A. Brand. 1989. Prediction of hip joint centre location from external landmarks. *Human Movement Science* 8:3-16.

Bell, A.L., D.R. Pederson, and R.A. Brand. 1990. A comparison of the accuracy of several hip centre location prediction methods. *Journal of Biomechanics* 23:717-21.

Bellemare, F., and N. Garzaniti. 1988. Failure of neuromuscular propagation during human maximal voluntary contraction. *Journal of Applied Physiology* 64:1084-93.

Bernstein, N. 1967. *Co-ordination and Regulation of Movements.* London: Pergamon Press.

Bernstein, N.A. 1996. On dexterity and development. In *Dexterity and Its Development,* eds. M.L. Latash and M.T. Turvey, 3-244. Mahwah, NJ: Erlbaum.

Betts, B., and J.L. Smith. 1979. Period-amplitude analysis of EMG from slow and fast extensors of cat during locomotion and jumping. *Electroencephalography and Clinical Neurophysiology* 47:571-81.

Biedermann, H., G. Shanks, W. Forrest, and J. Inglis. 1991. Power spectrum analyses of electromyographic activity: Discriminators in the differential assessment of patients with chronic low-back pain. *Spine* 16:1179-84.

Biewener, A.A., D.D. Konieczynski, and R.V. Baudinette. 1998. *In vivo* muscle force-length behavior during steady-speed hopping in tammar wallabies. *Journal of Experimental Biology* 201:1681-94.

Biewener, A.A., and T.J. Roberts. 2000. Muscle and tendon contributions to force, work and elastic energy savings: A comparative perspective. *Exercise and Sport Sciences Reviews* 28:99-107.

Bigland-Ritchie, B., R. Johansson, O.C.J. Lippold, S. Smith, and J.J. Woods. 1983. Changes in motoneurone firing rates during sustained maximal voluntary contractions. *Journal of Physiology* 340:335-46.

Bilodeau, M., A.B. Arsenault, D. Gravel, and D. Bourbonnais. 1990. The influence of an increase in the level of force on the EMG power spectrum of elbow extensors. *European Journal of Applied Physiology* 61:461-6.

Blemker, S.S., D.S. Asakawa, G.E. Gold, and S.L. 2007. Image-based musculoskeletal modeling: Applications, advances, and future opportunities. *Journal of Magnetic Resonance Imaging* 25:441-51.

Blemker, S.S., and S.L. Delp. 2005. Three-dimensional representation of complex muscle architectures and geometries. *Annals of Biomedical Engineering* 33:661-673.

Blemker, S.S., and S.L. Delp. 2006. Rectus femoris and vastus intermedius fiber excursions predicted by three-dimensional muscle models. *Journal of Biomechanics* 39:1383-91.

Bobbert, M.F., P.A. Huijing, and G.J. van Ingen Schenau. 1986a. A model of the human triceps surae muscle-tendon complex applied to jumping. *Journal of Biomechanics* 19:887-98.

Bobbert, M.F., P.A. Huijing, and G.J. van Ingen Schenau. 1986b. An estimation of power output and work done by the human triceps surae muscle-tendon complex in jumping. *Journal of Biomechanics* 19:899-906.

Bobbert, M.F., and G.J. van Ingen Schenau. 1990. Isokinetic plantar flexion: Experimental results and model calculations. *Journal of Biomechanics* 23:105-19.

Bobbert, M.F., and J.P. van Zandwijk. 1999. Dynamics of force and muscle stimulation in human vertical jumping. *Medicine and Science in Sports and Exercise* 31:303-10.

Bogduk, N., J.E. Macintosh, and M.J. Pearcy. 1992. A universal model of the lumbar back muscles in the upright position. *Spine (Phila Pa 1976)* 17:897-913.

Bogey, R.A., J. Perry, E.L. Bontrager, and J.K. Gronley. 2000. Comparison of across-subject EMG profiles using surface and multiple indwelling wire electrodes during gait. *Journal of Electromyography and Kinesiology* 10:255-9.

Bouisset, S., and B. Maton. 1972. Quantitative relationship between surface EMG and intramuscular electromyographic activity in voluntary movement. *American Journal of Physical Medicine* 51:285-95.

Bramble, D.M., and D.R. Carrier. 1983. Running and breathing in mammals. *Science* 219:251-6.

Brandon, S.C., and K.J. Deluzio. 2011. Robust features of knee osteoarthritis in joint moments are independent of reference frame selection. *Clinical Biomechanics (Bristol, Avon)* 26:65-70.

Braune, W., and O. Fischer. 1889. Über den Schwerpunkt des menschlichen Körpers, mit Rücksicht auf die Ausrüstung des deutschen Infanteristen [The center of gravity of the human body as related to the equipment of the German Infantry]. *Abhandlungen der mathematischphysischen Klasse der Königlich-Sächsischen Gesellschaft der Wissenschaften* 26:561-672.

Bresler, B., and F.R. Berry. 1951. *Energy and Power in the Leg during Normal Level Walking.* Prosthetic Devices Research Project, Series II, Issue 15. Berkeley: University of California.

Bresler, B., and J.P. Frankel. 1950. The forces and moments in the leg during level walking. *Transactions of the American Society of Mechanical Engineers* 72:27-36.

Brismar, T., and L. Ekenvall. 1992. Nerve conduction in the hands of vibration exposed workers. *Electroencephalography and Clinical Neurophysiology* 85:173-6.

Broker, J.P., and R.J. Gregor. 1990. A dual piezoelectric force pedal for kinetic analysis of cycling. *International Journal of Sport Biomechanics* 6:394-403.

Brooks, C.B., and A.M. Jacobs. 1975. The gamma mass scanning technique for inertial anthropometric measurement. *Medicine and Science in Sports* 7:290-4.

Brown, T.D., L. Sigal, G.O. Njus, N.M. Njus, R.J. Singerman, and R.A. Brand. 1986. Dynamic performance characteristics of the liquid metal strain gage. *Journal of Biomechanics* 19:165-73.

Buchanan, T.S., D.J. Almdale, J.L. Lewis, and W.Z. Rymer. 1986. Characteristics of synergic relations during isometric contractions of human elbow muscles. *Journal of Neurophysiology* 56:1225-41.

Buchanan, T.S., D.G. Lloyd, K. Manal, and T.F. Besier. 2004. Neuromusculoskeletal modeling: Estimation of muscle forces and joint moments and movements from measurements of neural command. *Journal of Applied Biomechanics* 20:367-95.

Buchanan, T.S., G.P. Rovai, and W.Z. Rymer. 1989. Strategies for muscle activation during isometric torque generation at the human elbow. *Journal of Neurophysiology* 62:1201-12.

Buchthal, F., and P. Rosenfalck. 1958. Rate of impulse conduction in denervated human muscle. *Electroencephalography and Clinical Neurophysiology* 10:521-6.

Buczek F., T. Kepple, S. Stanhope, and K.L. Siegel. 1994. Translational and rotational joint power terms in a six-degree-of-freedom model of the normal ankle complex. *Journal of Biomechanics* 27:1447-57.

Burgar, C.G., F.J. Valero-Cuevas, and V.R. Hentz. 1997. Finewire electromyographic recording during force generation: Application to index finger kinesiologic studies. *American Journal of Physical Medicine* 76:494-501.

Burgess-Limerick, R., B. Abernethy, and R.J. Neal. 1993. Relative phase quantifies interjoint coordination. *Journal of Biomechanics* 26:91-4.

Bustami, F.M. 1986. A new description of the lumbar erector spinae muscle in man. *Journal of Anatomy* 144:81-91.

Caldwell, G.E. 1995. Tendon elasticity and relative length: Effects on the Hill two-component muscle model. *Journal of Applied Biomechanics* 11:1-24.

Caldwell, G.E., W.B. Adams, and M.L. Whetstone. 1993. Torque/velocity properties of human knee muscles: Peak and angle-specific estimates. *Canadian Journal of Applied Physiology* 18(3):274-90.

Caldwell, G.E., and A.E. Chapman. 1989. Applied muscle modeling: Implementation of muscle-specific models. *Computers in Biology and Medicine* 19:417-34.

Caldwell, G.E., and A.E. Chapman. 1991. The general distribution problem: A physiological solution which includes antagonism. *Human Movement Science* 10:355-92.

Caldwell, G.E., and L.W. Forrester. 1992. Estimates of mechanical work and energy transfers: Demonstration of a rigid body power model of the recovery leg in gait. *Medicine and Science in Sports and Exercise* 24:1396-412.

Caldwell, G.E., L. Li, S.D. McCole, and J.M. Hagberg. 1998. Pedal and crank kinetics in uphill cycling. *Journal of Applied Biomechanics* 14:245-59.

Cappozzo, A., A. Cappello, U. Della Croce, and F. Pensalfini. 1997. Surface-marker cluster design criteria for 3D bone movement reconstruction. *IEEE Transactions on Biomedical Engineering* 44(2):1165-74.

Cappozzo, A., F. Catani, U. Della Croce, and A. Leardini. 1995. Position and orientation in space of bones during movement: Anatomical frame definition and determination. *Clinical Biomechanics* 10:171-78.

Cappozzo, A., F. Catani, A. Leardini, M.G. Benedetti, and U. Della Croce. 1996. Position and orientation in space of bones during movement: experimental artifacts. *Clinical Biomechanics* 11(2):90-100.

Cappozzo, A., F. Figura, M. Marchetti, and A. Pedotti. 1976. The interplay of muscular and external forces in human ambulation. *Journal of Biomechanics* 9:35-43.

Cappozzo, A., P.F. La Palombara, L. Luchetti, and A. Leardini. 1996. Multiple anatomical landmark calibration for optimal bone pose estimation. *Human Movement Science* 13:259-74.

Cappozzo, A., T. Leo, and A. Pedotti. 1975. A general computing method for the analysis of human locomotion. *Journal of Biomechanics* 8:307-20.

Carrière, L., and A. Beuter. 1990. Phase plane analysis of biarticular muscles in stepping. *Human Movement Science* 9:23-5.

Cavagna, G.A., N.C. Heglund, and C.R. Taylor. 1977. Mechanical work in terrestrial locomotion: Two basic mechanisms for minimizing energy expenditure. *American Journal of Physiology* 233:R243-61.

Cavagna, G.A., and M. Kaneko. 1977. Mechanical work and efficiency in level walking and running. *Journal of Physiology* 268:467-81.

Cavagna, G.A., L. Komarek, and S. Mazzoleni. 1971. The mechanics of sprint running. *Journal of Physiology* 217:709-21.

Cavagna, G.A., F.P. Saibene, and R. Margaria. 1963. External work in walking. *Journal of Applied Physiology* 18:1-9.

Cavagna, G.A., F.P. Saibene, and R. Margaria. 1964. Mechanical work in running. *Journal of Applied Physiology* 19:249-56.

Cavanagh, P.R. 1978. A technique for averaging center of pressure paths from a force platform. *Journal of Biomechanics* 11:487-91.

Cereatti, A., U. Della Croce, and A. Cappozzo. 2006. Reconstruction of skeletal movement using skin markers: Comparative assessment of bone pose estimators. *Journal of Neuroengineering and Rehabilitation* 3(7):3-7.

Chandler, R.F., C.E. Clauser, J.T. McConville, H.M. Reynolds, and J.W. Young. 1975. *Investigation of Inertial Properties of the Human Body.* AMRL Technical Report 74-137. Wright-Patterson Air Force Base, OH.

Chang, R., R.E.A. van Emmerik, and J. Hamill. 2008. Quantifying rearfoot-forefoot coordination in human walking. *Journal of Biomechanics* 41:3101-5.

Chao, E.Y.S. 1980. Justification for a triaxial goniometer for the measurement of joint rotation. *Journal of Biomechanics* 13:989-1006.

Chao E.Y.S., R.K. Laughman, E. Schneider, and R.N. Stauffer. 1983. Normative data of knee joint motion and ground reaction forces in adult level walking. *Journal of Biomechanics* 16:219-33.

Chapman, A.E. 1985. The mechanical properties of human muscle. *Exercise and Sport Sciences Reviews* 13:443-501.

Chapman, A.E. 2008. *Biomechanical Analysis of Fundamental Human Movements.* Champaign, IL: Human Kinetics.

Chapman, A.E., G. Caldwell, R. Herring, R. Lonergan, and S. Selbie. 1987. Mechanical energy and the preferred style of running. In *Biomechanics X-B*, ed. B. Jonsson, 875-9. International Series on Biomechanics. Champaign, IL: Human Kinetics.

Chapman, A.E., G.E. Caldwell, and W.S. Selbie. 1985. Mechanical output following muscle stretch in forearm supination against inertial loads. *Journal of Applied Physiology* 59(1):78-86.

Chapman, A.E., B. Vicenzino, P. Blanch, and P. Hodges. 2008. Is running less skilled in triathletes than runners matched for running training history? *Medicine and Science in Sports and Exercise* 40:557-65.

Chau, T. 2001a. A review of analytical techniques for gait data. Part 1: Fuzzy, statistical and fractal methods. *Gait and Posture* 13:49-66.

Chau, T. 2001b. A review of analytical techniques for gait data. Part 2: Neural network and wavelet methods. *Gait and Posture* 13: 102-20.

Chen, G. 2006. Induced acceleration contributions to locomotion dynamics are not physically well defined. *Gait and Posture* 23:37-44.

Chen, J.J., T.Y. Sun, T.H. Lin, and T.S. Lin. 1997. Spatiotemporal representation of multichannel EMG firing patterns and its clinical applications. *Medical Engineering and Physics* 19:420-30.

Cholewicki, J., and S.M. McGill. 1996. Mechanical stability of the in vivo lumbar spine: Implications for injury and chronic low back pain. *Clinical Biomechanics* 11:1-15.

Cholewicki, J., M.M. Panjabi, and A. Khachatryan. 1997. Stabilizing function of trunk flexor-extensor muscles around a neutral spine posture. *Spine (Phila Pa 1976)* 22:2207-12.

Clamann, H.P., and K.T. Broecker. 1979. Relation between force and fatigability of red and pale skeletal muscles in man. *American Journal of Physical Medicine* 58:70-85.

Clamann, H.P., and T.B. Schelhorn. 1988. Nonlinear force addition of newly recruited motor units in the cat hindlimb. *Muscle and Nerve* 11:1079-89.

Clarkson, D.B., C. Fraley, C. Gu, and J.O. Ramsay. 2005. *S+ Functional Data Analysis.* New York: Springer.

Clauser, C.E., J.T. McConville, and J.W. Young. 1969. *Weight, Volume and Center of Mass of Segments of the Human Body.* AMRL Technical Report 60-70. Wright-Patterson Air Force Base, OH.

Coffey, N., A.J. Harrison, O.A. Donoghue, and K. Hayes. 2011. Common Functional Principal Component Analysis: A new approach to analyzing human movement data. *Human Movement Science* 30(11): 1144-1166. doi:10.1016/j.humov.2010.11.005

Cole, G.K., B.M. Nigg, J.L. Ronsky, and M.R. Yeadon. 1993. Application of the joint coordinate system to three-dimensional joint attitude and movement representation: A standardization proposal. *Journal of Biomechanical Engineering* 115:344-9.

Connolly, S., D.G. Smith, D. Doyle, and C.J. Fowler. 1993. Chronic fatigue: Electromyographic and neuropathological evaluation. *Journal of Neurology* 240:435-8.

Contini, R. 1972. Body segment parameters, part II. *Artificial Limbs* 16:1-19.

Corazza, L., and T.P. Andriacchi. 2009. Posturographic analysis through markerless motion capture without ground reaction forces measurement. *Journal of Biomechanics.* 42:370-4.

Corazza, L., S. Mündermann, and T.P. Andriacchi. 2007. Accurately measuring human movement using articulated ICP with soft-joint constraints and repository of articulated model. In *IEEE International Conference on Computer Vision and Pattern Recognition.* 1-6.

Cram, J.R., G.S. Kasman, and J. Holtz. 1998. *Introduction to Surface Electromyography.* Gaithersburg, MD: Aspen.

Craven, P., and G. Wahba. 1979. Smoothing noisy data with spline functions: Estimating the degree of smoothing by method of generalized cross-validation. *Numerische Mathematik* 31:377-403.

Cronin, A., and D.G.E. Robertson. 2000. Two-segment model of the foot: Reduction of power imbalance. *Archives of Physiology and Biochemistry* 108:120.

Crowninshield, R.D., and R.A. Brand. 1981a. A physiologically based criterion of muscle force prediction in locomotion. *Journal of Biomechanics* 14:783-801.

Crowninshield, R.D., and R.A. Brand. 1981b. The prediction of forces in joint structures: Distribution of intersegmental resultants. *Exercise and Sport Sciences Reviews* 9: 159-81.

Cuturic, M., and S. Palliyath. 2000. Motor unit number estimate (MUNE) testing in male patients with mild to moderate carpal tunnel syndrome. *Electromyography and Clinical Neurophysiology* 40:67-72.

Cywinska-Wasilewska, G., J.J. Ober, and J. Koczocik-Przedpelska. 1998. Power spectrum of the surface EMG in post-polio syndrome. *Electromyography and Clinical Neurophysiology* 38:463-6.

Dainis, A. 1980. Whole body and segment center of mass determination from kinematic data. *Journal of Biomechanics* 13:647-51.

Dapena, J. 1978. A method to determine the angular momentum of a human body about three orthogonal axes passing through its center of gravity. *Journal of Biomechanics* 11:251-6.

Dapena, J., E.A. Harman, and J.A. Miller. 1982. Three-dimensional cinematography with control object of unknown shape. *Journal of Biomechanics* 15:11-19.

D'Apuzzo, N. 2001. Motion capture from multi image video sequences. In *Proceedings of the XVIIIth Congress of the International Society of Biomechanics*, July 8-13, Zurich, Switzerland. CD-ROM, paper 0106.

Davids, K., P.S. Glazier, D. Araujo, and R.M. Bartlett. 2003. Movement systems as dynamical systems: The role of functional variability and its implications for sports medicine. *Sports Medicine* 33:245-60.

Davis, R.B., S. Ounpuu, D. Tyburski, and J.R. Gage. 1991. A gait analysis data collection and reduction technique. *Human Movement Science* 10:575-87.

de Boer, R.W., J. Cabri, W. Vaes, J.P. Clarijs, A.P. Hollander, G. de Groot, and G.J. van Ingen Schenau. 1987. Moments of force, power, and muscle coordination in speed-skating. *International Journal of Sports Medicine* 8:371-8.

Debold, E.P., J.B. Patlak, and D.M. Warshaw. 2005. Slip sliding away: Load-dependence of velocity generated by skeletal muscle myosin molecules in the laser trap. *Biophysical Journal* 89:L34-L36.

de Boor, C. 1978. *A Practical Guide to Splines.* Springer-Verlag.

de Boor, C. 2001. *A Practical Guide to Splines.* Revised edition. New York: Springer.

De Leva, P. 1996. Adjustments to Zatsiorsky-Seluyanov's segment inertia parameters. *Journal of Biomechanics* 29:1223-30.

DeLisa, J.A., and K. Mackenzie. 1982. *Manual of Nerve Conduction Velocity Techniques.* New York: Raven Press.

Delp, S.L., F.C. Anderson, A.S. Arnold, P. Loan, A. Habib, C.T. John, E. Guendelman, and D.G. Thelen. 2007. OpenSim: Open-source software to create and analyze dynamic simulations of movement. *IEEE Transactions on Biomedical Engineering* 54:1940-50.

Delp, S.L., A.S. Arnold, and S.J. Piazza. 1998. Graphics based modeling and analysis of gait abnormalities. *Bio-medical Materials and Engineering* 8:227-40.

Delp, S.L., J.P. Loan, M.G. Hoy, F.E. Zajac, E.L. Topp, and J.M. Rosen. 1990. An interactive graphics-based model of the lower extremity to study orthopaedic surgical procedures. *IEEE Transactions on Biomedical Engineering* 37:757-67.

Deluzio, K.J., and J.A. Astephen. 2007. Biomechanical features of gait waveform data associated with knee osteoarthritis: An application of principal component analysis. *Gait and Posture* 25:86-93.

Deluzio, K.J., U.P. Wyss, P.A. Costigan, C. Sorbie, and B. Zee. 1999. Gait assessment in unicompartmental knee arthroplasty patients: Principle component modeling of gait waveforms and clinical status. *Human Movement Science* 18:701-11.

Demoulin, C., J.M. Crielaard, and M. Vanderthommen. 2007. Spinal muscle evaluation in healthy individuals and low-back-pain patients: A literature review. *Joint, Bone, Spine* 74:9-13.

Dempster, W.T. 1955. *Space Requirements of the Seated Operator: Geometrical, Kinematic, and Mechanical Aspects of the Body with Special Reference to the Limbs.* WADC Technical Report 55-159. Wright-Patterson Air Force Base, OH.

Derrick, T.R. 1998. Circular continuity of non-periodic data. *Proceedings of the Third North American Congress on Biomechanics*, 313-4. Waterloo, ON: American and Canadian Societies of Biomechanics.

Derrick, T.R., G.E. Caldwell, and J. Hamill. 2000. Modeling the stiffness characteristics of the human body while running with various stride lengths. *Journal of Applied Biomechanics* 16:36-51.

Dewald, J.P., P.S. Pope, J.D. Given, T.S. Buchanan, and W.Z. Rymer. 1995. Abnormal muscle coactivation patterns during isometric torque generation at the elbow and shoulder in hemiparetic subjects. *Brain* 118:495-510.

de Zee, M., and M. Voigt. 2001. Moment dependency of the series elastic stiffness in the human plantarflexors measured in vivo. *Journal of Biomechanics* 34:1399-1406.

Dickx, N., B. Cagnie, E. Achten, P. Vandemaele, T. Parlevliet, and L. Danneels. 2010. Differentiation between deep and superficial fibers of the lumbar multifidus by magnetic resonance imaging. *European Spine Journal* 19:122-8.

Diedrich, F.J., and W.H. Warren. 1995. Why change gaits? Dynamics of the walk to run transition. *Journal of Experimental Psychology: Human Perception and Performance* 21:183-201.

Donà, G., E. Preatoni, C. Cobelli, R. Rodano, and A.J. Harrison. 2009. Application of functional principal component analysis in race walking: An emerging methodology. *Sports Biomechanics* 8:284-301.

Donelan, J.M., R. Kram, and A.D. Kuo. 2002. Simultaneous positive and negative external mechanical work in human walking. *Journal of Biomechanics* 35:117-24.

Donoghue, O.A., A.J. Harrison, N. Coffey, and K. Hayes. 2008. Functional data analysis of the kinematics of running gait in subjects with chronic Achilles tendon injury. *Medicine and Science in Sports and Exercise* 40:1323-1335.

Dowling, J.J. 1997. The use of electromyography for the noninvasive prediction of muscle forces: Current issues. *Sports Medicine* 24:82-96.

Drake, J.D., and J.P. Callaghan. 2006. Elimination of electrocardiogram contamination from electromyogram signals: An evaluation of currently used removal techniques. *Journal of Electromyography and Kinesiology* 16:175-87.

Drillis, R., R. Contini, and M. Bluestein. 1964. Body segment parameters: A survey of measurement techniques. *Artificial Limbs* 8:44-66.

Duchateau, J., S. Le Bozec, and K. Hainaut. 1986. Contributions of slow and fast muscles of triceps surae to a cyclic movement. *European Journal of Applied Physiology* 55:476-81.

Dumitru, D., and J.C. King. 1999. Motor unit action potential duration and muscle length. *Muscle and Nerve* 22:1188-95.

Dumitru, D., J.C. King, and S.D. Nandedkar. 1997. Motor unit action potential duration recorded by monopolar and concentric needle electrodes: Physiologic implications. *American Journal of Physical Medicine* 76:488-93.

Durkin, J.L., and J. Dowling. 2003. Analysis of body segment parameter differences between four human populations and the estimation of errors of four popular mathematical models. *Journal of Biomechanical Engineering* 125:515-22.

Durkin, J.L., J. Dowling, and D.M. Andrews. 2002. The measurement of body segment inertial parameters using dual energy X-ray absorptiometry. *Journal of Biomechanics* 35:1575-80.

Eberstein, A., and B. Beattie. 1985. Simultaneous measurement of muscle conduction velocity and EMG power spectrum changes during fatigue. *Muscle and Nerve* 8:768-73.

Edman, K.A.P. 1975. Mechanical deactivation induced by active shortening in isolated muscle fibres of the frog. *Journal of Physiology* 246:255-75.

Edman, K.A.P. 1988. Double-hyperbolic force-velocity relation in frog muscle fibres. *Journal of Physiology* 404:301-21.

Edman, K.A.P., G. Elzinga, and M.I.M. Noble. 1978. Enhancement of mechanical performance by stretch during tetanic contractions of vertebrate skeletal muscle fibres. *Journal of Physiology* 466:535-52.

Eilers, P.H.C., and B.D. Marx. 1996. Flexible smoothing with B-splines and penalties, with comments. *Statistical Science* 11:89-121.

Elftman, H. 1934. A cinematic study of the distribution of pressure in the human foot. *Anatomical Record* 59:481-91.

Elftman, H. 1939a. Forces and energy changes in the leg during walking. *American Journal of Physiology* 125:339-56.

Elftman, H. 1939b. The function of muscles in locomotion. *American Journal of Physiology* 12:357-66.

Elftman, H. 1940. The work done by muscle in running. *American Journal of Physiology* 129:672-84.

Engsberg, J.R., and Andrews, J.C. 1987. Kinematic analysis of the talocalcaneal, talocrural joint during running support. *Medicine and Science in Sports and Exercise* 19:275-84.

Epifanio, I., C. Avila, A. Page, and C. Atienza. 2008. Analysis of multiple waveforms by means of functional principal component analysis: Normal versus pathological patterns in sit to stand movement. *Medical and Biological Engineering and Computing* 46:551-61.

Erdemir, A., S. McLean, W. Herzog, and A.J. van den Bogert. 2007. Model-based estimation of muscle forces exerted during movements. *Clinical Biomechanics* 22:131-54.

Etnyre, B.R., and L.D. Abraham. 1988. Antagonist muscle activity during stretching: A paradox re-assessed. *Medicine and Science in Sports and Exercise* 20:285-9.

Feldman, R.G., P.H. Travers, J. Chirico-Post, and W.M. Keyserling. 1987. Risk assessment in electronic assembly workers: Carpal tunnel syndrome. *Journal of Hand Surgery* 12:849-55.

Fenn, W.O. 1929. Work against gravity and work due to velocity changes in running. *American Journal of Physiology* 262:639-57.

Fenn, W.O. 1930. Frictional and kinetic factors in the work of sprint running. *American Journal of Physiology* 92:583-611.

Filligoi, G., and F. Felici. 1999. Detection of hidden rhythms in surface EMG signals with a non-linear time-series tool. *Medical Engineering and Physics* 21:439-48.

Finer, J.T., R.M. Simmons, and J.A. Spudich. 1994. Single myosin molecule mechanics: Piconewton forces and nanometre steps. *Nature* 368:113-9.

Finucane, S.D., T. Rafeei, J. Kues, R.L. Lamb, and T.P. Mayhew. 1998. Reproducibility of electromyographic recordings of submaximal concentric and eccentric muscle contractions in humans. *Electroencephalography and Clinical Neurophysiology* 109:290-6.

Fischer, O. 1906. *Theoretische Grundlagen für eine Mechanik der Lebenden Körper mit Speziellen Anwendungen auf den Menschen, sowie auf einige Bewegungs-vorange an Maschine* [Theoretical fundamentals for a mechanics of living bodies, with special applications to man, as well as to some processes of motion in machines]. Leipzig: B.G. Tuebner.

Fisher, N.I. 1993. *Statistical Analysis of Circular Data.* Cambridge, UK: Cambridge University Press.

Frantzell, A. and B. Ingelmark. 1951. Occurrence and distribution of fat in human muscles at various age levels. *Acta Societatis Medicorum Upsaliensis* 56:59-87.

Fregly, B.J., and F.E. Zajac. 1996. A state-space analysis of mechanical energy generation, absorption, and transfer during pedaling. *Journal of Biomechanics* 29:81-90.

Froese, E.A., and M.E. Houston. 1985. Torque-velocity characteristics and muscle fiber type in human vastus lateralis. *Journal of Applied Physiology* 59:309-14.

Fuglevand, A.J., D.A. Winter, A.E. Patla, and D. Stashuk. 1992. Detection of motor unit action potentials with surface electrodes: Influence of electrode size and spacing. *Biological Cybernetics* 67:143-53.

Full, R.J., and D.E. Koditschek. 1999. Templates and anchors: Neuromechanical hypotheses of legged locomotion on land. *Journal of Experimental Biology* 202:3325-3332.

Garner, B.A., and M.G. Pandy. 2003. Estimation of musculotendon properties in the human upper limb. *Annals of Biomedical Engineering* 31:207-220.

Gerard, M.J., T.J. Armstrong, A. Franzblau, B.J. Martin, and D.M. Rempel. 1999. The effects of keyswitch stiffness on typing force, finger electromyography, and subjective discomfort. *American Industrial Hygiene Association Journal* 60:762-9.

Gerber, H., and E. Stuessi. 1987. A measuring system to assess and compute the double stride. In *Biomechanics X-B*, ed. B. Jonsson, 1055-8. International Series on Biomechanics. Champaign, IL: Human Kinetics.

Gerbrands, J.J. 1981. On the relationships between SVD, KLT and PCA. *Pattern Recognition* 14: 375–81.

Gerilovsky, L., P. Tsvetinov, and G. Trenkova. 1986. H-reflex potentials shape and amplitude changes at different lengths of relaxed soleus muscle. *Electromyography and Clinical Neurophysiology* 26:641-53.

Gerilovsky, L., P. Tsvetinov, and G. Trenkova. 1989. Peripheral effects on the amplitude of monopolar and bipolar H-reflex potentials from the soleus muscle. *Experimental Brain Research* 76:173-81.

Gerleman, D.G., and T.M. Cook. 1992. Instrumentation. In *Selected Topics in Surface Electromyography for Use in the Occupational Setting: Expert Perspectives,* ed. G.L. Soderberg, 44-68. Washington, DC: National Institute for Occupational Safety and Health.

Gerritsen, K.G.M., A.J. van den Bogert, M. Hulliger, and R.F. Zernicke. 1998. Intrinsic muscle properties facilitate locomotor control—A computer simulation study. *Motor Control* 2:206-220.

Gerritsen, K.G.M., A.J. van den Bogert, and B.M. Nigg. 1995. Direct dynamics simulation of the impact phase in heel-toe running. *Journal of Biomechanics* 28:661-8.

Gervais, P., and F. Tally. 1993. The beat swing and mechanical descriptors of three horizontal bar release-regrasp skills. *Journal of Applied Biomechanics* 9:66-83.

Gilchrist, L.A., and D.A. Winter. 1997. A multisegment computer simulation of normal human gait. *IEEE Transactions on Rehabilitation Engineering* 5:290-9.

Gitter, J.A., and M.J. Czerniecki. 1995. Fractal analysis of the electromyographic interference pattern. *Journal of Neuroscience Methods* 58:103-8.

Glass, L. 2001. Synchronization and rhythmic processes in physiology. *Nature* 410:277-284.

Glitsch, U., and W. Baumann. 1997. The three-dimensional determination of internal loads in the lower extremity. *Journal of Biomechanics* 30:1123-31.

Godwin, A.A., Stevenson, J.M., Agnew, M.J., Twiddy, A.L., Abdoli-E, M., Lotz, C.A., 2009. Testing the efficacy of an ergonomic lifting aid at diminishing muscular fatigue in women over a prolonged period of lifting. *International Journal of Industrial Ergonomics* 39, 121–126.

Godwin, A., Takahara, G. Agnew, and M. Stevenson, J. 2010. Functional data analysis as a means of evaluating kinematic and kinetic waveforms. *Theoretical Issues in Ergonomics Science* 11(6):489-503.

Gollhofer, A., G.A. Horstmann, D. Schmidtbleicher, and D. Schonthal. 1990. Reproducibility of electromyographic patterns in stretch-shortening type contractions. *European Journal of Applied Physiology* 60:7-14.

Gordon, A.M., A.F. Huxley, and F.J. Julian. 1966. The variation in isometric tension with sarcomere length in vertebrate muscle fibres. *Journal of Physiology* 184:170-92.

Granzier, H.L., and S. Labeit. 2006. The giant muscle protein titin is an adjustable molecular spring. *Exercise and Sport Sciences Reviews* 34:50-3.

Green, P.J., and B.W. Silverman. 1994. *Nonparametric Regression and Generalised Linear Models: A Roughness Penalty Approach.* London: Chapman & Hall.

Gregor, R.J., and T.A. Abelew. 1994. Tendon force measurements and movement control: A review. *Medicine and Science in Sports and Exercise* 26:1359-72.

Gregor, R.J., P.V. Komi, R.C. Browning, and M. Jarvinen.1991. A comparison of the triceps surae and residual muscle moments at the ankle during cycling. *Journal of Biomechanics* 24:287-97.

Gregor, R.J., P.V. Komi, and M. Jarvinen. 1987. Achilles tendon forces during cycling. *International Journal of Sports Medicine* 8:9-14.

Grieve, D.W., S. Pheasant, and P.R. Cavanagh. 1978. Prediction of gastrocnemius length from knee and ankle joint posture. In *Biomechanics VI-A,* eds. E. Asmussen and K. Jorgensen, 405-12. Baltimore: University Park Press.

Griffiths, R.I. 1991. Shortening of muscle fibres during stretch of the active cat medial gastrocnemius muscle: The role of tendon compliance. *Journal of Physiology* 436: 219-36.

Grood, E.S., and W.J. Suntay. 1983. A joint coordination system for the clinical description of three-dimensional motions: Application to the knee. *Journal of Biomedical Engineering* 105:136-44.

Hagberg, M., and B.-E. Ericson. 1982. Myoelectric power spectrum dependence on muscular contraction level of elbow flexors. *European Journal of Applied Physiology* 48:147-56.

Hagemann, B., G. Luhede, and H. Luczak. 1985. Improved "active" electrodes for recording bioelectric signals in work physiology. *European Journal of Applied Physiology* 54:95-8.

Hägg, G. 1981. Electromyographic fatigue analysis based on the number of zero crossings. *Pflugers Archiv: European Journal of Physiology* 391:78-80.

Haken, H., J.A.S. Kelso, and H. Bunz. 1985. A theoretical model of phase transitions in human hand movements. *Biological Cybernetics* 51:347-56.

Hamill, J., G.E. Caldwell, and T.R. Derrick. 1997. A method for reconstructing digital signals using Shannon's sampling theorem. *Journal of Applied Biomechanics* 13:226-38.

Hamill, J., J.M. Haddad, B.C. Heiderscheit, R.E.A. van Emmerik, and L. Li. 2006. Clinical relevance of coordination variability. In: *Movement System Variability,* eds. K. Davids, S.J. Bennett, and K.M. Newell, 153-65. Champaign, IL: Human Kinetics.

Hamill, J., J.M. Haddad, and W.M. McDermott. 2000. Issues in quantifying variability from a dynamical systems perspective. *Journal of Applied Biomechanics* 16:407-18.

Hamill, J., R.E.A. van Emmerik, B.C. Heiderscheit, and L. Li. 1999. A dynamical systems approach to lower extremity running injuries. *Clinical Biomechanics* 14:297-308.

Hammelsbeck, M., and W. Rathmayer. 1989. Intracellular Na+, K+ and Cl– activity in tonic and phasic muscle fibers of the crab *Eriphia. Pflugers Archiv: European Journal of Physiology* 413:487-92.

Hanavan, E.P. 1964. *A Mathematical Model of the Human Body.* AMRL Technical Report 64-102. Wright-Patterson Air Force Base, OH.

Hannaford, B., and S. Lehman. 1986. Short time Fourier analysis of the electromyogram: Fast movements and constant contraction. *IEEE Transactions on Biomedical Engineering* 12:1173-81.

Hannah, R., S. Cousins, and J. Foort. 1978. The CARS-UBC electrogoniometer, a clinically viable tool. In *Digest of the 7th Canadian Medical and Biological Engineering Conference,* 133-4, Vancouver, BC.

Hannerz, J. 1974. An electrode for recording single motor unit activity during strong muscle contractions. *Electroencephalography and Clinical Neurophysiology* 37:179-81.

Harless, E. 1860. The static moments of the component masses of the human body. *Treatises of the Mathematics—Physics Class, Royal Bavarian Academy of Sciences* 8:69-96, 257-94. Trans. FTD Technical Report 61-295. Wright- Patterson Air Force Base, OH, 1962.

Harrington, M.E., A.B. Zavatsky, S.E.M. Lawson, Z. Yuan, T.N. Theologis. 2007. Prediction of the hip joint centre in adults, children and patients with cerebral palsy based on magnetic resonance imaging. *Journal of Biomechanics* 40:595-602.

Harris, G.F., and J.J. Wertsch. 1994. Procedures for gait analysis. *Archives of Physical Medicine and Rehabilitation* 75:216-25.

Harrison, A.J., W. Ryan, and K. Hayes. 2007. Functional data analysis of joint coordination in the development of vertical jump performance. *Sports Biomechanics* 6:196-211.

Harvey, R., and E. Peper. 1997. Surface electromyography and mouse use position. *Ergonomics* 40:781-9.

Hashimoto, S., J. Kawamura, Y. Segawa, Y. Harada, T. Hanakawa, and Y. Osaki. 1994. Waveform changes of compound muscle action potential (CMAP) with muscle length. *Journal of the Neurological Sciences* 124:21-4.

Hasson, C.J., and G.E. Caldwell. 2012. Effects of age on mechanical properties of dorsiflexor and plantarflexor muscles. *Annals of Biomedical Engineering* 40:1088-101.

Hasson, C.J., J.A. Kent-Braun, and G.E. Caldwell. 2011. Contractile and non-contractile tissue volume and distribution in ankle muscles of young and older adults. *Journal of Biomechanics* 44:2299-306.

Hasson, C.J., R.M. Miller, and G.E. Caldwell. 2011. Contractile and elastic ankle joint muscular properties in young and older adults. *PLoS One* 6(1):e15953.

Hastie, T., and R. Tibshirani. 1990. *Generalised Additive Models.* New York: Chapman and Hall.

Hatze, H. 1975. A new method for the simultaneous measurement of the moment on inertia, the damping coefficient and the location of the centre of mass of a body segment *in situ. European Journal of Applied Physiology* 34:217-26.

Hatze, H. 1979. A model for the computational determination of parameter values of anthropometric segments. NRIMS Technical Report TWISK 79. Pretoria, South Africa.

Hatze, H. 1980. A mathematical model for the computational determination of parameter values of anthropometric segments. *Journal of Biomechanics* 13:833-43.

Hatze, H. 1981. The use of optimally regularised Fourier series for estimating higher-order derivatives of noisy biomechanical data. *Journal of Biomechanics* 14:13-8.

Hatze, H. 1993. The relationship between the coefficient of restitution and energy losses in tennis rackets. *Journal of Applied Biomechanics* 9:124-42.

Hatze, H. 1997. A three-dimensional multivariate model of passive human joint torques and articular boundaries. *Clinical Biomechanics* 12:128-35.

Hatze, H. 1998. Validity and reliability of methods for testing vertical jumping performance. *Journal of Applied Biomechanics* 14:127-40.

Hatze, H. 2002. The fundamental problem of myoskeletal inverse dynamics and its application. *Journal of Biomechanics* 35:109-16.

Haxton, H.A. 1944. Absolute muscle force in the ankle flexors of man. *Journal of Physiology* 103:267-273.

Hay, J.G. 1973. The center of gravity of the human body. *Kinesiology III* 20-44.

Hay, J.G. 1974. Moment of inertia of the human body. *Kinesiology IV* 43-52.

Hayward, M. 1983. Quantification of interference patterns. In *Computer-Aided Electromyography,* ed. J.E. Desmedt, 128-49. New York: Karger.

Heiderscheit, B.C., J. Hamill, and R.E.A. van Emmerik. 2002. Variability of stride characteristics and joint coordination among individuals with unilateral patellofemoral pain. *Journal of Applied Biomechanics* 18:110-21.

Hermens, H.J., T.A.M. van Bruggen, C.T.M. Baten, W.L.C. Rutten, and H.B.K. Boom. 1992. The median frequency of the surface EMG power spectrum in relation to motor unit firing and action potential properties. *Journal of Electromyography and Kinesiology* 2:15-25.

Herzog, W. 1988. The relation between the resultant moments at a joint and the moments measured by an isokinetic dynamometer. *Journal of Biomechanics* 21:5-12.

Herzog, W., and T.R. Leonard. 1991. Validation of optimization models that estimate the forces exerted by synergistic muscles. *Journal of Biomechanics* 24:31-9.

Herzog, W., T.R. Leonard, and J.Z. Wu. 2000. The relationship between force depression following shortening and mechanical work in skeletal muscle. *Journal of Biomechanics* 33:659-68.

Herzog, W., and H.E.D.J. ter Keurs. 1988. Force-length relation of in-vivo human rectus femoris muscles. *Pflugers Archiv: European Journal of Physiology* 411:642-647.

Hides, J.A., M.J. Stokes, M. Saide, G.A. Jull, and D.H. Cooper. 1994. Evidence of lumbar multifidus muscle wasting ipsilateral to symptoms in patients with acute/subacute low back pain. *Spine (Phila Pa 1976)* 19:165-72.

Hilding, M.B., L. Ryd, S. Toksvig-Larsen, A. Mann, and A. Stenström. 1999. Gait affects tibial component fixation. *Journal of Arthroplasty* 14:589-93.

Hill, A.V. 1938. The heat of shortening and the dynamic constants of muscle. *Proceedings of the Royal Society B* 126:136-95.

Hill, D.A. 1967. *Schaum's Outline of Theory and Problems of Lagrangian Dynamics.* New York: McGraw-Hill.

Hinrichs, R.N. 1985. Regression equations to predict segmental moments of inertia from anthropometric measurements: An extension of the data of Chandler et al. (1975). *Journal of Biomechanics* 18:621-4.

Hodges, P.W., and B.H. Bui. 1996. A comparison of computer-based methods for the determination of onset of muscle contraction using electromyography. *Electroencephalography and Clinical Neurophysiology* 101:511-9.

Hof, A.L. 1998. In vivo measurement of the series elasticity release curve of human triceps surae muscle. *Journal of Biomechanics* 31:793-800.

Hof, A.L. 2009. A simple method to remove ECG artifacts from trunk muscle EMG signals. *Journal of Electromyography and Kinesiology* 19:e554-e555.

Hof, A.L., and J.W. van den Berg. 1981. EMG to force processing I: An electrical analogue of the Hill muscle model. *Journal of Biomechanics* 14: 747–758.

Holden, J.P., W.S. Selbie, and S.J. Stanhope. 2003. A proposed test to support the clinical movement analysis laboratory accreditation process. *Gait and Posture* 17:205-13.

Holt, K.G., J. Hamill, and R.O. Andres. 1990. The force-driven harmonic oscillator as a model for human locomotion. *Human Movement Science* 9:55-68.

Hu, Y., J.N. Mak, and K.D. Luk. 2009. Effect of electrocardiographic contamination on surface electromyography assessment of back muscles. *Journal of Electromyography and Kinesiology* 19:145-56.

Hubley-Kozey, C.L., K.J. Deluzio, J.A. McNutt, C.S.N. Landry, and W.D. Stanish. 2006. Neuromuscular alterations during walking associated with moderate knee osteoarthritis. *Journal of Electromyography and Kinesiology* 16:365-78.

Huijing, P.A. 1998. Muscle, the motor of movement: Properties in function, experiment and modeling. *Journal of Electromyography and Kinesiology* 8:61-77.

Huijing, P.A. 1999. Muscle as a collagen fiber reinforced composite: A review of force transmission in muscle and whole limb. *Journal of Biomechanics* 32:329-45.

Hurwitz, D.E., A.B. Ryals, J.P. Case, J.A. Block, and T.P. Andriacchi. 2002. The knee adduction moment during gait in subjects with knee osteoarthritis is more closely correlated with static alignment than radiographic disease severity, toe out angle and pain. *Journal of Orthopedics Research* 20:101-7.

Huxley, A.F. 1957. Muscle structure and theories of contraction. *Progress in Biophysics and Biophysical Chemistry* 7:255-318.

Huxley, A.F., and R.M. Simmons. 1971. Proposed mechanism of force generation in striated muscle. *Nature* 233:533-8.

Ikegawa, S., M. Shinohara, T. Fukunaga, J.P. Zbilut, and C.L.J. Webber. 2000. Nonlinear time-course of lumbar muscle fatigue using recurrence quantifications. *Biological Cybernetics* 82:373-82.

Inbar, G.F., J. Allin, O. Paiss, and H. Kranz. 1986. Monitoring surface EMG spectral changes by the zero crossing rate. *Medical and Biological Engineering and Computing* 24:10-8.

Ishida, Y., H. Kanehisa, J. Carroll, M. Pollock, J. Graves, and L. Ganzarella. 1997. Distribution of subcutaneous fat and muscle thicknesses in young and middle-aged women. *American Journal of Human Biology* 9:247-55.

Ito, M., Y. Kawakami, Y. Ichinose, S. Fukashiro, and T. Fukunaga.1998. Nonisometric behavior of fascicles during isometric contractions of a human muscle. *Journal of Applied Physiology* 85:1230-5.

Jackson, J.E. 1991. *A User's Guide to Principal Components.* New York: Wiley.

Jackson, K.M. 1979. Fitting of mathematical functions to biomechanical data. *IEEE Transactions on Biomedical Engineering* 26(2):122-4.

Jacobs, R., and G.J. van Ingen Schenau. 1992. Control of an external force in leg extensions in humans. *Journal of Physiology* 457:611-26.

Jensen, B.R., B. Schibye, K. Sogaard, E.B. Simonsen, and G. Sjøgaard. 1993. Shoulder muscle load and muscle fatigue among industrial sewing-machine operators. *European Journal of Applied Physiology* 67:467-75.

Jensen, R.K. 1976. Model for body segment parameters. In *Biomechanics V-B,* ed. P.V. Komi, 380-6. Baltimore: University Park Press.

Jensen, R.K. 1978. Estimation of the biomechanical properties of three body types using a photogrammetric method. *Journal of Biomechanics* 11:349-58.

Jensen, R.K. 1986. Body segment mass, radius and radius of gyration proportions of children. *Journal of Biomechanics* 19:359-68.

Jensen, R.K. 1989. Changes in segment inertial proportions between 4 and 20 years. *Journal of Biomechanics* 22:529-36.

Jensen, R.K., T. Treitz, and S. Doucet. 1996. Prediction of human segment inertias during pregnancy. *Journal of Applied Biomechanics* 12:15-30.

Johnson, S.W., P.A. Lynn, J.S.G. Miller, and G.L. Reed. 1977. Miniature skin-mounted preamplifier for measurement of surface electromyographic potentials. *Medical and Biological Engineering and Computing* 15:710-1.

Jonas, D., C. Bischoff, and B. Conrad. 1999. Influence of different types of surface electrodes on amplitude, area and duration of the compound muscle action potential. *Clinical Neurophysiology* 110:2171-5.

Jönhagen, S., M.O. Ericson, G. Nemeth, and E. Eriksson. 1996. Amplitude and timing of electromyographic activity during sprinting. *Scandinavian Journal of Medicine and Science in Sports* 6:15-21.

Juel, C. 1988. Muscle action potential propagation velocity changes during activity. *Muscle and Nerve* 11:714-9.

Kamen, G., S.V. Sison, C.C. Du, and C. Patten. 1995. Motor unit discharge behavior in older adults during maximal-effort contractions. *Journal of Applied Physiology* 79:1908-13.

Kameyama, O., R. Ogawa, T. Okamoto, and M. Kumamoto.1990. Electric discharge patterns of ankle muscles during the normal gait cycle. *Archives of Physical Medicine and Rehabilitation* 71:969-74.

Kamon, E., and J. Gormley. 1968. Muscular activity pattern for skilled performance and during learning of a horizontal bar exercise. *Ergonomics* 11:345-57.

Kang, W.J., J.R. Shiu, C.K. Cheng, J.S. Lai, H.W. Tsao, and T.S. Kuo. 1995. The application of cepstral coefficients and maximum likelihood method in EMG pattern recognition. *IEEE Transactions on Biomedical Engineering* 42:777-85.

Kaplan, M.L., and J.H. Heegaard. 2001. Predictive algorithms for neuromuscular control of human locomotion. *Journal of Biomechanics* 34:1077-83.

Karlsson, D., and R. Tranberg. 1999. On skin movement artifact: Resonant frequencies of skin markers attached to the leg. *Human Movement Science* 18:627-35.

Karlsson, S., J. Yu, and M. Akay. 1999. Enhancement of spectral analysis of myoelectric signals during static contractions using wavelet methods. *IEEE Transactions on Biomedical Engineering* 46:670-84.

Karlsson, S., J. Yu, and M. Akay. 2000. Time-frequency analysis of myoelectric signals during dynamic contractions: A comparative study. *IEEE Transactions on Biomedical Engineering* 47:228-38.

Kaufman, K.R., C. Hughes, B.F. Morrey, M. Morrey, and K.N. An. 2001. Gait characteristics of patients with knee osteoarthritis. *Journal of Biomechanics* 34(7):907-15.

Kawakami, Y., and R.L. Lieber. 2000. Interaction between series compliance and sarcomere kinetics determines internal sarcomere shortening during fixed-end contraction. *Journal of Biomechanics* 33:1249-55.

Kelso, J.A.S. 1995. *Dynamic Patterns: The Self-Organization of Brain and Behavior.* Cambridge, MA: MIT Press.

Kelso, J.A.S., J.P. Scholz, and G. Schöner. 1986. Nonequilibrium phase transitions in coordinated biological motion: Critical fluctuations. *Physics Letters A* 134:8-12.

Kepple T., K. Siegel, J. Winters, and S.J. Stanhope. 1998. The sensitivity of joint accelerations to net joint moments during normal gait. *Annals of Biomedical Engineering* 26:S-110.

Kilbom, A., G.M. Hägg, and C. Kall. 1992. One-handed load carrying: Cardiovascular, muscular and subjective indices of endurance and fatigue. *European Journal of Applied Physiology* 65:52-8.

Kilmister, C.W. 1967. *Lagrangian Dynamics: An Introduction for Students.* New York: Plenum Press.

Kim, H.J., J.W. Fernandez, M. Akbarshahi, J.P. Walter, B.J. Fregly, and M.G. Pandy. 2009. Evaluation of predicted knee-joint muscle forces during gait using an instrumented knee implant. *Journal of Orthopaedic Research* 27:1326-31.

Kim, M.J., W.S. Druz, and J.T. Sharp. 1985. Effect of muscle length on electromyogram in a canine diaphragm strip preparation. *Journal of Applied Physiology* 58:1602-7.

Kirkwood, R.N., E.G. Culham, and P. Costigan. 1999. Radiographic and non-invasive determination of the hip joint center location: Effect on hip joint moments. *Clinical Biomechanics* 14:227-35.

Knaflitz, M., and P. Bonato. 1999. Time-frequency methods applied to muscle fatigue assessment during dynamic contractions. *Journal of Electromyography and Kinesiology* 9:337-50.

Koh, T.J., and M.D. Grabiner. 1992. Cross talk in surface electromyograms of human hamstring muscles. *Journal of Orthopedic Research* 10:701-9.

Komi, P.V. 1990. Relevance of *in vivo* force measurements to human biomechanics. *Journal of Biomechanics* 23(Suppl. No. 1):23-34.

Komi, P.V., A. Belli, V. Huttunen, R. Bonnefoy, A. Geyssant, and J.R. Lacour. 1996. Optic fibre as a transducer of tendomuscular forces. *European Journal of Applied Physiology and Occupational Physiology* 72:278-80.

Koning, J.J., G. de Groot, and G.J. van Ingen Schenau. 1991. Speed skating the curves: A study of muscle coordination and power production. *International Journal of Sport Biomechanics* 7:344-58.

Kramer, M., V. Ebert, L. Kinzl, C. Dehner, M. Elbel, and E. Hartwig. 2005. Surface electromyography of the paravertebral muscles in patients with chronic low back pain. *Archives of Physical Medicine and Rehabilitation* 86:31-6.

Krogh-Lund, C. 1993. Myo-electric fatigue and force failure from submaximal static elbow flexion sustained to exhaustion. *European Journal of Applied Physiology* 67:389-401.

Kugler, P.N., and M.T. Turvey. 1987. *Information, Natural Law, and the Self-Assembly of Rhythmic Movement.* Hillsdale, NJ: Erlbaum.

Kumar, S., and A. Mital, eds. 1996. *Electromyography in Ergonomics.* London: Taylor & Francis.

Kwon, Y.-H. 1996. Effects of the method of body segment parameter estimation on airborne angular momentum. *Journal of Applied Biomechanics* 12:413-30.

Lakin G., M.H. Schwartz, and L. Schutte 1999. The effects of individual muscle forces on body mass center acceleration. *Gait and Posture* 9(2):117.

Lamontagne, M., R. Doré, H. Yahia, and J.M. Dorlot. 1985. Tendon and ligament measurement. *Medical Electronics* 6:74-6.

Lamoth, C.J.C., P.J. Beek, and O.G. Meijer. 2002. Pelvis-thorax coordination in the transverse plane during gait. *Gait and Posture* 16:101-14.

Lamoth, C.J.C., A. Daffertshofer, R. Huys, and P.J. Beek. 2009. Steady and transient coordination structures of walking and running. *Human Movement Science* 28(3):371-86.

Landjerit, B., B. Maton, and G. Peres. 1988. In vivo muscular force analysis during the isometric flexion on a monkey's elbow. *Journal of Biomechanics* 21:577-84.

Landry, S.C., K.A. McKean, C.L. Hubley-Kozey, W.D. Stanish, and K.J. Deluzio. 2007. Knee biomechanics associated with moderate knee osteoarthritis during gait at both a self-selected and a fast walking speed. *Journal of Biomechanics* 40(8):1754-61.

Lansdown, D.A., Z. Ding, M. Wadington, J.L. Hornberger, and B.M. Damon. 2007. Quantitative diffusion tensor MRI-based fiber tracking of human skeletal muscle. *Journal of Applied Physiology* 103:673-81.

Lanshammer, H. 1982a. On practical evaluation of differentiation techniques for human gait analysis. *Journal of Biomechanics* 15:99-105.

Lanshammer, H. 1982b. On precision limits for derivatives calculated from noisy data. *Journal of Biomechanics* 15:459-70.

Lariviere, C., D. Gagnon, and P. Loisel. 2000. The comparison of trunk muscles EMG activation between subjects with and without chronic low back pain during flexion-extension and lateral bending tasks. *Journal of Electromyography and Kinesiology* 10:79-91.

Latash, M.L., J.P. Scholz, and G. Schöner. 2002. Motor control strategies revealed in the structure of motor variability. *Exercise and Sport Sciences Reviews* 30:26-31.

Lee, R.G., P. Ashby, D.G. White, and A.J. Aguayo. 1975. Analysis of motor conduction velocity in the human median nerve by computer simulation of compound muscle action potentials. *Electroencephalography and Clinical Neurophysiology* 39:225-37.

Lehman, G.J., and S.M. McGill. 1999. The importance of normalization in the interpretation of surface electromyography: A proof of principle. *Journal of Manipulative and Physiological Therapeutics* 22:444-6.

Lemaire, E.D., and D.G.E. Robertson. 1989. Power in sprinting. *Track and Field Journal* 35:13-7.

Lemaire, E.D., and D.G.E. Robertson. 1990a. Validation of a computer simulation for planar airborne human motions. *Journal of Human Movement Studies* 18:213-28.

Lemaire, E.D., and D.G.E. Robertson. 1990b. Force-time data acquisition system for sprint starting. *Canadian Journal of Sport Sciences* 15:149-52.

LeVeau, B., and G. Andersson. 1992. Output forms: Data analysis and applications. In *Selected Topics in Surface Electromyography for Use in the Occupational Setting: Expert Perspectives,* ed. G.L. Soderberg. Washington, DC: National Institute for Occupational Safety and Health.

Lexell, J., K. Henriksson-Larsen, B. Winblad, and M. Sjostrom.1983. Distribution of different fiber types in human skeletal muscles: Effects of aging studied in whole muscle cross sections. *Muscle and Nerve* 6:588-95.

Li, L., and G.E. Caldwell. 1999. Coefficient of cross correlation and the time domain correspondence. *Journal of Electromyography and Kinesiology* 9:385-9.

Lieber, R.L., and J. Friden. 1997. Intraoperative measurement and biomechanical modeling of the flexor carpi ulnaris-to-extensor carpi radialis longus tendon transfer. *Journal of Biomechanical Engineering* 119:386-91.

Lipsitz, L.A. 2002. Dynamics of stability: The physiologic basis of functional health and frailty. *Journal of Gerontology* 57:B115-25.

Liu, M.M., W. Herzog, and H.H. Savelberg. 1999. Dynamic muscle force predictions from EMG: An artificial neural network approach. *Journal of Electromyography and Kinesiology* 9:391-400.

Loram, I.D., C.N. Maganaris, and M. Lakie. 2004. Paradoxical muscle movement in human standing. *Journal of Physiology (London)* 556:683-689.

Lu, G., J.S. Brittain, P. Holland, J. Yianni, A.L. Green, J.F. Stein, T.Z. Aziz, and S. Wang. 2009. Removing ECG noise from surface EMG signals using adaptive filtering. *Neuroscience Letters* 462:14-9.

Lu, T.-W., and J.J. O'Connor. 1999. Bone position estimation from skin marker coordinates using global optimization with joint constraints. *Journal of Biomechanics* 32:129-34.

MacDonald, D., G.L. Moseley, and P.W. Hodges. 2009. Why do some patients keep hurting their back? Evidence of ongoing back muscle dysfunction during remission from recurrent back pain. *Pain* 142:183-8.

Macintosh, J.E., F. Valencia, N. Bogduk, and R.R. Munro. 1986. The morphology of the human lumbar multifidus. *Clinical Biomechanics* 1:196-204.

Mansour, J.M., and M.L. Audu. 1986. The passive elastic moment at the knee and its influence on human gait. *Journal of Biomechanics* 19:369-73.

Marion, J.B., and S.T. Thornton. 1995. *Classical Dynamics of Particles and Systems.* 4th ed. New York: Harcourt Brace College.

Marks, R.J., II. 1993. *Advanced Topics in Shannon Sampling and Interpolation Theory.* New York: Springer-Verlag.

Marque, C., C. Bisch, R. Dantas, S. Elayoubi, V. Brosse, and C. Perot. 2005. Adaptive filtering for ECG rejection from surface EMG recordings. *Journal of Electromyography and Kinesiology* 15:310-5.

Martindale, W.O., and D.G.E. Robertson. 1984. Mechanical energy variations in single sculls and ergometer rowing. *Canadian Journal of Applied Sport Sciences* 9:153-63.

Marzan, T., and H.M. Karara. 1975. A computer program for direct linear transformation of the colinearity condition and some applications of it. In *American Society of Photogrammetry Symposium on Close Range Photogrammetry,* 420-76. Falls Church, VA: American Society for Photogrammetry.

Masuda, T., H. Miyano, and T. Sadoyama. 1985. A surface electrode array for detecting action potential trains of single motor units. *Electroencephalography and Clinical Neurophysiology* 60:435-43.

Masuda, T., H. Miyano, and T. Sadoyama. 1992. The position of innervation zones in the biceps brachii investigated by surface electromyography. *IEEE Transactions on Biomedical Engineering* 32:36-42.

Mathiassen, S.E., J. Winkel, and G.M. Hägg. 1995. Normalization of surface EMG amplitude from the upper trapezius muscle in ergonomic studies: A review. *Journal of Electromyography and Kinesiology* 5:197-226.

Maton, B., and D. Gamet. 1989. The fatigability of two agonistic muscles in human isometric voluntary submaximal contraction: An EMG study. II. Motor unit firing rate and recruitment. *European Journal of Applied Physiology* 58:369-74.

Mayagoitia, R.E., A.V. Nene, and P.H. Veltnik. 2002. Accelerometer and rate gyroscope measurement of kinematics: An inexpensive alternative to optical motion analysis systems. *Journal of Biomechanics* 35:537-42.

McClay, I., and K. Manal. 1997. Coupling parameters in runners with normal and excessive pronation. *Journal of Applied Biomechanics* 13:109-24.

McDermott, W.J., R.E.A. van Emmerik, and J. Hamill. 2003. Running training and adaptive strategies of locomotor-respiratory coordination. *European Journal of Applied Physiology* 89(5):435-44.

McFaull, S., and M. Lamontagne. 1993. The passive elastic moment about the *in vivo* human knee joint. In *Proceedings of the 14th Annual Conference, International Society of Biomechanics,* ed. S. Bouisset, 848-9. Paris: International Society of Biomechanics.

McFaull, S., and M. Lamontagne. 1998. *In vivo* measurement of the passive viscoelastic properties of the human knee joint. *Human Movement Science* 17:139-65.

McGill, S., and R.W. Norman. 1985. Dynamically and statically determined low back moments during lifting. *Journal of Biomechanics* 18:877-85.

McMahon, T.A. 1984. Mechanics of locomotion. *International Journal of Robotics Research* 3:4-28.

McMahon, T.A., G. Valiant, and E.C. Frederick. 1987. Groucho running. *Journal of Applied Physiology* 62:2326-37.

Meglan, D., and F. Todd. 1994. Kinetics of human locomotion. In *Human Walking,* eds. J. Rose and J.G. Gamble, 73-99. Baltimore: Williams & Wilkins.

Mehta, A., and W. Herzog. 2008. Cross-bridge induced force enhancement? *Journal of Biomechanics* 41:1611-1615.

Merletti, R., D. Farina, and A. Granata. 1999. Non-invasive assessment of motor unit properties with linear electrode arrays. *Electroencephalography and Clinical Neurophysiology Supplement* 50:293-300.

Merletti, R., A. Rainoldi, and D. Farina. 2001. Surface electromyography for noninvasive characterization of muscle. *Exercise and Sport Sciences Reviews* 29:20-5.

Mero, A., and P.V. Komi. 1987. Electromyographic activity in sprinting at speeds ranging from sub-maximal to supra-maximal. *Medicine and Science in Sports and Exercise* 19:266-74.

Miller, D.I. 1970. A computer simulation of the airborne phase of diving. PhD dissertation, Pennsylvania State University.

Miller, D.I. 1973. Computer simulation of springboard diving. In *Medicine and Sport, Volume 8: Biomechanics III,* eds. S. Cerquiglini, A. Venerando, and J. Wartenweiler, 116-9. Basel: Karger.

Miller, D.I., and W.E. Morrison. 1975. Prediction of segmental parameters using the Hanavan human body model. *Medicine and Science in Sports* 7:207-12.

Miller D.I., and R.C. Nelson. 1973. *Biomechanics of Sport.* Philadelphia: Lea & Febiger.

Miller, D.I., and E.J. Sprigings. 2001. Factors influencing the performance of springboard dives of increasing difficulty. *Journal of Applied Biomechanics* 17:217-31.

Miller, N.R., R. Shapiro, and T.M. McLaughlin. 1980. A technique for obtaining spatial kinematic parameters of segments of biomechanical systems from cinematographic data. *Journal of Biomechanics* 13:535-47.

Miller, R.H., Meardon, S.A., Derrick, T.R., and Gillette, J.C. 2008. Continuous relative phase variability during an exhaustive run in runners with a history of iliotibial band syndrome. *Journal of Applied Biomechanics* 24:262-270.

Miller, R.M., B.R. Umberger, and G.E. Caldwell. 2012a. Limitations to maximum sprinting speed imposed by muscle mechanical properties. *Journal of Biomechanics* 45:1092-7.

Miller, R.M., B.R. Umberger, and G.E. Caldwell. 2012b. Sensitivity of maximum sprinting speed to characteristic parameters of the muscle force-velocity relationship. *Journal of Biomechanics* 45:1406-13.

Mills, K.R., and R.T. Edwards. 1984. Muscle fatigue in myophosphorylase deficiency: Power spectral analysis of the electromyogram. *Electroencephalography and Clinical Neurophysiology* 57:330-5.

Milner-Brown, H.S., and R.G. Miller. 1990. Myotonic dystrophy: Quantification of muscle weakness and myotonia and the effect of amitriptyline and exercise. *Archives of Physical Medicine and Rehabilitation* 71:983-7.

Milner-Brown, H.S., R.B. Stein, and R.G. Lee. 1975. Synchronization of human motor units: Possible roles of exercise and supraspinal reflexes. *Electroencephalography and Clinical Neurophysiology* 38:245-54.

Minetti, A.E., and G. Belli. 1994. A model for the estimation of visceral mass displacement in periodic movements. *Journal of Biomechanics* 27:97-101.

Mineva, A., J. Dushanova, and L. Gerilovsky. 1993. Similarity in shape, timing and amplitude of H- and T-reflex potentials concurrently recorded along the broad skin area over soleus muscle. *Electromyography and Clinical Neurophysiology* 33:235-45.

Mirka, G.A. 1991. The quantification of EMG normalization error. *Ergonomics* 34:343-52.

Mitchelson, D.L. 1975. Recording of movement without photograph. *Techniques for the Analysis of Human Movement.* London: Lepus Books.

Miyatani, M., H. Kanehisa, M. Ito, Y. Kawakami, and T. Fukunaga. 2004. The accuracy of volume estimates using ultrasound muscle thickness measurements in different muscle groups. *European Journal of Applied Physiology* 91:264-272.

Mizrahi, J., and Z. Susak. 1982. In-vivo elastic and damping response of the human leg to impact forces. *Journal of Biomechanical Engineering* 104:63-6.

Mochon, S., and T.A. McMahon. 1980. Ballistic walking. *Journal of Biomechanics* 13:49-57.

Monti, R.J., R.R. Roy, J.A. Hodgson, and V.R. Edgerton.1999. Transmission of forces within mammalian skeletal muscles. *Journal of Biomechanics* 32:371-80.

Morgan, D.L. 1990. Modeling of lengthening muscle: The role of intersarcomere dynamics. In *Multiple Muscle Systems,* eds. J.M. Winters and S.L.-Y. Woo, 46-56. New York: Springer-Verlag.

Morimoto, S. 1986. Effect of length change in muscle fibers on conduction velocity in human motor units. *Japanese Journal of Physiology* 36:773-82.

Moritani, T., and M. Muro. 1987. Motor unit activity and surface electromyogram power spectrum during increasing force of contraction. *European Journal of Applied Physiology* 56:260-5.

Morrenhof, J.W., and H.J. Abbink. 1985. Cross-correlation and cross-talk in surface electromyography. *Electromyography and Clinical Neurophysiology* 25:73-9.

Morris, J.M., G. Benner, and D.B. Lucas. 1962. An electromyographic study of the intrinsic muscles of the back in man. *Journal of Anatomy* 96:509-20.

Moseley, G.L., P.W. Hodges, and S.C. Gandevia. 2002. Deep and superficial fibers of the lumbar multifidus muscle are differentially active during voluntary arm movements. *Spine (Phila Pa 1976)* 27:E29-E36.

Moss, R.F., P.B. Raven, J.P. Knochel, J.R. Peckham, and J.D. Blachley. 1983. The effect of training on resting muscle membrane potentials. In *Biochemistry of Exercise,* eds. H.G. Knuttgen, J.A. Vogel, and J. Poortmans, 806-11. 3rd ed. Champaign, IL: Human Kinetics.

Mundermann, A., C.O. Dyrby, and T.P. Andriacchi. 2005. Secondary gait changes in patients with medial compartment knee osteoarthritis: Increased load at the ankle, knee, and hip during walking. *Arthritis Rheumatism* 52(9):2835-44.

Mungiole, M., and P.E. Martin. 1990. Estimating segmental inertia properties: Comparison of magnetic resonance imaging with existing methods. *Journal of Biomechanics* 23:1039-46.

Muniz, A.M.S., and J. Nadal. 2009. Application of principal component analysis in vertical ground reaction force to discriminate between normal and abnormal gait. *Gait and Posture* 29(1):31-5.

Murphy, S.D., and D.G.E. Robertson. 1994. Construction of a high-pass digital filter from a low-pass digital filter. *Journal of Applied Biomechanics* 10:374-81.

Murtaugh, K., and D.I. Miller. 2001. Initiating rotation in back and reverse armstand somersault tuck dives. *Journal of Applied Biomechanics* 17:312-25.

Nagano, A, B.R. Umberger, M.W. Marzke, and K.G.M. Gerritsen. 2005. Neuromusculoskeletal computer modeling and simulation of upright, straight-legged, bipedal locomotion of *Australopithecus afarensis* (A.L. 288-1). *American Journal of Physical Anthropology* 126:2-13.

Neptune, R.R., S.A. Kautz, and F.E. Zajac. 2000. Muscle contributions to specific biomechanical functions do not change in forward versus backward pedaling. *Journal of Biomechanics* 33:155-64.

Neptune, R.R., S.A. Kautz, and F.E. Zajac. 2001. Contributions of the individual ankle plantar flexors to support, forward progression and swing initiation during walking. *Journal of Biomechanics* 34:1387-98.

Neptune R.R., F.E. Zajac, and S.A. Kautz. 2004. Muscle force redistributes segmental power for body progression during walking. *Gait and Posture* 19:194-205.

Nigg, B.M. 1999. General comments about modeling. In *Biomechanics of the Musculo-skeletal System*, eds. B.M. Nigg and W. Herzog, 435-445. Chichester, UK: Wiley.

Nigg, B.M., and W. Herzog. 1994. *Biomechanics of the Musculo-skeletal System.* Toronto: Wiley.

Nishimura, S., Y. Tomita, and T. Horiuchi. 1992. Clinical application of an active electrode using an operational amplifier. *IEEE Transactions on Biomedical Engineering* 39:1096-9.

Nordander, C., J. Willner, G.-A. Hansson, B. Larsson, J. Unge, L. Granquist, and S. Skerfving. 2003. Influence of the subcutaneous fat layer, as measured by ultrasound, skinfold calipers and BMI, on the EMG amplitude. *European Journal of Applied Physiology* 89:514-9.

Norman, R.W., G.E. Caldwell, and P.V. Komi. 1985. Differences in body segment energy utilization between world class and recreational cross country skiers. *International Journal of Sport Biomechanics* 1:253-62.

Norman, R.W., and P.V. Komi. 1987. Mechanical energetics of world class cross-country skiing. *International Journal of Sport Biomechanics* 3:353-69.

Norman, R., M. Sharratt, J. Pezzack, and E. Noble. 1976. Re-examination of the mechanical efficiency of horizontal treadmill running. In *Biomechanics V-B,* ed. P. Komi, 87-93. International Series on Biomechanics. Baltimore: University Press.

O'Connor, K.M., and M.C. Bottum. 2009. Differences in cutting knee mechanics based on principal components. *Medicine and Science in Sports and Exercise* 41(4):867-78.

O'Connor, K.M., and J. Hamill. 2004. The role of selected extrinsic foot muscles during running. *Clinical Biomechanics (Bristol, Avon)* 19:71-7.

Oh, S.J. 1993. *Clinical Electromyography: Nerve Conduction Studies.* 2nd ed. Baltimore: Williams & Wilkins.

Ohashi, J. 1995. Difference in changes of surface EMG during low-level static contraction between monopolar and bipolar lead. *Applied Human Science* 14:79-88.

Ohashi, J. 1997. The effects of preceded fatiguing on the relations between monopolar surface electromyogram and fatigue sensation. *Applied Human Science* 16:19-27.

Okada, M. 1987. Effect of muscle length on surface EMG wave forms in isometric contractions. *European Journal of Applied Physiology* 56:482-6.

Okamoto, T., and K. Okamoto. 2007. *Development of Gait by Electromyography: Application to Gait Analysis and Evaluation.* Osaka, Japan: Walking Development Group.

Okamoto, T., K. Okamoto, and P.D. Andrew. 2003. Electromyographic developmental changes in one individual from newborn stepping to mature walking. *Gait and Posture* 17:18-27.

O'Malley, J. 1992. *Schaum's Outline of Basic Circuit Analysis.* 2nd ed. New York: McGraw-Hill.

Onyshko, S., and D.A. Winter. 1980. A mathematical model for the dynamics of human locomotion. *Journal of Biomechanics* 13:361-8.

Padgaonkar, A.J., K.W. Krieger, and A.I. King. 1975. Measurement of angular acceleration of a rigid body using accelerometers. *Transactions of ASME, Journal of Applied Mechanics* 42:552-6.

Pain, M.T.G., and J.H. Challis. 2001. Whole body force distributions in landing from a drop. In *Proceedings of the XVIIIth Congress of the International Society of Biomechanics,* eds. R. Müller, H. Gerber, and A. Stacoff, 202. Zurich: International Society of Biomechanics.

Pan, Z.S., Y. Zhang, and P.A. Parker. 1989. Motor unit power spectrum and firing rate. *Medicine and Biological Engineering and Computing* 27:14-8.

Pandy, M.G. 2001. Computer modeling and simulation of human movement. *Annual Review of Biomedical Engineering* 3:245-73.

Pandy, M.G., and N. Berme. 1989. Quantitative assessment of gait determinants during single stance via a three-dimensional model. Part 1. Normal gait. *Journal of Biomechanics* 22:717-24.

Pandy, M.G., and F.E. Zajac. 1991. Optimal muscular coordination strategies for jumping. *Journal of Biomechanics* 24:1-10.

Pandy, M.G., F.E. Zajac, E. Sim, and W.S. Levine. 1990. An optimal control model for maximum-height human jumping. *Journal of Biomechanics* 23:1185-98.

Pattichis, C.S., I. Schofield, R. Merletti, P.A. Parker, and L.T. Middleton. 1999. Introduction to this special issue: Intelligent data analysis in electromyography and electroneurography. *Medical Engineering and Physics* 21:379-88.

Pavol, M.J., T.M. Owings, and M.D. Grabiner. 2002. Body segment parameter estimation for the general population of older adults. *Journal of Biomechanics* 35:707-12.

Perez, M.A., and M.A. Nussbaum. 2003. Principal component analysis as an evaluation and classification tool for lower torso sEMG data. *Journal of Biomechanics* 36(8):1225-9.

Perreault, E.J., C.J. Heckman, and T.G. Sandercock. 2003. Hill muscle model errors during movement are greatest within the physiologically relevant range of motor unit firing rates. *Journal of Biomechanics* 36:211-8.

Petrofsky, J. 2008. The effect of the subcutaneous fat on the transfer of current through skin and into muscle. *Medical Engineering and Physics* 30:1168-76.

Peters, B.T., J.M. Haddad, B.C. Heiderscheit, R.E.A. van Emmerik, and J. Hamill. 2003. Limitations in the use and interpretation of continuous relative phase. *Journal of Biomechanics* 36:271-4.

Pezzack, J.C. 1976. An approach for the kinetic analysis of human motion. Master's thesis, University of Waterloo, Waterloo, ON.

Pezzack, J.C., R.W. Norman, and D.A. Winter. 1977. An assessment of derivative determining techniques used for motion analysis. *Journal of Biomechanics* 10:377-82.

Piazza, S.J., and S.L. Delp. 1996. The influence of muscles on knee flexion during the swing phase of gait. *Journal of Biomechanics* 29:723-33.

Pierrynowski, M.R. 1982. A physiological model for the solution of individual muscle forces during normal human walking. PhD dissertation, Simon Fraser University, Burnaby, BC.

Pierrynowski, M.R., and J.B. Morrison. 1985. Estimating the muscle forces generated in the human lower extremity when walking: A physiological solution. *Mathematical Biosciences* 75:43-68.

Pierrynowski, M.R., R.W. Norman, and D.A. Winter. 1981. Mechanical energy analyses of the human during load carriage on a treadmill. *Ergonomics* 24:1-14.

Pierrynowski, M.R., D.A. Winter, and R.W. Norman. 1980. Transfers of mechanical energy within the total body and mechanical efficiency during treadmill walking. *Ergonomics* 23:147-56.

Pinter, I.J., R. van Swigchem, A.J. van Soest, and L.A. Rozendaal. 2008. The dynamics of postural Sway cannot be captured using a one segment inverted pendulum model: A PCA on one segment rotations during unperturbed stance. *Journal of Neurophysiology* 100:3197-208.

Plagenhoef, S. 1968. Computer programs for obtaining kinetic data on human movement. *Journal of Biomechanics* 1:221-34.

Plagenhoef, S. 1971. *Patterns of Human Motion: A Cinematographic Analysis.* Englewood Cliffs, NJ: Prentice Hall.

Pollard, C.D., B.C. Heiderscheit, R.E.A. van Emmerik, and J. Hamill. 2005. Gender differences in lower extremity coupling variability during an unanticipated cutting maneuver. *Journal of Applied Biomechanics* 21:143-52.

Proakis, J.G., and D.G. Manolakis. 1988. *Introduction to Digital Signal Processing.* New York: Macmillan.

Prilutsky, B.I., W. Herzog, and T.L. Allinger. 1997. Forces of individual cat ankle extensor muscles during locomotion predicted using static optimization. *Journal of Biomechanics* 30:1025-33.

Prodromos, C.C., T.P. Andriacchi, and J.O. Galante. 1985. A relationship between gait and clinical changes following high tibial osteotomy. *Journal of Bone and Joint Surgery. American volume* 67:1188-94.

Putnam, C.A. 1991. A segment interaction analysis of proximal-to-distal sequential motion patterns. *Medicine and Science in Sports and Exercise* 23:130-44.

Rack, P.M.H., and D.R. Westbury. 1969. The effects of length and stimulus rate on tension in the isometric cat soleus muscle. *Journal of Physiology* 204:443-60.

Ramey, M.R. 1973a. A simulation of the running broad jump. In *Mechanics in Sport,* ed. J.L. Bleustein, 101-12. New York: American Society of Mechanical Engineers.

Ramey, M.R. 1973b. Significance of angular momentum in long jumping. *Research Quarterly* 44:488-97.

Ramey, M.R. 1974. The use of angular momentum in the study of long-jump take-offs. In *Biomechanics IV,* eds. R.C. Nelson and C.A. Morehouse, 144-8. Baltimore: University Park Press.

Ramsay, J.O., and C.J. Dalzell. 1991. Some tools for functional data analysis (with discussion). *Journal of the Royal Statistical Society, Series B* 53:539-72.

Ramsay, J.O., G. Hooker, and S. Graves. 2009. *Functional Data Analysis with R and MATLAB.* New York: Springer.

Ramsay, J.O., and B.W. Silverman. 2002. *Applied Functional Data Analysis.* New York: Springer- Verlag.

Ramsay, J.O., and B.W. Silverman. 2005. *Functional Data Analysis.* 2nd ed. New York: Springer.

Ramsey, R.W., and S.F. Street. 1940. The isometric length-tension diagram of isolated skeletal muscle fibres of the frog. *Journal of Cellular and Comparative Physiology* 15:11-34.

Rau, G., C. Disselhorst-Klug, and J. Silny. 1997. Noninvasive approach to motor unit characterization: Muscle structure, membrane dynamics and neuronal control. *Journal of Biomechanics* 30:441-6.

Reber, L., J. Perry, and M. Pink. 1993. Muscular control of the ankle in running. *American Journal of Sports Medicine* 21:805-10.

Redfern, M.S. 1992. Functional muscle: Effects on electromyographic output. In *Selected Topics in Surface Electromyography for Use in the Occupational Setting: Expert Perspectives,* ed. G.L. Soderberg, 104-20. Washington, DC: National Institute for Occupational Safety and Health.

Reuleaux, F. 1876. *The Kinematics of Machinery: Outlines of a Theory of Machines.* London: Macmillan.

Reynolds, C. 1994. Electromyographic biofeedback evaluation of a computer keyboard operator with cumulative trauma disorder. *Journal of Hand Therapy* 7:25-7.

Risher, D.W., L.M. Schutte, and C.F. Runge. 1997. The use of inverse dynamics solutions in direct dynamic simulations. *Journal of Biomechanical Engineering* 119:417-22.

Robertson, D.G.E., and J.J. Dowling. 2003. Design and responses of Butterworth and critically damped digital filters. *Journal of Electromyography and Kinesiology* 13:569-73.

Robertson, D.G.E., and D. Fleming. 1987. Kinetics of standing broad and vertical jumping. *Canadian Journal of Sport Science* 12:19-23.

Robertson, D.G.E., and Y.D. Fortin. 1994. Mechanics of rowing. In *Proceedings of the Eighth Conference of the Canadian Society for Biomechanics,* eds. W. Herzog, B. Nigg, and A. van den Bogert, 248-9. Calgary, AB: Canadian Society for Biomechanics.

Robertson, D.G.E., J. Hamill, and D.A. Winter. 1997. Evaluation of cushioning properties of running footwear. In *Proceedings of the XVIth Congress of the International Society of Biomechanics,* eds. M. Miyashita and T. Fukunaga, 263. Tokyo: International Society of Biomechanics.

Robertson, D.G.E., and R.E. Mosher. 1985. Work and power of the leg muscles in soccer kicking. In *Biomechanics IX-B,* eds. D.A. Winter, R.W. Norman, R.P. Wells, K.C. Hayes, and A.E. Patla, 533-8. Champaign, IL: Human Kinetics.

Robertson, D.G.E., and V.L. Stewart. 1997. Power production during swim starting. In *Proceedings of the XVIth Congress of the International Society of Biomechanics,* eds. M. Miyashita and T. Fukunaga, 22. Tokyo: International Society of Biomechanics.

Robertson, D.G.E., and D.A. Winter. 1980. Mechanical energy generation, absorption and transfer amongst segments during walking. *Journal of Biomechanics* 13:845-54.

Roeleveld, K., D.F. Stegeman, H.M. Vingerhoets, and A. van Oosterom. 1997. Motor unit potential contribution to surface electromyography. *Acta Physiologica Scandinavica* 160:175-83.

Rolf, C., P. Westblad, I. Ekenman, A. Lundberg, N. Murphy, M. Lamontagne, and K. Halvorsen. 1997. An experimental *in vivo* method for analysis of local deformation on tibia, with simultaneous measures of ground forces, lower extremity muscle activity and joint motion. *Scandinavian Journal of Medicine and Science in Sports* 7:144-51.

Ronager, J., H. Christensen, and A. Fuglsang-Frederiksen. 1989. Power spectrum analysis of the EMG pattern in normal and diseased muscles. *Journal of the Neurological Sciences* 94:283-94.

Rosenblum, M., and J. Kurths. 1998. Analysing synchronization phenomena from bivariate data by means of the Hilbert transform. In *Nonlinear Analysis of Physiological Data,* eds. H. Kantz, J. Kurths, and G. Mayer-Kress, 91-100. Berlin: Springer.

Roy, B. 1978. Biomechanical features of different starting positions and skating strides in ice hockey. In *Biomechanics VI-B,* eds. E. Asmussen and K. Jørgensen, 137-41. Baltimore: University Park Press.

Ryan, W., A.J. Harrison, and K. Hayes. 2006. Functional data analysis in biomechanics: A case study of knee joint vertical jump kinematics. *Sports Biomechanics* 5:121-38.

Sacco, I.C., A.A. Gomes, M.E. Otuzi, D. Pripas, and A.N. Onodera. 2009. A method for better positioning bipolar electrodes for lower limb EMG recordings during dynamic contractions. *Journal of Neuroscience Methods* 180: 133-7.

Saitou, K., T. Masuda, D. Michikami, R. Kojima, and M. Okada. 2000. Innervation zones of the upper and lower limb muscles estimated by using multichannel surface EMG. *Journal of Human Ergology (Tokyo)* 29:35-52.

Sandberg, A., B. Hansson, and E. Stalberg. 1999. Comparison between concentric needle EMG and macro EMG in patients with a history of polio. *Clinical Neurophysiology* 110:1900-8.

Sanders, D.B., E.V. Stålberg, and S.D. Nandedkar. 1996. Analysis of the electromyographic interference pattern. *Journal of Clinical Neurophysiology* 13:385-400.

Schache, A.G., and R. Baker. 2007. On the expression of joint moments during gait. *Gait and Posture* 25(3):440-52.

Schneider, K., and R.F. Zernicke. 1992. Mass, center of mass and moment of inertia estimates for infant limb segments. *Journal of Biomechanics* 25:145-8.

Scholz, J.P., J.A.S. Kelso, and G. Schöner. 1987. Nonequilibrium phase transitions in coordinated biological motion: Critical slowing down and switching time. *Physics Letters* A123:390-4.

Scovil, C.Y., and J.L. Ronsky. 2006. Sensitivity of a Hill-based muscle model to perturbations in model parameters. *Journal of Biomechanics* 39:2055-63.

Seireg, A., and R.J. Arvikar. 1975. The prediction of muscular load sharing and joint forces in the lower extremities during walking. *Journal of Biomechanics* 18:89-102.

Selbie, W.S., and G.E. Caldwell. 1996. A simulation study of vertical jumping from different starting postures. *Journal of Biomechanics* 29:1137-46.

Shapiro, R. 1978. The direct linear transformation method for three-dimensional cinematography. *Research Quarterly* 49:197-205.

Sherif, M.H., R.J. Gregor, and J. Lyman. 1981. Effects of load on myoelectric signals: The ARIMA representation. *IEEE Transactions on Biomedical Engineering* 5:411-6.

Shiavi, R. 1974. A wire multielectrode for intramuscular recording. *Medical and Biological Engineering* 12:721-3.

Shiavi, R., L.Q. Zhang, T. Limbird, and M.A. Edmondstone. 1992. Pattern analysis of electromyographic linear envelopes exhibited by subjects with uninjured and injured knees during free and fast speed walking. *Journal of Orthopedic Research* 10:226-36.

Shorten, M.R., and D.S. Winslow. 1992. Spectral analysis of impact shock during running. *International Journal of Sport Biomechanics* 8:288-304.

Siegel, K.L., T.M. Kepple, and G.E. Caldwell. 1996. Improved agreement of foot segmental power and rate of energy during gait: Inclusion of distal power terms and use of 3D models. *Journal of Biomechanics* 29:823-7.

Siegel, K.L., T.M. Kepple, P.G. O'Connell, L.H. Gerber, and S.J. Stanhope. 1995. A technique to evaluate foot function during the stance phase of gait. *Foot and Ankle International* 16(12):764-70.

Siegel, K.L., T.M. Kepple, and S.J. Stanhope. 2004. Joint moment control of mechanical energy flow during normal gait. *Gait and Posture* 19:69-75

Siegel, K.L., T.M. Kepple, and S.J. Stanhope. 2007. A case study of gait compensations for hip muscle weakness in idiopathic inflammatory myopathy. *Clinical Biomechanics* 22:319-26.

Siegler, S., H.J. Hillstrom, W. Freedman, and G. Moskowitz. 1985. The effect of myoelectric signal processing on the relationship between muscle force and processed EMG. *Electromyography and Clinical Neurophysiology* 25:499-512.

Silder, A., B. Whittington, B. Heiderscheit, and D.G. Thelen. 2007. Identification of passive elastic joint moment-angle relationships in the lower extremity. *Journal of Biomechanics* 40:2628-35.

Sjøgaard, G., B. Kiens, K. Jorgensen, and B. Saltin. 1986. Intramuscular pressure, EMG and blood flow during low-level prolonged static contraction in man. *Acta Physiologica Scandinavica* 128:475-84.

Smith, A.J., D.G. Lloyd, and D.J. Wood. 2004. Pre-surgery knee joint loading patterns during walking predict the presence and severity of anterior knee pain after total knee arthroplasty. *Journal of Orthopaedic Research* 22:260-6.

Smith, G. 1989. Padding point extrapolation techniques for the Butterworth digital filter. *Journal of Biomechanics* 22:967-71.

Smith, R.M. 1996. Distribution of mechanical energy fractions during maximal ergometer rowing. PhD dissertation, University of Wollongong, Wollongong NSW, Australia.

Solomonow, M., R. Baratta, M. Bernardi, B. Zhou, Y. Lu, M. Zhu, and S. Acierno. 1994. Surface and wire EMG crosstalk in neighbouring muscles. *Journal of Electromyography and Kinesiology* 4:131-42.

Solomonow, M., R. Baratta, H. Shoji, and R. D'Ambrosia. 1990. The EMG-force relationships of skeletal muscle; dependence on contraction rate, and motor units control strategy. *Electromyography and Clinical Neurophysiology* 30:141-52.

Solomonow, M., C. Baten, J. Smit, R. Baratta, H. Hermens, R. D'Ambrosia, and H. Shoji. 1990. Electromyogram power spectra frequencies associated with motor unit recruitment strategies. *Journal of Applied Physiology* 68:1177-85.

Song P., P. Kraus, V. Kumar, and P. Dupont. 2001. Analysis of rigid-body dynamic models for simulation of systems with frictional contacts. *Journal of Applied Mechanics* 68:118-28.

Soutas-Little, R.W., G.C. Beavis, M.C. Verstraete, and T.L. Markus. 1987. Analysis of foot motion during running using a joint co-ordinate system. *Medicine and Science in Sports and Exercise* 19:285-93.

Sparrow, W.A., E. Donovan, R.E.A. van Emmerik, and E.D. Barry. 1987. Using relative motion plots to measure intra limb and inter limb coordination. *Journal of Motor Behavior* 19:115-29.

Spoor, C.W., and F.E. Veldpaus. 1980. Rigid body motion calculated from spatial coordinates of markers. *Journal of Biomechanics* 13:391-3.

Stefanyshyn, D.J., and B.M. Nigg. 1998. Contributions of the lower extremity joints to mechanical energy in running vertical and running long jumps. *Journal of Sports Sciences* 16:177-86.

Sternad, D., M.T. Turvey, and E.L. Saltzman. 1999. Dynamics of 1:2 coordination: Generalizing relative phase to n:m rhythms. *Journal of Motor Behavior* 31(3):207-33.

Stewart, C., N. Postans, M.H. Schwartz, A. Rozumalski, and A. Roberts. 2007. An exploration of the function of the triceps surae during normal gait using functional electrical stimulation. *Gait and Posture* 26:482-8.

Strogatz, S.H. 1994. *Nonlinear dynamics and chaos*. Reading, MA: Perseus Books.

Sung, P., A. Lammers, and P. Danial. 2009. Different parts of erector spinae muscle fatigability in subjects with and without low back pain. *Spine* 9:115-120.

Taelman, J., S. van Huffel, and A. Spaepen. 2007. Wavelet-independent component analysis to remove electrocardiography contamination in surface electromyography. *Conference Proceedings of the Annual International Conference of the IEEE Engineering in Medicine and Biology Society* 2007:682-5.

Taga, G. 1995. A model of the neuro-musculo-skeletal system for human locomotion. I. Emergence of basic gait. *Biological Cybernetics* 73:97-111.

Tang, A., and W.Z. Rymer. 1981. Abnormal force—EMG relations in paretic limbs of hemiparetic human subjects. *Journal of Neurology, Neurosurgery and Psychiatry* 44:690-8.

Tepavac, D., and E.C. Field-Fote. 2001. Vector coding: A technique for the quantification of intersegmental in multicyclic behaviors. *Journal of Applied Biomechanics* 17:259-70.

Thelen, D.G., and F.C. Anderson. 2006. Using computed muscle control to generate forward dynamic simulations of human walking from experimental data. *Journal of Biomechanics* 39:1107-15.

Thelen, D.G., F.C. Anderson, and S.L. Delp. 2003. Generating dynamic simulations of movement using computed muscle control. *Journal of Biomechanics* 36:321-8.

Thomsen, M., and P.H. Veltink. 1997. Influence of synchronous and sequential stimulation on muscle fatigue. *Medical and Biological Engineering and Computing* 35:186-92.

Thusneyapan, S., and G.I. Zahalak. 1989. A practical electrode-array myoprocessor for surface electromyography. *IEEE Transactions on Biomedical Engineering* 36:295-99.

Todorov, E. 2007. Probabilistic inference of multijoint movements, skeletal parameters and marker attachments from diverse motion capture data. *IEEE Transactions on Biomedical Engineering* 54:1927-39. PMID 18018688.

Trewartha, G., M.R. Yeadon, and J.P. Knight. 2001. Markerfree tracking of aerial movements. In *Proceedings of the XVIIIth Congress of the International Society of Biomechanics,* eds. R. Müller, H. Gerber, and A. Stacoff, 185. Zurich: International Society of Biomechanics.

Tsai, C.-S., and J.M. Mansour. 1986. Swing phase simulation and design of above knee prostheses. *Journal of Biomechanical Engineering* 108:65-72.

Turvey, M.T. 1990. Coordination. *The American Psychologist* 45(8):938-53.

Turvey, M.T., and C. Carello. 1996. Dynamics of Bernstein's level of synergies. In *Dexterity and Its Development,* eds. M.L. Latash and M.T. Turvey, 339-76. Mahwah, NJ: Erlbaum.

Umberger, B.R. 2010. Stance and swing phase costs in human walking. *Journal of the Royal Society Interface* 7:1329-40.

Umberger B.R., K.G.M. Gerritsen, and P.E. Martin. 2003. A model of human muscle energy expenditure. *Computer Methods in Biomechanics and Biomedical Engineering* 6:99-111.

van den Bogert, A., and A. Su. 2007. A weighted least squares method for inverse dynamics analysis. *Computer Methods in Biomechanics and Biomedical Engineering* 11:3-9.

van den Bogert, A.J., K.G.M. Gerritsen, and G.K. Cole. 1998. Human muscle modeling from a user's perspective. *Journal of Electromyography and Kinesiology* 8:119-24.

van der Helm F.C., H.E. Veeger, G.M. Pronk, L.H. van der Woude, and R.H. Rozendal. 1992. Geometry parameters for musculoskeletal modelling of the shoulder system. *Journal of Biomechanics* 25:129-44.

van Dieën, J.H., L.P. Selen, and J. Cholewicki. 2003. Trunk muscle activation in low-back pain patients, an analysis of the literature. *Journal of Electromyography and Kinesiology* 13:333-51.

Van Emmerik, R.E.A., and R.C. Wagenaar. 1996a. Effects of walking velocity on relative phase dynamics in the trunk in human walking. *Journal of Biomechanics* 29:1175-84.

Van Emmerik, R.E.A., and R.C. Wagenaar. 1996b. Dynamics of movement coordination and tremor in Parkinson's disease. *Human Movement Science* 15:203-35.

Van Emmerik, R.E.A., R.C. Wagenaar, A. Winogrodzka, and E.Ch. Wolters. 1999. Axial rigidity in Parkinson's disease. *Archives of Physical Medicine and Rehabilitation* 80:186-91.

van Ingen Schenau, G.J. 1989. From rotation to translation: Constraints on multi-joint movements and the unique action of bi-articular muscles. *Human Movement Science* 8:301-37.

van Ingen Schenau, G.J., and P.R. Cavanagh. 1990. Power equations in endurance sports. *Journal of Biomechanics* 23:865-81.

van Ingen Schenau, G.J., and A.J. van Soest. 1996. On the biomechanical basis of dexterity. In *Dexterity and Its Development*, eds. M.L. Latash and M.T. Turvey, 305-38. Mahwah, NJ: Erlbaum.

van Ingen Schenau, G.J., W.L.M. van Woensel, P.J.M. Boots, R.W. Snackers, and G. de Groot. 1990. Determination and interpretation of mechanical power in human movement: Application to ergometer cycling. *European Journal of Applied Physiology* 61:11-19.

van Soest, A.J., A.L. Schwab, M.F. Bobbert, and G.J. van Ingen Schenau. 1993. The influence of the biarticularity of the gastrocnemius muscle on vertical-jumping achievement. *Journal of Biomechanics* 26:1-8.

Vardaxis, V.G., and T.B. Hoshizaki. 1989. Power patterns of the lower limb during the recovery phase of the sprinting stride of advanced and intermediate sprinters. *International Journal of Sport Biomechanics* 5:332-49.

Vaughan, C.L. 1982. Smoothing and differentiating of displacement-time data: An application of splines and digital filtering. *International Journal of Bio-medical Computing* 13:375-86.

Vaughan, C.L. 1996. Are joint torques the holy grail of human gait analysis? *Human Movement Science* 15:423-43.

Vaughan, C.L., B.L. Davis, and J.C. O'Connor. 1992. *Dynamics of Human Gait.* Champaign, IL: Human Kinetics.

Veeger, H.E., L.S. Meershoek, L.H. van der Woude, and J.M. Langenhoff. 1998. Wrist motion in handrim wheelchair propulsion. *Journal of Rehabilitation Research and Development* 35:305-13.

Veigel, C., Molloy, J.E., Schmitz, S., and J. Kendrick-Jones. 2003. Load-dependent kinetics of force production by smooth muscle myosin measured with optical tweezers. *Nature Cell Biology* 5:980-6.

Vereijken, B., R.E.A. van Emmerik, H.T.A. Whiting, and K.M. Newell. 1992. Free(zing) degrees of freedom in skill acquisition. *Journal of Motor Behavior* 24:133-42.

Vigreux, B., J.C. Cnockaert, and E. Pertuzon. 1979. Factors influencing quantified surface EMGs. *European Journal of Applied Physiology* 41:119-29.

von Holst, E. 1939/1973. *The Behavioral Physiology of Animals and Man: Selected Papers of E. von Holst,* Vol. 1. Coral Gables, FL: University of Miami Press.

Vorro, J., and D. Hobart. 1981. Kinematic and myoelectric analysis of skill acquisition: I. 90cm subject group. *Archives of Physical Medicine and Rehabilitation* 62:575-82.

Wagenaar, R.C., and R.E.A. van Emmerik. 2000. Resonance frequencies of arms and legs identify different walking patterns. *Journal of Biomechanics* 33:853-61.

Wahba, G. 1985. A comparison of GCV and GML for choosing the smoothing parameter in the generalized spline smoothing problem. *Annals of Statistics* 13(4):1378-402.

Wahba, G. 1990. Spline Models for Observational Data. *Journal of the Society for Industrial and Applied Mathematics,* Vol.59.

Wallinga-De Jonge, W., F.L. Gielen, P. Wirtz, P. De Jong, and J. Broenink. 1985. The different intracellular action potentials of fast and slow muscle fibres. *Electroencephalography and Clinical Neurophysiology* 60:539-47.

Walthard, K.M., and M. Tchicaloff. 1971. Motor points. In *Electrodiagnosis and Electromyography*, ed. S. Licht. Baltimore: Waverly.

Walton, J.S. 1981. Close-range cine-photogrammetry: A generalized technique for quantifying gross human motion. PhD dissertation, Pennsylvania State University.

Wang, J.W., K.N. Kuo, T.P. Andriacchi, and J.O. Galante. 1990. The influence of walking mechanics and time on the results of proximal tibial osteotomy. *Journal of Bone and Joint Surgery. American volume* 72:905-9.

Webster, J.G. 1984. Reducing motion artifacts and interference in biopotential recording. *IEEE Transactions on Biomedical Engineering* 31:823-6.

Weiss, P. 1941. Self-differentiation of the basic patterns of coordination. *Comparative Psychology Monographs* 17(4).

Wells, R.P. 1988. Mechanical energy costs of human movement: An approach to evaluating the transfer possibilities of two-joint muscles. *Journal of Biomechanics* 21: 955-64.

Wheat, J.S., R.M. Bartlett, and C.E. Milner. 2002. Continuous relative phase calculation: Phase angle definition. pp. 322. *Proceedings of the 12th Commonwealth International Sport Conference,* Manchester, UK.

Wheat, J.S., and P.S. Glazier. 2006. Measuring coordination and variability in coordination. In *Movement System Variability,* eds. K. Davids, S.J. Bennett, and K.M. Newell, 167-81. Champaign, IL: Human Kinetics.

White, S.C., and D.A. Winter. 1985. Mechanical power analysis of the lower limb musculature in race walking. *International Journal of Sport Biomechanics* 1:15-24.

Whiting, W.C., and R.F. Zernicke. 1982. Correlation of movement patterns via pattern recognition. *Journal of Motor Behavior* 14:135-42.

Whittington, B., A. Silder, B. Heiderscheit, and D.G. Thelen. 2008. The contribution of passive-elastic mechanisms to lower extremity joint kinetics during human walking. *Gait and Posture* 27:628-34.

Whittlesey, S.N., and J. Hamill. 1996. An alternative model of the lower extremity during locomotion. *Journal of Applied Biomechanics* 12:269-79.

Wilkie, D.R. 1950. The relation between force and velocity in human muscle. *Journal of Physiology* 110:249-280.

Williams, K.R. 1985. The relationship between mechanical and physiological energy estimates. *Medicine and Science in Sports and Exercise* 17:317-25.

Williams, K.R., and P.R. Cavanagh. 1983. A model for the calculation of mechanical power during distance running. *Journal of Biomechanics* 16:115-28.

Winby, C.R., D.G. Lloyd, T.F. Besier, and T.B. Kirk. 2009. Muscle and external load contribution to knee joint contact loads during normal gait. *Journal of Biomechanics* 42:2294-300.

Windhorst, U., T.M. Hamm, and D.G. Stuart. 1989. On the function of muscle and reflex partitioning. *Behavioral and Brain Sciences* 12:629-81.

Winter, D.A. 1976. Analysis of instantaneous energy of normal gait. *Journal of Biomechanics* 9:253-7.

Winter, D.A. 1978. Calculation and interpretation of mechanical energy of movement. *Exercise and Sport Sciences Reviews* 6:183-201.

Winter, D.A. 1979a. *Biomechanics of Human Movement.* Toronto: Wiley.

Winter, D.A. 1979b. A new definition of mechanical work done in human movement. *Journal of Applied Physiology* 46:79-83.

Winter, D.A. 1980. Overall principle of lower limb support during stance phase of gait. *Journal of Biomechanics* 13:923-7.

Winter, D.A. 1983a. Moments of force and mechanical power in jogging. *Journal of Biomechanics* 16:91-7.

Winter, D.A. 1983b. Biomechanical motor patterns in normal walking. *Journal of Motor Behaviour* 15:302-30.

Winter, D.A. 1983c. Energy generation and absorption at the ankle and knee during fast, natural and slow cadences. *Clinical Orthopedics and Related Research* 175:147-54.

Winter, D.A. 1990. *Biomechanics and Motor Control of Human Movement.* 2nd ed. Toronto: Wiley.

Winter, D.A. 1991. *Biomechanics and Motor Control of Human Gait: Normal, Elderly and Pathological.* 2nd ed. Waterloo, ON: Waterloo Biomechanics.

Winter, D.A. 1996. Total body kinetics: Our diagnostic key to human movement. *Proceedings of the International Society of Biomechanics in Sports,* ed. J.M.C.S. Abrantes, 10. Fuchal, Portugal: International Society of Biomechanics in Sports.

Winter, D.A. 2009. *Biomechanics and Motor Control of Human Movement.* 4th ed. New York: Wiley.

Winter, D.A., J.A. Eng, and M.G. Ishac. 1995. A review of kinetic parameters in human walking. In *Gait Analysis: Theory and Application,* eds. R.L. Craik and C.A. Oatis, 252-70. St. Louis, MO: Mosby.

Winter, D.A., A.J. Fuglevand, and S. Archer. 1994. Crosstalk in surface electromyography: Theoretical and practical estimates. *Journal of Electromyography and Kinesiology* 4:15-26.

Winter, D.A., and A.E. Patla. 1997. *Signal Processing and Linear Systems for Movement Sciences.* Waterloo, ON: Waterloo Biomechanics.

Winter, D.A., A.O. Quanbury, and G.D. Reimer. 1976. Analysis of instantaneous energy of normal gait. *Journal of Biomechanics* 9:253-7.

Winter, D.A., and D.G.E. Robertson. 1979. Joint torque and energy patterns in normal gait. *Biological Cybernetics* 29:137-42.

Winter, D.A., and S.E. Sienko. 1988. Biomechanics of below-knee amputee gait. *Journal of Biomechanics* 21:361-7.

Winter, D.A., R.P. Wells, and G.W. Orr. 1981. Errors in the use of isokinetic dynamometers. *European Journal of Applied Physiology* 46:409-21.

Winter, D.A., and S. White. 1983. Moments of force and mechanical power in jogging. *Journal of Biomechanics* 16:91-7.

Winters, J.M., and L. Stark. 1985. Analysis of fundamental movement patterns through the use of in-depth antagonistic muscle models. *IEEE Transactions on Biomedical Engineering* 32:826-39.

Woittiez, R.D., P.A. Huijing, and R.H. Rozendal. 1983. Influence of muscle architecture on the length force diagram of mammalian muscle. *Pfluegers Archiv* 399:275-9.

Woltring, H.J. 1980. Planar control in multi-camera calibration for 3-D gait studies. *Journal of Biomechanics* 13:39-48.

Woltring, H.J. 1986. A FORTRAN package for generalized, cross-validatory spline smoothing and differentiation. *Advances in Engineering Software* 8:104-13.

Woltring, H.J. 1991. Representation and calculation of 3D joint movement. *Human Movement Science* 10:603-16.

Woltring, H.J., R. Huiskes, A. De Lange, and F.E. Veldpaus. 1985. Finite centroid and helical axis estimation from noisy landmark measurements in the study of human joint kinematics. *Journal of Biomechanics* 18:379-89.

Wood, G.A. 1982. Data smoothing and differentiating procedures in biomechanics. *Exercise and Sport Sciences Reviews* 10:308-62.

Wooten, M.E., M.P. Kadaba, and G.V.B. Cochran. 1990. Dynamic electromyography. I. Numerical representation using principal component analysis. *Journal of Orthopaedic Research* 8(2):247-58.

Wrigley, A.T., W.J. Albert, K.J. Deluzio, and J.M. Stevenson. 2006. Principal component analysis of lifting waveforms. *Clinical Biomechanics* 21:567-78.

Yang, J.F., and D.A. Winter. 1983. Electromyography reliability in maximal and submaximal isometric contractions. *Archives of Physical Medicine and Rehabilitation* 64:417-20.

Yeadon, M.R. 1990a. The simulation of aerial movement: II. A mathematical inertia model of the human body. *Journal of Biomechanics* 23:67-74.

Yeadon, M.R. 1990b. The simulation of aerial movement: III. The determination of the angular momentum of the human body. *Journal of Biomechanics* 23:75-84.

Yeadon, M.R. 1993. The biomechanics of twisting somersaults. Part I: Rigid body motions. *Journal of Sport Science* 11:187-98.

Yeadon, M.R., and M. Morlock. 1989. The appropriate use of regression equations for the estimation of segmental inertia parameters. *Journal of Biomechanics* 22:683-9.

Yoshihuku, Y., and W. Herzog. 1990. Optimal design parameters of the bicycle-rider system for maximal muscle power output. *Journal of Biomechanics* 23:1069-79.

Yu, B., and J.G. Hay. 1995. Angular momentum and performance in the triple jump: A cross-sectional analysis. *Journal of Applied Biomechanics* 11:81-102.

Zajac, F.E. 1989. Muscle and tendon: Properties, models, scaling, and application to biomechanics and motor control. *CRC Critical Reviews in Biomedical Engineering* 17:359-411.

Zajac, F.E., and M.E. Gordon. 1989. Determining muscle's force and action in multi-articular movement. *Exercise and Sports Sciences Reviews* 17:187-230.

Zajac, F.E., R.R. Neptune, and S.A. Kautz. 2002. Biomechanics and muscle coordination of human walking. Part I: Introduction to concepts, power transfer, dynamics and simulation. *Gait and Posture* 16:215-32.

Zajac, F.E., R.R. Neptune, and S.A. Kautz. 2003. Biomechanics and muscle coordination of human walking. Part II: Lessons from dynamical simulations, clinical implications and concluding remarks. *Gait and Posture* 17:1-17.

Zarrugh, M.Y. 1981. Power requirements and mechanical efficiency of treadmill walking. *Journal of Biomechanics* 14:157-65.

Zatsiorsky, V.M. 1998. *Kinetics of Human Motion*. Champaign, IL: Human Kinetics.

Zatsiorsky, V.M., and M.L. Latash. 1993. What is a "joint torque" for joints spanned by multi-articular muscles? *Journal of Applied Biomechanics* 9:333-6.

Zatsiorsky, V.M., and V.N. Seluyanov. 1983. The mass and inertia characteristics of the main segments of the human body. In *Biomechanics VIII-B*, eds. H. Matsui and K. Kobayashi, 1152-9. Champaign, IL: Human Kinetics.

Zatsiorsky, V.M., and V.N. Seluyanov. 1985. Estimation of the mass and inertia characteristics of the human body by means of the best predictive regression equations. *Biomechanics IX-B*, eds. D.A. Winter et al., 233-9. Champaign, IL: Human Kinetics.

Zhan, C., L. Yeung, and Z. Yang. 2010. A wavelet-based adaptive filter for removing ECG interference in EMGdi signals. *Journal of Electromyography and Kinesiology* 20:542-9.

Zipp, P. 1982. Recommendations for the standardization of lead positions in surface electromyography. *European Journal of Applied Physiology* 50:41-54.

Index

Note: The italicized *f* and *t* following page numbers refer to figures and tables, respectively.

About the Authors

D. Gordon E. Robertson, PhD, an emeritus professor and a fellow of the Canadian Society for Biomechanics, wrote *Introduction to Biomechanics for Human Motion Analysis.* He taught undergraduate- and graduate-level biomechanics at the University of Ottawa and previously at the University of British Columbia, Canada. He conducts research on human locomotion and athletic activities and authors the analogue data analysis software *BioProc3.*

Graham E. Caldwell, PhD, an associate professor and a fellow of the Canadian Society for Biomechanics, teaches undergraduate- and graduate-level biomechanics at the University of Massachusetts at Amherst and previously held a similar faculty position at the University of Maryland. He won the Canadian Society for Biomechanics  New Investigator Award and in 1998 won the Outstanding Teacher Award for the School of Public Health and Health Sciences at the University of Massachusetts at Amherst. He served as an associate editor for *Medicine and Science in Sports and Exercise.*

Joseph Hamill, PhD, is a professor and fellow of the Research Consortium, International Society of Biomechanics in Sports, Canadian Society for Biomechanics, American College of Sports Medicine, and National Academy of Kinesiology. He coauthored the popular undergraduate textbook *Biomechanical Basis of Human Movement.* He teaches undergraduate- and graduate-level biomechanics and is director of the Biomechanics Laboratory at the University of Massachusetts at Amherst. He serves on the editorial boards of several prestigious professional journals. He is adjunct professor at the University of Edinburgh in Scotland and the University of Limerick in Ireland and a distinguished research professor at Republic Polytechnic in Singapore.

Gary Kamen, PhD, is a professor and fellow of the American Alliance for Health, Physical Education, Recreation and Dance; American College of Sports Medicine; and National Academy of Kinesiology. He authored an undergraduate textbook on kinesiology, *Foundations of Exercise Science,* as well as a  primer on electromyography, *Essentials of Electromyography.* He was president of the Research Consortium of AAPHERD and teaches undergraduate and graduate courses in exercise neuroscience and motor control in the department of kinesiology at the University of Massachusetts at Amherst.

Saunders (Sandy) N. Whittlesey, PhD, a graduate of the University of Massachusetts at Amherst, is a self-employed technology consultant specializing in athletic training, sporting goods, and clinical applications.

Additional Contributors

Norma Coffey, PhD, a postdoctoral researcher in statistics at the National University of Ireland at Galway, has expertise is functional data analysis and worked extensively with the Biomechanics Research Unit at the University of Limerick. Her current area of research involves applying functional data analysis techniques to time-course gene expression data.

Timothy R. Derrick, PhD, a professor in the department of kinesiology at Iowa State University, has an extensive background in signal processing and conducts research on impacts to the human body particularly from the ground during running activities.

Kevin Deluzio, PhD, is a professor in the department of mechanical and materials engineering at Queen's University in Kingston, Canada, and held a similar position at Dalhousie University. He studies human locomotion to investigate the biomechanical factors of musculoskeletal diseases such as knee osteoarthritis. He is also interested in the design and evaluation of noninvasive therapies as well as surgical treatments such as total-knee replacement.

Andrew (Drew) J. Harrison, PhD, is a senior lecturer in biomechanics in the department of physical education and sport sciences at the University of Limerick in Ireland and a fellow of the International Society for Biomechanics in Sport. He is the director of the Biomechanics Research Unit at the University of Limerick. His research focuses on biomechanics of sport performance and sport injuries.

Thomas M. Kepple, PhD, is an instructor in the department of health, nutrition, and exercise sciences at the University of Delaware. He worked for many years as a biomechanist at the National Institutes of Health on motion capture technology and gait laboratory instrumentation.

Ross H. Miller, PhD, an assistant professor in the department of kinesiology at the University of Maryland, has published papers on static optimization and forward dynamics as well as methods on nonlinear techniques of data analysis.

Scott Selbie, PhD, is an adjunct professor at Queen's University, Canada, and at the University of Massachusetts at Amherst. He is a graduate of Simon Fraser University, Canada. He is the director of research at C-Motion, developers of the Visual3D software, and president of HAS-Motion in Canada.

Brian R. Umberger, PhD, is an associate professor teaching biomechanics at the undergraduate and graduate levels in the department of kinesiology at the University of Massachusetts at Amherst. In 2010, he received the Outstanding Teacher Award for the School of Public Health and Health Sciences at the University of Massachusetts at Amherst. In his research, he uses a combination of experimental, modeling, and simulation approaches to study the biomechanics and energetics of human locomotion.

Richard E.A. van Emmerik, PhD, is a professor in the kinesiology department at the University of Massachusetts at Amherst, where he teaches motor control at the undergraduate and graduate levels. In his research, he applies principles from complex and nonlinear dynamical systems to the study of posture and locomotion.